LIFE HISTORY EVOLUTION

LIFE HISTORY EVOLUTION

DEREK ROFF

University of California, Riverside

SINAUER ASSOCIATES, INC. • *Publishers*
Sunderland, Massachusetts U.S.A.

THE COVER
Cover shows an egret (*Casmerodius albus*) with its brood.
A primary feature of life history theory is the evolution of brood
size, which in this species includes siblicide. Also shown is the central
equation of life history theory, namely the characteristic equation.
Photograph © Joe McDonald / Visuals Unlimited.

LIFE HISTORY EVOLUTION

Sinauer Associates, Inc.
PO Box 407
23 Plumtree Road
Sunderland, MA 01375
U.S.A.

FAX 413-549-1118

Email: publish@sinauer.com, orders@sinauer.com

www.sinauer.com

Library of Congress Cataloging-in-Publication Data
Roff, Derek A., 1949-
     Life history evolution / Derek A. Roff.
         p. cm.
     Includes bibliographical references (p.).
     ISBN 0-87893-756-0
     1. Variation (Biology). 2. Evolutionary genetics. I. Title.
QH401 .R64 2001
576.5'4—dc21   2001049533

Printed in U.S.A.
5 4 3 2 1

TABLE OF CONTENTS

PREFACE

An obvious feature of the natural world is that it is made up of an enormous variety of different life histories. Even different populations of the same species can show considerable variation in life history patterns and parameters. This book describes how evolutionary biologists have attempted to account for this variation. My aim is to provide a framework within which such analyses can be understood and demonstrate by example the enormous explanatory and predictive power of present life history theory. I have kept mathematical derivations to a minimum and placed material that is useful but which can be skipped over in "boxes" within the body of the text. The analysis of life history variation is necessarily a quantitative endeavor and its theoretical foundation requires a mathematical structure. Therefore, the "minimal" amount of mathematical detail is still substantial. Nevertheless there is plenty of room for the "non-mathematical" biologist. What is important is to understand the assumptions underlying the various models and hence their potential limitations.

I could not have completed this work without the continued support and encouragement of my wife and colleague Dr. Daphne Fairbairn. Dr. Mats Bjorglund and Jeff Hutchings also read an earlier draft and gave very valuable advice and suggestions.

CHAPTER 1

An Overview

One of the most obvious features of the living world is that organisms differ enormously in their life histories. Such traits as the age at maturity, adult size, mortality rate, and age-specific fecundity rate all show wide variation. For example, the flatfish *Hippoglossoides platessoides* occurs on both sides of the Atlantic, but even though the populations are taxonomically the same species, the difference in life histories is profound. On the Grand Banks of Newfoundland, *H. platessoides* grows more slowly than off the coast of Scotland, reaches a much larger size (60 cm vs 25 cm), matures later (14 yrs vs 3 yrs), lives longer (30 yrs vs 6 yrs), and has a lower age-specific fecundity (Figure 1.1). This book is based on the proposition that such variation is explicable under the assumption that life history variation is primarily the result of natural selection and represents adaptation.

A critical tool for the study of the pattern and evolution of life histories is mathematical analysis. This statement is a self-evident fact to most population biologists. However, the use of mathematics in ecological investigations has had a much rockier road than its use in genetical analysis, and its general acceptance as an important tool dates only from the 1960s. (See Kingsland 1985 for an excellent historical survey of the rise of mathematical approaches in ecology from the work of Lotka in the 1920s to the studies of MacArthur up to 1970.) The importance of the mathematical approach to the understanding of genetic variation is amply illustrated by Provine's review of the history of population genetics (Provine 1971) and by the biographies of three of the most influential geneticists of this century: Fisher (Box 1978), Haldane (Clark 1984), and Wright (Provine 1986).

Even by the latter half of the 1940s, mathematical thinking had still not made a significant impact on ecological theory. Allee et al. (1949, p. 271) observed that "theoretical population ecology has not advanced to a great degree in terms of its impact on ecological thinking." An early antipathy to the use of mathematical analysis may account in part for the delay in the merging of the ecological and evolutionary perspectives in what is now commonly known as life history analysis. An influential factor encouraging the use of mathematical investigation into life history variation was Lamont Cole's 1954 paper "The population

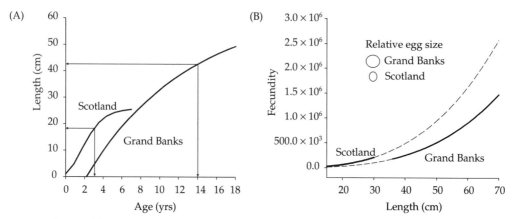

Figure 1.1   An example of variation in life history traits. (A) Variation in the growth curves (solid lines) and median maturity patterns (gray angles) of two separate populations of the flatfish *Hippoglossoides platessoides*. (B) Fecundity as a function of length in these two populations. The solid lines denote the range over which the fish actually reproduce; the dashed lines are extrapolated values for comparison with the other population. The difference in fecundity is due in part to differences in egg size as measured by egg diameter, shown drawn to a relative scale (Grand Banks = 3.0 mm, Scotland = 2.01 mm). Data from Bagenal (1955ab), Pitt (1966, 1967), and Roff (1982).

consequences of life history phenomena," which set out one of the basic mathematical frameworks by which the consequences of variation in life history traits can be analyzed. Cole's paper ushered in an era of research predicated on the integration of mathematics and biology in the study of the evolution of life history patterns.

In his review, Cole asked how changes in demographic attributes, such as the age at first reproduction, influenced the rate of increase of a population. Except for citations of its historical importance, Cole's paper gained widespread notice because of an apparent paradox with respect to the value of semelparity versus iteroparity: *"For an annual species, the absolute gain in intrinsic population growth which could be achieved by changing to the perennial reproductive habit would be exactly equivalent to adding one individual to the average litter size"* (Cole 1954, p. 118' the italics are Cole's). With the aid of hindsight, the resolution to this paradox is simple to the point of being trivial (see Chapter 4), but its importance lay in drawing attention to the value of mathematical analysis of life history phenomena. Cole's paper enunciated two important principles that are the basis of life history analysis:

> The birth rate, the death rate, and the age composition of the population, as well as its ability to grow, are consequences of the life-history features of the individual organisms. These population phenomena may be related

in numerous ways to the ability of the species to survive in a changing physical environment or in competition with other species. Hence it is to be expected that natural selection will be influential in shaping life-history patterns to correspond to efficient populations.

Thus natural selection is seen as maximizing some quantity, here termed "efficient populations," but elsewhere in the paper identified as the rate of population growth. This is not to be taken as indicating that Cole favored the idea of group selection. The tenor of his paper makes it clear that his use of population can be understood in modern terms to be equivalent to genotype. Thus Cole is making the point that selection favors those genotypes that have the highest rates of increase.

The second important principle put forward by Cole is that natural selection favors those patterns of birth, death, and reproduction that maximize the rate of increase. This observation was certainly not unique to Cole and can be traced back to Fisher (1930) and in verbal form to Darwin and Wallace. Andrewartha and Birch (1954) emphasized the importance of the potential for increase, devoting a whole chapter to the concept in their book *The Distribution and Abundance of Animals.* Birch later stressed the relationship between the genotype and its rate of increase, r: "Natural selection will tend to maximize r for the environment in which the species lives, for any mutation or gene combination which increases the chance of genotypes possessing them contributing more individuals to the next generation (that is, of increasing r) will be selected over genotypes contributing fewer of their kind to successive generations" (Birch 1960, p. 10).

Mathematical modeling has been, and continues to be, an important component of the analysis of life history variation (Stearns 1976, 1977, 1992; Parker and Maynard Smith 1990; Roff 1992a; Charnov 1993, McNamara 1993; Charlesworth 1994; Houston and McNamara 1999; Fox et al. 2001). The purpose of model construction is to address a particular aspect of the real world, ranging from a very detailed analysis of a very specific circumstance, to an assessment of a general proposition. All models necessarily are simplifications of reality. To ensure that the results are robust to the assumptions, Levins (1966, p. 423) recommended the use of several different models incorporating different assumptions: "Then, if these models, despite their different assumptions, lead to similar results we have what we can call a robust theorem which is relatively free of the details of the model. Hence our truth is the intersection of independent lies."

The analysis of life history evolution has developed along two lines, one concentrating on following evolutionary change via the underlying genetical determination of the traits, and another that focuses on the phenotype and, as indicated above, assumes that there is some phenotypic measure of fitness that is maximized. In recent years the barrier between these two approaches has been breaking down as more and more researchers appreciate the value of combining them.

A Framework for Analysis

Much of life history theory has developed without regard to the genetical basis of trait variation. But a full theory cannot ignore this obviously important component of evolutionary change. In Chapter 2 I present an outline of the quantitative genetic models presently available and discuss how these models can be used to generate the rationale for the phenotypic models upon which so much of the predictive basis of life history theory is based.

The general principle of heredity is that "like breeds like," which in more quantitative terms means that offspring resemble their parents. Observation suggests a more restrictive relationship in that, in general, the resemblance between mean offspring and mid-parent value can be described by a linear regression equation (Figure 1.2). This relationship is explainable by Mendelian genetics if we assume that the phenotypic trait value is equal to the sum of many genes of small effect. In this case the slope of the offspring/parent relationship is termed the heritability of the trait, h^2, and it is the quantitative genetic parameter that defines the degree to which selection will produce responses in the following generation. If we consider a subset or selected portion of parents, we can manipulate the offspring/parent regression to arrive at the familiar response equation

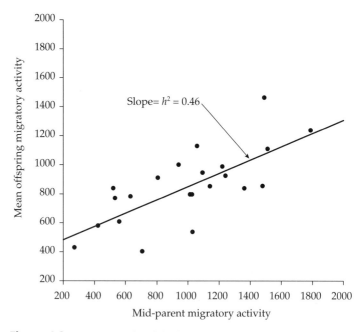

Figure 1.2   An example of the linear regression between mean offspring value and mid-parent value. The data shown are migratory activity, measured as half hours of night activity in the blackcap (*Sylvia atricapilla*). Data from Berthold and Pulido (1984).

$R = h_2S$, where R is the response to selection and S is the difference between the parental and population means. To this simple model we must add the complexities of linkage, dominance, and epistasis, but these change the details not the overall principle. If the response equation were all that we needed, then there would be little to occupy this book. The richness of evolutionary responses comes about because traits are not inherited or acted upon as independent entities. Rather, they are connected by virtue of having genes in common and by the net effect of selection on a trait being a consequence of selection acting on other traits, even if those traits are genetically independent. These two effects can be extracted from the expansion of the response equation to two traits: The response of trait X in standard deviation units, $R_{X,\sigma}$, is $R_{X,\sigma} = \beta_{X,\sigma} h^2_X + \beta_{Y,\sigma} h_{X}$-$h_Y r_A$, where Y is the second trait, the β's are the multivariate selection coefficients, and r_A is the genetic correlation between X and Y. The genetic correlation measures the effect of genes in common, whereas the selection coefficients are determined by both direct and indirect selection, the latter arising by virtue of a phenotypic correlation between traits.

Selection itself creates a surface over which trait values travel until they reach a combination at which selection ceases. This means that deviation in any direction away from this combination results in a negative selection coefficient and hence movement back to the equilibrium. Visually, we can think of this selection or fitness surface as a landscape in which there are one or several peaks at which selection ceases. Genetical theory, so far as it has been able to determine, predicts that a population will, except under precise and perhaps restrictive conditions (discussed in Chapters 2 and 3), "climb" up a peak, though it will not typically take the path of steepest ascent (that is, the shortest route). Thus if we can define the fitness surface, we should be able to predict the equilibrium trait combination. It is at this point that we arrive at the prediction that a population in a constant environment and not experiencing density-dependent or frequency-dependent effects will evolve such that its rate of increase, r, is maximized. Therefore, if we can define the phenotypic set of traits that maximize r, we have our desired "optimal" combination. Dropping any of the three restrictions changes the fitness criterion in ways that can be resolved. In this book I shall be most concerned with the effects of dropping the "constant environment" requirement, because, first, environments are patently not constant and, second, there are relatively few empirical and theoretical studies of the other effects on life history traits.

To calculate r we must solve the characteristic equation

$$\int_{x=\alpha}^{\infty} l(x)m(x)e^{-rx}dx = 1 \tag{1.1}$$

where α is the age at first reproduction, $l(x)$ is the probability of surviving to age x, and $m(x)$ is the production of female offspring at age x. If the population is stationary in the foregoing case, the equation reduces to

$$R_0 = \int_{x=\alpha}^{\infty} l(x)m(x)dx \qquad (1.2)$$

the expected lifetime production of female offspring (frequently simply referred to as expected lifetime fecundity). Without any restrictions the optimal life history would be to commence breeding at birth, live forever, and produce an infinite number of young. Obviously, this is not possible. This brings us to the second central tenant of life history theory, namely that trait combinations are restricted by virtue of trade-offs between traits, a subject dealt with in detail in the third chapter.

Trade-offs

As noted above, two central concepts in life history theory are that natural selection maximizes fitness and that trait combinations are constrained by trade-offs. Though the concept of a trade-off is intuitively simple it can be very difficult to disentangle causal factors unless manipulative experiments are feasible. A simple example of the problem is the distribution of choline between body and web in the spider *Nephila clavipes*. Choline is an important precursor to many physiologically important compounds (such as neurotransmitters) and ecologically important ones (such as web material) but does not appear to be synthesized by the spider (Higgins and Rankin 1999). All, or at least most of the choline, is obtained from the diet. Higgins and Rankin (1999) measured the amount of choline in the bodies and webs of spiders reared in the lab under a choline-restricted diet. There is a significant correlation between the amount of choline in the body and the amount in the web (Figure 1.3). Taken at face value this plot does not suggest a trade-off. However, this is because variation in acquisition confounds the variation in allocation. As the amount of choline is strictly limited, let us assume each spider has the same ability to utilize choline (an unlikely assumption, but one made for the present illustrative purposes). There thus must be a strict trade-off between body and web allocation, as illustrated by the single, example allocation isocline in Figure 1.3.

Trade-offs have both a genetic and phenotypic component and, as illustrated by the two-trait response equation, both can be important in shaping the evolutionary trajectory. Therefore, the measurement of trade-offs is necessary at both levels. Because of the problem described above, the two most appropriate means of measuring a trade-off are manipulative experiments (to estimate the phenotypic trade-off) and breeding experiments (to estimate the genetic correlation). In the final sections of Chapter 3 I provide examples of trade-offs pertinent to life history theory. There is an abundance of data showing that trade-offs occur, although many experiments are inconclusive because they lack statistical power. Further, most experiments, even if they establish the existence of a trade-off, lack

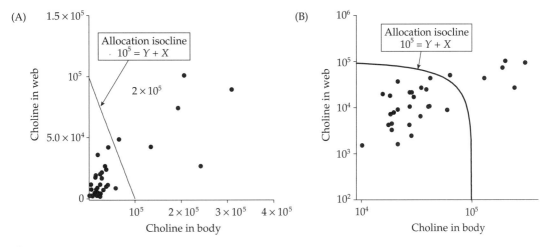

Figure 1.3 The distribution of choline (measured in units of radioactivity) in the body and web of the spider *Nephila clavipes*. The data are plotted both on an arithmetic scale (A) and on a log scale (B). Because of differences in acquisition there is a highly significant correlation between the two components and this "hides" the necessary trade-off in allocation (a single allocation isocline is shown). Data kindly supplied by Linden Higgins.

the resolution to mathematically describe it. This mathematical description is an essential in making qualitative predictions of life history variation, the subject of Chapters 4, 5, and 6.

Evolution in Constant Environments

I begin Chapter 4 with the general question of whether there is sufficient standing genetic variation for evolution in any environment to occur. The answer from breeding experiments and observations on rapid evolutionary change in natural populations is that there is generally an abundance of additive genetic variation in natural populations. The presence of large amounts of additive genetic variance in populations raises the question of what factors are responsible for its maintenance. Stabilizing selection is very commonly observed and is the form of selection most often predicted by the composite effect of trade-offs. However, stabilizing selection will erode variation, and hence we must look elsewhere for a mechanism. Various candidates are discussed in the chapter: mutation-selection balance, antagonistic pleiotropy (discussed in detail in Chapter 3), and frequency-dependent selection. All three are attractive candidates, but at present we simply lack the empirical basis to distinguish among them or decide if any can reasonably be invoked for any natural population.

Most theoretical studies of life history evolution have assumed that the environment is constant, as this greatly simplifies the mathematics. Although the assumption is clearly a simplification, it is not necessarily unrealistic in that environmental variation may be often so small that it can be disregarded. Further, it provides a useful starting point from which more complex analyses can proceed and provides a set of answers against which other models can be compared. Should it turn out that these more complex analyses arrive at basically the same answer, then we can continue to use the simpler modeling framework.

The general approach to the analysis of life history variation is to write the age-specific fecundity and survival functions, designated $m(x)$ and $l(x)$, respectively, as functions of age (x) and the trait or traits under study and find the trait combination that maximizes the fitness measure, typically r or R_0 (for cases in which it can be assumed that populations are more or less stationary). As an example, suppose we wish to estimate the optimal age at maturity given the two functions $m(x) = a + b\alpha$ for ($x = \alpha$) and $l(x) = e^{-Mx}$. In the first function, fecundity increases with age at maturity, α, with a minimum age/size below which fecundity is zero and is constant after maturity (e.g., growth ceases), and the second function designates a mortality rate, M, that is constant (Figure 1.4). For ease of presentation I shall use R_0 as the fitness measure:

$$R_0 = \int_{x=\alpha}^{\infty} (a + b\alpha)e^{-Mx}dx = \frac{1}{M}(a + b\alpha)e^{-M\alpha} \tag{1.3}$$

The optimal age at maturity is that which maximizes R_0, and is found by setting $\partial R_0 / \partial \alpha = 0$, which is when $\alpha = (1/M) - (a/B)$. A decreased mortality rate (M) or decreased rate of fecundity with age (b) favors a delayed age at maturation. This general method of analysis has been used successfully to predict the optimal age/size at maturity in a wide range of organisms. Additionally, it is the general theoretical framework for the analysis of conditions favoring continued growth after maturation, the evolution of clutch size and the joint evolution of propagule and clutch size.

Evolution in Stochastic Environments

In Chapter 5 I examine the opposite type of "world," namely one in which the environment is stochastic and hence only probability statements can be made about future environmental conditions. The intermediate and most realist scenario in which there are cues that can be used to predict forthcoming conditions is discussed in Chapter 6. Stochastic variation introduces two new possible mechanisms for the maintenance of genetic variation: temporal and spatial stochasticity. In its simplest form, temporal stochasticity is an erosive force because, at each generation, extreme phenotypes at one of the tails of the phenotypic distribution are eliminated. If, however, there is generation overlap, then temporal variation can preserve genetic variation. In contrast to temporal variation, the conditions under which spatial variation preserves genetic variation are very

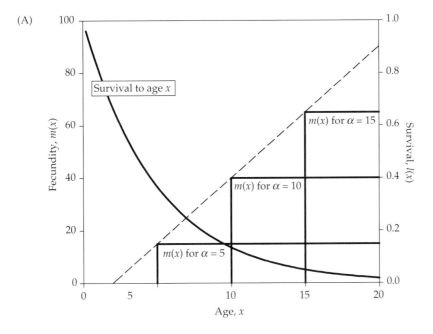

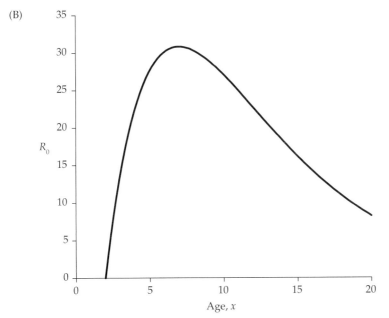

Figure 1.4 (A) A hypothetical life history in which $m(x) = a + b\alpha$, $x = \alpha$, $l(x) = e^{-Mx}$, where $a = -10$, $b = 5$, $M = 0.2$. (B) The relationship between R_0 and age. For the given parameters, the optimal age at maturity is 7.

broad, and it is likely that this phenomenon is a major determinant of the amount of standing genetic variance, although empirical data are sadly lacking.

The analogue for the fitness measure r (constant environment) in a temporally stochastic environment is the arithmetic r, or, equivalently, the geometric mean finite rate of increase, $\bar{\lambda}_{GM}$, which is calculated as

$$\ln(\lambda_{GM}) = \int P(y)\ln\{\lambda(y)\}dy \tag{1.4}$$

where $P(y)$ is the probability of environmental value y and $\gamma(y)$ is the finite rate of increase when the environment takes the value y. What this equation expresses is the potential importance of rare environments in which rates of increase are very low. For example, if there are just two environments the geometric mean is $\lambda_{GM} = \sqrt{\lambda_1^P \lambda_2^{1-P}}$, where P is the probability of λ_1. Suppose we have $\lambda_1 = 1.15$, $\lambda_2 = 0.05$, and $P = 0.95$. In the absence of the rare, poor environment, the population would grow at a rate of 15% per generation, but with the poor environment, the long-term growth rate is $\sqrt{1.5^{0.95}0.05^{0.05}} = 0.98$ and the population becomes extinct! The forgoing fitness measure is appropriate when there is no age structure in the population. With age structure we resort to matrix algebra and obtain a similar measure for the long-term growth rate. The important point about temporal variation is that its effects are multiplicative rather than additive, and this can have profound effects on the optimal life history.

An evolutionary response to the great impact that rare events can have on fitness is the production of a range of offspring. This response, termed "bet-hedging" or "spreading the risk," can be clearly seen in the case of the evolution of dormancy. In a seasonal environment, an organism that has several generations per year and passes the unfavorable period in a state of dormancy (hibernation, diapause, or aestivation) receives very high fitness benefits from the production of an extra generation but suffers very significant fitness decrements if unfavorable conditions commence before the production of the dormant life stage. As a consequence, selection can favor the production of mixed progeny, some attempting to complete development within the present season, and others entering dormancy. Although there are clear theoretical circumstances in which bet-hedging would evolve, the important issue is whether such parameter combinations occur frequently in the real world. I review the models examining the requirements for bet-hedging and the empirical data, reaching the conclusion that bet-hedging is very likely much less common than conventionally assumed.

In a spatially variable environment, one might suppose that selection would favor the evolution of a nondispersing life style. However, if sites become fully occupied, then it might be beneficial to migrate even if migration bears a cost, because it is better to compete with non-kin than with kin. Migration can have significant impacts on the optimal life history because the optimal value of r, for example, is a nonlinear function of the patch values. Specifically, if r is the fitness measure being maximized, then the characteristic equation is expanded to

$$\int p(h) \int l(x,h) m(x,h) e^{-rx} dx dh = 1 \qquad (1.5)$$

where $p(h)$ is the probability of habitat h occurring, $l(x,h)$ is the probability of survival to age x in habitat h, and $m(x,h)$ is the number of female births at age x in habitat h.

In the real world, environments are likely to vary in both space and time. One of the most important adaptations in such a world is migration (also known as dispersal, depending on one's preference). The effect of migration is to "smooth" out the temporal environmental fluctuations measured over the meta-population and hence increase the geometric mean rate of increase. I examine the extensive theoretical justification for this argument and empirical evidence in its support.

Evolution in Predictable Environments

In a constant environment, only one form typically has the highest fitness (frequency dependence providing a notable exception), and in a stochastic environment, either a single trait combination is most fit or selection favors the production of variable offspring (bet-hedging). If there are cues that give information on the state of the present or future environment, we can expect that organisms will evolve means of using such cues to develop into the optimal phenotype or adopt the appropriate behavior. Thus in predictable environments we should find single genotypes giving rise to different phenotypes, depending upon local conditions. This response is called phenotypic plasticity and the functional relationship between the phenotype and the environment is called the norm of reaction. In the final chapter I examine the circumstances favoring the evolution of norms of reaction.

The evolution of phenotypic plasticity can be understood and analyzed from either a character state perspective in which the same trait in different environments is considered to be different traits that are genetically correlated, or a reaction norm perspective in which the coefficients of the reaction norm are treated as traits. Both approaches are examined in Chapter 6.

There are many putative examples of adaptive phenotypic plasticity, but in many cases the evidence is anecdotal. I examine a number of cases in which there has been extensive theoretical and empirical investigation: the evolution of trait variation in seasonal environments with particular reference to reproductive effort, generation length, and dormancy; the evolution of parental care (desertion and defensive investment); the evolution of phenotypic plasticity in clutch and propagule size; reaction norms involving cues indicating mortality regimes with particular emphasis on the evolution of inducible defenses and the induction of alternative life history responses.

CHAPTER 2

A Framework for Analysis

For an organism to evolve we need two processes. First there must be a differential contribution by individuals of the parental generation to the next generation. Thus, for example, some individuals of the parental generation may simply fail to reproduce either because they do not survive to reproduce or they are unable to find mates. Alternatively, there may be a differential contribution among those that do reproduce. The second component we need for evolution to occur is that the traits under question are genetically determined, the precise definition of which is explained in this chapter. The process of evolution can thus be represented as

$$
\begin{array}{ccc}
\text{Step 1} & & \text{Step 2} \\
\text{Mean trait value} \Rightarrow & \text{Trait value of parents} \Rightarrow & \text{Trait value of offspring} \\
\text{in population} & \text{weighted by contribution} &
\end{array}
$$

Except under the most extraordinary circumstances, step 1 will result in a change in mean trait value. This occurs because even if all individuals in the population have an equal opportunity of contributing to the next generation, there will still be variation due to the effect of random sampling. Obviously, if the population size is small, there is a relatively high probability that step 1 will result in a large difference by chance alone. Thus the fact of observing a difference between the mean trait value of the population at large and that of the weighted mean trait value of parents is not sufficient grounds for concluding that some sort of natural selection has occurred. In this book we shall, however, be chiefly concerned with those factors that do result in step 1 being a selective episode (or series of episodes).

The evolution of traits is modulated both directly by selection and by the genetic basis of the traits. In this chapter I present an overview of quantitative genetic principles necessary to understand the influence of genetic architecture on the evolution of life history traits. Although genetic architecture does influence the evolutionary trajectories in most cases, it is not expected to determine the ultimate equilibrium trait combination. The determinant of this combination is the fitness surface, and for many purposes we can analyze life history variation using a strictly phenotypic approach. However, in the real world it is unlikely that equilibrium is ever achieved, because the biotic and abiotic environments are themselves continuously changing. Thus to understand the evolution of life history variation, we need both perspectives. The phenotypic perspective should serve to get us into the region of likely trait combinations, whereas the quantitative genetic perspective gives us insight into how this combination will fluctuate about some long-term average set.

A Single-Trait Model of Inheritance of Quantitative Traits

Even if step 1 does result in a selective change in mean trait value, this will not lead to a change in the next generation unless at least part of the variation is genetic and hence can be passed on. As discussed in the previous chapter, most life history traits show continuous variation; as a consequence, we need an approach that deals with continuous variation but can be reconciled with Mendelian genetics. This approach is quantitative genetics.

Mendelian Genetics and the "Regression to the Mean" Problem

At first glance quantitative and Mendelian genetics appear quite different, and the latter does not appear to be a solution to the question of continuous variation. To appreciate this point, consider the problem of "regression to the mean," as illustrated by Galton's data on the heights of parents and offspring in nineteenth-century England (Figure 2.1). From these data it is apparent that height is an inherited trait, provided we assume that there was not some overwhelming environmental component such that large parents "selected" environments promoting the growth of large offspring, whereas small parents "selected" the converse environment. Now suppose that the mean mid-parent height of contributors to the next generation is 74 inches. Then, assuming that the assumptions of linear regression analysis are fulfilled, the height of the offspring, and the parents of the next generation will be 21.52 + (0.69)(74) = 72.58, which is an overall reduction in height from the selected parents. If, on the other hand, the mean mid-parent height of contributors to the next generation is 64 inches, the height of the next generation will be 21.52 + (0.69)(64) = 65.68 inches, which is an overall increase in height from the selected parents. Assuming that there is no further selection of parents and that the population is sufficiently large that sampling error can be ignored, the recursive use of the linear regression equation brings the population mean eventually to 69.42 inches, regardless of the initial starting value. In other words, the value will regress over time to the point at which the regression line intersects the line of equality (Figure 2.1B). The apparent instability of the population mean following selection presents a challenge because it suggests that evolutionary change necessitates continuous selection. At first glance, Mendel's model of inheritance does not appear to offer a solution; it deals with discrete variation, whereas traits such as height are continuous.

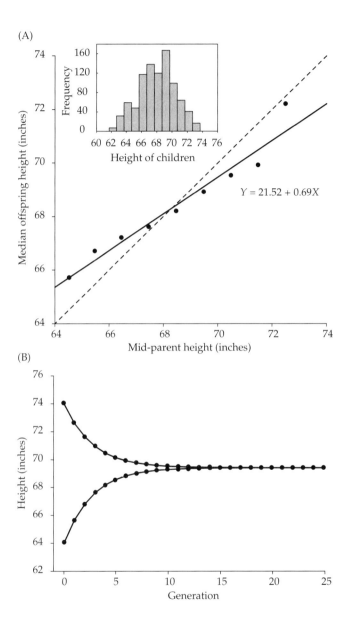

Figure 2.1   (A) Galton's data showing the median height of adult human offspring as a function of the mid-parent height. Each point is the median value of all children from parents with the particular mid-parent value. More correctly, the mean value of each family should be plotted separately. To correct for differences in height between men and women, Galton multiplied the heights of women by 1.08. The inset shows the distribution of the height of the children; note that it is approximately normal. Data taken from Table 11 of Galton (1889). (B) Two examples of the regression to the mean using the regression function derived from Galton's data.

The situation was resolved by Yule (1902, 1906), who realized that a continuous distribution will result from a large number of discrete loci that contribute additively to a trait value. The full details of how this conceptual framework eliminates the regression to the mean were later worked out by Fisher (1918). Under the simplest version of this scheme, each locus is inherited independently (unlinked), and the value of a trait is simply the sum of the contributions of each locus plus an environmental "error" term, which leads to a regression slope that is less than 1.

To see how we can move from Mendelian genetics to quantitative genetics, let us suppose that a particular trait X is determined by a single locus at which there are two alleles, one, denoted as A, contributing a value of $\frac{1}{2}$ to the trait, and the other, denoted as B, contributing a value of $-\frac{1}{2}$. (This does not imply any loss of generality because whatever the actual values of A and B are, we can always rescale to $\frac{1}{2}$ and $-\frac{1}{2}$. These values make the algebra much simpler.) The phenotypic value of the trait is then the sum of the allelic contributions plus a normally distributed random environmental error, e, which has a mean of zero. Thus the phenotypic values of the three genotypes are $AA = 1 + e$, $AB = 0 + e$, and $BB = -1 + e$.

As shown in Table 2.1, there are six types of matings. The mean offspring value is equal to the mid-parental genotypic value (because e has a mean of

Table 2.1   **The distribution of offspring values for a simple single-locus, two-allele model in which the phenotypic value is equal to the sum of the allelic values plus a normally distributed error e.**

Genotype of parents	Frequency of matings, f_i	Mid-parent value, X_i	Proportion of each type of offspring			Mean offspring value, Y_i
			AA	AB	BB	
AA, AA	p^4	$1 + \dfrac{e_{AA} + e_{AA}}{2}$	1	0	0	1
AA, AB	$4p^3q$	$\dfrac{1}{2} + \dfrac{e_{AA} + e_{AB}}{2}$	$\frac{1}{2}$	$\frac{1}{2}$	0	$\frac{1}{2}$
AA, BB	$2p^2q^2$	$0 + \dfrac{e_{AA} + e_{BB}}{2}$	0	1	0	0
AB, AB	$4p^2q^2$	$0 + \dfrac{e_{AB} + e_{AB}}{2}$	$\frac{1}{4}$	$\frac{1}{2}$	$\frac{1}{4}$	0
AB, BB	$4pq^3$	$-\dfrac{1}{2} + \dfrac{e_{AB} + e_{BB}}{2}$	0	$\frac{1}{2}$	$\frac{1}{2}$	$\frac{1}{2}$
BB, BB	q^4	$-1 + \dfrac{e_{BB} + e_{BB}}{2}$	0	0	1	-1

Note: The subscripts on the error term refer to the environmental values of the denoted parent. Environmental values with the same subscript are not necessarily equal (unlikely to be so).

zero), but the phenotypic mid-parent value is inflated or decreased by the contribution of the environmental component. (On average, of course, this change will be zero, but we plot the individual values, not the means.) If we knew the genotypic values of parent and offspring, the regression between mean offspring and mid-parent value would be 1 and the regression-to-the-mean problem would not arise. But the effect of the environmental contribution is to lower the slope. (This result will be true even if we measure only one offspring per family.) The regression to the mean arises because the offspring on parent regression is a consequence of both Mendelian genetic inheritance and environmental effects. This idea is shown in Figure 2.2, in which two hypothetical offspring-parent regressions are plotted. In both cases the phenotypic value of the offspring is determined by a single locus with two alleles plus a random normal

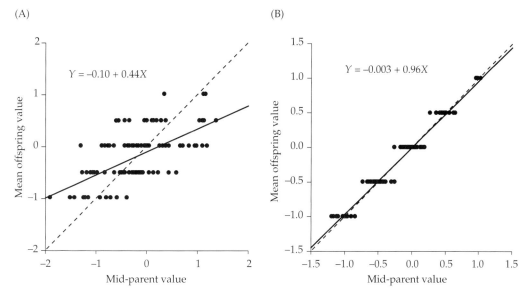

Figure 2.2 Hypothetical offspring on parent regressions using a single-locus, two allele genetic model in which the phenotype is the sum of the allelic values plus an environmental deviation (as shown in Table 2.1). The frequency of the *A* allele was set at 0.4 and the number of mean offspring values per mating category set at 100 times the frequency of the particular mating. (For example the mating *AA, AB* occurs with frequency $4p^3q$ and hence the number of mean offspring values [= families], to the nearest integer, from this mating is $(100)4(0.4)^3(0.6) = 15$.) The mean offspring value was set at the expected mean value (Table 2.1) and the mid-parent values for each determined as the expected mid-parent value + $\frac{1}{2}(e_M + e_F)$, where e_M and e_F are identically distributed normal variables with mean zero and variance 1 (A) or 0.01 (B). The solid lines show the fitted regression line.

deviate. Therefore, the underlying genetic relationships remain the same, but in one case (Figure 2.2B) the environmental variance is only 0.01, whereas in the second case (Figure 2.2A) the environmental variance is 1. In the former case the genetic relationship is clear, but in the second the pattern of inheritance is obscured by the environmental effect. Under Mendelian inheritance, one generation of selection changes the mean trait value, which thereafter, in the absence of further selection, remains the same and does not regress (Box 2.1).

Is there any way to adjust the regression model such that there is no regression to the mean? A simple solution is

$$\overline{Y}_{t+1} = (1-b)\overline{Y}_t + b\overline{X}_t \tag{2.1}$$

where $\overline{Y}_{t+1}$ is the population mean trait value in generation $t + 1$, $\overline{X}_t$ is the mean value of the parents that contribute to the next generation, and b is the slope of the offspring-parent regression. Under this formulation the intercept changes in each generation in such a manner that there is no longer a tendency to regress back from any new value. But does this model make sense under Mendelian

---

Box 2.1

In a Randomly Mating Population, the Mean Trait Value Does Not Regress

In a randomly mating population, the mean trait value should remain constant from generation to generation (assuming no drift due to sampling variance). Likewise, under Mendelian inheritance, if the genotypic value is shifted, then with no further selection it should remain at the new value. For example, in the present single locus case, if the proportion of the A allele is p (and $q = 1 - p$) then the population mean trait value is

$$p^2(1) + 2pq(0) + q^2(-1) = p^2 - q^2 \tag{2.2}$$

If we could determine genotypic value and select, say, only those greater than or equal to 0.0 (AA and AB genotypes), the mean trait value would shift to

$$\frac{p^2(1) + 2pq(0)}{p^2 + 2pq} = \frac{p}{p + 2q} \tag{2.3}$$

and the new proportion of A would be

$$\frac{p^2 + pq}{p^2 + 2pq} = \frac{1}{p + 2q} \tag{2.4}$$

From the Hardy-Weinberg principle, the proportion would stay constant thereafter (under random mating), and hence the mean trait value would remain at the selected value.

inheritance? To answer this question we must turn to the biometrical interpretation of the regression line. By standard linear regression theory, the slope of the regression equation is equal to the covariance between offspring and parental trait values divided by the variance of the mid-parent values. Taking the covariance term first, for simplicity, and without loss of generality, let the mean be set at zero (i.e., trait values are measured as deviations from the mean). Let $\bar{X}$ be the mid-parent value, X_m, X_f be the trait values of the two parents and $Cov(X,Y)$ be the covariance between X and the mean offspring value, $\bar{Y}$. Then, we have

$$\bar{X} = \frac{X_m + X_f}{2}$$

$$Cov\left(\bar{X}, \bar{Y}\right) = Cov\left(\frac{X_m + X_f}{2}, \bar{Y}\right) = \frac{1}{2}\left\{Cov\left(X_m, \bar{Y}\right) + Cov\left(X_f, \bar{Y}\right)\right\} \tag{2.5}$$

Now, because we have assumed that the environmental deviations are uncorrelated either among themselves or with the genetic value, the covariance between a parent and its offspring must represent only the covariance between genetic values. This remains true even if only one offspring per family is used. Further, because the offspring contain half the genes of each parent, the covariance between one parent and its offspring must equal one-half of the **additive genetic variance**, denoted as σ_A^2. The additive genetic variance is to be distinguished from the **total genetic variance**, denoted as σ_G^2. σ_A^2 is that component of the total genetic variance that contributes to the linear regression between offspring and parents. Thus the covariance between mid-parent and offspring is equal to one-half the additive genetic variance. If x and y are two independent random variables, the variance of the linear function $c_x x + c_y y$, where c_x and c_y are constants, is $c_x^2 \sigma_x^2 + c_y^2 \sigma_y^2$. Hence, assuming that the phenotypic variance in males is the same as that in females (denoted as σ_P^2), the **phenotypic variance** of the mid-parent values is $(\frac{1}{2})^2 \sigma_P^2 + (\frac{1}{2})^2 \sigma_P^2 = \frac{1}{2}\sigma_P^2$. The slope of the offspring on mid-parent regression is therefore equal to

$$b = \frac{\frac{1}{2}\sigma_A^2}{\frac{1}{2}\sigma_P^2} = \frac{\sigma_A^2}{\sigma_P^2} = h^2 \tag{2.6}$$

As indicated, this slope is given a specific symbol, h^2, called the heritability of the trait in the narrow sense, generally called simply **heritability.** (It is denoted as a squared parameter to symbolize that it is the ratio of variances.) Rearranging the offspring-on-mid-parent regression equation we get

$$\bar{Y}_{t+1} - \bar{Y}_t = h^2\left(\bar{X}_t - \bar{Y}_t\right) \tag{2.7}$$

The term on the left is the **response to selection,** *R,* and the term in parentheses on the right is the **selection differential,** *S.* Hence we arrive at the familiar response equation $R = h^2 S$. For purposes of comparison, it is often useful to write the selection differential in terms of phenotypic standard deviation units, usually designated by the letter *i.* Assuming truncation selection, the **intensity of selection** depends only on the proportion of the population selected ($i = z/p$, where *p* is the proportion selected and *z* is the ordinate at the point of truncation). Hence the intensity of selection can be readily obtained from statistical tables or from statistical computer packages. In terms of the selection intensity, $R = ih^2 \sigma_P = ih\sigma_A$. If we compare the response to selection of different traits, the response equation is problematic because it is not dimensionless and hence depends on the scale of measurement. A plausible solution is to consider the proportional change in the trait value, which can be written as

$$\frac{R}{\mu} = \frac{ih^2 \sigma_P}{\mu} \tag{2.8}$$

where μ is the mean trait value before selection. Houle (1992) suggested the term **evolvability** to denote the ratio

$$\frac{R}{i\mu} = \frac{\sigma_A^2}{\mu\sigma_P} \tag{2.9}$$

and an index of evolvability under various types of selection regime to be the coefficient of additive genetic variance, σ_A^2/μ. Although this is useful for comparing traits measured on the same scale there are problems with interpretation of evolvabilities measured on different traits (Roff 1997; Lynch and Walsh 1998).

The Constancy of Heritability

A problem indicated by the single-locus model is that the heritability (slope of the regression line) does not remain constant but changes each generation that selection is applied. For example, in the numerical example described in Box 2.2, when the environmental variance is 1, 25.87% of the population contributes to the next generation and the heritability changes from 0.32 to 0.31. Despite the fairly strong selection, the change in the heritability is quite small, but if the same selection were applied for many generations there would be a consistent erosion of the additive genetic variance. If, however, there are a large number of loci, then the change in allelic frequency at each locus will be greatly diminished and the rate of erosion greatly slowed.

For time periods of approximately 5–15 generations, the assumption of a constant heritability appears to be valid, given the intensities of selection typically used in artificial experiments (approximately 20% selected by truncation selection). This can be demonstrated by comparing the heritability estimated using the observed response to selection, termed the **realized heritability** (Box 2.3), with that estimated from pedigree analysis (e.g., offspring-parent regression, half-sib analysis; see Chapter 3 for an overview of such methods). A plot of realized heritabilities on estimates from other techniques is presented in Figure 2.3. The heritabilities were typically estimated over 5–15 generations of selection. Although there are more estimates lying below the line, there is certainly no overwhelming trend for the realized heritability to be markedly less than that estimated from sib

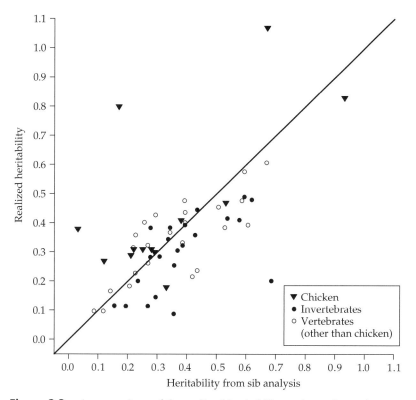

Figure 2.3 A comparison of the realized heritability estimated over less than 15 generations with the heritability estimated from pedigree analysis. Invertebrate data sources: *Drosophila*: Tantawy (1956), Clayton et al. (1957), Latter (1964), Sen and Robertson (1964), Frahm and Kojima (1966), Lopez-Fanjul and Hill (1973), van Dijken and Scharloo (1979), Sorenson and Hill (1982, 1983), Gallego and Lopez-Fanjul (1983), Gromko (1987), Gromko et al. (1991). *Tribolium*: Enfield et al. (1966), Martin and Bell (1960), Yamada and Bell (1969), Ruano et al. (1975). *Oncopeltus fasciatus*: Palmer and Dingle (1986). *Gryllus firmus*: Roff (1990a). *Wyeomyia smithii*: Hard et al. (1993), Bradshaw et al. (1996). Vertebrate data sources: Mouse: Eisen et al. (1970), Wilson et al. (1971), Rutledge et al. (1973), Cheung and Parker (1974), Lynch (1994). Chicken: Kinney (1969). Pigs and Japanese quail, Sheridan (1988, review).

Box 2.2
The Single-locus Model and the Mean Offspring on Mid-parent Regression

To see how well the equation developed to this point works in practice, consider the single-locus model previously described. In this model the variance in genotypic values can be readily obtained from the statistical relationship $\sigma_G^2 = \mu_2 - \mu^2$, where σ_G^2 is the genetic variance, μ_2 is the second moment (equal to the expected value of the squared values) and μ is the mean. Thus, if we let $q = 1 - p$,

$$\mu = (1)p^2 + (0)2pq + (-1)q^2 \quad = p - q$$
$$\mu_2 = (1^2)p^2 + (0^2)2pq + (-1^2)q^2 = p^2 + q^2$$
$$\sigma_G^2 = p^2 + q^2 - (p - q)^2 \quad\quad = 2pq$$

(2.10)

The phenotypic variance, σ_P^2 is therefore $\sigma_G^2 + \sigma_E^2 = 2pq + \sigma_E^2$, where σ_E^2 is the environmental variance. In the above model, because the effect of the alleles is strictly additive (no dominance), the genetic variance is also equal to the additive genetic variance (i.e., $\sigma_G^2 = \sigma_A^2$). The heritability of the trait is therefore

$$\frac{2pq}{2pq + \sigma_E^2}$$

(2.11)

Above we examined the consequence of selection when it was possible to distinguish the genotypes. In reality, we can only select on the basis of the phenotypic value and this, because of the environmental factor, may not be an accurate estimate of the genotypic value. Therefore, if we were to select as parents all individuals with phenotypic values greater than or equal to 0.5, we could, depending on the environmental variance, include some of each genotype in the selected group (Table 2.2). Table 2.3 shows the calculations required for the estimation of the mean value of offspring using the Mendelian model, the linear regression model using the theoretical heritability, and the linear regression model applied to the simulated data. There is fair agreement between the Mendelian model and the values predicted from linear regression model, though there is a large difference between the expected and estimated heritability for the larger environmental variance. The reason for this difference is due, at least in part, to the failure of the data to conform sufficiently well to the statistical requirement of linear regression that the offspring data be normally distributed about the regression line.

How could we alter the genetic model so that the assumptions of linear regression are better upheld? Obviously, what we require is that the genotypic values be more continuous. There are two ways we can envisage this happening. First, there could be many alleles of differing value at a single locus and, second, there could be many independent loci, with each locus contributing a small additive amount to the trait value. It is the latter model that is generally taken as the Mendelian interpretation of the offspring-parent regression. This model, with the inclusion of dominance, leads to the **infinitesimal model** described later in this chapter.

Table 2.2   **Probability that the phenotypic value of a given genotype will exceed 0.5, under two environmental variances, given an initial allele frequency for allele *A* of *p* = 0.4.**

Genotype "*IJ*"	Genotypic value G_{IJ}	Frequency of genotype $P_1(IJ)$	Probability[a] that phenotypic value > 0.5 $P_2(IJ)$		Mean trait value within genotypes $\bar{X}_{IJ}$	
			$\sigma_E^2 = 0.01$	$\sigma_E^2 = 1.0$	$\sigma_E^2 = 0.01$	$\sigma_E^2 = 1.0$
AA	1	0.16	1.00	0.69	1.00	1.51
AB	0	0.48	0.00	0.31	—	1.14
BB	−1	0.36	0.00	0.07	—	0.94

[a]The probability that the phenotypic value is greater than 0.5 is given by the cumulative normal distribution formula $P_2(IJ) = \dfrac{1}{\sigma_E \sqrt{2\pi}} \displaystyle\int_{0.5}^{\infty} e^{-\frac{1}{2}\left(\frac{x-G_{IJ}}{\sigma_E}\right)^2} dx$. The mean trait value within genotypes is computed using the formula for a truncated normal distribution, $G_{IJ} + \dfrac{\sigma_E e^{-\frac{1}{2}\left(\frac{G_{IJ}}{\sigma_E}\right)^2}}{P_2(IJ)\sqrt{2\pi}}$.

Table 2.3   **Heritability and mean offspring trait value when truncation selection is applied in which phenotypes greater than 0.5 contribute to the next generation.[a]**

Population genetics[b]		Regression method $\bar{Y}_{t+1} = (1 - h^2)\bar{Y}_t + h^2\bar{X}_t$				
		Theoretical[c]		Simulation		
σ_E^2	$\bar{Y}_{t+1}$	h^2,	$\bar{Y}_{t+1}$	$\hat{h}^2$,	$\bar{Y}_{t+1}$	$(n)^d$
0.01	1.00	0.98,	0.98	0.96,	0.71	(*n* = 9)
1.00	0.31	0.32,	0.28	0.44,	0.25	(*n* = 14)

[a]Genotypes are determined by the single-locus, two-allele model described in Table 2.1.

[b]The mean trait value in the offspring population, $\bar{Y}_{t+1}$, is equal to $(1)p'^2 + (0)2p'q' + (-1)q'^2$, where p' is the frequency of allele *A* in the offspring population, which is calculated from

$$\dfrac{P_1(AA)P_2(AA) + \frac{1}{2} P_1(AB)P_2(AB)}{P_1(AA)P_2(AA) + P_1(AB)P_2(AB) + P_1(BB)P_2(BB)}.$$ $P_1(\bullet)$ and $P_2(\bullet)$ are given in Table 2.2. $p' = 1$ when $\sigma_e^2 = 0.01$ and 0.653 when $\sigma_e^2 = 1$.

[c]$h^2 = \dfrac{\sigma_A^2}{\sigma_P^2}$, where $\sigma_A^2 = 2pq$, $\sigma_P^2 = 2pq + \sigma_E^2$. The mean trait value in the initial population, $\bar{Y}_t$, is calculated from $Y_t = (1)(0.4)^2 + (0)(2)(0.4)(0.6) + (-1)(.6)^2 = -0.2$. The mean trait value in the selected set of parents, $\bar{X}_t$, is $\bar{X}_t = \dfrac{\sum P_1(IJ)P_2(IJ)\bar{X}_{IJ}}{\sum P_1(IJ)P_2(IJ)}$ (see Table 2.2).

[d]The number of selected individuals.

Box 2.3
The Realized Heritability

Recall that $R = h^2 S$, from which an estimate of heritability can be obtained by rearrangement, $h^2 = R/S$. Response in a single generation is frequently very variable, and therefore, several generations of selection are generally required to obtain an accurate estimate of the realized heritability. If selection is continued over t generations we have

$$R_1 = \left(\overline{X}_1 - \overline{X}_0\right) = h^2\left(\overline{Y}_0 - \overline{X}_0\right) = h^2 S_1$$
$$R_2 = h^2\left(S_1 + S_2\right)$$
$$\cdots \qquad\qquad (2.12)$$
$$R_t = h^2\left(S_1 + S_2 + S_3 \cdots + S_t\right)$$

where $\overline{X}_i$ is the mean value of the trait measured in generation i, $\overline{Y}_i$ is the mean value of the selected individuals in generation i, R_i is the total or cumulative response (the difference between the mean value at the start of the experiment and that at ith generation), and S_i is the selection differential applied to the parents that gives rise to generation i. The sum $S_1 + S_2 + S_3 \cdots + S_t$ is the cumulative selection differential applied. Note that in total there are $t + 1$ generations. From the above it can be seen that the cumulative response is a linear function of the cumulative selection differential with a slope equal to the heritability of the trait. (Since the original trait value is a constant, one can replace R_i with $\overline{X}_i$.) Thus from a selection experiment the heritability can be estimated by a linear regression of response on cumulative selection differential (Falconer 1989; Figure 2.4). For a fuller discussion of the statistical problems associated with the estimation of the realized heritability, see Chapter 4 of Roff (1997).

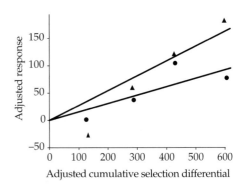

Figure 2.4   An example of the estimation of realized heritability. Females of the cricket *Gryllus firmus* were selected for high fecundity during the first seven days of adult life. Two lines were selected upwards, and for each line a control line was maintained. To account for variation among generations that was not a result of selection, the means of the selected lines were adjusted by subtracting the mean of the respective control line. The realized heritability is equal to twice the slope (because only females were selected) of the regression line forced through the origin (because of the adjustment using the control lines). Standard errors of the realized heritabilities were estimated using the formula of Hill, W. G. (1972). The estimates were 0.20 ± 0.05 and 0.34 ± 0.06, giving an average value of 0.27 ± 0.06. For a full description of this experiment, see Roff et al. (1999).

analysis; thus, this result supports the assumption of at least short-term constancy of the heritability.

Artificial selection experiments have shown that the heritability of a trait may be reduced after 20 generations or so of selection, given selection intensities roughly comparable to those in the single-locus model example—that is, approximately 20% (Figure 2.5). However, there are some notable exceptions in which continued response, indicative of retention of genetic variation, has occurred for many generations; for example, (1) 76 generations of selection for oil content in

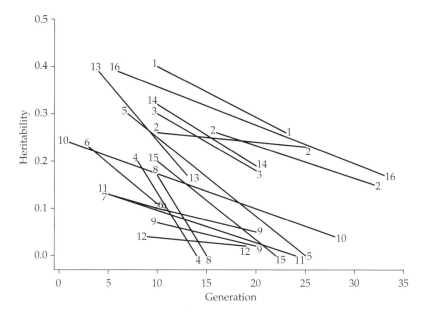

Figure 2.5 Some examples of declines in heritability observed during the course of artificial selection experiments. Heritabilities were estimated over a number of generations at the start of each experiment and then over differing periods after a period of selection (e.g., generations 1–10 and then 11–20): estimates are plotted on the last generation in each period (e.g., generation 10 and 20). Species (not shown if same as preceding), trait, % selected per generation (where available). Source: 1. *Tribolium casteneum*: pupal weight, 20% (Gall 1971); 2. fecundity, 20% (Ruano et al. 1975); 3. *D. melanogaster*: wing length (Robertson and Reeve 1952), 4. development time, 39% (Sang 1962); 5. mating speed, 20%, (Manning 1961); 6. copulation duration, 13%, (Gromko et al. 1991); 7. geotaxis, (Watanabe and Anderson 1976); 8. walking, (Choo 1975); 9. *Drosophila pseudoobscura*, (Dobzhansky et al. 1969); 10. body weight, 20%, (Frahm and Kojima 1966); 11. mating speed, 29%, (Kessler 1969); 12. *Drosophila subobscura*, light preference, (Kekic and Marinkovic 1974); 13. Mouse, growth rate, 23%, (Falconer 1960); 14. body weight, 21%, (Wilson et al. 1971); 15. litter weight, (Eisen 1972); 16. litter size index, 27%, (Schuler 1985).

maize (Dudley 1977); (2) 75 generations of selection for pupal weight in *Tribolium* (Enfield 1980); (3) 50 generations of selection for abdominal bristle number in *Drosophila melanogaster* (Jones et al. 1968); (4) 75 generations of selection for abdominal bristle number in *D. melanogaster* (Yoo 1980a; not all lines continued to respond to selection); (5) 60 generations of selection for ethanol resistance in *D. melanogaster* (Weber and Diggins 1990); and (6) 55 generations of selection for wing-tip height in *D. melanogaster* (Weber and Diggins 1990).

At first glance the experiments in which there was a significant loss of genetic variation might be taken as indicative of a low number of loci or alleles determining the trait. But the problem with many artificial selection experiments is that the experimental population is frequently very small, and hence there is a great potential for a loss of genetic variation through the effects of drift rather than selection. Within a population, the genetic variance at a single locus with two additive alleles (no dominance) at some time *t* is (Robertson, A. 1960a):

$$\sigma_t^2 = 2p_0q_0\left(1 - \frac{1}{2N_e}\right)^t \approx 2p_0q_0e^{-\frac{t}{2N_e}} \tag{2.13}$$

where p_0 is the initial allelic frequency and N_e is the effective population size. In the present case this is simply the census population size. (The difference between the effective and census population size will be described later.) Thus the variance is declining exponentially with time. An interesting feature of this equation is that it predicts that the total response to selection will be equal to the initial response times $2N_e$ (Robertson, A. 1970a). The range in effective population size in Figure 2.5 is from 6 to 700, but most values are approximately 50. For an effective population size of 50, the genetic variance will decline, on average, by a factor of 0.82 after 20 generations and by a factor of 0.74 after 30 generations. The actual decline in heritability will be much greater than this if the environmental variance remains constant. There are insufficient data for the experiments shown in Figure 2.5 to decide if the declines can be attributed to drift, but the point is that *a priori* a significant effect due to drift cannot be discounted.

Attempts to match long-term responses with theoretical predictions have frequently not been successful with respect to precise quantitative values. For example, selection for abdominal bristles in *D. melanogaster* has produced long-term responses but "the long-term behavior of these lines is bewilderingly complex" (Clayton and Robertson 1957, p. 166), "in general, agreement with these models was poor" (Jones et al. 1968, p. 265), and "the pattern of long-term response was diverse and unpredictable" (Yoo 1980a, p. 1). Possible reasons for the failure are the presence of lethal genes (Clayton and Robertson 1957; Yoo, 1980b), infertility of extreme females and heterozygosity (Clayton and Robertson 1957), presence of a few genes of large effect (Jones et al. 1968), and an overestimation of the effective population size (Frankham 1977).

Weber and Diggins (1990) compared the predicted and observed relative response after 50 generations of selection using both their own data and those obtained from the literature. Their starting point was the model of A. Robertson (1970a):

$$R(t) = ih_0^2 \sigma_{P0} \sum_{i=0}^{t-1} \left(1 - \frac{1}{2N_e}\right)^i \approx 2N_e ih_0^2 \sigma_{P0} \left(1 - e^{-\frac{t}{2N_e}}\right) \tag{2.14}$$

where $R(t)$ is the response after t generations of selection and h_0^2, σ_{P0} are, respectively, the initial heritability and initial phenotypic standard deviation. This equation takes into account loss of variation due to genetic drift but not nonadditive genetic effects, linkage, more than two alleles per locus, mutation, or a finite number of loci. Effects of linkage are unimportant (Robertson 1970b; Robertson and Hill 1983). Weber and Diggins (1990) incorporated mutation into their model, but it made very little quantitative difference to the predicted response. For population sizes less than 10 there was fair agreement with prediction, but thereafter the observed values fell below the predicted value (Figure 2.6). Nevertheless, there is the same overall sigmoidal increase in response that is

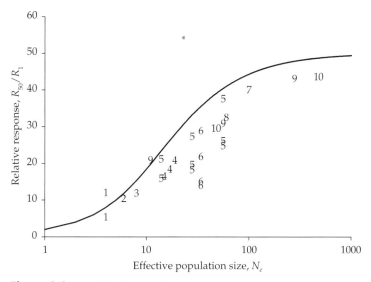

Figure 2.6 Predicted cumulative response relative to the initial response after 50 generations of selection in relation to population size, $R_{50}/R_1 = (1 - C^{50})/(1 - C)$, where $C = 1 - 1/2N_e$. Also plotted are the observed responses for various experiments, separate studies denoted by numbers. (Species used are *D. melanogaster*, 1, 2, 3, 5, 8, 9, 10; *Zea mays*, 6; *Mus domesticus*, 4, and *T. casteneum*, 7.) Data from Table 2 of Weber and Diggins (1990).

predicted by theory. In reviewing their analysis Webber and Diggins (1990, p. 593) noted that "populations that have actually been selected to the point of near-exhaustion of response (e.g., Reeve and Robertson 1953; Roberts 1966a,b; Enfield 1980; Yoo 1980a) indicate that Robertson's upper limit of $2NR_1$ [R_1 = initial response] cannot be attained except in very small populations. For populations of 10 it is realistic; for populations of 50 it is not remotely possible." One reason for the failure of the foregoing models is that the loss of genetic variation is almost certainly underestimated by the assumption of an infinite number of loci, as demonstrated by the simulation models of Hospital and Chevalet (1993).

A significant challenge to the theory of the long-term response is that reversed or relaxed selection frequently demonstrates the presence of additive genetic variance even when little or no further progress is being made in the original direction of selection (Reeve and Robertson 1953; Robertson 1955; Dawson 1965a; Clayton and Robertson 1957; Rathie and Barker 1968; Wilson et al. 1971; Eisen 1972; Lerner and Dempster 1951; Dickerson 1955; Kaufman et al. 1977; Yoo et al. 1980). Such a result could occur because (1) artificial selection in one direction is opposed by natural selection; (2) the alleles favoring the change in the direction of selection are dominant, inbreeding then causing a depression of the response (see later in this chapter for a discussion of inbreeding depression); (3) overdominance (considered unlikely by Falconer, 1989). The question of the maintenance of genetic variation in the face of selection is discussed in more detail in Chapter 4. The brief review given here highlights the simplicity of the present theoretical foundation. Application of such a theory to natural populations in which one or all of the complicating factors are likely to be found must be done with considerable caution, or skepticism. Nevertheless, it is only with field tests that the worth of the model can be properly tested. In particular, the very strong selection intensities that are applied continuously for many generations are probably not representative of the situation in the wild, where there may be bouts of strong selection followed by longer periods of weak selection (see Chapter 2).

To sum up: The regression to the means problem can be solved by invoking a Mendelian model in which there are many independent loci with two or more alleles per locus. The phenotypic value is given as the sum of all the allelic values plus an environmental effect that is normally distributed with a mean of zero. Only a few loci are actually required to produce an effectively continuously varying, normally distributed phenotype. With only a few loci and alleles, directional selection can erode genetic variation. Attempts to use the model to predict long-term response to artificial selection have not been quantitatively very successful, though the models do give reasonable qualitative answers.

The Mendelian model we have made use of so far ignores three phenomena that are most clearly present: linkage, dominance, and epistasis. How does the inclusion of such features affect our simple statistical model?

Linkage

To derive the genomic variance, we assumed that the loci were independent. Such would be the case only if each locus were on a separate chromosome. This case is unlikely, and hence we must inquire whether linkage is an important concern. (Its relative unimportance has been noted above, but no rationale given.) First, consider a randomly mating population. According to the **Hardy-Weinberg law**, in a randomly mating population under no selection, genotypic frequencies will come into equilibrium (i.e. p^2, $2pq$, q^2) after one generation. However, while this is true for the individual loci, it is not true for the multi-locus genotypes. Suppose a population consisting of individuals homozygous at two loci, say, *AABB* (not to be confused with the previous use of these symbols), and another population consisting entirely of the genotype *aabb* are randomly mixed. In the first generation, only two parental crosses are possible, giving rise to only three offspring genotypes, although at equilibrium nine genotypes will be found (Figure 2.7). The distribution of the individual loci will be at Hardy-Weinberg equilibrium (e.g., frequencies of 0.25, 0.5, and 0.25 for *AA*, *Aa*, and *aa*, respectively). In the second generation of random mating all nine genotypes will appear, but they will not occur at the frequency that is expected at equilibrium (Figure 2.7). The equilibrium frequencies will be approached asymptotically. The phenomenon of nonrandom association of loci is called **linkage disequilibrium** or **gametic phase disequilibrium.** The term is somewhat misleading since the disequilibrium does not imply that the loci are linked in the sense of being on the same chromosome. It is intuitively obvious that in the absence of retarding forces loci, whether linked or not, will eventually attain linkage equilibrium, the time taken decreasing with the crossover rate. It can be shown that (Crow and Kimura 1970, p. 48; Falconer 1989, p. 19):

$$P_t(AB) - p_A q_B = \{P_0(AB) - p_A q_B\}(1-c)^t \tag{2.15}$$

where $P_t(AB)$ is the frequency of genotypes containing at least one *A* and one *B* at time t, p_A is the frequency of the *A* allele, p_B is the frequency of the *B* allele, and c is the crossover rate. (Thus $p_A p_B$ is the equilibrium probability of an individual containing at least one *A* and one *B*.) For unlinked loci, $c = 0.5$, and thus for unlinked loci the disequilibrium value is halved each generation. In a randomly mating population, linkage equilibrium will eventually be achieved and maintained thereafter. Therefore, the foregoing assumption of independence of loci is not violated at equilibrium in a randomly mating population.

This property of independence will be lost in a population undergoing selection. For example, under directional selection for a large phenotypic value there will be a build-up of a correlation between alleles of large effect. However, linkage disequilibrium has little effect on the additive genetic variance maintained by mutation and stabilizing selection (Turelli and Barton 1990).

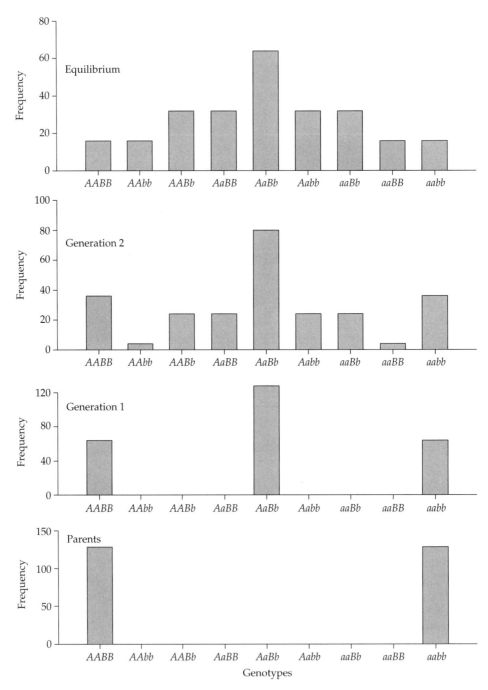

Figure 2.7   The distribution of genotypes with random mating given an initial cross of *AABB* × *aabb*. Notice that, unlike a single-locus model, equilibrium is not achieved after one generation of random mating.

Dominance

A standard feature of single-locus Mendelian models is the phenomenon known as **dominance,** in which the phenotype of the heterozygote resembles one homozygote more than the other. Mathematically, this condition can be represented by designating the genotypic value of the heterozygote as d. Given the present values of the homozygous genotypes in the single locus model (AA, BB) as 1 and –1, then if $d = 1$ there is complete dominance of the A allele, whereas if $d = -1$ there is complete dominance of the B allele. If d is less than –1 or greater than +1, the heterozygote is more extreme in phenotype than the relevant homozygote, a condition known as **overdominance.**

Complete dominance does not necessarily mean that the effect of the alleles is not additive. Suppose, for example, the effect is additive and produces a titre of hormone or some other regulatory substance, dependent upon the sum of the allelic values (plus an environmental deviation). Further, suppose that the phenotype produced depends not upon the hormone titre per se but upon one of two possible developmental pathways, and that the pathway depends upon a threshold value of the hormone. If the hormone exceeds the threshold, one particular morph is produced, whereas if the threshold is not exceeded, the alternate morph is produced. The production of dimorphic variation by this mechanism is well known, whether the variation is due to a single locus or the additive effect of multiple loci. It occurs in a wide range of taxa and traits (Roff 1996a).

The apparent presence of dominance may result from the particular choice of scale. Consider, for example, the following two cases: (1) $AA = 10$, $Aa = 3.16$, $aa = 1$, and (2) $AA = 1$, $Aa = 0.5$, $aa = 0$. In the first case AA appears to be dominant over aa, but in the second case there is apparent additivity. However, the second is merely the logarithm (base 10) of the first. Dominance can also be a function of the environment. An example of this is pesticide and herbicide resistance. In wild populations, resistance is frequently associated with a single locus (Taylor, A. D. 1986; Roush and McKenzie 1987; Mallet 1989; Ffrench-Constant 1994; McKenzie and Batterham 1994; Jasieniuk et al. 1996; Taylor and Feyereisen 1996). Figure 2.8 shows a hypothetical example in which the alleles act additively, moving the dose response curve of the three genotypes a fixed distance along the dosage axis. If the application of the pesticide is below the first threshold value, the mutant allele appears to be dominant, because genotypes AB and BB both survive. On the other hand, if the application is such that the dosage is above the second threshold value, the mutant allele appears to be recessive, because only BB genotypes survive.

The foregoing example is built on the proposition that dominance is a physiological phenomenon. This view of dominance as a being largely a result of metabolic variation was first proposed by Wright (1929, 1934) and has gained general acceptance (Bourguet 1999). This does not mean that dominance cannot evolve, since metabolic pathways can be changed quantitatively, for example, by modification of the amount of enzyme activity. Thus, in the case of Figure 2.8, a new mutation could arise that would shift the curves right or left, thereby altering the dominance pattern.

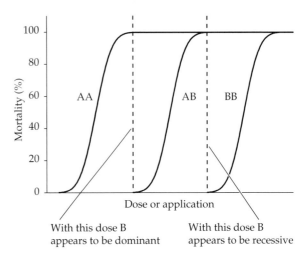

Figure 2.8   A example of dominance appearing as a consequence of a change in the environment. The three sigmoidal curves represent the resistance of three genotypes (*AA, AB, BB*) to the amount of pesticide/herbicide received. If the dosage is applied at the first vertical line, then genotypes *AB* and *BB* both have 0% mortality, whereas genotype *AA* has 100% mortality. Thus, at this dosage, allele *B* is dominant. However, at the second vertical line, genotypes *AA* and *AB* have 100% mortality and genotype *BB* has 0% mortality; at this dosage, allele *B* is recessive.

How does dominance affect the offspring on mid-parent regression? There are six different genotypic mid-parent values. In four cases the mean offspring value is equal to the mid-parent value, whereas in two cases the mean offspring values differ by an amount d or $d/2$ (Figure 2.9). The relationship between parents and offspring can thus been seen to be made up of two components, an additive component and a dominance deviation. Thus, even without any environmental dominance, the heritability (or slope of regression) of the trait will be less than 1. The additive genetic variance is equal to

$$\sigma_A^2 = 2pq\left[1 + d\left(q - p\right)\right]^2 \tag{2.16}$$

and the total phenotypic variance, in the absence of environmental effects, is

$$\sigma_P^2 = \sigma_A^2 + \left(2pq\right)^2 = \sigma_A^2 + \sigma_D^2 \tag{2.17}$$

The second term is known as the **variance of the dominance deviations,** or simply the **dominance variance.** It is important to note that, even when there is complete dominance, the variation can still be divided statistically into an additive variance and a dominance variance. Therefore, the addition of dominance to

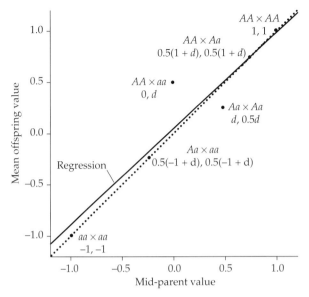

Figure 2.9   Mean offspring value on mid-parent value for the single-locus model with dominance. The genotypic values are $AA = 1$, $Aa = d$, $aa = -1$. The mid-parent and mean offspring values are shown below the cross indicator. For the purposes of illustration, $d = 0.5$.

the model does not change the utility of the biometrical model of inheritance. In some experimental designs it is only possible to estimate the total genetic variance: The ratio of the total genetic variance to the phenotypic variance is called the **heritability in the broad sense** and is an upper limit of heritability in the narrow sense.

Epistasis

Dominance represents the interaction within a locus; **epistasis** is the interaction among loci. Epistasis is a potential problem because it destroys the assumption of independence among loci and hence the simple addition of variances across loci to obtain the component variances. Nevertheless, there will still exist an overall regression between offspring and parents. (For the method of estimating variance components in a two-locus epistatic model, see Crow and Kimura, 1970, pp. 124–128). Statistically, the genetic variance can be decomposed in an additive component, a dominance component, and an epistatic or interaction component, which can be further divided into additive × additive interactions (σ^2_{AA}), additive × dominance interactions (σ^2_{AD}), and dominance × dominance interactions (σ^2_{DD}). The covariance between mean offspring and mid-parent is no longer simply the additive genetic variance but equal to $\frac{1}{2}\sigma^2_A + \frac{1}{4}\sigma^2_{AA}$ (Crow and Kimura 1970). Thus the slope of the regression is an upwards-biased estimate of the heritability as previously defined. In general, σ^2_{AA} will be small relative to σ^2_A. This is illustrated in Figure 2.10, which shows a plot of the relationship between heritability and allele frequencies for the preceding two models.

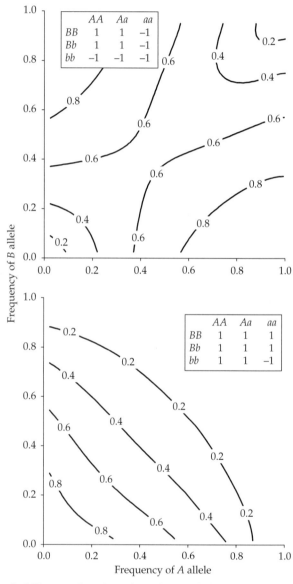

Figure 2.10   Contour plot of heritability as a function of two loci with dominance and epistatic effects. The genotypic values are shown in the boxes within the plots. There is no environmental variance.

Despite the large epistatic effects, much of the variance appears in the additive component. Thus, to a first approximation, the inclusion of epistasis does not change the overall model. The potential difficulties introduced by epistasis are, however, such that most mathematical models of quantitative genetics explicitly ignore it.

The Infinitesimal Model

Epistatic effects greatly complicate quantitative genetic theory and hence they are generally assumed to be absent. As shown above, even if present, the decomposition into additive and nonadditive components typically results in the largest component of variation coming from additive effects. If we ignore epistasis, the effects of different loci are independent, and the genotypic value, G, of a trait determined by n loci can be decomposed into

$$G = \mu + \sum_{i=1}^{i=n} \sum_{j=1}^{j=2} (A_{i,j} + D_{i,j}) = \mu + A + D \tag{2.18}$$

where μ is the population mean, and $A_{i,j}$ and $D_{i,j}$ are, respectively, the additive and dominance deviations due to locus i. The second subscript, j, is needed to account for the two alleles per locus in a diploid organism. From the central limit theorem A and D will become bivariate normal when the number of loci, n, is large (Bulmer 1985a, p. 123), and, as a consequence, the relationship between offspring and parents can be described using standard statistical methods. The variance of the genotypic value is simply

$$\sigma_G^2 = \sigma_A^2 + \sigma_D^2 \tag{2.19}$$

where the subscripts refer to the relevant components of Equation (2.18). As discussed above, in general, there will be environmental influences, and hence the general relationship for the variance of the phenotypic value is

$$\sigma_P^2 = \sigma_A^2 + \sigma_D^2 + \sigma_E^2 \tag{2.20}$$

It follows from this that the regression of the mean offspring value, X, on the mid-parent value, Y, is as previously derived: $Y = (1 - h^2)\mu + h^2 X$. As is expected from the single locus case, the slope of the regression of mean offspring on mid-parent is equal to the heritability of the trait.

The Importance of Nonadditive Genetic Variance

In large randomly mating populations the course of evolutionary change in a single trait considered in isolation can be largely described in terms of the additive genetic variance and phenotypic variance. These variances can be assumed to a first approximation to remain constant. However, assortative mating or drift resulting from small population size can have significant effects on the trait means and variances.

POPULATION BOTTLENECKS. Although in very large populations nonadditive genetic variance can be ignored, at least in the short term, drastic reductions in population size can bring nonadditive effects into prominence. To understand how this comes about, consider an infinitely large population that passes through a single generation in which the population size is reduced to N individuals. For simplicity we shall focus on a single locus with two alleles A_1 and A_2 at frequencies p and q (= $1 - p$), respectively.

Two factors affect the additive genetic variance after passage through the population bottleneck. The first is the allele frequency itself: Due to drift, the allelic frequency moves away from its initial position and, depending on its initial position, a change in p in any particular population can generate an increase or decrease in additive genetic variance and a consequent change in heritability (see Equations [2.16] and [2.17]). The second factor is the magnitude of the non-additive genetic contribution. From Equation (2.16) it can be seen that the probability of heritability increasing varies with the degree of dominance (d) and the allele frequency (p). Populations passing through a single bottleneck of size 2 may experience large increases in heritability due to the shift in allelic frequency, but only if $d > 0$. The latter result is dependent upon the assumption of a single locus. If the trait is determined by a large number of loci, then the expected additive genetic variance will always decrease. There will, however, still be a finite probability of an increase in additive genetic variance after passing through a bottleneck (Box 2.4).

The above analysis ignored epistatic effects, which require by definition at least two loci. As with dominance, increases in heritability can occur when epistasis is present. From his examination of an epistatic model involving two loci and no dominance, Goodnight (1987, 1988) concluded, for the particular epistatic model he examined, that: (1) "a single founder event, regardless of size, will lead to an increase in the additive genetic variance, provided the epistatic genetic variance in the ancestral population is at least 1/3 of the additive genetic variance" (Goodnight 1987, p. 446); (2) "When founder events are of size 2, regardless of linkage, more than half of the epistatic genetic variance in the ancestral population is converted to additive genetic variance in the new colony. With free recombination, fully 75% of the epistatic genetic variance is converted to additive genetic variance each generation that the population size remains at two individuals" (Goodnight 1987, p. 447); (3) The amount of conversion decreases rapidly with population size, 50% being converted when $N = 1$, approximately 10% when $N = 5$, and 6% when $N = 16$.

Regardless of the model chosen, any changes in allelic frequency other than fixation need not necessarily result in a permanent change of the additive genetic variance. If the original allelic frequency is a product of selection, then the shift

---

Box 2.4
Estimating changes in variance due to a population bottleneck

The probability of an increase given a bottleneck of size N can be obtained using the multinomial probability function (Sokal and Rohlf 1995; Mitchell-Olds 1991):

$$P(n_1, n_2, n_3) = \frac{N!}{n_1! n_2! n_3!} p_1^{n_1} p_2^{n_2} p_3^{n_3} \tag{2.21}$$

where n_i ($i = 1,2,3$) is the number of genotypes (A_1A_1, A_1A_2, and A_2A_2, respectively), $N = n_1 + n_2 + n_3$, $p_1 = p^2$, $p_2 = 2pq$, and $p_3 = q^2$. The frequency of A_1 after the bottleneck, p', is $p' = (n_1 + 0.5n_2)/N$. The probability of an increase in heritability is a function of the allele frequency, the environmental variance, and the bottleneck size. The expected variances can be obtained by analysis of the moments of p', from which we obtain (Willis and Orr 1993):

$$E\left(\sigma_A^2\right) = 2\mu_1\left(1 + 2d + d^2\right) - 2\mu_2\left(1 + 6d + 5d^2\right) + 8\mu_3\left(d + 2d^2\right) - 8\mu_4 d^2$$
$$E\left(\sigma_G^2\right) = 2\mu_1\left(1 + 2d + d^2\right) - 2\mu_2\left(1 + 6d + 3d^2\right) + 8\mu_3\left(d + d^2\right) - 4\mu_4 d^2 \tag{2.22}$$

where μ_i is the ith moment. (In other words, they are the expected values of p', p'^2, p'^3 and p'^4. For their derivation, see p. 335 of Crow and Kimura, 1970). These moments are functions of the bottleneck size. If there is no dominance ($d = 0$), the expected genetic variances are $E\left(\sigma_A^2\right) = E\left(\sigma_G^2\right) = 2\mu_1 - 2\mu_2$.

---

in genetic variance will only be temporary, selection eventually restoring the initial conditions. This may not occur if random drift breaks up adaptive gene complexes leading to establishment of new combinations (Wright's shifting-balance process; Goodnight, 1995). A second important point to be remembered is that there must necessarily be a reduction in genetic variation in an absolute sense if the bottleneck is very small (Nei et al. 1975; Maruyama and Fuerst 1985). For example, suppose that the bottleneck consists of a single male and female; in this case the maximum number of alleles that can pass through the bottleneck is four. This loss of variation is seen in a comparison of the mean number of alleles per locus of source and founder populations (Figure 2.11), in a positive correlation between population size and electrophoretic variation in the plants *Scabiosa columbaria* and *Salvia pratensis* (Bijlsma et al. 1994), the loss of electrophoretic variability following a bottleneck caused by vulcanism in wild populations of the rodent *Ctenomys maulinus* (Gallardo and Kohler 1994; Gallardo et al. 1995) and a loss of genetic variation in red squirrels living in woodland fragments

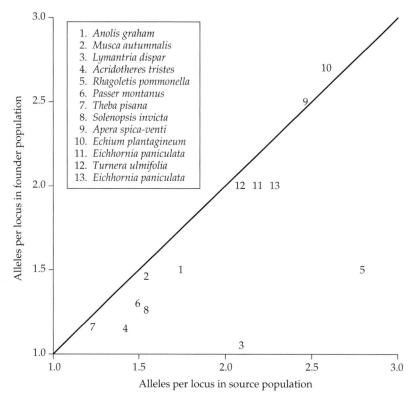

Figure 2.11 Relationship between allelic variation in source and founder populations. Data sources: Barrett and Husband (1990), Howard (1993), Ross et al. (1993).

(Wauters et al. 1994). The loss of alleles could negate the first effect, because after the bottleneck part of the genetic variability is lost, it may therefore not be possible to return to the original conditions (for example, one locus might be fixed).

INBREEDING DEPRESSION. In the above analyses, mating was assumed to be random. **Inbreeding** is defined as the mating of two related individuals (e.g., full-sibs, cousins, and so on). Nonrandom mating may result in inbreeding (or the converse, outbreeding). The probability that two parents will be related even if mating is at random will increase as population size is decreased. To make use of the concept of inbreeding we define genes as being identical by descent if they have originated from the replication of the same gene. This should not be confused with identity of effect. Two alleles at a diallelic locus may be the same in the sense that both are A_1 or both A_2 but not be identical by descent because they have originated from the replication of two different A_1s or A_2s. The probability that the two genes at a locus are identical by descent is termed the **coefficient of inbreeding,** F. A base population is defined as that population in which $F = 0$, that is, in which there is no inbreeding. Typically, the starting population is arbitrarily defined as having an inbreeding coefficient of zero. The estimation of the inbreeding coefficient for autosomal loci can be calculated following a simple addition rule. (For details, see Crow and Kimura, 1970, pp. 69–73 or Falconer, 1989, pp. 85–97). With random mating, the rate of inbreeding is inversely proportional to population size (Box 2.5).

The expected change in the mean for a single locus with dominance is

$$\mu_t = \mu_0 - 2p_0 q_0 dF_t \tag{2.23}$$

From the above equation it can be seen that, when there is dominance, the expected trait value declines linearly with the degree of inbreeding. When there is epistasis but no dominance, there will be no inbreeding depression, and when both epistasis and dominance are present, the relationship between inbreeding depression and F may be linear or quadratic, though more likely the latter (Crow and Kimura 1970, pp. 77–85). Because inbreeding leads to homozygosity there will also be a decline in genetic variance that is proportional to the inbreeding coefficient. The latter phenomenon does not depend on dominance. So inbreeding reduces genetic variation and, if there is dominance present, it also reduces trait means. The latter phenomenon is termed **inbreeding depression,** as the

---

Box 2.5
Population size and change in the inbreeding coefficient

Consider the change in the inbreeding coefficient over time. Given random mating with selfing possible, the probability that any two genes will be identical by descent is $1/(2N)$; thus $F_1 = 1/(2N)$. In the second generation there are two ways in which genes may be identical by descent; first, because they originate by replication in the previous generation, and, second, because they originate from the same gene in the first generation. From this we can write

$$F_2 = \frac{1}{2N} + \left(1 - \frac{1}{2N}\right)F_1 \tag{2.24}$$

The rate of inbreeding, ΔF, is equal to $1/(2N)$ and arises from self-fertilization. If self-fertilization is not permitted, the increment is not much affected, as the effect is to simply push the process back to the grandparental generation. Note that the rate of inbreeding is inversely proportional to population size. Applying the same argument as used to derive F_2 gives the recursive equation

$$F_t = \frac{1}{2N} + \left(1 - \frac{1}{2N}\right)F_{t-1} = 1 - \left(1 - \frac{1}{2N}\right)^t \tag{2.25}$$

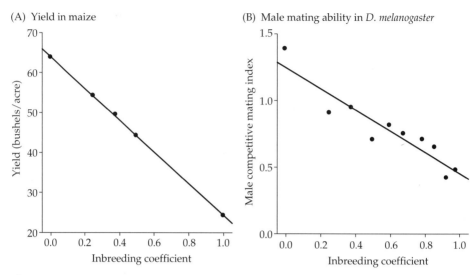

Figure 2.12 Two examples of inbreeding depression in life history traits. (A) data from Crow and Kimura (1970); (B) data from Sharp (1984).

reduction in trait means is typically found in fitness-related traits such as fecundity (Figure 2.12). In the analysis of inbreeding depression the usual statistic computed is the **coefficient of inbreeding depression**, δ, $\delta = 1 - X_I/X_O$ where X_I is the trait value from inbred progeny and X_O is the trait value from outcrossed progeny.

Empirical data from nondomestic species show overwhelming evidence for significant inbreeding depression in both plants and animals (Roff 1997). An inbreeding depression of only 10% may appear small, but when applied to a series of traits, such as survival at various life cycle stages and fecundity, that contribute to fitness, the combined multiplicative effect can produce an enormous depression. Estimation of inbreeding depression under different conditions indicates that the degree of depression is increased under field or stressful conditions in both plants (Allard 1965; Antonovics 1968; Schemske 1983; Mitchell-Olds and Walle 1985; Dudash 1990; Schmitt and Ehrhardt 1990; Johnston 1992; Eckert and Barrett 1994, McCall et al. 1994) and animals (Chen 1993; Jimenez et al. 1994; Miller 1994; Bijlsma et al. 1999). These results demonstrate the importance of circumstance in the determination of fitness differences between inbred and outbred individuals.

Asymmetry of Response

According to the simple theory developed above, the response to selection does not depend upon the direction of selection. However, it is frequently observed that the same selection intensity will produce a different response depending on the direction applied (for example, selecting for increased or decreased fecundity). This does not represent a conceptual challenge to the theory, because, as described below, there are several reasons why such asymmetry is expected. The asymmetry in response does represent a modeling challenge in so far as it should be incorporated into genetic models that consider the evolution of life history traits.

Factors causing asymmetry in response can be divided in to three broad categories: (1) measurement artifacts, (2) finite population size, and (3) genetic asymmetry, which is of particular significance for life history theory.

MEASUREMENT ARTIFACTS. Measurement artifacts may themselves be divided into three groups: (i) scalar effects, (ii) differences in selection differentials, and (iii) use of surrogate measures.

(i) *Scalar effects*: Scalar effects arise when the trait is measured on an inappropriate scale. For example, suppose the trait is log normally distributed. If measurements are made on an arithmetic scale, the distribution will by highly asymmetrical and the basic assumption underlying the equation $R = h^2 S$ will be violated. The supposed section intensities, computed based on a normal distribution, will be in error, and hence the formula will have no meaning. Asymmetrical responses due to scalar effects will disappear once measurements are transformed to the appropriate scale.

(ii) *Differences in selection differentials*: Obviously, if the supposed selection differential is not correct, then there will be an apparent asymmetrical response. A factor that can lead to an incorrect assessment of the selection differential is the scalar effect introduced when the phenotypic variance changes in a different fashion in the two divergent directions of selection. The phenotypic variance of body weight, for example, is typically correlated with the mean; therefore, due to the greater variance in the direction of larger body size, selection for an increased body size may produce a greater response than selection for a decreased body size. A second way in which selection differentials may be incorrect is if the effects of natural selection are not considered. For example, suppose we select for increased and decreased adult weight, and fecundity is correlated with body size. In each group we select two females to be parents of the next generation. Consider two females in the "large body size" group with weights X_1, X_2 and corresponding fecundities f_1 and f_2. Ignoring effects due to the differential fecundities, the selection differential is $\mu - (X_1 + X_2)/2$, where μ is the population mean weight. Because each female contributes differentially to the next generation, the correct selection differential is $\mu - (f_1 X_1 + f_2 X_2)/(f_1 + f_2)$, which clearly could be quite different from the uncorrected value.

(iii) *Use of surrogate measures*: If the method of selecting on the trait is based on a surrogate measure rather than the trait itself and there is a nonlinear relationship between the two traits, the response of the surrogate measure may be symmetric but that of the target trait, which responds due to its genetic correlation with the surrogate trait, may be asymmetric. For example, Baptist and

Robertson (1976) selected for body size in *Drosophila melanogaster* not by direct measurement of size but by having the flies walk through a series of slits of diminishing diameter. Thus the actual trait being selected was not body size but a combination of body size and willingness to perform the required behavior. Baptist and Robertson found a negative correlation between size and activity, which they suggested accounted for the slower response in the direction of increased body size. However, other researchers selecting for body size (thorax length, weight, wing length, and so on) by direct measurement of the trait have also observed asymmetrical response (Robertson and Reeve 1952; Tantawy 1956; Martin and Bell 1960) and thus other factors are probably also important.

FINITE POPULATION SIZE.   Measurement artifacts represent experimental problems but in no way affect the validity of the simple biometrical model. The second category, finite population size, does affect the theory, because it is assumed that population size is large enough that each population behaves in an entirely deterministic manner. This is most assuredly not the case for natural populations or for experimental populations. We have already encountered the possible influence of finite population size in causing a decline in heritability and changes in heritability when populations pass through bottlenecks. Asymmetry in response can arise due to (i) drift per se, in which case it is of no particular concern for the general theory or (ii) inbreeding depression, which is of considerable importance for life history theory (Box 2.6).

GENETIC ASYMMETRY.   Additive genetic variance is a function of allele frequencies. If the distribution of allelic values is skewed, then selection in one direction will have a greater effect on allelic frequencies than selection in the other direction. Therefore, there will be a differential change in heritability and hence a differential response to selection. This response can be illustrated with the single-locus model (Figure 2.13). If the allele frequency at a locus is precisely 0.5 and there is no dominance, then selection in either direction will lead to the same change in the heritability. If, on the other hand, the allele frequency is, say, 0.6, then "downward" selection will lead to a greater change in h^2 than "upward" selection. The same phenomenon occurs when there is dominance, except that the point of symmetrical change is at an allele frequency of 0.25. If there is dominance, there must still be eventual asymmetry in the response, as the function relating h^2 and allele frequency is not itself symmetrical (Figure 2.13). The overall effect of genetic asymmetry will depend upon the combined action of all the loci controlling the trait. Asymmetry in response could result from asymmetry in the frequency of a few of genes with large effect or from a large number of genes, each with very small effects.

 From artificial selection experiments we have seen that strong selection will slowly erode additive genetic variance and hence reduce heritability. As is discussed in greater detail in Chapter 3, the same argument can be advanced to predict that fitness-related traits will have low heritabilities and relatively large

---

Box 2.6
Asymmetry of selection response
due to drift and/or inbreeding depression

(i) *Drift:* In an artificial selection experiment, if only two selected lines are used, one "up" and one "down," then it is very difficult to discount the possibility of drift being responsible for different rates of divergence from the base population. The variance in selection response, σ_R^2, after t generations of selection is approximately (Nicholas 1980):

$$\sigma_R^2 \approx \frac{\sigma_P^2}{N}\left(th^2 + 2p\right) + 2\sigma_c^2 \qquad (2.26)$$

where N is the number of parents, p is the proportion selected (so number actually measured is N/p), and σ_c^2 is the common environmental variance. Assuming that σ_c^2 is negligible, the variability in the response is proportional to the phenotypic variance and inversely proportional to population size. Since the response to selection is proportional to σ_p, the coefficient of variation in the response to selection is independent of the phenotypic variance and inversely proportional to $\sqrt{N}$. Thus, for population sizes much less than about 100, there is likely to be considerable variation in the observed response to selection, and asymmetry in response could easily arise solely to drift.

(ii) *Inbreeding depression*: In a small population there will be increased breeding between related individuals, possibly leading to a depression of the mean value of the trait (see above). As a consequence, the trait will appear to respond more quickly in the direction of decreased mean than in the direction of increased mean. (Since this effect is a consequence of dominance effects, it is similar to the effect caused by genetic asymmetry, described in the next section.) The effects of asymmetry due to small population size causing inbreeding depression can be accounted for in an artificial selection experiment by use of a control population.

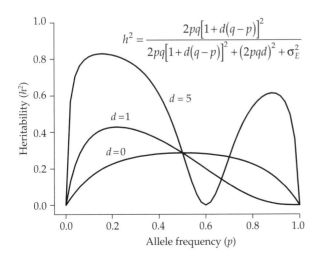

$$h^2 = \frac{2pq\left[1 + d(q-p)\right]^2}{2pq\left[1 + d(q-p)\right]^2 + \left(2pqd\right)^2 + \sigma_E^2}$$

Figure 2.13 Heritability as a function of allelic frequency for the single-locus, two-allele genetic model. Heritability is given by the function, where σ_E^2 is the environmental variance, here set at 1.

dominance variance. From this we can further predict that asymmetry of response will be common for traits such as fecundity and development time, traits that are closely related to fitness. Because selection will have tended to drive those alleles with positive effects on fitness close to fixation, we predict that the asymmetry of the response will be manifested as a slow response in the direction of increased fitness (Falconer 1989; for a mathematical description of the process, see Kojima 1961). Frankham (1990) analyzed data from 30 studies of bidirectional selection on traits that are probably related to fitness; these included both life history traits (fecundity, development time, age to maturity) and behavioral traits (mating speed, mating competence, feeding rate). A significantly higher proportion of studies (80%) showed asymmetry of response towards lower fitness. Thus the heritability of fitness-related traits is significantly more often higher when measured in the direction of reduced fitness than when measured in the direction of increased fitness—the direction in which natural selection will have tended to push the organism and hence deplete additive genetic variance; Figure 2.14. On the other hand, body size, which is not invariably a fitness-related trait (though in some cases it could be, as with mating success and body size), does not show any directionality in asymmetry of response. Asymmetry in this case could be due to other factors such as drift.

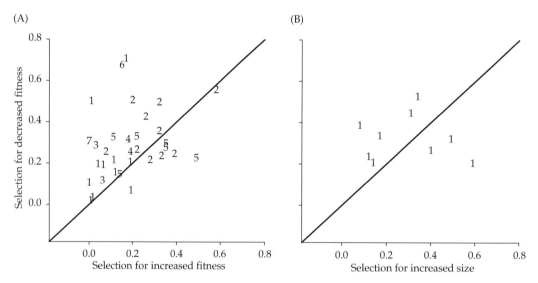

Figure 2.14   Realized heritability estimates from bi-directional selection. (A) Estimates in the plot are for traits directly connected to life history, whereas those in (B) are for morphological traits, which are typically only indirectly related to fitness via their relationship with fecundity, mating success, survival, and so on. Data from Frankham (1990). Data codes: 1, *Drosophila*; 2, Mouse; 3, Japanese quail; 4, Chicken; 5, *Tribolium*; 6, *Gambusia*; 7, *Cyprinus*.

Estimating Heritability

OFFSPRING ON PARENT REGRESSION. In the theory developed above, the parameter termed heritability plays a central role. How can it be estimated? In this section I present some of the more common methods. A regression of mean offspring on mid-parent is obviously one method of estimating heritability. If the number of offspring per family is constant, then heritability (equal to the slope of the regression line, b) can be analyzed using simple linear regression methods. There is no bias introduced by assortative mating, and hence it is worthwhile to mate parents assortatively to increase the range of values. The estimated standard error is the usual standard error of the slope. If the number of offspring per family is not constant, then the mean value for each value is estimated with differing precision. This is not a problem if the number is reasonably large (>10) but may cause problems if sample sizes are small and very variable (Bohren and McKean (1961). In this case heritability is estimated from the following (Kempthorne and Tandon 1953; Bulmer 1985a, p. 79):

$$\hat{h}^2 = \frac{\sum w_i (x_i - \bar{x})(y_i - \bar{y})}{\sum w_i (x_i - \bar{x})^2} \tag{2.27}$$

where the w_i are weights estimated as

$$w_i = \left[\frac{\hat{\sigma}_y^2 \left(1 - \frac{1}{2} h^2\right)}{n_i} \right] \tag{2.28}$$

$\hat{\sigma}_y^2$ is the estimated phenotypic variance among the offspring means and n_i is the number of individuals in family i. Because h^2 occurs on both the left- and right-hand side of Equation (2.27), an iterative procedure must be adopted to find the value of h^2 that satisfies both sides of the equation. The standard error of the heritability estimate is $1 / \left[\sum w_i (x_i - \bar{x})^2 \right]$.

If the variance in males differs from that of females, then separate heritabilities should be estimated for each sex using sons on sires and daughters on dams. Heritabilities estimated from the regressions of sons on dams and daughters on sires must be corrected to take into account the different variances. For the former, $h^2 = b \hat{\sigma}_M / \hat{\sigma}_F$, where b is the slope of the regression and $\hat{\sigma}_M$, $\hat{\sigma}_F$ are the estimated phenotypic standard deviations in males and females, respectively. As described below, the standard error is considerably inflated when only a single parent is used; it is, therefore, worthwhile to search for a transformation that will remove the differences in variance. A typical reason for differences in variance is that the sexes vary in size and there is a relationship between the mean and the variance. In such a case, a logarithmic transformation may be sufficient to remove the difference.

If we assume a constant family size, the standard error of heritability is equal to (Latter and Robertson, A. 1960b):

$$SE\left(h^2\right) = \sqrt{\frac{h^2\left(1-h^2\right)+\left(2-h^2\right)/n}{nN/(n+2)-3}} \qquad (2.29)$$

from which suitable sample sizes can be readily determined. The standard error is very sensitive to the number of families (N) and family size (n) when n is less than about 6. If the total number of organisms to be measured is fixed (i.e., $N(n + 2) =$ constant) then the optimum number of offspring per family is approximately

$$n = \sqrt{\frac{2-h^2}{h^2\left(1-h^2\right)}} \qquad (2.30)$$

For h^2 between 0.25 and 0.75, a total family size of 4 (2 parents + 2 offspring) is optimal when the total number of individuals is fixed. However, the potential for a great increase in the standard error if by chance family size is reduced suggests a more conservative family size between 6 and 10 offspring.

A regression of mean offspring on one parent is used to estimate heritability if the phenotypic variance in the trait under consideration differs between males and females, or if there are likely to be effects attributable to one parent. The latter is particularly likely in mammals where maternal effects are common. The slope of the regression of mean offspring value on one parent is equal to one-half of the heritability. Thus, to obtain the heritability, the slope must be multiplied by two. Similarly, the standard error of the heritability is equal to twice the standard error of the regression slope. Use of a single parent considerably increases the confidence region of the heritability estimate. Differences between the heritabilities estimated using fathers or mothers may be attributable to non-Mendelian factors such as maternal effects, but the relative imprecision of the estimate makes it difficult to detect significant differences. Assortative mating biases the estimate of heritability, the appropriate correction being (Falconer 1989, p. 178) $h^2 = 2b/(1 + r)$, where r is the phenotypic correlation between parents.

To obtain a preliminary estimate of the standard error given a fixed family size one can use Equation (2.22) multiplied by $\sqrt{2}$. Thus the optimum family size remains the same, but the associated standard error is increased.

HALF-SIB AND FULL-SIB DESIGNS. Pedigree information other than offspring and parents can be used to estimate the covariance between relatives of any degree. The two most frequently used relationships are those between full-sibs and half-sibs:

$$Cov\left(Full\text{-}sibs\right) = \frac{\sigma_A^2}{2} + \frac{\sigma_D^2}{4} + \frac{\sigma_{AA}^2}{4} + \frac{\sigma_{AD}^2}{8} + \frac{\sigma_{DD}^2}{16}$$

$$Cov\left(Half\text{-}sibs\right) = \frac{\sigma_A^2}{4} + \frac{\sigma_{AA}^2}{16}$$

(2.31)

If we ignore epistatic variance (which is typically done), then the covariance between half-sibs is the additive genetic variance; hence, heritability can be estimated without difficulty. The usual half-sib mating design is to mate several females with a single male, the number of sires ranging anywhere from 10 to hundreds. Because standard errors can be depressingly large, at least 20 sires and three dams per sire should be used. (For a full discussion of estimation methods and sampling numbers, see Roff, 1997.) In addition to the half-sib estimate of heritability, obtained from the sire intraclass correlation, it is also possible to estimate heritability using the dam intraclass correlation, which provides a full-sib estimate. The problem with the full-sib estimate is that even if one ignores the epistatic variance there is potentially a bias introduced by dominance variance. Additionally there is the possibility that the dam estimate will be inflated by maternal effects. For example, early growth may be strongly influenced by maternal care. A full-sib estimate is also available from the mean offspring on mid-parent regression. A comparison of the full-sib estimate with that obtained from the sire intraclass correlation or the slope of the regression might reveal the presence of maternal and/or nonadditive genetic effects but the standard errors of estimates are generally so large that meaningful discrimination is difficult.

A full-sib design in which each female is mated to a separate male is experimentally easier than the half-sib design, but the problem of inflation due to dominance variance makes the method suspect unless there exists evidence that dominance variance is negligible. In general, morphological traits seem to be relatively free of dominance variance, but life history traits frequently show large nonadditive variance components (reviewed in Roff, 1997). Even when nonadditive genetic variance might be ignored, the estimated heritability can still be inflated as a consequence of common environment effects. Thus full-sibs should, in general, be separated into several cages/blocks so that a nested ANOVA can be used to isolate effects due to common environment.

The variance components in either the half-sib or full sib designs can be estimated from analysis of variance (Table 2.4) or directly using a variance component estimation procedure available in a number of computer statistical packages. The advantage of the latter method is that techniques such as restricted maximum likelihood can be used, which are not so sensitive to imbalance in the data as is ANOVA. However, the estimates typically do not differ much unless the imbalance is very severe.

Standard error estimates are fairly complex, particularly for the half-sib design, and the reader is referred to Roff (1997, pp. 42–45) for these. Arveson and

---

Table 2.4   Estimation relationships for full-sib and half-sib estimates of heritability

Estimate	Relationship	ANOVA estimate
Full-sib	$\dfrac{2\sigma^2_{AF}}{\sigma^2_{AF}+\sigma^2_{AP}}$	$\dfrac{2\left(MS_{AF}-MS_{AP}\right)}{MS_{AF}+(k-1)MS_{AP}}$
Half-sib, sire estimate	$\dfrac{4\sigma^2_{AS}}{\sigma^2_{AS}+\sigma^2_{AD}+\sigma^2_{AP}}$	$\sigma^2_{AS}=4\left(MS_{AS}-\left[MS_{AP}+k_2\sigma^2_{AD}\right]\right)/k_3$ $\sigma^2_{AD}=\left(MS_{AD}-MS_{AP}\right)/k_1$ $\sigma^2_{AP}=MS_{AP}$
Half-sib, dam estimate	$\dfrac{2\sigma^2_{AD}}{\sigma^2_{AS}+\sigma^2_{AD}+\sigma^2_{AP}}$	See above.

Note: *AF*, Among Families; *AP*, Among Progeny; *AS*, Among Sires; *AD*, Among Dams
MS, Mean Squares.

For balanced designs: k = number of families, $k_1 = k_2$ = number of offspring per dam family, k_3 = number of progeny per sire. For unbalanced designs, see formulas in Roff (1997).

If parents are mated assortatively, the heritability estimates have to be corrected. The full-sib correction (including the dam estimate in the half-sib design) is $h^2 = \left(-1+\sqrt{1+4rH^2}\right)\Big/2r$, where H^2 is the uncorrected heritability (Falconer 1989, p. 179). The half-sib correction is $h^2 = \left(-1+\sqrt{1+\left(2r+r^2\right)H^2}\right)\Big/\left(2\left[2r+r^2\right]\right)$ (Bulmer 1985, p. 129).

Schmitz (1970) suggested the jackknife procedure as a method of estimating variance components and their standard errors. Simons and Roff (1994) tested the utility of the jackknife as a method for the full-sib design and found it to be superior to the approximate formula. The jackknife method has also been used for the half-sib design (Fox et al. 1999), though its performance has not been checked with simulation.

As with the mean offspring on mid-parent regression, for the full-sib and half-sib designs a family size of between 6 to 10 is advisable, unless one can be assured of maintaining family size at 4 or above. If the total sample size (nN) is fixed, the optimal family size for the full-sib design is approximately equal to $h^2/2$ (Robertson 1959a). There is little difference between the optimal design for the full-sib and mean offspring on mid-parent regression (Hill and Nicholas 1974). Robertson (1959a, 1960b) investigated the optimal sample size for the half-sib design when the total number of individuals to be measured is kept constant. The recommendations from this study are:

1. If the magnitude of h^2 is known, then:
   a) If only the sire component is to be used, then the number of offspring per dam (k_1) should be set at $k_1 = 4/h^2$.

b) If both the sire and dam components are to be used, then $k_1 = 2/h^2$ with 3 or 4 dams per sire.

2. If no *a priori* estimate of h^2 is available, then:
   a) If only the sire component is to be used, $20 < k_1 < 30$.
   b) If both components are desired, then $k_1 = 10$ with 3 or 4 dams per sire.

Family sizes as low as 2 to 3 should be avoided (Robertson 1959a). If there is no dominance variance nor maternal effects, the full-sib design is preferable to the half-sib because the standard error of the former is approximately $4\sqrt{(h^2/T)}$ while that of the latter is $4\sqrt{(2h^2/T)}$, where T is the total number of individuals measured. If h^2 is less than 0.25, then, for a given number of individuals measured, the half-sib method is more accurate than the mean offspring on mid-parent regression, the reverse being true for $h^2 > 0.25$ (Robertson 1959a).

There are a large number of other potential designs, from some of which dominance and epistatic variances components can be estimated. For a review of different breeding designs, see Becker (1992), Roff (1997), and Lynch and Walsh (1998).

ESTIMATING HERITABILITY IN THE WILD. Most heritability estimates are made in the laboratory, though what we are really interested in is the heritability in wild populations. In some taxa, such as birds, it may be possible to measure both parents and offspring in the wild, but more typically it is only possible to collect individuals from the field and raise their offspring in the laboratory. Under these conditions a lower bound of heritability in nature, h^2_{Nature} (Riska et al., 1989) is

$$h^2_{Nature} = \frac{b^2 \sigma_P^2}{r_A^2 \sigma_A^2} \tag{2.32}$$

where b is the slope of the mean offspring on mid-parent regression, r_A is the genetic correlation between the two environments, σ_P^2 is the phenotypic variance of the parents under the natural conditions, and σ_A^2 is the additive genetic variance of the offspring under the laboratory conditions. If only a single parent is used, Equation (2.32) is multiplied by 4. The concept of genetic correlation is discussed later in this chapter; at present it is sufficient to note that like all correlation coefficients genetic correlation lies between zero and one. Since the value of r_A is generally not known, a minimum estimate of h^2_{Nature} is made by assuming that $r_A = 1$. The additive genetic variance is calculated from the offspring data using the full-sib design described above. There is no fully satisfactory method of placing confidence limits on h^2_{Nature}. Riska et al. (1989) suggested the use of the bootstrap method, while Simons and Roff (1994) estimated the upper and lower 95% confidence bounds by substituting the upper and lower bounds of the regression slope. Recently, methods of estimating heritability in natural populations using molecular markers have been proposed. These methods are beyond the scope of the present discussion and the reader is instead referred to Ritland (1996, 2000) and Mousseau et al. (1998).

A Multiple-Trait Model of Inheritance of Quantitative Traits

The model presented above is not without its problems; provided that neither epistasis nor selection is too strong (but what constitutes "too strong" is yet to be resolved satisfactorily) this model should serve as a useful tool to study quantitative genetic phenomena. Even if long-term quantitative predictions are unreliable (also yet to be determined), the qualitative responses are in accord with theory (e.g., Figure 2.6). Selection in natural populations is unlikely to act on a single trait, and even if it does, there will still be repercussions throughout the genome as a consequence of genes affecting several traits. Thus we need to expand the model to include several traits simultaneously. I first consider the simple case of two traits.

The Genetic Correlation

An appropriate statistical measure of association between two variables is the correlation. When applied to genotypic values this correlation is called the **additive genetic correlation** or, more generally, the **genetic correlation**. ("Additive" is used because the correlation applies only to the additive component of genetic variation.) A genetic correlation can arise from two causes; linkage disequilibrium and pleiotropy. Linkage disequilibrium has been described above in considering two or more loci determining the same trait. Exactly the same arguments hold when the two loci determine different traits. In this circumstance, a correlation can arise because of physical linkage or because selection builds up the association. Such correlations are transitory and disappear after a few generations of random mating. In life history theory, genetic correlation by linkage disequilibrium has not been considered an important factor except in the case of the evolution of mate preference (Fisher 1930; Lande 1981; Kirkpatrick 1982; Heisler 1984; Pomiankowski 1988; Pomiankowski and Iwasa 1993). Unless otherwise stated, genetic correlations discussed in this book are assumed to be caused by pleiotropy. **Pleiotropy** refers to the influence of a locus (gene) on two or more traits. Two traits that share no loci in common will have a zero genetic correlation, whereas traits that have no independent loci will have a genetic correlation of ±1.

Generally, genetic correlations are not directly observable. What is observed is the phenotypic correlation. Thus we must relate the phenotypic correlation to the genetic correlation. As shown in Box 2.7, this relationship is

$$r_P = r_A \sqrt{h_X^2 h_Y^2} + r_E \sqrt{\left(1 - h_X^2\right)\left(1 - h_Y^2\right)}.$$

If the two heritabilities are very small, the phenotypic correlation is determined primarily by the environmental correlation, whereas if they are both high,

---

Box 2.7

Decomposing the phenotypic correlation into the genetic and environmental correlations

From standard probability theory the phenotypic correlation, r_P, between two traits, X and Y, is

$$\frac{\sigma_{PXY}}{\sigma_{PX}\sigma_{PY}} \tag{2.33}$$

where σ_{PXY} is the covariance between X and Y, and σ_{PX}, σ_{PY} are the phenotypic standard deviations of X and Y, respectively. The phenotypic correlation is made up of two components, a component that is ascribable to the additive action of the two sets of overlapping genes, and a component that comprises the correlation of environmental effects plus the nonadditive genetic effects. The first is equivalent to the correlation between breeding values (equal to the genotypic value in the present context) and is called the genetic correlation, whereas the second is termed, somewhat misleadingly, the environmental correlation. In the formulas that follow I shall use the subscript P to designate phenotypic values, A to designate additive genetic values, and E to designate environmental values. Because environmental and genetic effects are assumed to be uncorrelated, the phenotypic covariance between X and Y is the sum of the genetic and environmental covariances, $\sigma_{PXY} = \sigma_{AXY} + \sigma_{EXY}$, and from algebraic manipulation we have

$$r_P = r_A \frac{\sigma_{AX}\sigma_{AY}}{\sigma_{PX}\sigma_{PY}} + r_E \frac{\sigma_{EX}\sigma_{EY}}{\sigma_{PX}\sigma_{PY}} \tag{2.34}$$

If we note that $h^2 = \dfrac{\sigma_A^2}{\sigma_P^2}$ and that $\sigma_E^2 = \sigma_P^2 - \sigma_A^2 = \left[1 - \dfrac{\sigma_A^2}{\sigma_P^2}\right]\sigma_P^2$, the above can be rewritten as

$$r_P = r_A \sqrt{h_X^2 h_Y^2} + r_E \sqrt{\left(1 - h_X^2\right)\left(1 - h_Y^2\right)} \tag{2.35}$$

---

it is the genetic correlation that is most important. Note that the phenotypic correlation does not by itself give any idea of the importance of the genetic relationship between the two traits.

We now have two genetic parameters, the heritability, which describes the relationship of a particular trait between parent and offspring, and the genetic correlation that describes the association between two different traits as a consequence of shared genes. Now suppose we choose a particular group of parents contingent on the values of two traits. There will be a response due to the direct effect of selection plus an indirect response due to the genetic correlation between the two traits. Because we have assumed normal distributions for both genetic and environmental effects, the total response can be obtained from standard statistical theory (Young and Weiler 1960; Bell and Burris 1973). The expected response of traits X and Y in phenotypic standard deviation units is:

$$R_{X,\sigma} = \beta_{X,\sigma} h_X^2 + \beta_{Y,\sigma} h_X h_Y r_A$$
$$R_{Y,\sigma} = \beta_{Y,\sigma} h_Y^2 + \beta_{X,\sigma} h_X h_Y r_A \qquad (2.36)$$

where

$$\beta_{X,\sigma} = \frac{i_X - r_p i_Y}{1 - r_p^2}, \qquad \beta_{Y,\sigma} = \frac{i_Y - r_p i_X}{1 - r_p^2} \qquad (2.37)$$

The first terms on the right of Equation (2.36) determine the response of the trait due to selection directly on that trait (i.e., the coefficient β plays the same role as standardized S), whereas the second terms determine the response of the trait due to its genetic correlation with the other trait. If selection is applied only to trait X, then $i_Y = i_X r_p$ and the response is simply

$$R_{X,\sigma} = i_X h_X^2 \qquad (2.38)$$

while the correlated response of trait Y (denoted as CR_Y) is

$$R_{Y,\sigma} = CR_{Y,\sigma} = i_X r_A h_X h_Y \qquad (2.39)$$

Substituting the variance and covariances in Equation (2.36) gives the response in unstandardized units. This form can be written concisely using matrices:

$$\begin{pmatrix} R_1 \\ R_2 \end{pmatrix} = \begin{pmatrix} \sigma_{A11}^2 & \sigma_{A12} \\ \sigma_{A21} & \sigma_{A22}^2 \end{pmatrix} \begin{pmatrix} \beta_1 \\ \beta_2 \end{pmatrix} \qquad (2.40)$$

where, for simplicity, I have substituted numerical designators for the two traits (e.g., X = 1, Y = 2). The diagonal elements in the matrix immediately to the right of the equality sign are the additive genetic variances, and the off-diagonal elements are the covariance ($\sigma_{A12} = \sigma_{A21}$). The vector on the extreme right is known as the **selection gradient vector** and can be decomposed into the phenotypic variance/covariance matrix and a vector of selection differentials,

$$\begin{pmatrix} \beta_1 \\ \beta_2 \end{pmatrix} = \begin{pmatrix} \sigma_{P11}^2 & \sigma_{P12} \\ \sigma_{P21} & \sigma_{P22}^2 \end{pmatrix}^{-1} \begin{pmatrix} S_1 \\ S_2 \end{pmatrix} \qquad (2.41)$$

The above formulation for the response of two traits can be expanded immediately to any number of traits by simply increasing the size of the matrices and vectors. In general, the equation is written in shorthand notation by using bold font:

$$\Delta \bar{z} = \mathbf{G} \mathbf{P}^{-1} \mathbf{S} \qquad (2.42)$$

where $\Delta \bar{z}$ is the vector of mean responses, **G** is the additive genetic variance-covariance matrix ("additive" is usually omitted), $\mathbf{P}^{-1}$ is the inverse of the phenotypic variance-covariance matrix, and **S** is the vector of selection differentials. Though z is frequently used to denote the traits, for consistency with the notation in the rest of this book I shall use uppercase X when only a single trait is being discussed.

The Constancy of the Genetic Correlation

Evidence was presented above that under strong selection heritability can be expected to remain more or less constant for approximately 10–15 generations. Bohren et al. (1966, p. 55) argued that the genetic correlation will remain constant over even shorter periods:

> The additive genetic variance of any character will be made up of contributions from the separate loci. These contributions will change as the gene frequencies are altered by selection or by random drift and they will not all change in the same way, depending on the gene frequencies at the loci concerned. But the genetic covariance (if the genetic correlation is not close to 1) will either be made up of a much smaller number of terms, if all loci contribute to the covariance with the same sign, or will be made up of positive and negative contributions from different loci. In either case the *proportional* [authors' italics] change in the genetic covariance is likely to be greater than in the genetic variances themselves. *It must be therefore expected that the static description of a population in terms of additive genetic variances and covariances will be valid in prediction over a much shorter period for correlated responses than it will be for direct responses* [my italics].

This prediction is rather depressing but is supported by simulation analyses (Bohren et al. 1966, Parker et al. 1969, 1970a,b; Slatkin and Frank 1990). As with heritability, a critical test is the comparison of the genetic correlation estimated from a pedigree analysis with that obtained from the observed correlated response of a trait. The results of such a comparison are rather more heartening than the theoretical analyses and over periods as long as 50 generations the two estimates can be quite concordant (Figure 2.15). Another test of the constancy of the genetic correlation is the prediction of the correlated response. The primary problem here is that the confidence limits about such a prediction are so large as to preclude very little. With this caveat, the overall results support the previous finding that predictions over time periods of about 20 generations are reasonably reliable (Roff 1997, pp. 178–182).

Simultaneous Selection on Two or More Traits

The model developed above predicts the change in trait values when selection acts on several traits simultaneously, as might be expected in natural populations. Several points should be noted.

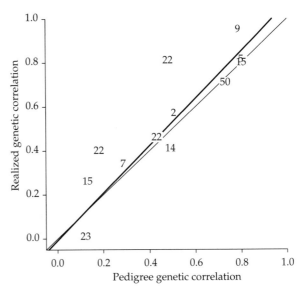

Figure 2.15   Estimates of genetic correlation from pedigree analysis and from correlated response to selection. The numbers indicate the number of generations over which selection was done. Regression and 1:1 lines shown. Data from Table 5.2 in Roff (1997).

First, the response is a function not only of the genetic variances and covariances but also the phenotypic variances and covariances. This is true even in the simplest case of selection on a single trait, because heritability is the ratio of additive genetic to phenotypic variance. With multiple traits the single phenotypic variance is replaced by the phenotypic variance-covariance matrix (**P**) and the additive genetic variance with the genetic variance-covariance matrix. The dependence of both the magnitude and the direction of the response on **P** is illustrated in Figure 2.16 in which the heritabilities and genetic correlations are kept constant and the phenotypic and environmental correlations varied. For a detailed discussion of how negative responses may be generated when all selection differentials and correlations are positive, see Deng and Kibota (1995).

Second, it is possible for there to be no response to selection in spite of both genetic variance for the trait and a genetic correlation with another trait under selection. This is readily seen by expanding Equation (2.36) in terms of variances and covariances to give the response in measurement units,

$$R_X = \frac{\sigma_{AX}^2\left(\sigma_{PY}^2 S_X - \sigma_{PXY} S_Y\right) + \sigma_{AXY}\left(\sigma_{PX}^2 S_Y - \sigma_{PXY} S_X\right)}{\sigma_{PX}^2 \sigma_{PY}^2 - \sigma_{PXY}} \qquad (2.43)$$

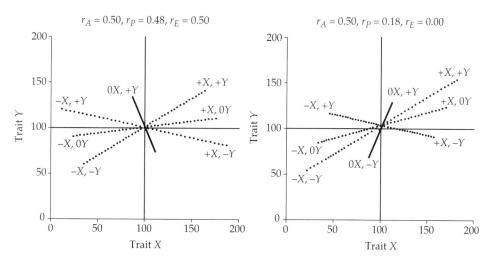

Figure 2.16   Direct and correlated responses predicted for two different combinations of correlations differing in the phenotypic and environmental correlations. Each dot represents the response after one generation of selection in which 20% of the population was selected. For each combination of selection, 20 generations of selection are shown. In all cases $h_X^2 = 0.5$, $h_Y^2 = 0.25$.

with a similar term for R_Y. The additive genetic variance in X could be substantial and the additive genetic covariance between X and Y also very large, but if the parenthetical terms, which are functions only of the phenotypic variances and covariances and the selection differentials, are each zero, then there will be no response in X. Thus a trait may be under selection but show no response (Box 2.8).

Finally, for the two-trait case, if the genetic correlation is exactly ±1, the **G** matrix is singular and evolutionary response is constrained (Lande 1979; Maynard Smith et al. 1985; Via and Lande 1985). The reason is clear when one visualizes the linear regression between the additive genetic values of the two traits. According to linear regression theory, the two traits will covary with some normally distributed "error variance," ε, about the regression line. The additive genetic correlation is the correlation between the additive genetic values of the two traits. When the correlation is exactly ±1, all the values lay along a single line and ε is zero. So long as ε is not zero, then any combination of trait values is possible, even if unlikely. (This means simply that it will take longer for selection to drive the trait combination to that point.) However, with a correlation of exactly ±1, the combination of values is constrained to move only along the regression line and cannot attain the optimal combination unless by chance it happens to pass through this combination. The probability that two traits will have a genetic correlation of exactly ±1 seems remote, for in this case they would represent no more than mathematical transformations of each other (for an example, however,

---

Box 2.8
Selection without response

A pertinent case for life history theory can be illustrated by relating the selection differential to the selection gradient and phenotypic variances and covariances:

$$S_X = \beta_X \sigma_{PX}^2 + \beta_Y \sigma_{PXY} \qquad (2.44)$$

If there is no additive genetic variance for trait Y, trait Y could be phenotypically correlated with X (i.e., $\sigma_{PXY} \neq 0$) but cannot be genetically correlated (i.e., $\sigma_{AXY} = 0$). Suppose that there is no direct selection on trait X (i.e., $\beta_X = 0$) but there is selection on trait Y (i.e., $\beta_Y > 0$). From Equation (2.36) it can be seen that a positive selection differential is generated ($S_X > 0$) whenever the phenotypic covariance is positive. But, given that $R_X = \sigma_{AX}^2 \beta_X + \sigma_{AXY}^2 \beta_Y$, it is apparent that there will be no response to this selection, even if there is additive genetic variance for trait X. A possible case in which this situation can arise is when trait X is a heritable trait such as fecundity and Y is a trait such as nutritional status, which might have zero heritability. Thus well-nourished individuals have high fecundity, which gives them an apparent selective advantage, but this is not realized because selection is acting only on the environmentally determined component of fecundity. The mechanism described above, or ones conceptually similar, have been proposed for the observation of directional selection without evolutionary response in breeding date in birds (Price et al. 1988), clutch size in birds (Price and Lia 1989), and tarsus length in birds (Alatalo et al. 1990; Thessing and Ekman 1994).

---

see Roff 1994a). What effect will a genetic correlation that is not ±1 have on the evolutionary trajectory? Two examples of bivariate distributions of additive genetic values are shown in Figure 2.17. In the first case, the genetic correlation is 0.6 and as a result the axes of the binormal distribution are rotated relative to the trait axes. We can define two new "traits" relative to the axes of the binormal distribution, that are linear functions of the original traits. These traits will be uncorrelated with each other. The axis that runs along the region of greatest spread is known as the major axis and the other axis is called the minor axis. The equation of the principal axis can be written $Y = \mu_Y + b_Y(X - \mu_X)$, where μ_Y, μ_X are the trait means (shown as zero in the plots) and b_Y is the slope of the principal axis. The slope b_Y, is then equal to

$$b_Y = \frac{\sigma_{AYX}}{\left(\lambda_Y - \sigma_{AY}^2\right)} \qquad (2.45)$$

where

$$\lambda_Y = \frac{1}{2} \left\{ \sigma_{AY}^2 + \sigma_{AX}^2 + \sqrt{\left(\sigma_{AY}^2 + \sigma_{AX}^2\right)^2 - 4\left(\sigma_{AY}^2 \sigma_{AX}^2 - \sigma_{AYX}^2\right)} \right\} \qquad (2.46)$$

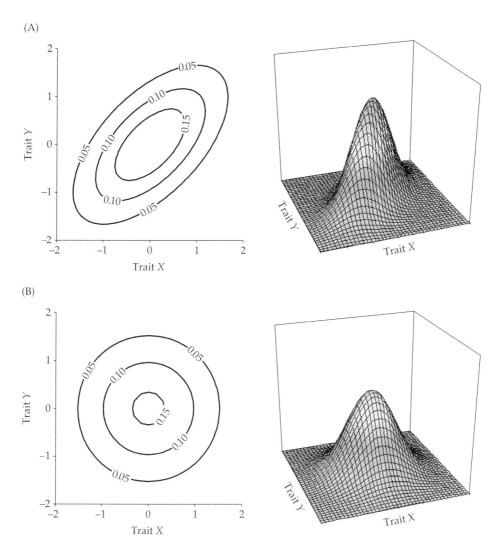

Figure 2.17 Two examples of bivariate distributions of additive genetic values. (A) The left the genetic correlation is 0.6 and, as a result, the axes of the binormal distribution are rotated relative to the trait axes. (B) Two traits that are uncorrelated.

The slope, b_X, of the minor axis (that which is orthogonal to the major axis) is simply $b_X = -1/b_Y$. The quantities λ_Y, λ_X are known as the eigenvalues and are a measure of the amount of variation along the rotated axes. When the two traits are uncorrelated with each other (Figure 2.17B), there is no rotation required and the eigenvalues do not differ. The major and minor axes are the eigenvectors (also called the latent roots or characteristic vectors). Now, *all other things being equal*, because the greatest amount of additive genetic variance is in the direction of the major axis, the evolutionary trajectory will tend to be strongly biased in this direction (Kirkpatrick and Lofsvold 1992; Bjorklund 1996; Schluter 1996; Merila and Bjorklund 1999), as illustrated in Figure 2.18. But this will not prevent evolution from eventually attaining the equilibrium combination. This conclusion depends upon the evolutionary trajectory being determined primarily by the **G** matrix. However, as described above, this trajectory is also a function of the phenotypic variance-covariance matrix and it is possible that the direction of the bivariate phenotypic distribution will negate the directional effect of the **G** matrix. This is a question that can only be resolved empirically: Interspecific and intraspecific comparisons appear to indicate that genetic constraints can be significant over long periods of time (Schluter 1996; Merila and Bjorlund 1999. But see Badyaev and Martin, 2000, for an example in which growth trajectories appear not to be strongly constrained).

The above two-trait case can be extended immediately to any number of cases, though visualization is not possible. Evolutionary response will be constrained in any direction in which the eigenvalue is zero, which means that in this direction there is no additive genetic covariance. (Thus in the two-trait case, when the eigenvalue is zero, the genetic correlation is ±1 and there is no genetic

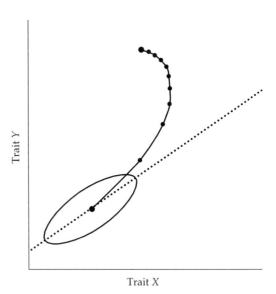

Figure 2.18 A hypothetical trajectory of two traits whose major axis of genetic variation lies away from the optimal combination (large dot at left). Initially, the direction of evolution is strongly influenced by the major axis of genetic variation, but the trajectory eventually bends towards the optimal combination. Modified from Schluter (1996).

variation except along the regression line.) Except for this condition, selection can, under this model, move the set of traits to any combination, given enough time (Zeng 1988; Kirkpatrick and Lofsvold 1992). The number of eigenvectors equals the number of traits and, as before, response to selection will be biased by the relative sizes of the eigenvalues, moving more in the direction dictated by the largest values. An example is presented in Figure 2.19, in which the bivariate distributions of five morphological traits of the cricket, *Gryllus pennsylvanicus* is plotted. For the purposes of illustration I have plotted the mean values for 39

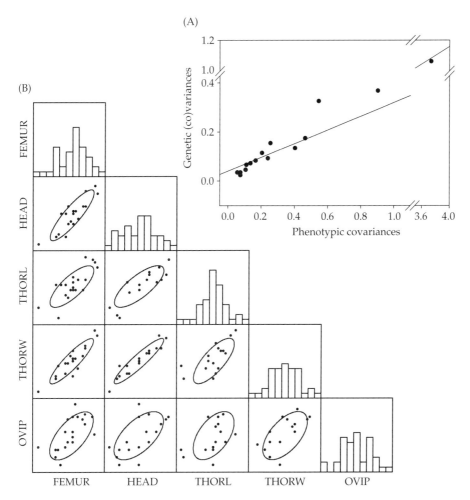

Figure 2.19 (A) A bivariate plot of genetic covariances versus phenotypic covariances in the cricket *Gryllus pennsylvanicus*. (B) Bivariate plot of family mean values for five traits, with 68% contour. The histograms show the distribution of the trait designated at the bottom of the column.

families, which approximates the genetic covariance distribution (Via 1984a,b; Roff and Preziosi 1994). The eigenvalues of the **G** matrix are 1.28, 0.22, 0.016, 0.008, and 0.004, respectively, with the first eigenvalue explaining 84% of the variation. The first principal component of morphological traits is typically strongly biased towards general size, and hence selection will be most effective at simply changing overall size rather than, say, shape (Bjorklund 1996). Such a response will be reinforced if the **G** and **P** matrices covary positively. This is certainly the case for the *G. pennsylvanicus* data (Figure 2.19) and is true, in general, for morphological traits (Figure 2.20). Although there is an overall trend for phenotypic and genetic correlations involving life history traits to covary positively, there are a considerable number of instances in which the signs differ between the two types of correlation (Figure 2.20), and, therefore, the phenotypic correla-

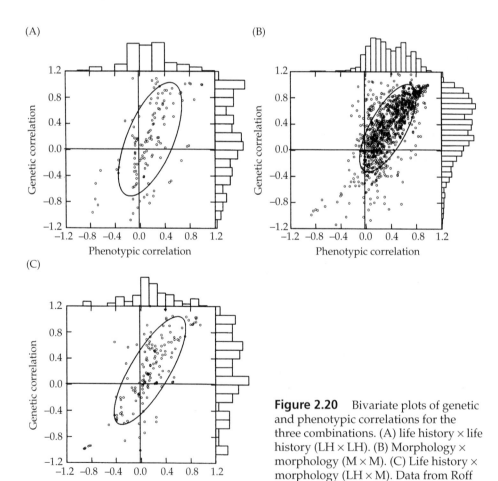

Figure 2.20 Bivariate plots of genetic and phenotypic correlations for the three combinations. (A) life history × life history (LH × LH). (B) Morphology × morphology (M × M). (C) Life history × morphology (LH × M). Data from Roff (1996a).

tion cannot be taken as a reliable indicator of the genetic correlation in the absence of other confirmatory data (Roff 1996b). The distribution pattern of correlations involving a life history and a morphological trait is intermediate between the former two combinations. The important point in the present context is that, because of the differing directions of the phenotypic and genetic covariances, the responses of life history traits to selection might be quite erratic.

In Chapter 3 I take up the issue of genetic variation in trade-offs and consider in more detail how trade-offs that involve several traits can constrain evolution though no single genetic correlation is –1.

Experimental Observations on Response to Joint Selection

Given constant **G** and **P** matrices, one can predict the evolutionary trajectory when selection is simultaneously applied on two or more traits. However, there are a number of reasons why this may be wishful thinking for artificial selection experiments: (1) The strength of selection is typically sufficient to erode at least additive genetic variances, and possibly additive genetic covariances (expected but not well demonstrated), (2) genetic drift could alter the response both by the erosion of genetic variation and also by generating additive genetic variation (most likely in the case of life history traits), (3) asymmetry of responses, particularly in life history traits, indicates that the simple biometrical model is lacking in some respects, and (4) the estimation of the genetic parameters is difficult and the standard errors generally associated with the estimates are so large that there will be considerably uncertainty in predictions. Under these circumstances the failure to observe a good match between the trajectories predicted using the above model and those found in selection experiments must be evaluated with considerable caution. In particular, the problem of obtaining precise parameter estimates is not to be dismissed lightly. Given these caveats, what are the findings of artificial selection on two traits simultaneously?

Sen and Robertson (1964) used two types of joint selection on the abdominal and sternopleural bristles of *D. melanogaster*: (1) *index selection,* in which the highest 10 of 40 individuals of each sex were selected using the index, Abdominal score + 1.5 (sternopleural score), and (2) *independent culling selection,* in which the highest 20 of 40 individuals of each sex were first selected according to their abdominal score, and then from these the highest 10 of each sex according to their sternopleural score. There was a possible decrease in the heritability of abdominal bristle number but not in that of sternopleural bristles. (For data, see Table 5.8 in Roff, 1997.) The estimate of r_A in the base population was approximately 0.1 and there was no evidence from the selection experiments of a significant decline after 12 generations. If anything, there may have been an increase, the estimate from both methods of selection being 0.21 (SE = 0.08). Sheridan and Barker (1974) also used joint selection on bristle number in *D. melangaster*, in this case on the coxals and sternopleurals. They used all four possible combinations and found, as in the previous study, a slight decline in the heritabilities and a general trend for the genetic correlation to increase regardless of the type of selection.

There are few experiments that have simultaneously selected on two traits; three examples are presented in Figure 2.21. The general pattern in all three experiments was the same. Simultaneous selection for two traits in the direction of the genetic correlation (into the quadrants +,+ or –,–) was successful, but selection in which the change in trait values was contrary to the genetic correlation (into the quadrants +,– or –,+) tended to produce very erratic responses (Figure 2.21). Nordskog (1977, p. 576) called selection into the –X, +Y (large egg weight, small body size) quadrant incompatible antagonistic selection because not only is it contrary to the sign of the genetic correlation but also because "very small chickens don't

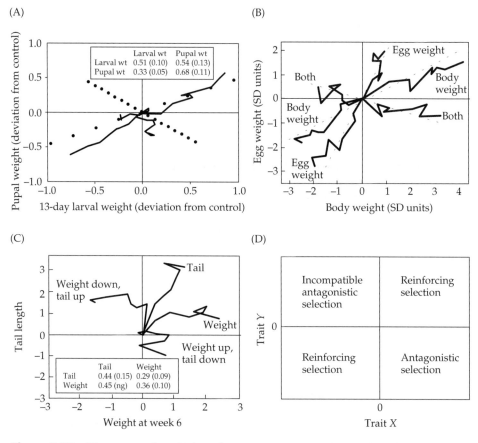

Figure 2.21   Three examples of joint selection on two traits. (A) Predicted (• = generation) response and observed (solid lines) response in *Tribolium casteneum*. For clarity, the traces have been individually centered at zero. Heritabilities and genetic correlations (SE) are shown in the box; data from Bell and Burris (1973). (B) Responses in the chicken; data from Nordskog (1977). (C) Responses in mice, with heritabilities and genetic correlations (SE) shown in the box; data from Rutledge et al. (1973).

naturally lay very large eggs." Note, however, that the overall response is qualitatively what is predicted (dotted lines). Rutledge et al. (1973), from their analysis of selection for body weight and tail length in mice, concluded that "In contrast to single-trait-selection responses, the responses to index selection were not consistent with current theory. . . . Our results indicate that in the dynamic situation of antagonistic selection, the genetic correlation may be more powerful in impeding component responses than predicted from presently available theory."

Two points emerge from the above three examples. First, the quantitative response may be different from that predicted by the genetic parameters from the base population (as, for example, selection in *T. casteneum*). However, given the relatively large standard errors, such a failure cannot be taken very seriously. For example, in the experiment of Bell and Burris, 1973 (Figure 2.21A), the observed response is approximately twice as great as that predicted. But the confidence region of the heritability estimates is certainly large enough to accommodate this discrepancy.

The second point is that symmetry in response may not be obtained. We have already seen that asymmetry in response to selection on a single trait is actually to be expected under certain circumstances, and hence we should not be surprised to find it in the case of two-trait selection. *A priori* more erratic responses would be expected when selection was antagonistic.

Taken as a whole I find the results reassuring and indicative that, though a simplification, the biometrical model of quantitative genetics does sufficiently capture the essential features of the inheritance of quantitative traits to be useful. This is certainly not to say that there is not considerable room for improvement, but in its present form and with suitable additions already available in the literature (such as the addition of finite population size, mutation, and dominance variance) we can use this approach to investigate problems in the evolution of life history traits.

Fitness and Fitness Components

The quantitative genetic model presented above states that, in general, genetic architecture is not a barrier to selection driving trait values to any particular combination. However, as is discussed in Chapter 3, trade-offs involving several traits can result in prohibited directions of evolution, even though the component genetic correlations are all greater than –1. In the absence of prohibited directions, if we wish to know what the ultimate equilibrium point will be, we do not need to measure the **G** matrix but we do have to measure the vector of selection differentials. More specifically, we need to construct the fitness surface, which requires the combination of the separate fitness components. Some traits have an obvious relationship to fitness: thus we might expect selection to favor increased fecundity and increased survival. But what of development time? Is there any advantage for an organism to develop faster or mature earlier? Before proceeding to answer this question I shall address the question "given that we can assign a fitness to a phenotype, what will be the change in mean fitness over time?"

Changes in Fitness under Directional Selection

Under the quantitative genetic model, natural selection will always act to increase the mean fitness of a population (Lande 1976). This can be demonstrated readily for a single trait; extension to the multivariate case is immediate. The average phenotypic value at generation t before selection, $\overline{X}(t)$, is given by the sum of the product of the phenotypic values and their proportional representation in the population:

$$\overline{X}(t) = \int Xp(X,t)dX \tag{2.47}$$

where $p(X,t)$ is the frequency of trait value X at generation t. This frequency is assumed to be normal with mean $\overline{X}(t)$ and variance σ_P^2. Letting the fitness of an individual with phenotype X, be $W(X)$, the mean fitness of individuals in the population, $\overline{W}$, is

$$\overline{W} = \int p(X,t)W(X)dX \tag{2.48}$$

and the mean phenotype after selection is

$$\overline{X}_W(t) = \frac{1}{\overline{W}} \int Xp(X,t)W(X)dX \tag{2.49}$$

The change in mean fitness with the change in the mean phenotype is given by

$$\frac{\partial \overline{W}}{\partial \overline{X}(t)} = \int \frac{\partial p(X,t)}{\partial \overline{X}(t)} W(X)dX \tag{2.50}$$

If we note that, for any function $f(X)$, $\dfrac{\partial e^{f(X)}}{\partial X} = e^{f(X)} \dfrac{\partial f(X)}{dX}$, the above equation can be solved to give

$$\frac{\partial \overline{W}}{\partial \overline{X}(t)} = \int \frac{X - \overline{X}(t)}{\sigma_P^2} p(X,t)W(X)dX$$

$$= \frac{\overline{W}}{\sigma_P^2} \left[\overline{X}_W(t) - \overline{X}(t) \right] \tag{2.51}$$

The response to selection, $\Delta \overline{X}(t)$, is equal to $h^2 \left[\overline{X}_W(t) - \overline{X}(t) \right]$ and hence

$$\Delta \overline{X}(t) = \frac{h^2 \sigma_P^2}{\overline{W}} \frac{\partial \overline{W}}{\partial \overline{X}(t)}$$

$$= h^2 \sigma_P^2 \frac{\partial \ln \overline{W}}{\partial \overline{X}(t)} \tag{2.52}$$

All of the terms on the right are positive, and hence the average phenotype will, under the assumptions of this model, evolve in the direction that increases the mean fitness of the population.

Fisher's Fundamental Theorem

Suppose that the trait is itself fitness. In this case $\frac{\partial \overline{W}}{\partial \overline{X}(t)} = \frac{\partial \overline{W}}{\partial \overline{W}} = 1$. Noting that $\sigma_A^2 = h^2 \sigma_P^2$ and substituting in Equation (2.51), we have

$$\Delta \overline{W} = \frac{\sigma_{AW}^2}{\overline{W}}$$

$$\frac{\partial \overline{W}}{\partial t} = \sigma_{AW}^2 \tag{2.53}$$

where σ_{AW}^2 is the additive genetic variance in fitness. In words, the above equation can be stated as "the change in fitness caused by natural selection is equal to the additive genetic variance in fitness" (Frank and Slatkin 1992, p. 93), although the exact meaning of Fisher's original proposition was not so clearly stated (Price 1972; Ewens 1989, 1992). Two important points to be remembered are that it refers only to additive genetic variance and assumes that the environment does not change.

Hard and Soft Selection

Wallace (1968, 1975) defined two types of selection: hard selection and soft selection. **Hard selection** is selection that operates without regard to the frequency or density of conspecifics. In population genetics it is equivalent to a fixed selection coefficient, and in life history theory it would be equivalent to probabilities of death, fecundity rates, and so forth, probabilities that are independent of density and frequency, though they could vary over time or space. Hard selection is both density- and frequency-independent. In **soft selection** the individual selection coefficients are dependent on density and/or frequency. For simplicity I shall consider the definition of fitness in several distinct circumstances: hard selection under constant conditions, hard selection in a temporally stochastic environment, hard selection in a spatially stochastic environment, density-dependent selection, frequency-dependent selection.

Defining Fitness when Selection is Hard and the Environment Constant

Given that we have proven that under at least some circumstances natural selection will increase fitness, we are lead to inquire what exactly is fitness. To approach this subject I shall first consider the very simple case in which there are a number of clones. Suppose that the generation interval is fixed and nonoverlapping, so that we can write for each clone, $N(t + 1) = N(t)\lambda$, where $N(t)$ is the number in a particular clone that survive to reproduce and λ is the number of offspring that each individual of that clone contributes to the next generation. The ratio

$$\frac{N(t+1)}{N(t)} = \lambda \tag{2.54}$$

measures the rate at which a given clone expands (or contracts). Over time the clone that has the largest λ will predominate in the population, and hence, by definition, it will be fittest genotype. Clones are unlikely to have exactly the same generation interval, and, therefore, we need a parameter that measures the rate at which the population size of a given clone increases over time. A clonal population growing in an unlimited, homogeneous and constant environment follows the simple exponential growth function:

$$\frac{dN(t)}{dt} = (\text{Births} - \text{Deaths})N(t) = rN(t) \tag{2.55}$$

where r is variously referred to as the **Malthusian parameter**, the **intrinsic rate of increase,** or simply the **rate of increase**. This equation also applies to any population that has achieved a stable age distribution. Solving Equation (2.55) gives

$$N(t) = N(0)e^{rt} = N(0)\lambda^t \tag{2.56}$$

where $\lambda = e^r$ and is known as the **finite rate of increase**. Suppose there are two clones with population sizes, $N_1(t)$ and $N_2(t)$, respectively, the first with an intrinsic rate of increase of r_1 and the second with r_2, where $r_1 > r_2$. The ratio of population sizes after some time t, given that both clones start with the same population size is

$$\frac{N_1(t)}{N_2(t)} = \left(\frac{\lambda_1}{\lambda_2}\right)^t = e^{(r_1 - r_2)t} \tag{2.57}$$

It is clear that as time progress the above ratio will increase, clone 1 becoming numerically more and more dominant in the combined population. This conclusion does not depend upon the two clones beginning with the same population size. Differences in starting condition simply accelerate or retard the rate at which clone 1 increases in frequency relative to clone 2.

For the two clones described above, an appropriate measure of fitness is r, since the frequency of the clone with the highest value of r will increase toward unity. Thus any mutation in a set of clones that increases r by changing rates of birth or death will increase in frequency in the population. There are no difficulties in assigning r as a measure of fitness in the above circumstances. Difficulties arise, however, when sexual reproduction and age structure are introduced (Pollack and Kempthorne 1970, 1971). Suppose we have a random mating population in which a mutation arises that increases birth rate or decreases death rate. Since the mutant will initially be rare in the population, its fate can be ascertained by considering the birth and death rates of the heterozygote alone (Charlesworth 1974; Charlesworth and Williamson 1975; Christiansen and Fenchel 1977; Reed and Stenseth 1984). If the heterozygote's rate of increase is enhanced, the mutation will increase in frequency in the population, but its ultimate fate depends upon the relative birth and death rates of the homozygotes and heterozygotes bearing the mutant allele. If the homozygote carrying both mutant alleles has a higher birth rate and/or a lower death rate than the heterozygote, the mutant allele will eventually be fixed in the population; otherwise the population will reach a stable polymorphism.

The general assumption, stemming from the work of Fisher (1930), has been that r can be associated with genotypes that follow particular life histories, and that selection will favor that genotype with the highest value of r. In other words, fitness can be equated with r and selection will drive the trait values to the combination that maximizes r. Remember that this conclusion is arrived at under the assumption of no environmental variation and no density dependence. (These complications will be tackled later.) Using normal distribution theory, Lande (1982) showed that provided the **G** matrix remains constant (i.e., weak selection) and there is a nearly stable age distribution, then populations will evolve towards the maximum value of r. More recently, Charlesworth (1993, 1994) beginning from a Mendelian model of multivariate traits has shown that evolution will maximize r regardless of the strength of selection or demographic equilibrium.

The Characteristic Equation and Fitness Components

In a population that is at a stable age distribution $\sum_{x=\alpha}^{\omega} e^{-rx} l(x) m(x) = 1$, where r is the rate of increase, $l(x)$ is the probability of surviving to age x, $m(x)$ is the number of female births at age x and α, ω are the age at first reproduction and death, respectively (Box 2.9).

Changes in the age at maturity, survival rates or fecundity schedule will alter r and hence affect fitness. Thus development time, by changing the age at maturity, is a fitness-related trait. A reduction in development time (i.e., α) would by itself increase r, but changes in development time may affect other components of the characteristic equation, such as survival rate. Hence, the overall change in r could be positive, negative, or zero. The components of the characteristic equa-

Box 2.9
Derivation of the characteristic equation

The number of newborns at time t to females of age x is equal to

$$\begin{pmatrix} \text{Number of newborns at} \\ \text{time } t \text{ to females of age } x \end{pmatrix} = \begin{pmatrix} \text{Number of newborns} \\ \text{at time } t-x \end{pmatrix} \begin{pmatrix} \text{Probability of} \\ \text{survival to } x \end{pmatrix} \begin{pmatrix} \text{Number of newborns} \\ \text{from a female of age } x \end{pmatrix} \quad (2.58)$$

$$= N(t-x)l(x)m(x)$$

By convention, the first age at reproduction is designated by the Greek letter α and the end of the life by the Greek letter ω. Hence the total number of newborns is

$$N(t) = \sum_{x=\alpha}^{\omega} N(t-x)l(x)m(x) \quad (2.59)$$

If the population is growing exponentially, $N(t) = N(t-x)e^{rx}$. Rearranging yields $N(t-x) = N(t)e^{-rx}$ and substituting in Equation (2.49) gives

$$N(t) = \sum_{x=\alpha}^{\omega} N(t)e^{-rx}l(x)m(x) \quad (2.60)$$

Dividing both sizes by $N(t)$ gives the characteristic equation

$$\sum_{x=\alpha}^{\omega} e^{-rx}l(x)m(x) = 1 \quad (2.61)$$

tion and those traits that affect the components of the characteristic equation are fitness components. The effect of selection on the fitness components can only be assessed by a consideration of how they interact to alter r. A fuller discussion of fitness components and the concept of fitness trade-offs is deferred until the next chapter.

The characteristic equation in discrete and integral form is frequently written as

$$\sum_{x=1}^{\infty} e^{-rx}l(x)m(x) = 1$$
$$\int_{0}^{\infty} e^{-rx}l(x)m(x) = 1 \quad (2.62)$$

Note that, whereas the integral form is initiated at age 0, the discrete version is subscripted initially at 1 (Goodman 1982).

Fisher (1930) modified the characteristic equation to produce a variable he termed **reproductive value**

$$V(x) = \frac{e^{rx}}{l(x)} \int_{x}^{\infty} e^{-rx}l(y)m(y)dy \quad (2.63)$$

The reproductive value of an individual of age x is a measure of the extent to which it contributes to the ancestry of future generations (Taylor et al. 1974). It is a measure of the expected contribution of zygotes to the population from age x. Because the population is increasing at rate e^{-r} the future contribution at age y is discounted by e^{-ry}. Its discrete version is

$$V(x) = \frac{e^{r(x-1)}}{l(x)} \sum_{i=x}^{\infty} e^{-ri} l(x) m(x) \qquad (2.64)$$

In most analyses the −1 is omitted from the first exponent, leading to a reproductive value that is off by a factor of e^r at every age (Goodman 1982). Williams (1966) postulated that natural selection maximizes r by maximizing reproductive value at every age. This conjecture was supported by the analyses of Goodman (1974) and Schaffer (1974a, 1979a), but a semantic misunderstanding led Caswell (1980) to question it, leading to a flurry of rebuttals (Schaffer 1981; Yodzis 1981). A correct statement of the principle is: "Reproductive value at each age is maximized relative to reproductive effort at that age, although not necessarily with respect to effort at other ages (therein lies the misinterpretation of Schaffer's work in Caswell [1980]" (Caswell 1982a, p. 1220).

Defining Fitness when Selection is Hard and the Environment Temporally Stochastic

There is no analytical treatment of what natural selection will maximize under the quantitative genetic model. However, since in the case of a stable environment, all perspectives—clonal, single locus, multilocus, biometrical—lead to the same conclusion, it is reasonable to suppose that if several of the foregoing approaches lead to the same fitness measure for stochastic environments, then it is likely to be valid under other approaches. Thoday (1953) suggested that an appropriate measure of fitness is the **probability of persistence.** If clone A has a persistence time of 100 generations and clone B a persistence time of 200 generations then, on average, after a sufficiently long time, only clone B will still be extant. Thus persistence time is an index of fitness, but a better one (or at least easier to work with) suggested by Cohen (1966) but actually going back to Haldane and Jayakar (1963a), is the geometric mean of the finite rate of increase, which is equivalent to the arithmetic mean of r. The rationale for this measure can be understood by considering the growth of a population in a stochastic environment (Lewontin and Cohen 1969). Population size (of a clone) at time t is given by

$$N(t+1) = N(0)\lambda_1 \lambda_2 \lambda_3 \cdots \lambda_t = N(0) \prod_{i=1}^{t} \lambda_i = N(0) e^{\sum_{i=1}^{t} r} \qquad (2.65)$$

Now assume that λ_i is a random, uncorrelated variable with mean $\bar{\lambda}$. The expected population size at time t is then given by the product of the initial population size, $N(0)$, times the expectation of the product $\lambda_1\lambda_2\lambda_3\cdots\lambda_t$. Because the λs are uncorrelated, the expected value of the product is equal to the product of the expected values, giving

$$E\{N(t)\} = N(0)E\left\{\prod_{i=1}^{t}\lambda_i\right\} = N(0)\prod_{i=1}^{t}E\{\lambda_i\} = N(0)\overline{\lambda}^t \tag{2.66}$$

At first glance the above result would suggest that the appropriate measure of fitness is $\overline{\lambda}$, which is the **arithmetic mean of the finite rates of increase** (i.e., $\overline{\lambda} = \sum_{i=1}^{t}\lambda_i$) not the **geometric mean**, which is given as

$$\overline{\lambda}_g = \left(\prod_{i=1}^{t}\lambda_i\right)^{\frac{1}{t}} \tag{2.67}$$

Hastings and Caswell (1979) used the above result to argue for the arithmetic mean rather than the geometric mean as a measure of fitness. However, the behavior of populations in a temporally randomly varying environment has the curious property that the expectation of population size will grow without bound whenever $\overline{\lambda} > 0$, but the probability of extinction within a few generations can be virtually certain (Lewontin and Cohen 1969; Levins 1969; May 1971, 1973; Turelli 1977). This paradoxical behavior can be illustrated with a simple example: Suppose that λ can take two values, 0 or 3, with equal frequency. The expected value of λ is $(0 + 3)/2 = 1.5$, and hence the expected population size increases without bound as t increases. For example, starting from a single female, after 10 generations $E\{N(10)\} = 1.50^{10} = 57.7$ but either $N(10) = 59{,}049$ or $N(10) = 0$ and the probability that the population persists for the 10 generations is $(0.5)^{10} = 0.00098$, a very small probability indeed. A more realistic example is displayed in Figure 2.22, in which the finite rate of increase is a uniform random variable between 0 and 2.2, giving a mean rate of increase of 1.1. Despite the positive arithmetic mean rate of increase, no population persisted more than 500 generations, and the median persistence time was only 15 generations (Figure 2.22).

As illustrated in Box 2.10, the geometric mean is always smaller than the arithmetic and the two are related by the approximation

$$E(\ln X) \approx E(X) - \frac{Var(X)}{2E(X)} \tag{2.68}$$

(Caswell 1989), where $E(\ln X)$ is the geometric mean, $E(X)$ is the arithmetic mean, and $Var(X)$ is the variance.

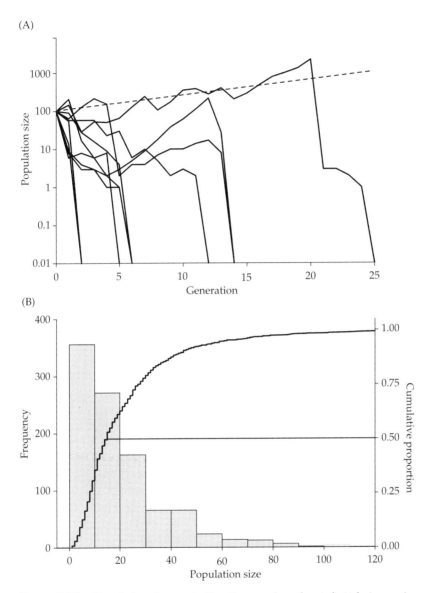

(A)

(B)

Figure 2.22   Simulation demonstrating the paradox of an infinitely increasing expected population size but a finite and short time to extinction. Each population was initiated with 100 individuals and subjected to the recursive equation $N(t + 1) = \lambda(t)N(t)$, where $\lambda(t)$ was a uniform random variable between 0 and 2.2. (A) The expected population size grows exponentially (dashed line), but none of the ten randomly selected populations shown persist for more than 25 generations. (B) 1000 runs of the simulation show the highly skewed distribution of time to extinction, with the median time being only 15 generations.

Box 2.10
A comparison of arithmetic and geometric fitnesses

Consider two clones living in an environment that comprises two types of year, "good" and "bad," each occurring with equal frequency. In "good" years genotype A has a finite rate of increase of 2 and in a "bad" year a finite rate of increase of 0.5, while genotype B has finite rates of increase of 1 and 1.1, respectively. The arithmetic averages of A and B are 1.25 and 1.05, respectively, but the geometric averages are 1 and 1.1. Thus genotype B has the higher long-term fitness, although it has a smaller arithmetic finite rate of increase. Genotype A increases more than genotype B in "good" years but suffers a greater reduction in "bad" years. The relatively high fitness of genotype B resides in the fact that, although it has a smaller arithmetic average, it also has a smaller variance in its finite rate of increase.

The long-run growth rate of a population in a temporally stochastic environment, $\bar{r}_G$ or $\bar{\lambda}_G$, is variously given by the following approximations:

$$\bar{r}_G = E(\ln \lambda) \approx \ln \bar{\lambda} - \frac{\sigma_\lambda^2}{2\bar{\lambda}^2} \quad \text{Lewontin and Cohen (1969)}$$

$$\bar{r}_G = E(\ln \lambda) \approx \bar{r} - \tfrac{1}{2}\, \sigma_r^2 \quad \text{Turelli (1977)} \tag{2.69}$$

$$\bar{\lambda}_G = e^{E(\ln \lambda)} \approx \bar{\lambda} - \frac{\sigma_\lambda^2}{2\bar{\lambda}} \quad \text{Lacey et al. (1983)}$$

where $\bar{r}$ and $\bar{\lambda}$ are the arithmetic average values of r and λ, respectively, and σ_r^2, σ_λ^2 are the respective variances. The parameter $\bar{r}_G$ is the expected rate of growth of the logarithm of population size:

$$E\{\ln N(t)\} \approx \ln N(0) + \left(\bar{r} - \tfrac{1}{2}\sigma_r^2\right)t = \ln N(0) + \bar{r}_G t \tag{2.70}$$

Lande (1993) examined the dynamics of the population model governed by the simple rules $dN/dt = rN$ when $1 < N < K$ and 0 for $N = K$, where K is the carrying capacity. (For an excellent review of different mathematical models, see Hanski 1999.) For this model the logarithm of the average time to extinction, $\ln T$, is approximately

$$\ln T \approx \left(\bar{r} - \tfrac{1}{2}\sigma_r^2\right)\frac{2}{\sigma_r^2}\ln K - \ln\left[\left(\bar{r} - \tfrac{1}{2}\sigma_r^2\right)^2 \frac{2}{\sigma_r^2}\right] \approx \bar{r}_G \frac{2}{\sigma_r^2}\ln K - \ln\left(\bar{r}_G^2 \frac{2}{\sigma_r^2}\right) \tag{2.71}$$

A given geometric rate of increase can result from an infinite set of combinations of $\bar{r}$ and σ_r^2 and, in general, the persistence time for these combinations will not

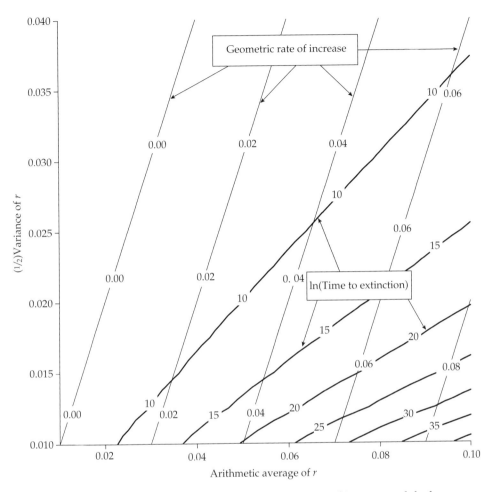

Figure 2.23   Isoclines of equal values of the geometric rate of increase and the logarithm of the predicted time to extinction for a carrying capacity (K) of 100 as determined from the equation $\ln T = \bar{r}_G \dfrac{2}{\sigma_r^2} \ln K - \ln\left(\bar{r}_G^2 \dfrac{2}{\sigma_r^2} \right)$, where $\bar{r}_G = \bar{r} - \frac{1}{2}\sigma_r^2$.

be the same (Figure 2.23; Orzack 1997). All things being equal, we might expect that natural selection would favor trait combinations that jointly maximized $\bar{r}_G$ and persistence time.

   The introduction of age structure into the analysis greatly complicates the algebra, although the general finding remains that fitness can be defined as the long-run average of a genotype (Tuljapurkar 1982, 1989, 1990; Lande and Orzack 1988; Orzack and Tuljapurkar 1989). The flavor of the results can be illustrated with a simple example from Tuljapurkar (1989). Consider a population with two age classes, described by the model $N_{t+1} = X_{t+1}N_t$, where

$$N_t = \begin{pmatrix} N_t(1) \\ N_t(2) \end{pmatrix}, \qquad X_t = \begin{pmatrix} m_1 F_t & m_2 F_t \\ S & 0 \end{pmatrix} \qquad (2.72)$$

in which m_i is the fecundity at age i, S is the probability of surviving from age 1 to age 2, and F_t is an independent random variable with $(1/F_t)$ having a gamma distribution with probability density function $g(w) = (n^n/(n-1)!)w^{n-1}\exp(-nw)$. The parameter n measures the variance, with variance increasing as n approaches zero. The long-run growth rate of this population with the parameter values fixed at their average values (i.e., deterministic case) is

$$r = \ln(\overline{\lambda}) = \ln\left(\left[\frac{m_2 S}{m_1}\right]\left[\frac{m_1 N(1)}{m_2 N(2)}\right]^*\right) \qquad (2.73)$$

where the term in brackets with the superscript * is the value once the population has achieved a stable age distribution. The average growth rate, a, of the population is given by

$$a \approx \overline{r} - \left(\frac{1}{2n\overline{\lambda}^2 C^2}\right)\left(m_1 + \frac{m_2 p}{\overline{\lambda}}\right)^2 \qquad (2.74)$$

where

$$C = 2 - \frac{n}{n-1}\frac{m_1}{\overline{\lambda}} \qquad (2.75)$$

As in the case of the population without age structure, the long-term average growth rate is equal to the "arithmetic" average growth rate (the growth rate at the fixed mean parameter values) minus an amount that is a function of the variance in growth rate (n). Increases in variance decrease the growth rate and decrease fitness.

The above models are phenotypic and do not take into account the genetic basis of the trait. The appropriateness of the geometric mean as the correct measure of fitness has been investigated in a population genetics context for both discrete generations (Haldane and Jayakar 1963a; Gillespie 1973; Karlin and Liberman 1974, 1975) and age-structured populations (Charlesworth 1994). These analyses (Box 2.11) have shown that the important parameter is the geometric means of the genotypes, providing support for the phenotypic models.

Defining Fitness When Selection Is Hard and the Environment Spatially Stochastic

An insufficient consideration of population dynamics in a spatial environment lead to incorrect assessments of the appropriate measure of fitness in an environ-

---

Box 2.11

Fitness and the geometric mean in population genetics

Following Gillespie (1973), let us consider a diploid population with genotypes AA, AB, BB with fitnesses in generation t of $W_{AA}(t)$, $W_{AB}(t)$, $W_{BB}(t)$ respectively. The frequency of allele A, p_A, (with $q_A = 1 - p_A$) satisfies the stochastic difference equation

$$\Delta p_A = \frac{p_A q_A \left[p_A \left(W_{AA} - W_{AB} \right) + q_A \left(W_{AB} - W_{BB} \right) \right]}{1 - \left[p_A^2 \left(1 - W_{AA} \right) + 2 p_A q_A \left(1 - W_{AB} \right) + q_A^2 \left(1 - W_{BB} \right) \right]} \tag{2.76}$$

where, for notational convenience, the generation indicators have been dropped. The above can be approximated in the region of the origin by

$$\Delta p_A = \left(\frac{\left(W_{AB} - W_{BB} \right) t}{1 - W_{AB} t} \right) p_A t \tag{2.77}$$

The origin will be unstable if, as generations increase, the probability that p_A is greater than some arbitrarily small value equals unity. From Equation (2.64) this statement is equivalent to the condition for the maintenance of all three genotypes in the population being

$$E \left\{ \ln \frac{W_{AB}}{W_{BB}} \right\} > 1,$$

$$E \left\{ \ln \frac{W_{AB}}{W_{AA}} \right\} > 1 \tag{2.78}$$

Thus, if the geometric mean fitness of a homozygote exceeds that of the other two genotypes, the relevant allele will increase to fixation.

---

ment in which habitat quality varied (Stearns and Koella 1986; Roff 1992a). The correct approach was independently, and almost simultaneously, published by Houston and McNamara (1992) and Kawecki and Stearns (1993). The environment is modeled as a set of patches in which the survival or fecundity (female births) rates differ, and are designated as $l(x,h)$, $m(x,h)$, respectively. Here x is age and h the patch designator. If there were no migration among patches, selection would favor the set of trait values that maximized the rates of increase in each patch. Now suppose that there is migration among patches and the organism is unable to alter its life history according to the type of patch in which it finds itself (i.e., the individual is not phenotypically plastic). The population as a whole increases at some rate r and, by direct analogy with the first case considered (hard selection in a constant environment), selection will favor the combination of trait values that maximize r, not the individual patch r. For a population in stable age distribution this means the value of r that satisfies

$$\int p(h) \int l(x,h) m(x,h) e^{-rx} dx dh = 1 \tag{2.79}$$

where $p(h)$ is the probability of habitat h occurring, $l(x,h)$ is the probability of surviving to age x in habitat h, and $m(x,h)$ is the number of female births at age x in habitat h.

If populations are nonoverlapping, temporal stochasticity will be a stronger selective factor than spatial heterogeneity. For example, if there is a probability that λ will equal zero, then in a temporally varying environment the population goes extinct, whereas in the spatial environment only a portion of the population will be affected.

Defining Fitness When the Population Is Density-Regulated

For a density-regulated population there are two scenarios that need to be considered with respect to the definition of fitness. To examine these, suppose that the population is regulated by mortality in some early juvenile stage.

In one scenario all genotypes suffer the same mortality rate. That is, there is no genetic variation for density-dependent survival through this stage, though there could be variation for density-independent survival. In this scenario fitness can be defined by setting $r = 0$ in the characteristic equation to give

$$\int l(x)m(x)e^{-0x}dx = \int l(x)m(x)dx = R_0 \tag{2.80}$$

where R_0 is known as the **net reproductive rate** and is the expected number of female offspring produced by a female over her lifetime. This definition of fitness is also suitable if the particular fitness components under study are genetically uncorrelated with traits that are components of the density-regulation. For example, there could be genetic variation for larval survival under different densities, and hence selection could operate on this variation. However, if it is uncorrelated with factors that determine the age at maturity, then the latter can be analyzed using R_0.

In the second scenario there is genetic variation for the ability to survive under different densities, and either these traits are of direct interest or are genetically correlated with those traits under study. MacArthur (1962) suggested that in density-regulated populations the appropriate measure of fitness is the carrying capacity that can be attributed to a genotype. This suggestion is intuitively appealing, and a rough justification for its use is as follows. Suppose there are two clones that in isolation equilibrate at population sizes K_1 and K_2, where $K_1 < K_2$, and are otherwise identical. The two clones are mixed together in equal proportion to make a population initially of size K_1. Since the total population size is equal to the equilibrium population size for clone 1, this clone will show no response, but clone 2 is below its equilibrium and hence will increase in numbers. But now the total population size is above the equilibrium size for clone 1, and hence clone 1 must decline in numbers until the total population size is restored to K_1. Since clone 2 can always increase at this level (i.e., population size $= K_1$), clone 1 will continue to decline, eventually being eliminated and the population increasing in size to K_2.

Though it is not universally true that maximization of population size is the appropriate fitness measure in MacArthur's model (Green 1980), the typical finding with other more general population models is that population size is maximized (Anderson 1971; Hastings 1978; Iwasa and Teramoto 1980; Turelli and Petri 1980; Desharnais and Constantine 1983; Benton and Grant 2000). Difficulties arise when there is age structure in the population, and in this case population size may not be maximized (de Jong 1984). The important variable is the stage at which the density-dependence occurs, called the **critical age group** by Charlesworth (1972). In a density-regulated population "selection tends to maximize the number of individuals in the critical age group subject to the assumption of weak selection" (Charlesworth 1994, p. 149). Because selection maximizes only the number in the critical age group, it is not necessarily true that population size as a whole will be maximized. To accommodate density effects, Equation (2.67) can be rewritten as

$$\int l(x,d)m(x,d)dx = R_0(d) \tag{2.81}$$

where d designates that the life history components are functions of density. The tendency for density in the critical age group to increase has been termed by Leon and Charlesworth (1978) the "ecolological version of Fisher's fundamental theorem of natural selection."

Recently, simulation modeling has been used relatively extensively to examine fitness measures when there is density-dependence, stochastic fluctuations, and age structure. For such complex circumstances the concept of **invasibility** has been promoted, using as a measure the Lyapunov exponent (Ferriere and Gatto 1995; Grant 1997). Not surprisingly, fitness measures such as r or R_0 fare poorly as predictors of the probability that a mutant type can invade the population (Benton and Grant 2000). In circumstances in which there is both density-dependence and stochastic variation, a simulation approach is perhaps the only presently viable approach.

r- and K-Selection

Density-independent (= hard) selection in an unlimited environment maximizes r, whereas density-dependent selection maximizes population size over a particular age group. These two processes are sometimes referred to as r- and K-selection. Although these terms were used more popularly in the past, they are still frequently used, though considerable confusion has surrounded both the concept and its application.

Pianka (1970) credited Dobzhansky (1950) with originating the basic idea, though not the terms. In his paper Dobzhansky compared the tropical and temperate regions and suggested that populations in the temperate zones are primarily limited by abiotic factors and rarely achieve densities at which intra-specific competition prevails. In contrast, populations in the supposedly constant conditions of the tropics typically attain densities where density-dependence becomes a principal factor in population regulation. MacArthur and Wilson (1967) coined the terms r- and K-selection to describe these two conditions. The terms are derived from the standard model of population growth in ecology, the logistic equation in which the per capita growth rate declines linearly with density relative to some carrying capacity, K,

$$\frac{dN}{dt} = rN\left(1 - \frac{N}{K}\right) = rN - \frac{rN^2}{K}$$

$$N(t) = \frac{K}{1 + e^{c-rt}}$$

(2.82)

where c is a constant of integration dependent on the initial population size. MacArthur and Wilson, drawing on an earlier study by MacArthur (1962), proposed that alleles could be assigned values of r and K. In an unsaturated environment those genotypes with the highest values of r would prevail, whereas in a saturated environment those that could maintain the largest population size—those with the highest K value—would have the advantage. Thus far there is no real problem with the definition since these conform reasonably well to the definitions given in the previous sections of this chapter.

Since the analysis by MacArthur and Wilson there have been a great number of further analyses (Anderson 1971; King and Anderson 1971; Roughgarden 1971; Charlesworth 1971; Charlesworth and Giesel 1972a,b; Clarke 1972; Armstrong and Gilpin 1977; Turelli and Petri 1980; Asmussen 1983; Desharnais and Costantio 1983). Most analyses were based on the hypothesis, suggested by Gadgil and Bossert (1970), that there is a trade-off between r and K. This hypothesis may not be correct; it has received some support in experiments on *Drosophila melanogaster* (Mueller and Ayala 1981; Mueller et al. 1991), but not in those on the rotifer *Asplanchna brightwelli* (Snell 1977), the cladocera, *Bosmina longirostris* (Kerfoot 1977) or the bacteria *Escherichia coli* (Luckinbill 1984). The foregoing theoretical analyses all demonstrated that K will be maximized, and that if the heterozygote has the highest K, a stable polymorphism can be generated. Although several analyses have indicated that this relationship may not be universally true (Green 1980; de Jong 1982), the overall conclusion is that in a density-regulated population, selection maximizes population size (or the critical age cohort). To this extent the concept of r- and K-selection has been a useful theoretical concept. The problem arises when the concept is applied without refinement to natural populations, most particularly when these definitions are used to define the life history traits of organisms.

Pianka (1970) suggested that r-selected organisms could be characterized by rapid development, high rate of increase, early reproduction, small body size, and semelparity (single reproductive episode). K-selective organisms were proposed to have the opposite characteristics, the first two traits being tradedoff for high population size and competitive ability. No rationale for these categories was given. The categorization ignores the fact that in ectotherms, and to a lesser extent endotherms, development time, size at maturity and fecundity are intercorrelated (see Chapter 4). Thus, depending on the exact functional relationships between traits, the rate of increase may be maximized by late reproduction. In support of his categorization, Pianka presented data on the body lengths of vertebrates and insects. To compare vertebrates and insects is to compare apples and oranges. Differences between taxa as widely separated as vertebrates and insects are undoubtedly due to a multiplicity of causes. A correct analysis would be between different genotypes or at least between different populations of the same species.

Mertz (1975, p. 3) objected to the notion of an *r-K* continuum because, "*r* and *K* are usually thought of in terms of the logistic equation, and classical logistic theory is plainly incompatible with complex life history considerations," a view echoed by others (Gill 1972; Wilbur et al. 1974; Whittaker and Goodman 1979; Caswell 1982b; Hall 1988; see also the experimental tests of Pianka's categories by Barclay and Gregory 1981, 1982). Unfortunately, the perceived inadequacies of the *r-K* concept led others to invent yet more terms: *b* and *d* selection (Hairston et al. 1970), α selection (Gill 1974), *h* and *T* selection (Demetrius 1977).

The real problem that Pianka's 1970 paper caused was that workers began classifying species on the basis of their life history characteristics rather than by a demonstration of the type of selection operating (reviewed in Parry 1981). Thus the terms *r* and *K* have become to apply either to density-dependent selection or to presumed characteristics associated with *r*- and *K*-selection.

I concur with the view of Mueller (1988a, p. 787): "Given the substantial problems with the verbal theory of *r*- and *K*-selection, it is reasonable to ask if the formal theories that assume logistic fitness functions can be used to predict the evolution of phenotypes other than density-dependent rates of population growth. The answer is probably no. To develop a theory that accounts for the evolution of body size or competitive ability, models with the relevant ecological phenomena must be developed." The theoretical analysis of Mueller (1988a) is a fine example of this approach, grounded firmly in the biology of a particular organism, *Drosophila melanogaster*, and designed to address the processes observed during density-dependent selection in this species (Mueller and Sweet 1986; Joshi and Mueller 1988; Mueller 1988b, 1990, 1991).

To summarize, the concept of *r*- and *K*-selection has been useful in helping to formalize the definition of fitness in density-regulated populations, but attempts to transfer the concept to actual populations without regard to the realities of the complexities in life history have probably been detrimental rather than helpful. The terms *r*- and *K*-selection should be interpreted strictly in terms of models of density-dependence (Boyce 1984; Elgar and Catterall 1989) and, given the confusion that now surrounds the issue, it may be preferable to avoid use of the terms altogether.

Defining Fitness When Selection Is Frequency-Dependent

Frequency-dependent selection occurs when the fitness of a phenotype or genotype varies with the phenotypic or genotypic composition of the population (Ayala and Campbell 1974; Gromko 1977; DeBenedictis 1978). The important differences between frequency-dependent selection and those described above is that in the latter there is typically only a single most fit genotype at equilibrium, whereas with frequency-dependent selection there must necessarily be several phenotypes maintained in the population at equilibrium.

In the case of a single locus genetic model with nonoverlapping generations, we can assign a selection coefficient to each genotype. This selection coefficient represents the relative contribution of each genotype to the next generation (i.e., fecundity × survival, where these terms include components such as the probability of breeding at all). A characteristic finding for such models is that frequency-dependent selection can readily maintain genetic variation at a single locus (Wright 1948; Haldane and Jayakar 1963b; Clarke and O'Donald 1964; Clarke 1964; Anderson 1969). Slatkin (1979) in a general analysis of frequency- and density-dependent selection concluded that "at the equilibrium reached . . . the distribution of a normally distributed character does not depend on the underlying genetic model as long as the model imposes no constraints on the mean and variance." This means that, with regard to equilibrium conditions, we need consider only a phenotypic or clonal model.

Suppose there exist two clones with rates of increase r_1 and r_2, which are functions of the relative frequencies. Thus we can write $r_1 = f[N_1/(N_1 + N_2)]$, $r_2 = g[N_2/(N_1 + N_2)]$, where $f(\bullet)$ and $g(\bullet)$ denote functions and N_i is the population size of clone i. Clearly, at equilibrium one solution is $r_1 = r_2 = 0$, which is equivalent to finding the two population sizes at which $f(\bullet)$ and $g(\bullet)$ equal zero. Another possible solution is for r_1 and r_2 to be equal but not equal to zero, in which case the populations are expanding (or contracting) at the same rate and thereby maintaining the same relative frequency. This is probably an unlikely scenario, particularly if we consider a distribution of genotypes that would be necessary to expand this approach to a quantitative trait. Unlike the density-dependent case, frequency-dependent selection will not maximize population size (Slatkin 1978).

The analysis of frequency-dependent selection using the type of phenotypic model outlined above typically focuses not upon fitness per se but upon some component of fitness, assuming that maximization of this component will lead to maximization of fitness. This approach has given rise to the adoption of a particular method of analysis known as **game theory** (Maynard Smith 1982; Riechert and Hammerstein 1983). Such analyses make two assumptions: The first is that particular trait values (most frequently, behaviors) will persist in a population provided no mutant adopting an alternate value (behavior) can invade (Maynard Smith and Price 1973). Such stable combinations are termed **evolutionarily stable strategies** (ESS). The concept of the ESS is not unique to game theory. The maximization of r, R_0 or the numbers in the critical age group are all ESSs within the context in which they are appropriate. Second, for each type there must be an assigned gain or loss in fitness when this type interacts with another individual. From this payoff matrix we compute the expected payoff for each trait value (behavior). For a trait to be maintained in polymorphic condition (i.e., for it to be evolutionarily stable), their fitnesses must be equal. For a game involving two states—the game that is the one most usually studied—we can write the payoff matrix as

$$\text{Payoff to} \downarrow \quad \begin{matrix} A & B \end{matrix}$$
$$\begin{matrix} A \\ B \end{matrix} \begin{bmatrix} W_{AA} & W_{AB} \\ W_{BA} & W_{BB} \end{bmatrix} \tag{2.83}$$

where W_{IJ} is the payoff to individual with trait I when interacting with an individual displaying trait J. To ascertain if both trait values will be maintained in the population, we assume that the proportion of A types in the population is p. This could represent separate types in the population or individuals adopting the type A value. We then test for equality of the expected payoffs,

$$pW_{AA} + (1-p)W_{AB} = pW_{BA} + (1-p)W_{BB} \tag{2.84}$$

Upon rearrangement, this equation gives

$$p = (W_{BB} - W_{AB}) / (W_{AA} + W_{BB} - W_{AB} - W_{BA}) \tag{2.85}$$

Since its conception, game theory, or as it is more popularly (and incorrectly) known, ESS theory, has undergone considerable mathematical refinement (e.g., Riley 1979; Gadgil et al. 1980; Hines 1980; Cressman and Dash 1987; Vickers and Cannings 1987; Rand et al. 1994; Kisdi and Meszena 1995; see also reviews by Rapoport, 1985, and Vincent and Brown, 1988). However, these do not appear to have greatly influenced the application of the approach to actual case studies. For a detailed and lucid description of the application of game theory to real data, see Maynard Smith (1982). The most important theoretical question is the extent to which the method is valid in a sexual population. With simple Mendelian models, such as single-locus, two allele models, an ESS may not be possible (Auslander et al. 1978; Maynard Smith 1981), but more complex genetic models do permit populations at least to approach, if not attain, the predicted equilibrium (Eshel 1982; Bomze et al. 1983; Hines 1987; Taper and Case 1992; Roff 1998a).

The Fitness Surface

TRADE-OFFS AND OPTIMAL TRAIT VALUES. As discussed above, the quantitative genetic equation does not appear to limit the long-term change in trait combinations. Further, selection maximizes a quantity that depends upon the environmental conditions (constant, temporally or spatially variable) or population processes (density-independent, density-dependent, or frequency-dependent). Even though selection might be a continuous process, it is evident that trait values do not show a continuous directional change; why is this so? The simplest answer is that, rather than being directional, in general selection is balancing or stabilizing, maintaining the trait combinations at some intermediate set. There are numerous examples of stabilizing selection in both laboratory stocks and natural populations (see Table 9.1 in Roff 1997). How can a peaked fitness surface come about? The answer is that the fitness surface is composed of the interaction between several traits for which the directions of directional selection are in opposition. For example, in the waterstrider *Aquarius remigis*, daily fecundity increases with body size. Hence there is directional selection for increased body size, but survival (longevity) decreases with body size. Thus there is directional selection for decreased body size (Figure 2.24). Let these two relationships be $Y_{DF} = a_{DF} + b_{DF}X$ and $Y_L = a_L - b_L X$, where the subscripts DF and L, denote daily fecundity and longevity, respectively. Lifetime fecundity, Y_{LF}, is equal to daily fecundity times longevity, giving

$$Y_{LF} = Y_{DF}Y_L = (a_{DF} + b_{DF}X)(a_L - b_L X)$$
$$= a_{DF}a_L + (a_L b_{DF} - a_{DF}b_L)X - b_{DF}b_L X^2 \qquad (2.86)$$

which is a parabola. (The more bumpy shape in Figure 2.24 is a consequent of fitting with a spline function.) With no constraint imposed by the **G** matrix, and with the absence of other selective factors acting on body size, selection should eventually drive body size to the turning point of the fitness curve. The observed body sizes of males and females are indeed quite close to the observed fitness maxima (Figure 2.24): Preziosi and Fairbairn (2000) speculated, based on a theoretical analysis (Reeve and Fairbairn 2001), that the deviations are a consequence of the system not yet having attained equilibrium. Fluctuating selection might also cause the observed mean to deviate from the long-term optimum (Ferguson and Fairbairn 2000).

GENETIC CONSTRAINTS ON THE FITNESS SURFACE. The prediction of the optimal combination of trait values thus boils down to estimating the fitness surface. It means determining the trade-offs between traits. However, there is a problem. To understand its nature let us examine the two-trait case, illustrated in Figure 2.25. The trade-off between traits X and Y is a represented by concave curve, $Y = f(X)$. Assume that fitness is a linear combination of the two traits, $W = aX + bY$. Therefore, there are linear contours of equal fitness given by the relationship $Y = (W - aX)/b$. Geometrically, it is obvious that the optimal combination is that at which a fitness contour is tangent to the trade-off curve (Figure 2.25). The actual values, X^*, Y^*, can be found using the calculus as shown in the legend to Figure 2.25. Suppose the initial combination lies below the trade-off curve. Selection will drive the trait combinations towards X^*, Y^*, with the actual trajectory dependent upon the **G** matrix. Since there are higher values of fitness beyond the point

Figure 2.24   (A) Stabilizing selection in the water strider *Aquarius remigis*, resulting ▶ from the combined effect of directional selection favoring (B) increased body size with respect to daily fecundity but (C) decreased body size with respect to longevity. Solid lines are cubic spline fits to the data, and the dashed lines are bootstrapped standard errors. The arrow indicates the observed body size. From Preziosi and Fairbairn (1997).

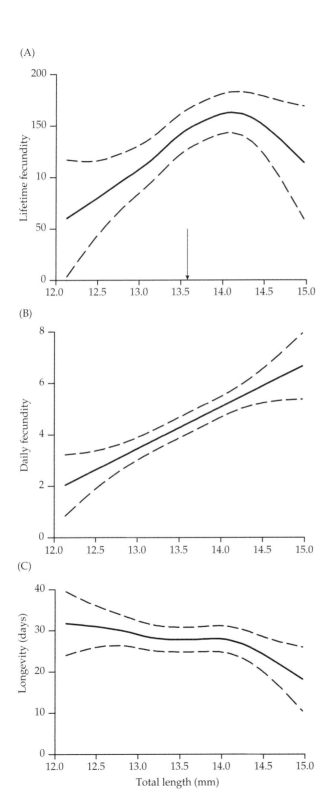

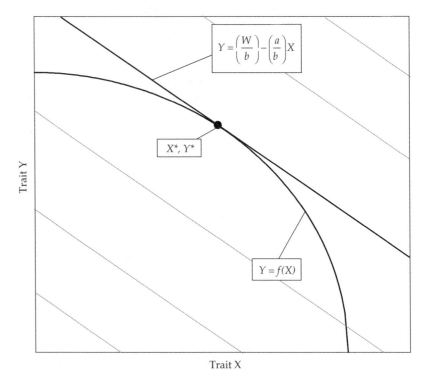

$$Y = \left(\frac{W}{b}\right) - \left(\frac{a}{b}\right)X$$

X^*, Y^*

$Y = f(X)$

Trait Y

Trait X

Figure 2.25   Hypothetical trade-off between two traits X and Y superimposed over the fitness contours. The optimal combination of X,Y is the point at which a fitness contour is tangent to the trade-off curve. To find the optimal combination using calculus, we proceed as follows:

$$W = aX + bY = aX + bf(X), \quad \frac{\partial W}{\partial X} = a - b\frac{\partial f(X)}{\partial X} = 0 \text{ when } \frac{\partial f(X)}{\partial X} = -\frac{a}{b}$$

X^*,Y^*, if there remain additive genetic variance and covariance, the trait values would be driven beyond the trade-off value. One way that this will be prevented from happening is if the genetic correlation between the two traits is –1 at the point X^*,Y^*. This idea was independently made by Charnov (1989a) and Charlesworth (1990). Charlesworth (1990) assumed that the trade-off function represented a genetic constraint; his example is illuminating. A hypothetical organism lives for three years, breeding for the first time at the end of the first year. All potential eggs are produced prior to first reproduction and hence $m(1) + m(2) + m(3) = $ constant, where $m(x)$ is the number of female offspring produced at age x. This is the first functional constraint. Two further constraints are assumed: that there is a concave-down trade-off function relating fecundity at age 1 with survival, $s(1)$ from ages 1 to 2, and a similar function for age 2, qualitatively the same as shown in Figure 2.25. The particular numerical form chosen was $s(x) = 1 - m(x)^2$. Population size was assumed fixed, with density-dependence occurring in the immature stage. The appropriate fitness measure is thus lifetime reproduction, $R_0 \propto m(1) + s(1)m(2) + s(2)m(3)$. Given these constraints, the signs of the genetic correlations between traits can be determined (see Charlesworth 1990 for the algorithm):

$$
\begin{pmatrix}
 & s(2) & m(1) & m(2) & m(3) \\
s(1) & <0 & \boxed{-1} & >0 & >0 \\
s(2) & & >0 & \boxed{-1} & 0 \\
m(1) & & & \boxed{<0} & \boxed{<0} \\
m(2) & & & & \boxed{0}
\end{pmatrix}
\tag{2.87}
$$

where correlations between functionally constrained traits are boxed. The genetic correlations between the two functionally defined trade-offs relating fecundity and survival are both –1. The constraint on the number of eggs produced $\left(\sum m(x) = \text{constant}\right)$ does not result in a genetic correlation of –1 between the component pairs, but they are either 0 [$m(2)$ vs $m(3)$], or are negative [$m(1)$ vs $m(2)$ and $m(1)$ vs $m(3)$]. The reason for this is that the constraint is not a line but a three-dimensional surface. Consider the correlation between any two fecundities, keeping the value of the third fixed, say, $m(1)$ vs $m(2)$ with $m(3)$ fixed at m^*, the functional constraint is now $m(1) + m(2) - c = 0$, where c equals the constant plus m^*. Because of the fixed sum, an increase in one fecundity, say, $m(1)$, will result in a reduction in the other, $m(2)$, and a genetic correlation of –1. Now suppose we set $m(3)$ at another fixed value. There will still be a single constraint line but the intercept of that line will be changed. If we combine both values of $m(3)$, we still get an overall decline in $m(2)$ with an increase in $m(1)$, but the correlation will now be greater than –1. Moving over all possible combinations of $m(x)$—by definition $m(x) \geq 0$—gives the same result, namely that the genetic correlation between any two fecundities will lie between 0 and –1. The question of genetic correlations constrained to a surface is considered in more detail in the next chapter.

Correlations between survival from ages x to $x + 1$ and fecundity at ages other than x are positive or zero. One might have hypothesized costs of reproduction that predicted negative trade-offs in these cases, but such trade-offs are not explicitly specified by the assumed set of functional constraints. Those that are explicitly specified do indeed produce negative genetic correlations. The set of genetic correlations does, therefore, reflect the underlying functional constraints. The requirement that the genetic correlation between two traits be –1 to be a constraint seems very stringent; there are probably few instances when such correlations occur. One obvious instance is when two activities cannot be done concurrently and hence time is the constraining factor, giving $\sum_{i}^{n} T_i = c$, where T_i is the time spent on the ith activity and c is the total time available. Another instance where the correlation could be –1 occurs when there are constraints imposed by physics. The low clutch weight of *Anolis*, lizards is a possible example.

Anoline lizards are characterized by a clutch size of 1 egg and, as a consequence, clutch weight relative to body size is extremely low in this genus compared to a more typical iguanid genus such as *Sceloporus* (Figure 2.26). The low
clutch size is partly compensated by multiple broods, a pattern possible in the
tropical environment typical of the genus *Anolis*, but one that may not be available to their ecologically temperate counterparts in the new world, *Sceloporus*,
which lay larger clutches but only 1–3 clutches per season (Tinkle 1972, 1973;
Tinkle and Ballinger 1972; Ballinger 1973). Anoline lizards are highly arboreal,
spending the majority of their time climbing among bushes and trees. (For a
comparative analysis of climbing ability and morphology, particularly pad-area,
see Irschick et al., 1996.) Andrews and Rand (1974) speculated that aboreality in
anoles may place a mechanical constraint on clutch weight. They argued that the
efficacy of the subdigital lamellae, which provide the grip required for climbing,
may be greatly diminished by weight, and, as a consequence, the requirements
of climbing may limit clutch size.

An important component of the argument of Andrews and Rand is the difference in the relationship between clutch mass and female length of the two lizard
genera. The hypothesis verbally suggested by Andrews and Rand is that in *Anolis* mean clutch weight will increase as the square of female length, whereas in
Sceloporus mean clutch weight will increase as the cube of female length. A linear
regression on these data after logarithmic transformation supports the hypothe-

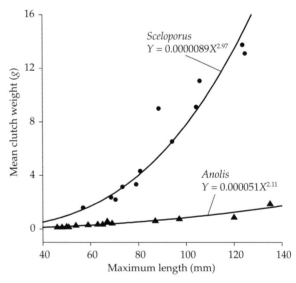

Figure 2.26   Relationship between mean clutch
weight and maximum length in an arboreal genus of
lizard (*Anolis*) and a terrestrial genus (*Sceloporus*). Data
from Andrews and Rand (1974).

sis (Figure 2.26), the slope for *Anolis* being 2.11 ($SE = 0.184$, $r = 0.957$, $n = 14$, $P < 0.001$), and that for *Sceloporus* being 2.97 ($SE = 0.269$, $r = 0.961$, $n = 12$, $P < 0.001$). There is very little variance about the *Anolis* regression line, suggesting that gravity might be acting as a fundamental constraint in this case, leading to a genetic correlation between maximum length and mean clutch weight of +1. This relationship acts as a trade-off despite the positive correlation, because to increase clutch weight there must be an increase in size, which presumably carries some cost. I do not wish to argue this example too strongly, but to suggest that it is in the realm of the environment where stringent constraints (e.g., time, gravity) may occur.

A genetic correlation of ±1 essentially means that we really have a single trait that has two phenotypic manifestations. An example is the environmental threshold model of Hazel et al. (1990) for dimorphic traits (Roff 1994a). Many traits occur as dimorphic, rather than continuously distributed characters: for example, pupal color in swallowtail butterflies (Hazel 1977), shell shape in acorn barnacles (Lively 1986a), cyclomorphosis in zooplankters (Dodson 1989), paedomorphosis in amphibia (Semlitsch 1985), dental dimorphism in some species of fish (Skúlason et al. 1989; Meyer 1990; Schluter 2000), wing dimorphism in insects (Harrison 1980; Roff 1986), sex ratio in turtles (Bull et al. 1982), and diapause in insects (Mousseau and Roff 1989). Although only two phenotypes are discernable, dimorphic variation may be due to the additive effect of many loci, the particular manifestation of the trait being a function of a threshold of sensitivity. According to this **threshold model**, a continuously varying character underlies the expression of the trait, with individuals with values lying above the threshold developing into one morph, and individuals lying below the threshold developing into the other morph (Figure 2.27).

The threshold model was developed specifically to address the question of discrete states in a fixed environment. However, the proportion of each morph in a population frequently varies both with genotype and environment. For example, in wing dimorphic insects both temperature and genotype determine the proportion of macropterous individuals (fully winged, flight capable) (Harrison 1980; Roff 1986). Similarly, the proportion of males in some reptile species is determined by incubation temperature of the eggs as well as by genotype (Bull et al. 1982; Janzen 1992). The effect of environment can also be accommodated with a threshold model (Hazel et al. 1990). The threshold model, as typically presented, assumes a fixed threshold and a continuously distributed underlying trait. However, from a mathematical perspective it could equally well be assumed that the value of the underlying trait is fixed and that the threshold value is genetically variable. Which mathematical model we assume is based upon mathematical convenience and does not necessarily specify a particular mechanism. When considering the expression of a dimorphic trait across an environmental variable—such as photoperiod, temperature or density that varies in a continuous fashion—it is most convenient to assume a variable threshold. According to the model proposed by Hazel et al. (1990), each geno-

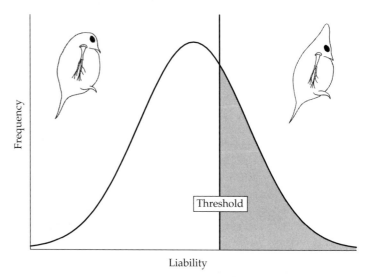

Figure 2.27   The threshold model and the generation of dimorphic variation when genetic determination is polymorphic. According to this model there is a normally distributed underlying trait termed the **liability**. If the liability exceeds a critical threshold during development, one morph is produced (e.g., the crested form of *Daphnia*, right), whereas if the liability lies below the threshold, the alternate morph is produced (e.g., the round-head morph of *Daphnia*, left).

type has a switch point, or threshold, along the environmental gradient at which the phenotypic expression of the genotype abruptly shifts from one morph to the other. I have termed this model the **environmental threshold** (ET) **model** (Roff 1994a). Under this model the genetic correlation between conditions (e.g., two temperatures) is +1 (Box 2.12).

According to the ET model, after several generations of selection the norms of reaction curves should be shifted, but not changed in shape. Because the norm of reaction curve is predicted to be cumulative normal, it can be linearized by transformation to z, where z is the abscissa on the unit normal corresponding to the proportion of one morph. Selection for, say, increased macroptery is predicted to shift the reaction curve to the left, but it should remain parallel to the reaction curve of the unselected population. This prediction is supported by selection on proportion macroptery in the cricket *Dianemobius fascipes* and the small brown planthopper *Laodelphax striatellus* (Roff 1994a).

It may be uncommon for two traits to be so tied together that there is no variation about the regression line. But, as traits are accumulated, the total range of variation may become sufficiently constrained that, for all practical purposes, we will find that the interaction among all n component traits, $x_1, x_2, x_3, \ldots, x_n$ is such that there is a function, f, for which $f(x_1, x_2, x_3, \ldots, x_n) = c$, where c is a constant.

Box 2.12
The genetic correlation in the environmental threshold model

To apply standard quantitative genetic theory to the ET model we make the usual assumption that the character, in this case the switch point, is normally distributed in the population. As a consequence, the relationship between the proportion of a particular morph and the value of the environmental variable (equal to the norm of reaction for the population) will follow a cumulative normal. Since the distribution of thresholds in each environment along the gradient is, by definition, the same, it follows that whereas the proportion of a morph will vary across environments, the heritability of the trait will not. Thus, if we designate one environment as x and the other as y, we have $h_x^2 = h_y^2$, and for the phenotypic variances, $\sigma_x^2 = \sigma_y^2$. The genetic correlation across environments along the gradient will be +1. This can be demonstrated in two ways. First, by changing the environment, we do not change the underlying character, only its expression on the 0–1 scale. Since the method of estimation corrects for this change, we are in fact measuring the same trait independently of the environment, and therefore, the genetic correlation must be +1. The second method of demonstration considers the effect of selection on the reaction norm. Selection will shift the distribution of switch points, thereby shifting the reaction norm by the same amount. From rearrangement of Equation (2.39) the genetic correlation between x and y, is

$$r_A = \left(\frac{CR_y}{R_x} \right) \left(\frac{h_x \sigma_x}{h_y \sigma_y} \right) \tag{2.88}$$

From the considerations above, $\sigma_y = \sigma_x$, $h_x = h_y$, and the correlated response of y is the same as the response of x, $R_x = R_y$. Substituting in the foregoing equation gives $r_A = 1$.

Such functions do present us with what Pease and Bull (1988) call the problem of dimensionality, for to omit one trait is to miss the trade-off. The omission of important trade-offs can have significant consequences for predicting variation among traits. An example is shown in Figure 2.28, in which the failure to take into account the variation in egg size among different species of flatfish obscures a general relationship between fecundity, measured in terms of total egg volume, and body size.

Charlesworth (1993) showed that, at equilibrium under any strength of selection, one of three conditions must be satisfied: (1) all the components of **G** are zero, (2) the gradient vector of the intrinsic rate of increase with respect to the mean trait, $\nabla r_{\bar{z}}$, is zero, or (3) the determinant of **G** is zero. The many estimates of additive genetic variance and covariance in both laboratory and wild populations (Mousseau and Roff 1987; Weigensberg 1996; Roff 1996b; Hoffman 2000) have demonstrated that the first condition is highly unlikely. If the component traits of **G** are life history traits, then the derivatives of r must be positive and

(A)

(B)

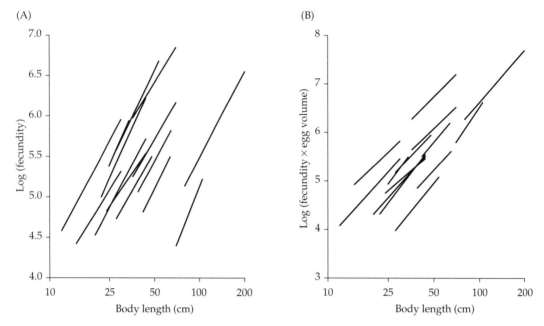

Figure 2.28   An example of how the omission of a trade-off can distort relationships among traits. (A) Fecundity as a function of body length (length at maturity to maximum length) for 14 species of flatfish. (B) The effect of including variation in egg size. Note that the spread of the lines is greatly diminished. Data from Roff (1982).

thus condition 2 cannot be satisfied. This leaves condition 3 as the necessary requirement. For the two-trait case, this requirement is equivalent to a genetic correlation of ±1; with more traits, the condition requires that at least some of the correlations be negative. At the same time, it also means that some combinations will show positive correlations. (Compare the hypothetical age-structured model discussed above.)

Summary

The evolution of traits showing quantitative variation can be modeled using a Mendelian framework in which the trait value is determined by the action of many genes. This action can be divided into additive and nonadditive components, the first of which is the major determinant of the linear regression between mean offspring and mid-parent values. The ratio of the additive genetic variance to the total phenotypic variance is termed heritability (in the narrow sense) and is equal to the slope of the regression of mean offspring on mid-parent values. From this basis can be developed an evolutionary model for several traits, the structure of which consists of three matrices: the additive genetic variance-covariance matrix, the phenotypic variance-covariance matrix, and the vector of selection differentials. The direction of evolution is modulated by the action of all three matrices. If the genetic correlation between any two traits is ± 1, then the evolution of the two traits is constrained to occur only along a single line. If the traits are constrained in higher dimensions (e.g., a plane) there is still potentially a strong constraint on the direction of evolution but the genetic correlation between any two pairs of traits will not be ± 1. A fundamental component in the analysis of the evolution of life history evolution/variation is the definition and measurement of the fitness surface. There is no single measure that is best for all occasions. In the absence of density or frequency dependence, a suitable measure is r or R_0. Temporal and spatial variation affect the method of estimating these two parameters (geometric average, population-wide value). With density dependence, population size or the size of the age-component subjected to the density dependence is frequently but not universally maximized. The analysis of frequency-dependent selection requires a game-theory approach.

CHAPTER 3

Trade-offs

All other things being equal, selection will act to maximize the $l(x)$ and $m(x)$ functions and, in particular environments will minimize the age at maturity. Selection will be prevented from doing so by the existence of trade-offs between the $l(x)$ and $m(x)$ functions deriving either from a direct trade-off between these fitness components or indirectly via an indirect trade-off with a third variable. A direct trade-off could result from a physiological constraint in which increased reproduction lowered life span. An indirect trade-off could result if increased reproduction required increased energy intake, which necessitated increased foraging, which exposed the organism to increased predation and hence reduced its survival.

The foundation of life history theory is the premise that trait combinations are constrained by trade-offs among traits. From the perspective of life history theory trade-offs can be conveniently divided into two groups; first, those that involve traits that enter directly into the characteristic equation, namely, age at maturity, fecundity, and survival. I shall refer to these as **life history** or **direct fitness traits**. The second category are those trade-offs that act indirectly on fitness via one of the preceding traits. For example, body size does not directly enter into the characteristic equation but it may be important if it determines, in part, fecundity, mating success, survival, or age to maturity. A relationship between body size and a life history trait is not by itself a trade-off because selection will, all other things being equal, simply favor an increasing or decreasing body size. The trade-off arises when body size is related to two life history traits so that there arises a trade-off between the two life history traits. So, for example, a large body size may increase fecundity but also increase development time and hence decrease survival. Suppose, for illustrative purposes,

$$Fecundity = a + b\big(Body\ size\big)$$
$$Survival = c - d\big(Body\ size\big)$$

(3.1)

Equations (3.1) can be rearranged so that body size is the dependent variable. Equating the two relationships and further rearranging gives

$$Fecundity = \left(a + \frac{bc}{d}\right) - \left(\frac{b}{d}Survival\right)$$

(3.2)

Thus there is a trade-off between fecundity and survival, mediated by their separate relationships with body size. A positive correlation between fecundity and body size and a negative correlation between body size and survival will necessarily lead to a negative correlation between fecundity and survival (Box 3.1).

In this chapter I shall first present a theoretical framework within which trade-offs can be understood, second a discussion on methods of measuring trade-offs, and third some examples of trade-offs that involve life history traits.

Theoretical Considerations

Trade-offs and Missing Variables

A point that is drummed into every student of science is that correlation does not imply causation. In the present context the important consideration is that the omission of a trait from a proposed set of interacting traits may lead to the erro-

---

Box 3.1

The production of a trade-off by the interaction of two component functions

For simplicity and without loss of generality, we rescale fecundity and survival such that $b = d = 1$: *Fecundity* $= a + X$, *Survival* $= c - X$, where X is body size. The covariance between survival and body size, σ_{SX}, is

$$\sigma_{SX} = E\{(c - X)X\} = E\{cX\} - E\{X^2\} = c\mu_X - \left(\sigma_X^2 + \mu_X^2\right)$$

(3.3)

The covariance between fecundity and survival, σ_{FS}, can be obtained in a similar manner:

$$\begin{aligned}
\sigma_{FS} &= E\{(F - \mu_F)(S - \mu_S)\} = E\{(a + X - \mu_F)(c - X - \mu_S)\} \\
&= E\{a(c - X) + cX - X^2 - (c - X)\mu_F - (a + X)\mu_S + \mu_F\mu_S\} \\
&= a\mu_S + c\mu_X - \left(\sigma_X^2 + \mu_X^2\right) - \mu_S\mu_F - \mu_F\mu_S + \mu_F\mu_S \\
&= \mu_S(a - \mu_F) + \sigma_{SX} = \sigma_{SX} - \mu_S\mu_X
\end{aligned}$$

(3.4)

Now, since the covariance between survival and body size is negative ($\sigma_{SX} < 0$) and both means are positive ($\mu_S > 0$ and $\mu_X > 0$), then the covariance between fecundity and survival must be negative ($\sigma_{FS} < 0$).

neous conclusion that there is no trade-off between traits. To illustrate this idea, consider the three-trait relationship $Z = X - Y$, where for simplicity the data have been scaled so that all coefficients are unity and the intercept equal to zero. An example of such a relationship would be the functional relationship between fecundity and the two traits, body size and egg size. If fecundity is constrained, for example, by geometric constraints then we might find *Fecundity* = *c*(*Body size*)/*Egg size*, where *c* is a constant. Upon log transformation, we have ln(*Fecundity*) = ln*c* + ln(*Body size*) − ln(*Egg size*). Now suppose we hypothesize that fecundity, Z, declines as egg size, Y, increases, but fail to take into account variation in body size, X. The covariance between X and Y, σ_{XY}, can be obtained as follows:

$$X = Z + Y$$
$$\sigma_X^2 = \sigma_Z^2 + \sigma_Y^2 + 2\sigma_{ZY} \tag{3.5}$$
$$\sigma_{ZY} = \tfrac{1}{2}\left(\sigma_X^2 - \sigma_Z^2 - \sigma_Y^2\right)$$

If the variance in X (body size) is small relative to the variances in Z and Y, then we shall observe the predicted trade-off. (For an example see Guntrip et al. 1997.) However, if the variance in X is large the trade-off could either be absent or even be positive, suggesting the absence of a trade-off (Figure 3.1; for an example, see Atkinson and Begon 1987).

A positive correlation can also arise due to the evolution of an optimal reaction norm that depends upon the missing variable. Genoud and Perrin (1994) provide a particularly good example of this for the case of the trade-off between litter size and offspring weight (Box 3.2).

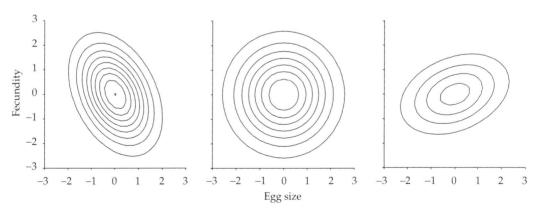

Figure 3.1 Examples of the bivariate normal distributions (lines show contours of equal probability) for two traits which are themselves constrained by a third variable: $X = Z + Y$, $\sigma_{ZY} = \tfrac{1}{2}\left(\sigma_X^2 - \sigma_Z^2 - \sigma_Y^2\right)$. For clarity only, specific life history traits are shown (X is body size, Z is fecundity, Y is egg size). Parameter values used, reading from left to right: $\sigma_X^2 = 1.6$, $\sigma_Y^2 = 1$: $\sigma_X^2 = 1.6$, $\sigma_Y^2 = 1.6$: $\sigma_X^2 = 3.2$, $\sigma_Y^2 = 1.6$.

Box 3.2
Example of a positive correlation arising due to the presence of a missing variable

Suppose that for a female of given "quality", Q, there is a linear trade-off between litter size, X, and weaning weight, W:

$$W = c_0 + Q - c_1 X \tag{3.6}$$

where the c_i are constants. Assume also that offspring survival, S_0, increases with offspring mass at weaning, (this is only approximate, because the relationship would almost certainly be curvilinear: see Chapter 4), and that (also approximately) female survival, S_F, decreases with litter mass:

$$S_0 = -c_2 + c_3 W$$
$$S_F = c_4 - c_5 WX \tag{3.7}$$

Assuming that generations are discrete, reproductive value is given as

$$V(t) = S_0 X + S_F V(t+1) \tag{3.8}$$

From standard calculus, the litter size that maximizes reproductive value, X^* is found by setting $\dfrac{\partial V(t)}{\partial X} = 0$. (This procedure is discussed in greater detail in Chapter 4.)

$$\frac{\partial V(t)}{\partial X} = S_0 + X \frac{\partial S_0}{\partial W}\frac{\partial W}{\partial X} + \frac{\partial S_0}{\partial (WX)}\left(X \frac{\partial W}{\partial X} + W\right)V(t+1) \tag{3.9}$$

And $\dfrac{\partial V(t)}{\partial X} = 0$ when

$$-c_2 + c_3\left(c_0 + q - c_1 X^*\right) - X^* c_1 c_3 - c_5\left(c_0 + q - 2c_1 X^*\right)V(t+1) = 0 \tag{3.10}$$

from which we obtain the optimal litter size to be

$$X^* = a + bQ \tag{3.11}$$

where

$$a_1 = \frac{c_2}{2c_1\left(c_5 V(t+1) - c_3\right)} + \frac{c_0}{2c_1}, \quad b_1 = \frac{1}{2c_1} \tag{3.12}$$

As can be seen from Equation (3.11), the optimal litter size increases linearly with female quality. We can now compute the optimal offspring weaning weight, W^*, as a function of litter size and female quality, which is done by substituting Equation (3.6) into Equation (3.11):

$$W^* = c_0 + \frac{X^* - a_1}{b} - c_1 X^* \tag{3.13}$$

Substituting the expressions for a_1 and b_1 we get

$$W^* = \frac{c_2}{c_3 - c_5 V(t+1)} + c_1 X^* \tag{3.14}$$

The optimal litter size increases with the optimal weaning weight. Thus, providing females are capable of responding appropriately to their quality, in the field litter size and offspring weight should vary positively although the underlying trade-off is negative (Figure 3.2). For further discussion of this model, see Genoud and Perrin (1994).

(A)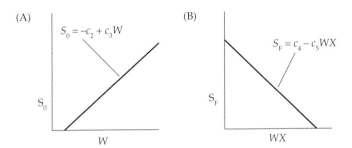

(C)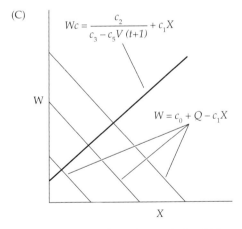

Figure 3.2 A hypothetical example in which a trade-off is masked by a missing third variable. (A) Offspring survival, S_0, is an increasing linear function of weaning weight, W. (B) Female survival is a decreasing linear function of reproductive investment, measured as the product of litter number and weaning weight, WX. (C) Weaning weight, W, declines linearly with litter size, X, for females of fixed quality, Q. $W = c_0 + Q - c_1 X$. The trade-off lines for three values of Q are shown. Fitness is maximized by a female varying her litter size and weaning weight of offspring according to her quality such that $W = c_2 / \left[c_3 - c_5 V(t+1) \right] + c_1 X$, which produces the appearance that there is no trade-off.

Phenotypic Versus Genetic Expression of Trade-offs

A trade-off clearly has no selective effect unless it is expressed at the level of the phenotype. The "conventional wisdom" that a trade-off has no direct evolutionary consequence unless it is expressed at the genetic level in the form of a genetic covariance between the traits is correct only if selection is acting on a single trait. To illustrate, consider the two-trait response equation for trait X:

$$R_{X,\sigma} = \frac{(i_X - r_p i_Y)h_X^2 + (i_Y - r_p i_X)h_X h_Y r_A}{1 - r_p^2} \tag{3.15}$$

where $R_{X,\sigma}$ is the response in standard deviation units. Now if the genetic correlation is zero ($r_A = 0$), the second term in the numerator is zero but there still can be an influence of the phenotypic correlation by virtue of its appearance in the first term. If there is no selection on trait Y, then $i_Y = i_X r_P$ (Chapter 2) and $R_{X,\sigma} = i h_X^2$, which is the usual direct response equation. If there is selection on trait Y (and $r_A = 0$), then, letting $i_Y = c i_X r_P$,

$$R_{X,\sigma} = \frac{i_X (1 - c r_p^2) h_X^2}{1 - r_p^2} \tag{3.16}$$

and the selection on trait Y affects the response of trait X by virtue of its phenotypic correlation. Thus in a world in which selection is acting on multiple traits their responses depend not only upon the "genetical" foundations of the trade-offs but also upon their phenotypic expression.

If the phenotypic and genetic correlations were of opposite sign, then the effect of the trade-off will at least be muted, if not reversed in the short term. It must be remembered that the fitness surface is generated by the phenotypic trade-offs whereas the evolutionary trajectory along that surface is a function of the heritabilities and genetic correlations. Thus the determination of the phenotypic trade-offs gives us where selection will ultimately drive the trait combinations, given that there are no absolute genetic constraints as described in the previous chapter.

The genetic and phenotypic correlations may not match in sign if there is a large environmental correlation, as can be readily seen from the equation relating the three correlations:

$$r_P = r_A h_X h_Y + r_E \sqrt{(1 - h_X^2)(1 - h_Y^2)} \tag{3.17}$$

As before, the covariance due to a third variable may produce apparently anomalous results, in this case giving genetic and phenotypic covariances of different sign. To see this result, consider again the functional relationship $Z = X - Y$. Form standard statistical theory, the covariance between Z and X can be derived as $\sigma_Y^2 = \sigma_X^2 + \sigma_Y^2 - 2\sigma_{XZ}$. Rearranging gives

$$\sigma_{XZ} = \frac{1}{2}\left[\sigma_X^2 + \sigma_Z^2 - \sigma_Y^2\right] = \frac{1}{2}\left[\sigma_X^2 + \sigma_X^2 + \sigma_Y^2 - 2\sigma_{XY} - \sigma_Y^2\right] = \sigma_X^2 - \sigma_{XY} \quad (3.18)$$

So if the covariance between X and Y, σ_{XY}, is small relative to the variance in X, σ_X^2, there will be a positive correlation between X and Z. However, if the covariance between X and Y is large, the correlation could be reversed. So, if the genetic and environmental covariances differed in magnitude, the genetic and phenotypic correlations could be of opposite sign. A plausible example of this occurrence is the relationship of development time and adult size and growth rate, discussed in detail in Chapter 4. In the example described in the previous section concerning the correlation between fecundity and egg size we could easily find that body size showed a very large environmental, and hence phenotypic, variance but only a small additive genetic variance. As a consequence, the genetic correlation between fecundity and egg size would be negative, showing the trade-off relationship, whereas the phenotypic correlation would be positive.

The Partitioning of Resources: A Fundamental Trade-off

It is not unreasonable to suppose that in many, if not most, cases energy is a limiting resource and that allocation to one trait means deprivation of another, which I shall refer to as the **Partition of Resources Model**. An example of this model is shown in Figure 3.3, in which ovary weight (which is proportional to fecundity) is plotted against the weight of the major flight muscles of the sand cricket, *Gryllus firmus*. Dorsal-longitudinal flight muscle weight is an index to flight capability, lower weights indicating histolysis of the muscles and a shift in reproduction (Crnokrak and Roff 2000; Stirling et al. 2001).

James (1974) analyzed the partition-of-resources model for the relatively simple case in which the total amount of resources, T, and the proportion, P, of T allocated to one trait are determined only by additive genetic variance (i.e., they have heritabilities of 1). Because the proportion P can itself be regarded as a trait, I have denoted it with an uppercase symbol. (Obviously P is not normally distributed, but we can use the threshold model and assume that P is determined by a normally distributed liability, as described in Chapter 2.) Letting the two "competing" traits be X and Y we have

$$\begin{aligned} X &= PT + e_X \\ Y &= (1 - P)T + e_Y \end{aligned} \quad (3.19)$$

where e_X, e_Y are uncorrelated environmental effects. Because we have assumed that T and P are entirely determined by additive effects, the covariance between X and

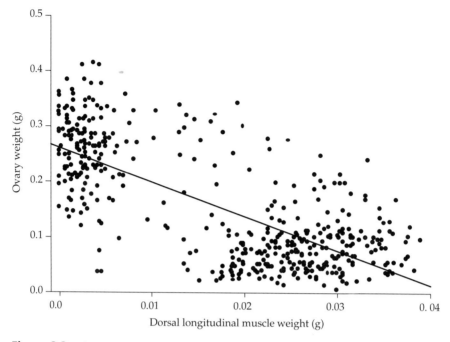

Figure 3.3   An example of a trade-off that is hypothesized to involve the partitioning of resources between two traits, in this case, flight capability and fecundity in the sand cricket *Gryllus firmus*. Data from Roff et al. 2002.

Y is also determined entirely by additive effects and arises solely from the partition of resources. Following the same procedure as in the previous sections, we have

$$\sigma_{XY} = Cov\{PT,\ (1-P)T\}$$
$$= \mu_P(1-\mu_P)\sigma_T^2 - \mu_T^2\sigma_P^2 + \mu_T(1-2\mu_P)\sigma_{PT} \tag{3.20}$$
$$= \mu_P(1-\mu_P)\sigma_T^2 - \mu_T^2\sigma_P^2$$

A striking observation from the above equation is that the partition of resources does not necessarily lead to a negative covariance between the component traits: The covariance between X and Y will be positive whenever $\mu_P(1 - \mu_P)\sigma_T^2 > \mu_T^2\sigma_P^2$. Thus if the mean of the total resources to be allocated is small enough, no trade-off will occur. What will happen to the genetic correlation if selection is imposed on one or both component traits? James (1974) assumed that P and T are uncorrelated and thus

$$\sigma_{XP} = \mu_T\sigma_P^2 \qquad \sigma_{YP} = -\mu_T\sigma_P^2$$
$$\sigma_{XT} = \mu_P\sigma_T^2 \qquad \sigma_{YP} = (1-\mu_Y)\sigma_T^2 \tag{3.21}$$

Making the quantitative genetic assumption that the variances and covariances remain constant we can analyze the effect of selection on the component traits by considering selection on a linear combination of X and Y, $Z = aX + bY$. Changes in mean trait values are then related to the variances and covariances according to the following:

$$\Delta\mu_P \propto \frac{\sigma_{PZ}}{\sqrt{a^2\sigma_X^2 + 2ab\sigma_{XY} + b^2\sigma_Y^2}}$$

$$\Delta\mu_T \propto \frac{\sigma_{TZ}}{\sqrt{a^2\sigma_X^2 + 2ab\sigma_{XY} + b^2\sigma_Y^2}} \qquad (3.22)$$

$$\Delta\sigma_{XY} = \left(1 - 2\mu_P\right)\sigma_T^2\Delta\mu_P - 2\mu_T\sigma_P^2\Delta\mu_T$$

If we use the above relationships, the changes in the genetic correlation for various types of selection can be enumerated (Table 3.1). Selection will change the value of the genetic correlation but the direction of change is dependent on the type of selection.

Noordwijk and de Jong (1986) analyzed the same model as described above, but unaware of the earlier work, called it the **acquisition and allocation model**. The work was motivated by the observation of many examples of positive phenotypic correlations between life history traits when negative correlations might be expected. In the same year, James Riska (1986) apparently unaware of the paper by James published an analysis of a model, which he called the **variable-proportion model**, which is the same as the partition of resources model. Noordwijk and de Jong (1986) present a very elegant pictorial representation of this model (Figure 3.4), and in a later paper (de Jong and Noordwijk 1992) provide a detailed genetic analysis that is not so constrained as that done by James. The view obtained by examining the phenotypic correlation can be quite different

---

Table 3.1 **Effects of different selection regimes on the genetic correlation between the component traits X and Y in the partition of resources model.**

Type of selection (a,b)[a]		$P < 0.5$	$P > 0.5$
For X	(1,0)	Parameter dependent	Decreases
For $X + Y$	(1,1)	Decreases	Decreases
For $X - Y$	(1,–1)	Increases	Decreases
Against X	(–1,0)	Parameter dependent	Increases
Against $X + Y$	(–1,–1)	Increases	Increases
Against $X - Y$	(–1,1)	Decreases	Increases

[a] $Z = aX + bY$

Figure 3.4 A schematic illustration of the partition of resources or acquisition-allocation model. (A) The trade-off curves for three values of *P* (solid lines), and three values of *T* (dashed lines). (B–C) The effect of two different pairwise distributions of *P* and *T*, which result in positive (B) or negative (C) covariation between *X* and *Y*. Figure modified from Van Noordwijk and de Jong (1986).

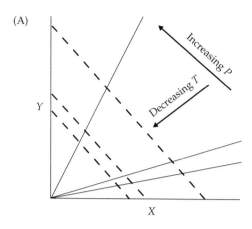

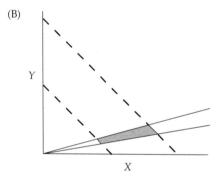

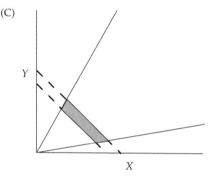

from that obtained by examining the genetic correlation. In the above model, suppose that there is a genetic correlation of –1 between the two traits but there is considerable phenotypic variation in *T* (resource acquisition) but not in *P* (resource allocation.) Despite the genetic correlation of –1, the phenotypic correlation can be positive (Figure 3.4B),

Houle (1991) examined a somewhat more complex version of the partition-of-resources model. He assumed (1) directional selection on resource acquisition (*T*), (2) stabilizing selection on resource allocation (*P*), (3) n_T acquisition loci, n_P allocation loci, (4) a reversible mutation rate of m_T, where mutant alleles decrease acquisition by an amount a_T, (5) a parameter *H* that measures the "hardness" of genetic effects, such that when $H = 1$ resource acquisition is determined only by the absolute value of an individuals genotypes, whereas when $H = 0$, only the relative ability is affected. (Thus, there is a gradation from hard to soft selection.) From his model Houle derived the conditions for positive covariance. The equations are rather complex and depend on all the foregoing parameters but all share the feature that a positive covariance between acquisition and allocation due to mutation-selection balance requires that the number of acquisition loci must exceed the number of allocation loci. At present there are few data on the possible number of each type of loci (and operationally defining them is itself difficult), and no general conclusion can be reached as to the likelihood of positive covariances arising via this process.

The Genetic Architecture of Trade-offs

GENETIC CORRELATIONS WHEN VARIATION IS CONSTRAINED. In the last chapter I discussed the case of a constraint involving two traits, and the general requirement for evolution to be barred from a particular direction. It is unlikely that a trade-off will involve only two traits and the general condition is mathematically obscure. Therefore, here I examine the case of a constraint involving three traits, which gives some idea of the potential complexity of the issue and how examining genetic correlations alone may be rather uninformative. Mathematical details are outlined in Box 3.3.

As illustrated in Figure 3.5, the correlations can take on a wide variety of combinations. Thus the finding that a genetic correlation is not –1 is not sufficient grounds to conclude that evolution is not constrained. It seems very likely that suites of traits will be constrained by physiological and/or ecological factors to vary as a complex function. Looking for evolutionary constraints by examining the genetic correlation structure is probably an unprofitable enterprise, as the standard errors are large enough so that in many cases, the existence of complex constraint surfaces cannot be discounted. More profitable is a research program that is organized about the nature of the constraints, with the estimation of the genetic variance-covariance matrix as a secondary goal.

GENETIC CORRELATIONS AND PLEIOTROPY. Houle (1991) assumed that genetic variation for the trade-off was maintained by mutation-selection balance, a phenomenon that is discussed in greater detail in Chapter 4. Consider a locus that influences two life history traits. Mutations that increase the fitness contribution of each trait will be quickly fixed in the population, whereas mutations that are deleterious to both traits will be quickly lost from the population. Those mutations that increase fitness due to one trait while decreasing it due to the other trait would, at least temporarily, be left segregating in the population (Prout 1980). According to this view, the pleiotropic mutations are not individually at equilibrium, but there is a constant flux of mutations that are eventually removed. An alternative mechanism is that it is the antagonistic pleiotropy itself that maintains the genetic variation. The initial reasoning (Hazel 1943; Falconer 1981) was that alleles affecting two traits that had a positive effect on fitness would be quickly fixed, whereas alleles that had a negative effect would be lost, leaving alleles that had a positive effect via one trait and a negative effect via the other. It was assumed that such alleles would be maintained, though the theoretical basis for such an assumption was not examined. Recent analyses (Rose 1982, 1985; Curtsinger et al. 1994) have shown that this scenario is insufficient to ensure genetic variation at equilibrium except by mutation-selection balance. If there is

(A)

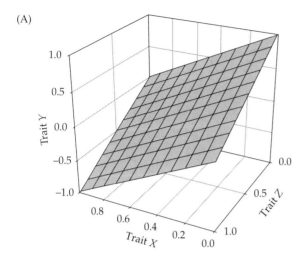

Figure 3.5   An example of a three-trait constraint function and the correlations so generated. (A) The constraint surface, $Y = 1 - X - Z$. (B,C) The correlations under the restriction that $\sigma_Z = 1$.

(B)

(C)

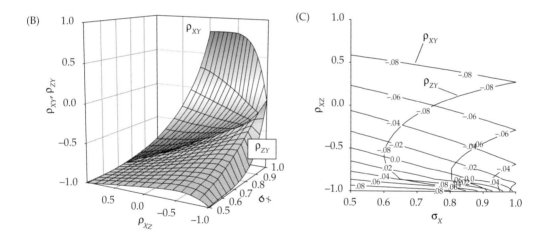

Box 3.3
Correlations between traits constrained to a plane

Let the three traits, X, Y, Z, be related to fitness, W, by the linear relationship

$$W = a + bX + cY + dZ \tag{3.23}$$

where the lowercase letters indicate coefficients. Rearranging gives (Figure 3.5)

$$Y = (W - a - bX - dZ)/c \tag{3.24}$$

Now if we assume that Y is normally distributed, the variance of Y depends on both X and Y. Without loss of generality, let the means of X and Z be zero. The expected value of Y, $E(Y)$ is therefore

$$E(Y) = (W - a)/c \tag{3.25}$$

and the variance of Y, σ_Y^2, is

$$\sigma_Y^2 = E(Y - \mu_Y)^2 = (1/c^2)E\left[(W - a - bX - dZ) - (W - a)\right]^2$$
$$= (1/c^2)E(-bX - dZ)^2 = (1/c^2)E(b^2X^2 + dZ^2 + bdXZ)$$
$$= \left(\frac{b}{c}\right)^2\sigma_X^2 + \left(\frac{d}{c}\right)^2\sigma_Z^2 + 2\frac{bd}{c^2}\sigma_{XZ} \tag{3.26}$$

The covariance between X and Y is

$$\sigma_{XY} = E\left[(Y - \mu_Y)(X - \mu_X)\right] = E\left[(Y - \mu_Y)X\right] = E(XY)$$
$$= (1/c)E\left[X(W - a - bX - dZ)\right] = (1/c)E(-bX^2 - dXZ)$$
$$= -\left(\frac{b}{c}\right)\sigma_X^2 - \left(\frac{d}{c}\right)\sigma_{XZ} \tag{3.27}$$

The correlation between X and Y, ρ_{XY} is

$$\rho_{XY} = \frac{\sigma_{XY}}{\sigma_X\sigma_Y} = \frac{-(b\sigma_Z^2 + d\sigma_{XZ})}{\sigma_X(b^2\sigma_X^2 + d^2\sigma_Z^2 + 2bd\sigma_{XZ})^{1/2}} \tag{3.28}$$

By the same rationale, the correlation between Z and Y is

$$\rho_{ZY} = \frac{\sigma_{ZY}}{\sigma_X\sigma_Y} = \frac{-(d\sigma_Z^2 + b\sigma_{XZ})}{\sigma_Z(b^2\sigma_X^2 + d^2\sigma_Z^2 + 2bd\sigma_{XZ})^{1/2}} \tag{3.29}$$

The sign of the correlations depends upon the sign and magnitude of the covariance between X and Z (σ_{XZ}). In the simple case in which $b = d = 1$, the above equations can be simplified to

$$\rho_{XY} = \frac{-(\sigma_X + \rho_{XZ}\sigma_Z)}{(\sigma_X^2 + \sigma_Z^2 + 2\rho_{XZ}\sigma_X\sigma_Z)^{1/2}}, \quad \rho_{ZY} = \frac{-(\sigma_Z + \rho_{XZ}\sigma_X)}{(\sigma_X^2 + \sigma_Z^2 + 2\rho_{XZ}\sigma_X\sigma_Z)} \tag{3.30}$$

dominance variance of sufficient magnitude, then equilibrium can be achieved. The basic model is outlined in Box 3.4.

The important message from the analysis of Curtsinger et al. (1994) is that, for trade-offs to occur that are represented by genetic correlations greater than −1

Box 3.4
Outline of the two-trait model with antagonistic pleiotropy

Suppose there are a large number of loci affecting trait X and trait Y and there is one locus that jointly affects both traits. The single and joint effects of this one locus are shown in Table 3.2. When $d_X = d_Y = \frac{1}{2}$, the heterozygote lies exactly halfway between the two homozygotes and there is strict additivity. When both are greater than 0.5, there is **deleterious reversal** so that dominance favors the homozygote with the lower fitness. When both are greater than 0.5 there is **beneficial reversal** with dominance favoring the fitter homozygote. If the $d_X < 0.5$, $d_Y < 0.5$ or $d_X > 0.5$, $d_Y < 0.5$, there is **parallel dominance** in that one allele is dominant for both fitness components. There are two ways in which the trade-off determines overall fitness. First, the effects could

Table 3.2   Antagonistic pleiotropy generated by the effects of a single locus.

	Genotype		
	AA	*AB*	*BB*
Trait X selection coefficients	1	$1 - d_X S_X$	$1 - S_X$
Trait Y selection coefficients	$1 - S_Y$	$1 - d_Y S_Y$	1
Additive fitness	$2 - S_Y$	$2 - (d_X S_X + d_Y S_Y)$	$2 - S_X$
Multiplicative fitness	$1 - S_Y$	$(1 - d_X S_X)(1 - d_Y S_Y)$	$1 - S_X$

be additive, as might be expected in the case of fecundity at different ages. Second, the effects could be multiplicative, as might occur if the traits were fecundity and survival. If the fitness of the heterozygote is less than one of the homozygotes variation will be lost but the trade-off will persist with a genetic correlation of –1. For both alleles to be maintained in the population and hence a genetic correlation greater than –1, the fitness of the heterozygote must exceed that of both homozygotes, which requires

$$\text{Additive case} \quad 2 - (d_X S_X + d_Y S_Y) > 2 - S_X \text{ and } 2 - S_Y$$
$$\text{Multiplicative case} \quad (1 - d_X S_X)(1 - d_Y S_Y) > 1 - S_X \text{ and } 1 - S_Y \tag{3.31}$$

where S_X, S_Y are the selection parameters (Table 3.2). For the case of additivity of allelic effects ($d_X = d_Y = 0.5$) there is no equilibrium when fitness effects are additive, and when fitness is multiplicative, the requirement is $2S_X/(2 + S_X) < S_Y < 2S_X/(2 - S_X)$, a relationship that is unlikely unless selection is strong (Figure 3.6). The effect of variation in all four parameters is not, unfortunately, easy to visualize. However, a general pattern can be discerned by examining the pair-wise plots of the two dominance coefficients and the two selection coefficients (Figure 3.7A). Stable polymorphisms, and hence genetic correlations greater than –1, are most likely to occur, as is intuitively obvious, in the region of beneficial reversal and they are least likely in the region of deleterious reversal. Stable polymorphisms are also most likely when selection is strong on both traits, particularly if fitness is multiplicative (Figure 3.7B).

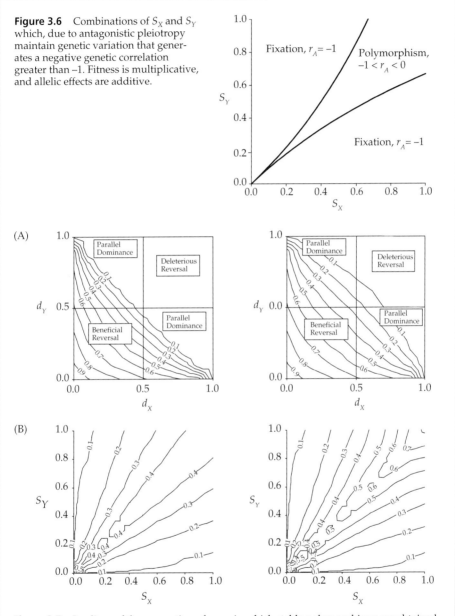

Figure 3.6 Combinations of S_X and S_Y which, due to antagonistic pleiotropy maintain genetic variation that generates a negative genetic correlation greater than –1. Fitness is multiplicative, and allelic effects are additive.

Figure 3.7 Isoclines of the proportion of cases in which stable polymorphisms are obtained when all four parameters in the antagonistic pleiotopy genetic model are varied. (A) The pairwise combinations of d_X and d_Y with S_X and S_Y varied over their full range (0–1). (B) The alternate set of combinations. For the additive fitness model shown on the left in each part, a polymorphism occurs when $2 - (d_X S_X + d_Y S_Y)$ is greater than both $2 - S_X$ and $2 - S_Y$. For the multiplicative fitness model, shown on the right in each part, a polymorphism occurs when $(1 - d_X S_X)(1 - d_Y S_Y)$ is greater than $1 - S_X$ and $1 - S_Y$.

and maintained by antagonistic pleiotropy, there will most likely be directional dominance. Using the formula presented in the previous chapter, we can derive the relative amounts of dominance and additive genetic variance contributed by the single diallelic locus:

$$\text{Trait } X, \qquad \frac{\sigma_D^2}{\sigma_A^2} = \frac{pq(1-C_x)^2}{2(C_x p + q)^2}$$

$$\text{Trait } Y, \qquad \frac{\sigma_D^2}{\sigma_A^2} = \frac{pq(1-C_y)^2}{2(C_y q + p)^2} \qquad (3.32)$$

where $C_i = d_i/(1 + d_i)$ and p is the frequency of allele A. If there is no epistasis and the frequency is the same at all loci that are pleiotropic, the above ratios apply also to the multilocus case. Curtsinger et al. (1994) determined the average ratio for the case of multiplicative fitness by evaluation over a "hyper" lattice of equally spaced values of d_x, d_y, S_x and S_y. If, for a given combination, a stable polymorphism is possible, the equilibrium allele frequency p is readily found by solving the equation (Crow and Kimura 1970, p. 182):

$$p = \frac{W_{AB} - W_{BB}}{(W_{AB} - W_{BB}) + (W_{AB} - W_{AA})} \qquad (3.33)$$

where W_{IJ} is the fitness of genotype with alleles IJ. Averaging over all combinations, Curtsinger et al. (1994) obtained a mean ratio of 0.51. However, ratios are generally lognormal rather than normal, and the present case is no exception (Figure 3.8). Therefore, a more appropriate measure of central tendency is the median. For multiplicative fitness the median value is considerably smaller at 0.18. (The mean ratio I obtained was 0.55 due to differences in the spacing of combinations.) For additive fitness, the mean was 0.64 and the median 0.21. Contours of equal ratios are shown in Figure 3.8A for the case of multiplicative fitness. Note that the contours for trait X are a mirror image of trait Y, which is a result of the symmetry in parameter values. The largest ratios are found in the quadrants of parallel dominance, whereas the ratio for the combinations of beneficial reversal vary roughly from 0.2 to 0.6, with a clear trend towards the lower ratio.

The conditions for polymorphism when there is more than a single diallelic locus are more stringent than with a single locus but, to date, the exact requirements have not been worked out (Curtsinger et al. 1994). With this restriction in mind, we can predict the following: The probability that genetic variation in trade-offs is maintained by antagonistic pleiotropy increases as the ratio of dominance to additive genetic variance increases, with a minimum requirement of approximately 0.25. More than 40% of morphological traits and more than 65% of life history traits satisfy this requirement (Figure 3.9) with 54% of life history

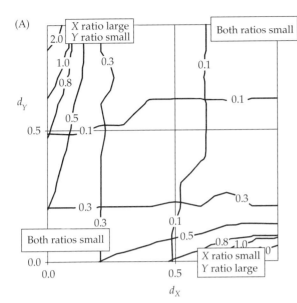

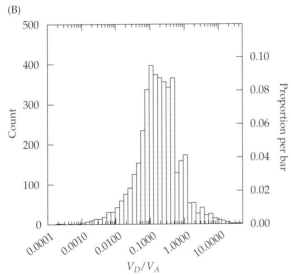

Figure 3.8 (A) Contours of median values of the ratio of dominance to additive genetic variance for those combinations for which both alleles are maintained in the population when fitness is multiplicative. Vertical contour lines show values for trait X, while horizontal lines show values for trait Y. (B) The distribution of ratios plotted on a log scale.

traits having ratios greater than 0.5 and 35% with ratios greater than 2. These data suggest that genetic variation in trade-offs, which will thus lead to correlations less than –1, can very well be sustained by antagonistic pleiotropy. This does not mean that this is the main, or even necessarily a frequent, mechanism by which variation is maintained, but it does mean that it is a mechanism worthy of more detailed investigation.

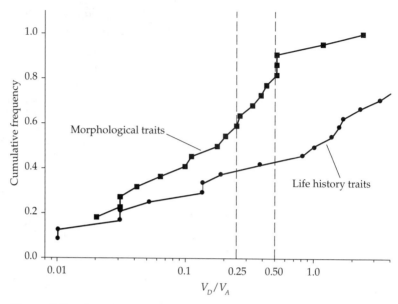

Figure 3.9 Cumulative plots of the ratio of dominance to additive genetic variance for morphological and life history traits. Data sources: Hill (1964); Dawson (1965b); Jinks et al. (1969); Travis et al. (1987); Clark (1990); Cooper et al. (1991); Harding et al. (1991); Legner (1991); Suh and Mukai (1991); Antolin (1992); Mullin et al. (1992); Shukla and Khanna (1992); Price and Burley (1993).

The Further Importance of Nonadditive Genetic Variance and Trade-offs

Inbreeding or passage through a population bottleneck affects a trade-off because there can be a change in mean trait value and amounts of additive genetic variance (Roff and DeRose 2001). Trade-off functions can take any shape, but in order to determine the phenotypic and genetic correlations between the component traits, the function must be transformed so that it is linear. The phenotypic linear regression between any two traits, X, Y, can be written as

$$Y = \mu_Y + r_P \frac{\sigma_{P,Y}}{\sigma_{P,X}} \left(X - \mu_X \right) \tag{3.34}$$

where r_P is the phenotypic correlation between traits and $\sigma_{P,X}$, $\sigma_{P,Y}$ are the phenotypic standard deviations. Assuming no environmental covariance between the two traits, the equation can be written as

$$Y = \mu_Y + \frac{\sigma_{A,XY}}{\sigma_{P,X}^2} \left(X - \mu_X \right) \tag{3.35}$$

where $\sigma_{A,XY}$ is the additive genetic covariance between X and Y. First, consider the trade-off function involving a life history trait, Y, and a morphological trait, X (e.g., fecundity on body size). Now, under inbreeding the mean of the life history trait, μ_Y, will be decreased as a linear function of the inbreeding coefficient, F, but the mean value of the morphological trait, μ_X, because it has little or no dominance variance, will remain the same (Crow and Kimura 1970, p. 80). Inbreeding reduces the additive genetic covariance between Y and X and the additive genetic variance of X by a proportion F (Crow and Kimura 1970, p. 100) and hence the slope of the above function, $\sigma_{A,XY}/\sigma_{P,X}^2$, will show little or no change. Thus inbreeding should shift the trade-off function by a change in the intercept but not the slope. If both traits have significant dominance variance, as might be expected if both were life history traits, then both μ_Y and μ_X will decrease with inbreeding. However, the effect of inbreeding on the slope will depend on how inbreeding affects $\sigma_{P,X}$, which now contains both additive and dominance variance (Crow and Kimura 1970, p. 343).

The above analysis shows that changes in means and variances brought about by inbreeding can produce shifts in a trade-off function provided the genetic variance of at least one of the components of the function consists of both additive and dominance genetic variance. Thus the evolution of trade-offs can be sensitive to inbreeding effects.

Measuring Trade-offs

The measurement of trade-offs can be divided into four categories (Reznick 1985):

1. *Phenotypic correlations:* The correlation between two traits measured at the level of the phenotype, and involving no manipulation of the organism or its environment.

2. *Experimental manipulations:* The direct manipulation of a single factor while keeping all other factors constant, or at least randomly assigned.

3. *Genetic correlations from sib analysis:* Estimation of the genetic correlation between two traits using covariation between individuals within and among families, or covariation between clones or inbred lines.

4. *Genetic correlations from selection experiments:* Estimation of the genetic correlation between two traits using correlated changes in one trait in response to selection on another.

The first two categories measure only the phenotypic association between traits whereas the second two address the issue of whether the trade-off arises from the joint additive action of genes. Reznick (1985) argued that categories 1 and 2 are flawed because of the problem of inferring causation from correlation and because they do not demonstrate that the trade-offs are under genetic control. As discussed above, if selection is acting on both traits in a trade-off, then both the phenotypic and genetic correlations are important in modulating the evolution of the two traits. A phenotypic correlation is a component of the set of trade-offs that together determine fitness: If we view it simply as such, then there is no problem. Problems may arise if we ignore potentially confounding "third variables." Such a failure could lead to ascribing an incorrect causal mechanism for the trade-off. In the case of genetic correlations, the "cause" is the genes, and so this problem does not arise. Nevertheless, if the object is to construct the fitness surface, we need all of the phenotypic correlations, or more specifically the phenotypic trade-off functions, in addition to the genetic variances and covariances. (The genetic parameters are required in order to determine if there are prohibited directions of evolutionary change.)

Phenotypic Correlations

There are many examples of correlations based on data from unmanipulated situations, but because of the problem of inferring causation from correlation, the interpretation of such data is difficult if one is interested in determining the causal reasons for the apparent trade-off (Partridge and Harvey 1985; Reznick 1985; Noordwijk and de Jong 1986; Pease and Bull 1988). Reznick (1985) examined the evidence for a cost of reproduction (i.e., a trade-off between fitness components associated with reproduction) using the above four categories. Data based on phenotypic correlations demonstrated a cost to reproduction in 22 of 33 cases (67%), whereas 85% (17 of 20) of experimental manipulations and 100% (10) genetic analyses confirmed a cost to reproduction. These results suggest that the phenotypic correlations are arising in many cases as a function of both the variable measured and other unmeasured variables.

As indicated above, the observation of a particular phenotypic correlation is no guarantee that even the sign of the genetic correlation will be the same. Phenotypic correlations between morphological traits appear to be generally reliable (Cheverud 1988; Koots and Gibson 1994, 1996; Roff 1996b; Reusch and Blankenhorn 1998; Bayaev and Hill 2000), but those between two life history traits are extremely unreliable (Roff 1996b). In part the problem with the latter category is the difficulty of obtaining precise estimates of the genetic correlations; this issue is discussed below.

The measurement of phenotypic covariation plays a vital role in the analysis of selection in natural populations. Considerable attention in the last chapter was given to the $\mathbf{G}$ matrix in the multivariate response equation $\Delta\mu_X = \mathbf{G}\mathbf{P}^{-1}\mathbf{S} = \mathbf{G}\beta$, but there will be no change unless there is selection, the directional forces of which are specified by the vector β (Lande 1982). This vector is the set of partial regression coefficients of relative fitness on the characters and hence can be estimated by linear regression methods. Relative fitness is typically a fitness component such as survival or fecundity. If the phenotypic traits follow a multivariate normal distribution, as assumed in quantitative genetic theory, the vector β is equivalent to the average gradient of the relative fitness surface weighted by the phenotypic distribution, for which reason Lande and Arnold (1983) designated it the **directional selection gradient**. Stabilizing selection can be assessed by a multiple regression model including quadratic terms (Lande and Arnold 1983). With two traits the regression model would be

$$W = a + \beta_1 X_1 + \beta_2 X_2 + \beta_3 X_1^2 + \beta_4 X_2^2 + \beta_5 X_1 X_2 + error \qquad (3.36)$$

The coefficients β_3, β_4 describe the shape of the surface while the crossproduct coefficient, β_5, accounts for selection for covariation between the two traits (in other words, **correlational selection**). For a discussion of these equations, see Fairbairn and Reeve (2001). The estimation of β is the estimation of the trade-off between the trait value and fitness component, typically some index of fecundity or survival.

Because the method of analysis is based on regression methods, it is subject to all the requirements of regression. In particular, exclusion of traits can have a major impact on estimates of coefficients. This problem is illustrated in Figure 3.10. Part A shows the fitness surface for two traits in which fitness is a quadratic function of these traits. (Note that the effect of correlational selection is to rotate the axis of the ellipse.) Parts B–C show the result of taking the data used to construct the two-dimensional surface and ignoring one trait. There is now considerable scatter in the fitness function, and although there is still stabilizing selection present in both traits, that on X would almost certainly be missed in most "real" data.

Another major problem with estimating the phenotypic relationships is that each coefficient in the equation is tested separately and hence the problem of multiple tests rears its ugly head. With just two independent variables, there are five coefficients (see above) and hence after Bonferroni correction, the required probability for significance at the 5% level is 1%. The number of required tests increases as a power function of the number of coefficients; and hence very quickly the probability of being able to declare any coefficient significant becomes negligible. There is no easy solution to this problem. With a large number of traits it is probably better to approach the problem as a multiple regression problem and use stepwise regression to find the best model that describes the relationship between fitness and the traits. This approach has the advantage that it is more robust and one still ends up with a description of the selective process. (See, for example, Ferguson and Fairbain 2000.) An alternate approach that shows considerable promise is that of path analysis (Kingsolver and Schemske 1991; Sinervo and DeNardo 1996; Abell 1999; Gomez 2000; Scheiner et al. 2000).

Experimental Manipulation

THE IMPORTANCE OF EXCLUDING CONFOUNDING VARIABLES. It is clearly preferable to examine a putative trade-off by manipulating the traits in question, at the same time keeping all other factors constant or randomly assigned to each group. For example, the costs of increasing brood size may be obscured if "better quality" females have larger broods. This problem can be overcome by randomly assigning broods, a relatively easy task with organisms such as birds. Similarly, the effect of reproduction on, say, longevity can be examined by allowing some

Figure 3.10   (A) A hypothetical fitness surface generated by the equation: *Fitness* = $10 + X + 3Y - X^2 - 3Y^2 - 2XY$. Fitness surface plotted as a function of *X* (B) or *Y* (C) alone. The solid line shows the fitted quadratic equation.

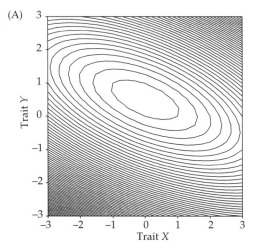

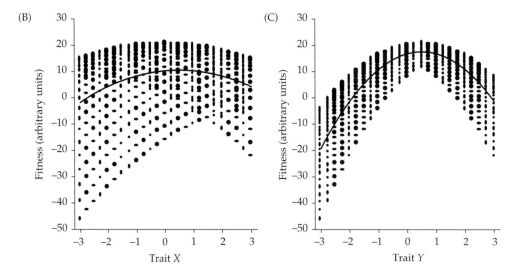

individuals to mate, while others are kept virgin. Great care must be taken to ensure that the manipulation involves only a single change. Increased longevity of virgins, for example, might be a consequence of a lack of interaction with the male rather than the physiological stress associated with reproduction. In an elegant study, Partridge and Farquah (1981) examined the cost of sexual activity in male fruitflies, *Drosophila melanogaster*, by maintaining two types of controls: one in which the males were kept isolated from females, and a second in which males were kept with newly mated females, which are sexually unreceptive. There was no significant difference in longevity between the two categories of control, but a difference between virgin and mated males could be statistically detected when the correlation between body size and longevity was factored out.

Tatar et al. (1996) manipulated reproductive effort in the bean beetle, *Callosobruchus maculates,* by two types of treatments. First, some females were allowed access to mates for a short period early in adult life and some kept virgin, and, second, mated females were provided with oviposition sites (beads) for varying amounts of time. Unmated females or females without oviposition sites resorb eggs. Using a relatively large sample size of 200 females, Tatar et al. (1996) were able to demonstrate that early reproduction decreased later survival. These studies illustrate the power of experimental manipulation, the care needed to isolate possible confounding factors, and the importance of adequate sample sizes.

TRADE-OFFS ARE CONTINGENT ON CONDITIONS. In many branches of science a single experiment can refute a hypothesis. This is not the case in evolutionary biology. Demonstrating that at a particular time and place an organism does not exhibit a trade-off between two traits is insufficient grounds for rejecting the general hypothesis that trade-offs exist. There is no reason to suppose that trade-offs will be manifested in the same way either among different organisms or by the same organism in different circumstances. Trade-offs are often a consequence of physiological or ecological factors. In either case, a trade-off, particularly one involving two life history traits, is likely to be apparent only when the organism is subjected to a particular set of conditions.

Reznick (1985) noted that costs to reproduction may be more evident under conditions of stress, a hypothesis examined in more detail by Bell and Koufopanou (1985). They divided the phenotypic correlations obtained from nongenetic studies into those obtained under laboratory conditions and those from natural populations. In the former category only 5 of 29 cases produce evidence of a cost, whereas 9 of 14 cases in the latter category indicate a cost to reproduction, supporting the stress hypothesis. Reproduction is an energetically expensive activity (Roff 1992); if enough resources are provided, the organism may be able to meet all the demands placed upon it, but if resources are in short supply, then some functions must necessarily be relatively neglected. Consider, for example, an organism that is given just enough food to meet maintenance costs. To reproduce, this individual must sacrifice body tissue. Initially, this may not cause any harm since storage tissues may be able to meet the demands of maintenance and reproduction, but once the storage tissues are depleted, the organism must either cease reproduction or degrade its essential body components. Failure to provide energy for maintenance will not only cause a reduction in weight but a general deterioration in physical condition. Continued reproduction under such conditions will likely increase the individual's susceptibility to disease or prevent it from adjusting its internal milieu when challenged by adverse environmental conditions, thereby increasing the likelihood of death.

It should be apparent that the above scenario is not independent of ecological conditions. A mammal stressed by lactation may be capable of maintaining its condition in a thermally benign environment but succumb to hypothermia when the temperature is lowered. Also, trade-offs leading to a cost to reproduction may be generated by the particular ecological circumstances even if the organism is not physiologically stressed. For example, the presence of eggs may increase the visibility of an animal and hence make it more vulnerable to predators (i.e., a trade-off between fecundity and survival). The trade-off in this case is contingent upon the presence of predators. Considerations of physiology and ecology must be important factors in the design of experiments seeking to examine the presence and cause of trade-offs. Equally, interpretation of results should be given due consideration. If one reflects on the energetic costs of reproduction—the behavioral components and the ecological circumstances under which they occur—it is inconceivable that there are no trade-offs. The question is not whether there are trade-offs, but under what circumstances and in what ways trade-offs are exhibited.

THE IMPORTANCE OF CONSIDERING BOTH TYPE I AND TYPE II ERRORS.   No experiment can demonstrate that no trade-off exists. In a trivial sense this is true because the appropriate measurements may not have been taken. For example, a survival cost to reproduction may have been tested, whereas the actual cost is a reduction in future fecundity. More importantly, no experiment can ever conclude that there is no difference between two groups. It can provide only a probability statement about the likelihood of any specified difference being observed.

Suppose we wish to test the hypothesis that there is a trade-off between reproduction and lifespan such that reproduction decreases lifespan: One method is to compare a group that has been allowed to mate and reproduce, with another group that has been kept virgin. Assuming that lifespan is normally distributed (or can be transformed to be so), the probability that the difference between the two groups is no greater than expected by chance can be assessed using a t-test. Let the number in each group be 20. For such a sample size the 5% level of significance is 2.02. There is nothing sacrosanct about the 5% level of significance; it is entirely arbitrary. Suppose in one experiment we obtained a value of $t = 2.03$ and in another, $t = 2.01$. It would be silly to build a theory based on the "significance" of the former experiment while accepting no difference in the latter.

Suppose t is much smaller than the 5% level of significance. In this case we cannot reject the null hypothesis of no difference between groups. However, not rejecting the null hypothesis is a far cry from accepting it. Consider the following situation: The mean lifespan of the group permitted to reproduce is 8 time units, that of the virgin group is 12 time units, and the standard deviation of both groups is the same at 10 units. The t value is 1.26, which is well below the critical value: In fact, the probability of obtaining this result by chance alone is approximately 0.20. But it is obvious that the difference between 12 and 8 is large, the former being 1.5 times as large as the latter. The problem is that the variances are large and the sample sizes relatively small. The relevant question in the present circumstances is: "What difference is biologically meaningful, and does the present test reject this difference?"

Suppose that, in fact, the samples do come from different statistical populations (i.e., there is a real effect on lifespan). Further, suppose that the observed difference and standard deviations are actually equal to the true values. What difference could be detected? Simple algebraic manipulation of the formula for *t* gives the difference to be 6.39. In other words, a significant difference will be evident only when the difference in means is in excess of 6.39. If on biological grounds we can state that a difference less than 6.39 is biologically meaningful, then our test proves only that the difference is less than 6.39 *and not that there is no biologically meaningful difference between the two groups.* This example illustrates the importance of deciding a *priori* what differences are biologically important and then, on the basis of preliminary estimates of the parameters, deciding upon the appropriate sample size. In the present case, suppose that we decide that a difference of 1 is the minimum difference likely to have biological significance, and the minimum sample size (obtained by rearrangement of the formula for *t*) is 816, which is considerably larger than 20! If this sample size cannot be achieved, then one might question whether it is worth undertaking the experiment. Of course, there may be circumstances in which no a *priori* estimates can be made. The important point is that the statistical test tells us the minimal difference between the two samples, not that the difference is negligible simply because it is nonsignificant.

In practice we do not know the "truth," and our estimation of required sample size must include the probability of accepting no difference when the truth is that the means are different (Type II error), and the probability of rejecting the null hypothesis when the truth is that the two means are not different (Type I error). In the simple case where the difference is simply a difference in means, appropriate formulas are available (e.g., Zar 1999). Peterman (1990) provides a very useful look at power analysis from an ecological perspective. The foregoing discussion may seem obvious, but a surprisingly large number of studies fail to consider the power of a statistical test. A good example of this failing can be seen in experiments in which brood size is manipulated to test for effects on future survival and fecundity (Roff 1992a). As an illustration I shall consider data from a brood manipulation study on house sparrows conducted by Hegner and Wingfield (1987).

In their experiment Hegner and Winfield (1987) reduced by 2 babies the brood size of 9 pairs, increased by 2 babies those of 8 pairs, and left 11 pairs with their original brood size as "controls." Because the reasons for variation in clutch size among the unmanipulated pairs is unknown, it is debatable whether these are really an appropriate control group. A better control would be to have taken the babies from the 11 pairs and then distributed the brood sizes among the 11 pairs at random while keeping the same distribution of brood sizes. For example, suppose we have five control pairs and their original brood sizes are 5, 2, 3, 6, 7. These broods are randomly assigned to parents, giving perhaps the sequence 3, 6, 5, 7, 2. This method then randomizes any differences in quality among the parents. For the present purposes I shall only consider the reduced and enlarged groups. Two possible trade-offs that could be considered are (1) a trade-off between present and future brood size, and (2) a trade-off between brood size and survival of the parents. According to the first trade-off, the subsequent brood size of parents with enlarged broods will be smaller than that of the brood size from parents with reduced broods. A simple test of this first hypothesis is a two-sample *t*-test of the subsequent brood sizes. (More complex tests utilizing the fact that brood sizes were increased or decreased rather than randomly assigned are possible, but for the purposes of demonstrating the general statistical perspective the simple *t*-test is appropriate.) From the 13 pairs for which data on subsequent brood sizes were available, Hegner and Winfield (1987) observed a mean brood size of 5.0 (SD = 1, *n* = 5) in the group that had received reduced broods and a mean brood size of 2.9 (SD = 1.5, *n* = 8) in the group that had received enlarged broods. The estimated common variance is $(4 * 1^2 + 7 * 1.5^2)/(4 + 7) = 1.7954$, giving a standard error of 0.7639 and $t = (5 - 2.9)/0.7639 = 2.749$. The one-tailed probability of observing a deviation as large as this is 0.0095, and hence we conclude that pairs that had reduced broods subsequently produced statistically significant larger broods than those that had enlarged broods. (Hegner and Wingfield used a Kruskal-Wallis test, including the control group, and found no significant effect.) Suppose the difference had been 0.7639, giving a *t* value of 1.00 and a one-tailed probability of 0.1694. We could not now reject the null hypothesis, although a difference of almost one offspring is surely potentially a large difference in fitness.

At this point it is necessary to examine the power of the test. For simplicity I shall assume a common sample size of 6 per group. How can we determine the power of the test? The appropriate equation (Zar 1999) is

$$t_{\beta(1),v} \leq \frac{\delta}{\sqrt{\dfrac{2SD^2}{n}}} - t_{\alpha(1),v} \tag{3.37}$$

where δ is the difference between the means, α is the probability of a type I error, β is the probability of a type II error. (Note that t_β is always one-tailed whereas t_α may be one- or two-tailed, depending on the hypothesis.) v is the degrees of freedom, SD is the standard deviation, and *n* is the size of each group. The type I error is conventionally taken as 0.05. Substituting the appropriate values in the above equation we get

$$t_{\beta(1),10} \leq \left(0.7639/\sqrt{2 * 1.795/6}\right) - 1.812 = -0.824 \tag{3.38}$$

which has an associated probability of 0.79. Thus the power of the test (the probability of correctly rejecting the null hypothesis) is 0.21, which is extremely low. What difference could be detected given the above values? Rearranging the above equation we have

$$\delta \geq \left(t_{\alpha(1),v} + t_{\beta(1),v}\right)\sqrt{\frac{2SD^2}{n}} \tag{3.39}$$

There is no conventional value for the type II error; I shall use 0.10, meaning that I require a 90% probability of correctly rejecting the null hypothesis. Substituting the parameter values gives

$$\delta \geq (1.372 + 1.812)\sqrt{2 * 1.795/6} = 2.46 \tag{3.40}$$

Thus we require a difference of at least 2.46 offspring, which is an enormous fitness difference, given that the average brood size of the house sparrow is only 4 babies (Hegner and Wingfield 1987). That such a difference was actually found is attributable to the very large manipulation (± 2). What sample size would be required to detect a difference as small as 0.7639? Again rearrangement of the above equation gives

$$n \geq \frac{2SD^2}{\delta^2}\left(t_{\alpha(1),v} + t_{\beta(1),v}\right)^2 \tag{3.41}$$

which can only be solved iteratively since $v = 2n - 2$. Solving the above gives $n = 54$, or a total sample size of 108 broods, which is considerably larger than that actually used (12).

Hegner and Wingfield (1987) found no difference in survival rates among the three groups. However, the detection of differences in survival rates is far more difficult than the detection of differences in brood size (Roff 1992a; see also Graves, 1991, for a comment). Consider the problem of comparing the survival rates of birds that raised enlarged broods with those that raised normal sized broods. (I shall assume here that these controls were correctly constituted.) A typical test for such data is the χ^2 contingency test. For the two groups considered here, the test would be one-tailed and, regardless of the sample size, there would be one-degree of freedom. Power analysis for this test is not as simple as that for the t-test and the reader is referred to Zar (1999) for full details. An alternative test is a test for two proportions, for which the reader is referred to Zar (1999) or Fleiss (1981). The scope of the problem of detecting statistical differences in survival can be seen by considering the set of χ^2 values obtained when the exact proportions are observed in a sample. Suppose the survival probability of birds raising normal sized broods is p and that of birds raising enlarged broods is cp ($0 < c < 1$). The four cell frequencies in the contingency table are pn, $(1 - p)n$, cpn, $(1 - cp)n$, from which the χ^2 value is readily computed. Figure 3.11 shows the set of χ^2 values of the full range of c and sample sizes from 10 to 100. What is apparent is that there is a very large set of combinations for which the expected χ^2 is not larger than the critical value of 2.706. Even for samples as large

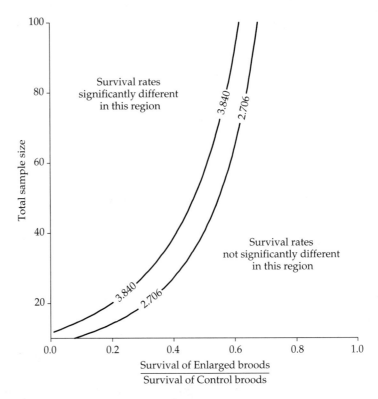

Figure 3.11 Response surface for χ^2 values as a function of the proportional change in survival of enlarged broods, assumed to be 0.5 (a reasonable value for passerines), and total sample size. Note that the χ^2 values are calculated assuming that the exact proportions are observed.

as 100, a survival decline of 30% will not be detected. The detection of trade-offs between survival and some other trait will generally be far more difficult than that between fecundity and some other trait.

Pedigree Analysis

It is important to note at the outset that the quantitative genetic theory, being based on linear regression theory, assumes a linear correlation between traits. There are, however, many circumstances in which the trade-off will be nonlinear. It is, therefore, necessary to linearize any relationship prior to the estimation of correlations.

TRAITS EXPRESSED IN THE SAME INDIVIDUAL. There are a number of experimental techniques to estimate the genetic correlation, the most obvious of which, deriving from the theory presented in Chapter 2, is the offspring-on-parent regression.

There are two such estimates, one obtained using trait X in the offspring and trait Y in the parents and the other using trait Y in the offspring and trait X in the parents. The genetic correlation is given by

$$r_A = \frac{\sigma_{XY}}{\sigma_{XX}\sigma_{YY}} \qquad (3.42)$$

where σ_{XY} is the covariance across traits (i.e., covariance between offspring and parent for different traits) and the denominator is the product of the covariances within traits (i.e., covariance between offspring and parent for the same trait). The latter should, of course, be equal to the additive genetic variances. An overall estimate is obtained using the arithmetic average of the two separate estimates (Falconer 1989, p. 317; Bulmer 1985a, p. 93). Becker (1992) also suggests computing the geometric average but does not provide a rationale for such a choice.

The estimation of confidence limits for the genetic correlation is difficult, and even in the restricted cases where estimation methods have been worked out, the statistical behavior is not well understood (Robertson 1959b, 1960b; Van Vleck and Henderson 1961; Van Vleck 1968; Hammond and Nicholas 1972; Grossman and Norton 1974; Becker 1992). As an estimate of the standard error, Falconer (1989) suggested the approximate formula derived by Reeve (1955) and Robertson (1959b):

$$SE(r_A) \approx \frac{1-r_A^2}{\sqrt{2}} \sqrt{\frac{SE(h_X^2)SE(h_Y^2)}{h_X^2 h_Y^2}} \qquad (3.43)$$

where $SE(h_X^2)$, $SE(h_Y^2)$ are the estimated standard errors of the heritabilities, h_X^2, h_Y^2, respectively. However, this approximation produces confidence intervals that are smaller than the required 95% (approximately 80–90% instead of 95%; Roff and Preziosi 1994). An empirical evaluation by Koots and Gibson (1996) suggests that it may be much worse in some cases. For the ANOVA methods described below, the jackknife method gives exact confidence intervals (Roff and Preziosi 1994) and will probably also do so in the present case. Other possible methods are the delta technique (Bulmer 1985a, p. 94) or the bootstrap (Aastveit 1990).

As with the estimation of heritability, genetic correlations can also be readily estimated using either full-sib or half-sib data. In the former case there is a potential bias due to dominance variance and common environment effects. The necessary additive genetic variance estimates are made as in the estimation for heritability. The covariance terms can be estimated from covariance analysis. (Formulas are given in Becker 1992; Roff 1997.) As described above, the standard error can be estimated with the jackknife. Bias due to common environmental effects can be eliminated by use of a nested design.

Box 3.5
Use of the family mean to estimate the genetic correlation

This estimate is an approximation because the variance and covariance terms contain a fraction of the within-family error term $Cov_m = Cov_{among} + (1/n)Cov_{within}$, where Cov_m is the covariance based on family means and n is the family size. The potential advantage of the family-mean method is that, because they are simply standard product-moment correlations, the usual statistical tests for correlations can be applied and confidence intervals computed. Specifically, if the number of families is N, the confidence interval is computed by first transforming the correlation to the z scale, $z = \frac{1}{2}\ln[(1+r)/(1-r)]$. Then the standard error is approximately $\sqrt{N-3}$. Confidence limits on r_m can be estimated by computing the confidence limits on z and back transforming. (For more details, see Sokal and Rohlf 1995.) Note that the family size, n, does not appear in the formula, and thus any bias due to an insufficiently large family size will not be reflected in the confidence interval.

The family mean correlation is related to the genetic correlation according to the formula (Roff and Preziosi 1994):

$$r_m = \frac{r_A + \dfrac{1}{n}\left(\dfrac{2r_P}{h_X h_Y} - r_A\right)}{\sqrt{\left[1 + \dfrac{1}{n}\left(\dfrac{2}{h_X^2} - 1\right)\right]\left[1 + \dfrac{1}{n}\left(\dfrac{2}{h_Y^2} - 1\right)\right]}} \tag{3.44}$$

Unlike the estimated standard error of r_m, which contains only N, the family mean correlation contains only n, the family size. For family sizes less than 20 there is a considerable bias in the family mean correlation as an approximation to the genetic correlation. Further, the estimated 95% confidence intervals are far from correct when family sizes are small (Roff and Preziosi 1994). The bias and error in the confidence interval will decline as the genetic correlation approaches the phenotypic correlation. However, given that these are the parameters we are attempting to estimate, it would certainly be unwise to make any a priori assumptions. The formula given above can be used as a guide for the bias resulting from a particular family size.

An alternative approach suggested by Via (1984a,b) for the estimation of the genetic correlation is to use the Pearson product-moment correlation between family means, for which the usual methods of estimating confidence intervals on correlations can be applied (Box 3.5).

The standard error of the genetic correlation is approximately proportional to the geometric mean of the standard errors of the heritabilities. Consequently, the optimal design for the estimation of the heritabilities is also the optimal design for the estimation of the genetic correlation. Whether the standard error of the genetic correlation exceeds that of the heritabilities depends upon the value of the standard error. Consider the case in which the two heritabilities are equal. Using the approximate formula (Equation [3.43]) we have

$$Est(SE) = SE(h^2)\left(\frac{1-r_A^2}{h^2\sqrt{2}}\right) \qquad (3.45)$$

As r_A approaches ± 1, the standard error decreases to zero. When r_A is very small, its standard error is approximately $0.71/h^2$ times the standard error of the heritability, which can lead to a considerable increase. For example, if h^2 equals 0.25, the standard error of the genetic correlation is 2.8 times that of the heritability. Figure 3.12 shows the standard errors for the three major breeding designs, assuming that the total number of individuals that can be measured is 2000 and a heritability of 0.25. Note that, as with heritability, the smallest standard errors are obtained with the full-sib design. Despite the large number of individuals measured, the standard errors can be very large. In the absence of prior informa-

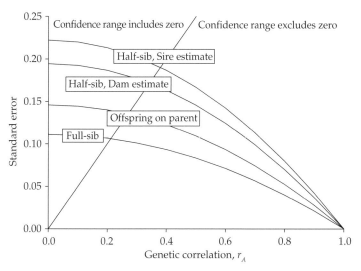

Figure 3.12   The standard error of the genetic correlation, estimated using the approximate formula of Robertson (1959b), $Est\{SE\} = SE(h^2)\left[(1-r_A^2)/(h^2\sqrt{2})\right]$, as a function of breeding design. The approximate sample size is 2000 individuals. (Note that the equivalent figure [Figure 3.3] given in Roff 1997 is in error.) Heritability for both traits was set at 0.25. The actual sample sizes were: Half-sib, 50 sires, 4 dams per sire, 10 offspring per dam; Offspring on parent, 167 families, 10 offspring per dam (total number = $167 \times 10 + 167 \times 2 = 2004$); full-sib, 200 families, 10 offspring per family.

tion on the genetic correlation, very large sample sizes are required to ensure reasonably small confidence limits.

TRAITS EXPRESSED IN DIFFERENT INDIVIDUALS (DIFFERENT ENVIRONMENTS). The above genetic correlations assume that the traits are all "contained" within the same individual—for example, traits of age and size at first maturity, or fecundity at ages t and $t + i$. There may also be trade-offs between environments. For example, the ability of a herbivore to grow rapidly on one particular host plant species may be traded off against the ability to grow on another type of host species. The experimental problem in this case is that, in general, the same individual cannot be subjected simultaneously to two separate environments. If the organism can be cloned, then the problem is resolved, but there are large numbers of organisms for which this is not possible. A common solution in such cases is to use the family means. As discussed above, these may be badly biased unless family sizes are large. Further, unless a full-sib design is used, information must necessarily be lost when a more complex design such as a half-sib design is collapsed to a full-sib design. Fry (1992), building on an analysis by Yamada (1962), showed that the mixed-model analysis of variance could be used to estimate the genetic correlation between environments. In this model, "families" are random effects because we wish to make inferences about the population as a whole, but the "environments" (habitats, hosts, morphs, and so on) are fixed because we restrict attention only to the specific environments used. The statistical model is $M_{ij} = \mu + \alpha_i + \beta_j + (\alpha\beta)_{ij}$, where M_{ij} is the expected mean of the jth family in the ith environment, μ is the population mean, α_i is the effect due to the ith environment, β_j is the effect due to the jth family and $(\alpha\beta)_{ij}$ is the effect due to environment by family interaction. The variance in β is equal to the covariance between M_{1j} and M_{2j} (more generally, any two environments). Consequently, the genetic correlation is estimated from the variance components as

$$r_A = \frac{\sigma_\beta^2}{\sigma_\beta^2 + \sigma_{\alpha\beta}^2} \tag{3.46}$$

For a full-sib breeding design σ_β^2 is estimated as $\sigma_\beta^2 = (MS_{AF} - MS_{FxE})/2n$, where n is the family size per environment. For a half-sib design, a nested ANOVA is used and the two estimates are

$$\sigma_\beta^2(Dam) = \left(MS_{AF(D)} - MS_{SxE(D)}\right)\!\big/(2n)$$

$$\sigma_\beta^2(Sire) = \left(MS_{AF(S)} - MS_{FxE(S)} - 2n_D\sigma_\beta^2(Dam)\right)\!\big/(2nn_D) \tag{3.47}$$

where the S and D refer to sire and dam components, respectively, and n_D is the number of dams per sire. In all cases, the variances are best estimated from separate one-way ANOVAs in each environment. Standard errors can be approximated using the jackknife (Windig 1997; Fox et al. 1999). For a discussion of restricted maximum likelihood approaches, see Shaw and Fry (2001).

A test for a genetic correlation different from 0 is given by the F-test constructed as the ratio of the mean square among families to the interaction mean square. A two-tailed test is conducted by using the 2.5th percentile to test for a negative correlation and the 97.5th percentile to test for a positive correlation. Fry cautions that tests for negative genetic correlations should be viewed carefully because variance components are frequently estimated by restricting variance components to be positive. Of course, if the trade-off is negative, it is an easy matter, for the purposes of testing, to reverse the sign of one of the components to produce a positive correlation. This test is exact only for a balanced design and assumes that:

1. Errors are independent and normally distributed with constant variance.

2. The family means, M_{ij} are normally distributed with the same variance in each environment. These two assumptions imply that heritabilities do not differ between environments. If the variances do differ between the environments, they should both be standardized to a common variance. (For example, 1; for an alternative approach based on family means, see Dutilleul and Potvin, 1995.) This standardization is necessary for the statistical testing of the interaction, but not for the estimation of the genetic correlation if the denominator in the appropriate formula comes from separate analyses of variance.

3. If there are more than two environments, then the $Cov(M_{ij}, M_{i'j})$ is the same for very pair of environments i and i'. This is automatically satisfied for two environments.

The hypothesis that $r_A < 1$ cannot be tested using the ANOVA, and Windig (1997) recommends use of the jackknife or bootstrap. The jackknife method of testing employs a one-sample t-test using the standard error estimated from the jackknife (Knapp et al. 1989; Roff 1997; Roff and Bradford 2000). Ratio estimators are frequently not normally distributed and so the following test might also be used (Fox et al. 1999). The test for $r_A = 1$ can be written as $Cov/Prod = 1$, where Cov is the covariance and Prod is the square root of the two variances (calculated separately). Rearranging the equation we obtain $Cov–Prod = 0$. Letting $X = Cov – Prod$, we can estimate X for the original sample and then estimate its standard error using the jackknife procedure as before. We can now test the hypothesis that $r_A < 1$ indirectly by testing the hypothesis that $X < 0$.

Selection Experiments

If two traits are genetically correlated, selection on one trait will result in a correlated change in the other. The presence of a correlated response is evidence of a genetic correlation between the traits and hence a trade-off (given the appropriate sign). However, a genetic correlation can arise as a result of linkage disequilibrium, which is transitory and hence not an indication of a trade-off unless the linkage disequilibrium is maintained in the population (e.g., by sexual selection). Linkage disequilibrium will in fact be generated by selection and hence could potentially bias the interpretation of selection experiments. This is a problem that has not been well studied.

There are two experimental approaches to estimating the genetic correlation from selection. In both cases selection is done on one trait and the direct and correlated responses measured. Recall that

$$CR_Y = ir_A h_X h_Y \sigma_{P,Y} \tag{3.48}$$

If trait X is selected and the responses of both X and Y are measured, then the genetic correlation can be estimated by rearrangement of the above equation:

$$r_A = \frac{CR_Y}{ih_X h_Y \sigma_{P,Y}} \tag{3.49}$$

If separate experiments are done, one in which X is selected and the other in which Y is selected, an estimate of the genetic correlation can be obtained without the necessity of estimating the heritabilities:

$$r_A = \frac{CR_Y \sigma_{A,X}}{R_X \sigma_{A,Y}} = \frac{CR_X \sigma_{A,Y}}{R_Y \sigma_{A,X}} = \frac{CR_Y CR_X}{R_X R_Y} \tag{3.50}$$

So far as I know, the statistical properties of the above two methods of estimating the genetic correlation have not been investigated.

Examples of Trade-offs Between Life History Traits Determined from Observation Alone

Reproduction and Survival

SOME EXAMPLES. It is certainly easy to imagine circumstances in which a trade-off between reproductive effort and survival might occur. Reproductive individuals might be more visible, slower or less maneuverable than nonreproductives. The energy diverted into reproduction might reduce immunocompetence and hence increase the risk of death from disease. Finally, activities specifically associated with reproduction, such as courtship, incubation, or calling or searching for mates, may increase the risk of predation (Aleksiuk 1977; Burk 1982; Sargeant et al 1984; Endler 1987; Trail 1987; Zuk and Kolluru 1998). Frequently, attempts to demonstrate this increased risk have been based not on a comparison of breeding and nonbreeding individuals but by showing that the predator locates its prey by the sexual advertisement of the prey, or by comparing differences in behavior and survival of males and females.

Reproductive behavior can increase the risk of being preyed upon in four ways. First, parasites or predators may home in on the mating call of the male or female. Taped calls have demonstrated that both parasites and predators can locate males via their accoustical signals. Various dipteran parasites locate their insect prey by homing in on their acoustical signal. Some examples are crickets (Cade 1975; Mangold 1978; Zuk et al. 1995), katydid (Burk 1982), cicada (Soper et al. 1976), and stink bug (Harris and Todd 1980). Similarly, cats (Walker 1964) and herons (Bell 1979) have been observed to orient to and capture male crickets using their call. Several species of frogs have been recorded as being located and preyed upon by possums (Ryan et al. 1981) and bats (Tuttle and Ryan 1981; Ryan et al. 1981, 1982) using their calls. Females may attract males using pheromones; at least one tachinid fly, *Trichopoda pennipes,* uses this cue to locate females on which to lay its eggs (Harris and Todd 1980). Predators might also act as satellites of a calling male and prey upon incoming females, as is found in the gecko *Hemidactylus tursicus* preying on females of the decorated cricket *Gryllodes supplicans* (Sakaluk and Belwood 1984).

Second, instead of homing-in on the mating call of the prey, a predator may attract males by mimicking the call of the female. The bolas spider, *Mastophora* sp., catches insects by means of a sticky ball suspended on the end of short vertical thread that is attached to a single horizontal line. The spider rests on the horizontal line holding the vertical thread with one front leg and swings the bola at passing insects. The incidence of male noctuiid moths captured and the direction of approach of prey (upwind) suggest that the spider may incorporate a chemical that mimics the male-attracting pheromone emitted by female noctuiid moths (Eberhard 1977, 1980a). More convincing evidence of female call mimicry comes from the study of Lloyd (1965) on female fireflies of the genus *Photuris*. In this genus the females are carnivorous and attract males of the genus *Photinus* by mimicking the response flashes of the female, a phenomenon demonstrated by the use of artificial calls (Lloyd 1965; Lloyd and Wing 1983).

Third, the increased activity of an animal may increase its susceptibility to predation either by making it more visible or simply by increasing the likelihood that predator and prey meet. The former situation occurs in digger wasps, *Palmodes laeviventris,* which provision their nests with more female than male mormon crickets, *Anabrus simplex* (Gwynne and Dodson 1983). The male mormon cricket calls from the protection of a burrow and so is less likely to be preyed upon by a visually hunting predator. In contrast, the territorial behavior of male digger wasps increases their vulnerability to predators. Significantly more males than females are taken by robberflies (Gwynne and O'Neill 1980). This bias arises because males patrol their territory and approach intruders; if the intruder happens to be the predatory robberfly, the wasp may patrol no more.

The situation of increased vulnerability due to increased activity is illustrated by mortality rates in Mecoptera. In the mecopteran *Hylobittacus apicalis*, the female does not hunt but accepts a nuptial gift of prey from the male. To hunt for prey the male must forage among the herbs and this makes it more likely than the female to fly into spider webs, as evidenced by a highly significant male-biased ratio within the webs (Thornhill 1980). However, nuptial feeding does not occur in the related mecopteran *Bittacus strigosus*, and females hunt in the same manner as the males; in this species there is no sex-biased mortality from web-building spiders (Thornhill 1980).

Fourth, parental defense of young may also logically place a reproductive individual more at risk than a nonreproductive individual. Male three-spined sticklebacks, *Gasterosteus aculeatus*, defend their eggs and young against fish that could potentially eat both the young or the parent: this defence could therefore increase the mortality of guarding male sticklebacks (Pressley 1981; Giles 1984). Similarly, mobbing behavior of birds defending their young can be dangerous, as evidenced by numerous accounts of kills occurring during this behavior (reviewed by Curio and Rengelmann 1985).

THE IMPORTANCE OF CIRCUMSTANCE AND THE PROBLEM OF MISSING VARIABLES. The simple comparison between survival rates of reproducing and nonreproducing organisms is likely to be uninformative because there is almost certainly a third missing variable that is determining whether an organism breeds and its potential survival. For example, reproduction may be cued to accumulated resources, and only individuals with adequate supplies will initiate reproduction. This being the case, it is quite conceivable that reproducers are in much better physical condition than the nonreproducers and that their survival rates even with the stress of reproduction will be greater than the nonreproducers. In the magpie *Pica pica*, the number of breedings is correlated with both clutch size and fledgling production, but these are also correlated with territory quality (Högstedt 1980,1981). As a consequence, the females that have the best territories have the highest survival rates and the largest reproductive success. As noted by Högstedt, unless territory quality can be accounted for, trade-off between reproduction and survival cannot be properly assessed.

Because they are relative benign, laboratory conditions may obscure any relationship between reproduction and survival. This is suggested by more negative correlations being observed from field than lab studies (Figure 3.13). Variation in conditions in the wild might be expected to have a profound impact on a trade-off. The analysis of Tinbergen et al. (1985) on the great tit is particularly illuminating in demonstrating the importance of ecological conditions in modulating the probability of survival associated with breeding. Analysis of data for the years 1957 to 1978 revealed that the survival cost of reproduction was determined by the size of the beech crop. When the crop is poor, a negative correlation between parent survival and number of fledglings produced is found, but in years of high beech seed yield, no correlation is obtained (Figure 3.14). The importance of the beech crop is that it provides winter food for the birds: a low crop presumably leads to increased starvation, which is exacerbated by the previous breeding effort. Clobert et al. (1987) suggested that female starlings raising

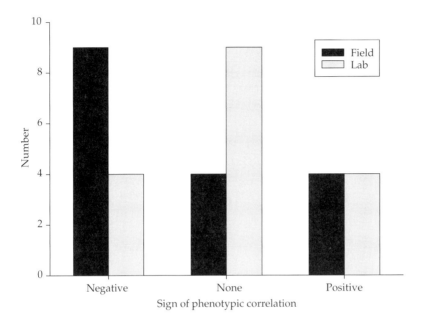

Figure 3.13 Distribution of the sign of the phenotypic correlation between reproduction (all or none) and survival for field and lab studies. Data from Tables 6.2 and 6.3 of Roff (1992a).

large broods had an increased mortality rate only during a particularly cold winter, though the analysis is not as convincing as Tinbergen's.

A similar hypothesis was advanced by Fairbairn (1977) to account for the relatively high mortality of female deer mice that attempted to breed in spring. Fairbairn suggested, based on work by Sadleir (1974), that lack of food in spring causes high mortalities among breeding females because they cannot support both lactation and thermoregulation. This hypothesis is supported by a study of Fordham (1971) in which supplementary food was given to a population of deer mice from February to September. The proportion of early breeding females increased, as did their survival, suggesting that food supply is critical to the breeding and survival of females in spring.

An inability to respond to an environmental challenge because of the stresses of reproduction is suggested by the observation of Festa-Bianchet (1989) that in bighorn sheep, *Ovis canadensis*, ewes that had produced a lamb at 2 years of age were more likely to die during a pneumonia epizootic than ewes that had not lambed at 2 years. In the common lizard, *Lacerta vivipara*, Sorci et al. (1996) found a negative correlation between adult survival and clutch mass but no effect of hematazoan parasite load. Effects of parasites on offspring survival have been found in birds (see below). In humans there is a negative correlation between

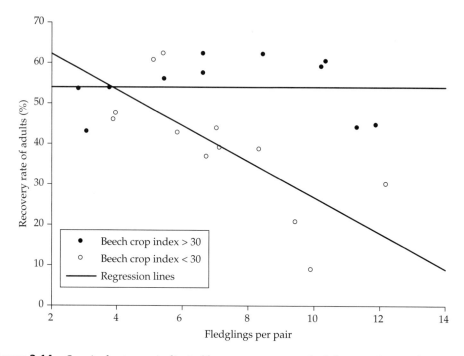

Figure 3.14   Survival rates, as indicated by recovery rates, of adult great tits as a function of the number of fledglings produced and the beech crop index. There is no significant correlation when the beech crop index is less than 30. Modified from Tinbergen et al. (1985).

longevity and fecundity that persists after correction for confounding factors such as history, religion, geography, and so forth (Thomas et al. 2000). The causative reasons for this trade-off are not known but are most likely connected with the large physiological cost of childbirth.

In all the above examples it is the survival of the parent that decreases with reproductive effort. There may additionally be increased mortality of offspring if reproductive effort is increased in terms of the initial offspring production but is insufficient to provide for the enlarged brood. This appears to be the case in small mammals, where survival to weaning is negatively correlated with litter size (Table 3.3). The lowered survival is almost certainly due to a decrease in resources obtained from the mother, as indicated by a decreased initial size and a decreased weaning weight and growth rate (Table 3.3). These indirect fitness effects have also been observed in *Mus musculus* (Parkes 1926; Konig et al. 1988), *Phenacomys longicaudus* (Hamilton 1962), *Dicrostonyx groenlandicus* (Hasler and Banks 1975), *Clethrionomys gapperi* (Innes and Millar 1979), *Spermophilus richardsonii* (Michener 1989), and *Crocidura russula* (Genoud and Perrin 1994) but not in

Table 3.3   Effects of natural variation in litter size on survival (*S*) and traits that may account for the effect (mass at birth, M_B, mass at weaning, M_W, and growth rate, *G*).

Species	S	M_B	M_W	G	Reference
Peromyscus leucopus	–	0	?	?	Millar (1975, 1978); Hill, R.W. (1972)
P. maniculatus	–	–	0	0	Linzey (1970); Myers and Master (1983)
P. polionotus	–	–	–	–	Kaufman and Kaufman (1978)
Sigmodon hispidus	–	0	–	–	Kilgore (1970); Randolph et al. (1977); Mattingly and McClure (1982)
Microtus californicus	–	0	–	?	Krohne (1981)
Neotoma lepida	–	?	0	–	Cameron (1973)
Cavia porcellus	–	–	–	–	Wright and Eaton (1929)

Notes: – = negative effect. 0 = no significant effect. ? = not measured. If studies differ in what was found, the results are given sequentially according to reference.

Microtus pennsylvanicus (Innes and Millar 1979) or *Dipodomys desertii* (Butterworth 1961).

Present and Future Reproduction

Bell and Koufopanou (1985) list 11 studies (8 species) in which the phenotypic correlation between early and late fecundity were estimated under laboratory conditions. In only one case was the phenotypic correlation negative. Positive correlations were found in the invertebrates *Aelosoma, Cypridopsis, Daphnia, Gargaphia, Philodina, Platyias* and the vertebrate, the domestic chicken. The single negative correlation occurred in the invertebrate *Pristina* (see Table 3 of Bell and Kofopanou 1985).

However, phenotypic correlations estimated for 10 field populations are negative in 9 cases. The single positive correlation occurred in the pied flycatcher *Ficedula hypoleuca* (Harvey et al. 1985). But a negative correlation has been found for three other bird species: the great tit, *Parus major*, (Kluyver 1963); the Eastern bluebird, *Sialia sialis* (Pinkowski 1975); and the house sparrow, *Passer domesticus*, (McGillivray 1983). Negative correlations between early and late fecundity and between fecundity and survival have been found in two plant species: meadow grass, *Poa annua* (Law 1979a). Since the Law uses family means, the correlation obtained is a measure of the genetic correlation between early and late fecundity; it is mentioned here as it is one of those listed by Bell and Kofopanou 1985); *Senecio leridendron* (Smith and Young 1982); and a single mammalian species, the female red deer, *Cervus elaphus* (Clutton-Brock et al. 1983). Three studies of primates report a negative correlation between present and future reproduction. In Japanese macaques, 95% of females that lose infants reproduce the following year, while only 8.6% of females not losing their infants reproduce the next year

(Tanaka et al. 1970). In rhesus monkeys two studies observed that the date and probability of parturition depended upon the immediate prior reproductive history (Drickamer 1974; Wilson et al. 1978).

Examples of Trade-offs Between Life History Traits Determined from Experimental Manipulation

Reproduction and Survival

Intuitively we might expect a trade-off between reproductive effort and survival or between present and future reproduction. A simple experimental approach is to compare the survival rates and/or future fecundities of presently breeding and nonbreeding organisms. Although this approach can establish that a cost exists it does not permit the quantitative assessment of the trade-off and hence should only be viewed as a first step. A decreased survival of reproductives could result from increased visibility, either optically or pheromonally. Alternatively, the stress of reproduction could reduce immunocompetence and hence make the individual more susceptible to disease (Moreno 1993; Sheldon and Verhulst 1996; Zuk 1996; Demas et al. 1997; Zuk and Johnsen 1998), which itself could cause death or could so weaken the organism that it is likely to be unable to escape a predator. The foraging of bumblebees, which is an important component of colony survival and reproduction, decreases immunocompetence compared to workers prevented from foraging (Konig and Schmid-Hempel 1995). A number of experiments examining predation rates on gravid and nongravid invertebrates have demonstrated that increased visibility is a prime factor in the increased mortality of reproductives, while brood manipulation experiments, examined in a later section, have shown that reproductives are less able to cope with parasites.

EXPERIMENTS DEMONSTRATING INCREASED PREDATION RISK BECAUSE OF INCREASED VISIBILITY. Copepods carry their eggs in sacs at the posterior of their body and may be both more visible and less maneuverable than females without eggs. Winfield and Townsend (1983) presented ovigerous (egg bearing) and non-ovigerous *Cyclops vicinus* individually to two fish species, bream (*Abramis brama*) and roach (*Rutilus rutilus*). With increasing number of trials, roach showed an increasing ability to capture ovigerous cyclops but remained relatively incompetent at capturing cyclops without eggs. On the other hand, bream were proficient at capturing both types of cyclops. In both species the survival of ovigerous females was less than that of nonovigerous females, but was statistically significant only in roach. The reaction distance of both bream and roach is greater for the ovigerous females, reflecting the greater visibility of these. A similar finding is reported by Hairston et al. (1983) for another copepod, *Diaptomus sanguineus* being preyed upon by sunfish, *Lepomis macrochirus*. Increased visibility probably also accounts for the higher susceptibility of egg-bearing females of the copepod, *Eurytemora hirundoides* (Vuorinen et al. 1983) and *Eudiaptomus gra-*

cilis (Svensson 1992). Logerwell and Ohman (1999) used stickleback as predators on six species of marine copepod and found that in all species a higher percentage of ovigerous females were taken, although in only two species were the results statistically significant. The foregoing studies examined only the binary situation, gravid versus nongravid. They suggest, but do not demonstrate, that the larger the clutch, the lower the survival rate. This was confirmed by Svensson (1995) using the copepod *Eudiaptomus gracilis*, and the fish predators, zebra fish and roach.

In the spring and fall, *Daphnia* populations may produce resting eggs that are carried in pigmented envelopes called ephippia. Mellors (1975) found a significantly higher proportion of ephippial female *Daphnia galeata mendotae* in the stomachs of pumpkinseed sunfish, *Lepomis gibbosus,* and perch, *Perca flavescens,* than in the water column. He tested the hypothesis that ephippial female *Daphnia* are more susceptible to predators by exposing equal numbers of ephippial and nonephippial *Daphnia pulex* to either sunfish or a second predator, the red-spotted newt, *Diemictylus viridescens*. In all cases, the proportion of nonephippial females in the test group increased as a result of selective predation on ephippial females, but the difference was significant only in the groups exposed to bluegills at the two highest light levels (0.13 and 0.86 lx). The lack of significance at the lowest light level may be due to relatively few *Daphnia* being eaten (9.8 per trial, compared to 14.8 and 13.1 at the two higher light levels).

The importance of the type of predator and the environmental conditions is highlighted by the analysis of Koufopanou and Bell (1984), who also used *Daphnia pulex*, but a wider variety of predators than did Mellors. Against a light background, guppies, *Poecilia reticulata*, selectively preyed upon egg-bearing females, but against a dark background, no selectivity was observed. The results for other fish species also suggest selective feeding, though statistical significance was achieved in only one case (stickleback). A visually hunting invertebrate predator, the backswimmer *Notonecta* sp. took more egg-bearing females than expected by chance, but a tactile feeder, *Hydra pseudooligactis,* appeared not to differentiate.

Females carry eggs or young for periods much longer than the duration of copulation. Nevertheless, in some organisms, such as some insects and amphibia, copulation or some type of amplexus may be extended and potentially put the couple at risk due to decreased mobility or increased visibility. This idea has been experimentally tested by Arnqvist (1989) using the water strider *Gerris odontogaster* and a potential predator, the backswimmer, *Notonecta lutea*. Survival rates of male and females in pairs were compared to those of female-female pairs. Waterstriders in copula are significantly more likely to be taken by a backswimmer than two separate females. Furthermore, it is the female not the male that is subject to the increased vulnerability (Arnqvist 1989). This makes sense, as the males "rides" on top of the female and the backswimmer attacks from below. Fairbairn (1993) has observed a similar phenomenon in the gerrid *Aquarius remigis*; pairs in copula were more likely to be taken by frogs, though in these experiments both males and females were consumed.

Reproductive females may be more visible to predators by virtue of different pheromones emitted. For example, Cushing (1985) showed that prairie deer mice, *Peromyscus maniculatus bairdii*, are more likely to be captured by a weasel, *Mustela nivalis*, when in estrous than out of estrous. These experiments followed a similar protocol as used in the invertebrate experiments previously described, in that two mice, one in estrous and one diestrous, were introduced into an arena (1.83 m × 1.83 m) containing a single weasel. Each trial ended when the weasel caught one of the mice. To examine the hypothesis that the weasel was preferentially finding the estrous mouse because of its odor, Cushing (1985) repeated the experiments using diestrous mice that had been painted with urine from an estrous mouse. Under these conditions there was no difference in mortality rate between the estrous and painted diestrous females.

Increased susceptibility to predation may be a result of increased visibility of a gravid female or decreased ability to escape. Gravid females of the Australian agamid species *Amphibolurus nuchalis*, the common lizard *Lacerta vivipara*, the western fence lizard *Sceloporus occidentalis*, and the snow skink *Niveoscincus microlepidotum* show a significant reduction in running speed (Garland 1985; Van Damme et al. 1989; Sinervo et al. 1991; Olsson et al. 2000), and in the first species preliminary data suggest reduced endurance (Garland and Else 1987). Although an "obvious" explanation for the reduced running speed is the increased burden of the eggs, the actual cause may be physiological not mechanical (Olsson et al. 2000). Very disparate running speed results were obtained for the skink *Lampropholis guichenoti*. In one population there was the expected decline in running speed in gravid females but in a second, geographically separate population no difference was found (Qualls and Shine 1997).

Shine (1980) tested for a trade-off between reproduction and survival in the lizard *Leiolopisma coventryi* by exposing a gravid female and male of approximately equal body size to a predator, the white-lipped snake, *Drysdalia coronoides*. Eight of 16 lizards were taken by the snakes, of which 7 were gravid females, a statistically significant difference. Although this experiment demonstrates that gravid females may be more vulnerable than males, it does not follow that gravid females are more vulnerable than nonbreeding females. Bauwens and Thoen (1981) pointed out that a lizard can shift its behavior, supporting this hypothesis with data on the European lacertid, *Lacerta vivipara*. Gravid females of this species have reduced sprint speeds (Van Damme et al. 1989) but increase their reliance on crypsis when gravid (Bauwens and Thoen 1981). This point is reinforced by the study of Schwarzkopf and Shine (1992) on the water skink, *Eulamprus tympanum*. Gravid females of this species shift their behavior from flight to crypsis and in open enclosures the susceptibility of gravid females to two natural predators, kookaburras and common blacksnakes, was no different from that of either nongravid females or males.

BROOD SIZE MANIPULATION EXPERIMENTS EXAMINING SURVIVAL AND/OR FUTURE REPRODUCTION. Researchers working with birds have a great propensity to move newly hatched young among nests. Although this has lead to numerous studies on the effect of reduced or enlarged brood sizes on trait characteristics, there are unfortunately rather a large number for which the resulting sample sizes have produced a study that possesses little statistical power. In particular, estimation of survival rates of the parents requires larger sample sizes than is frequently found. However, analysis of a suite of experiments does reveal some general patterns. Figure 3.15 shows the results for studies in which were estimated the consequences of enlarged broods on future reproduction, on future survival of the parents, and on survival of the offspring (generally to fledging). The most obvious feature is that among species there is no general trend for the cost of raising an enlarged brood to increase with the relative increase of the brood. Within studies, when an effect of enlarged brood size is detected, there is generally an increasing cost with increased enlargement of the brood.

The evidence for a trade-off between present and future fecundity is mixed, although where I was able to calculate the appropriate ratio, with one exception, it was smaller in the subsequent brood. There appears to somewhat stronger evidence for a more general trade-off between fecundity and parental survival, and very strong evidence for a decrease in the survival of the offspring, at least up to the age of fledging. Further evidence for the last comes from a review of 58 studies (Roff 1992a; Figure 3.16). Evidence from manipulations of clutch or brood size in birds indicates that larger clutches or broods lead to increased number of fledglings. Although experimentally enlarged clutches typically produce an increased number of fledglings (Lessells 1986; Ydenberg and Bertram 1989; Dijkstra et al. 1990; Roff 1992; Vanderwerf 1992), most studies also demonstrate that mass at fledging and survival to fledging are reduced (Figure 3.16). In 6 of 12 studies the later survival of the fledglings from enlarged broods was reduced. Given the difficulties of detecting a change in survivorship, this percentage suggests that a decrease in survival is a typical phenomenon. Enlarged clutches produce more offspring, but these are typically underweight at fledging (Figure 3.16). Their weight probably reduces their chances of survival, at least if they fledge late in the season and conditions during the winter months are more severe than average. This hypothesis is further supported by a significant association between mortality from hatching to fledging and mass at fledging (Figure 3.16, $\chi^2 = 5.21$, $P < 0.05$). It is reasonable to suppose that such detrimental effects will continue for a period following fledging.

There are a number of ways in which a brood enlargement experiments have been conducted:

1. Increase some broods while using unmanipulated broods as controls.

2. Randomly assign brood sizes to parents.

3. Increase or decrease brood sizes while keeping track of the original clutch size.

The ability of parents to raise their young is likely to be reflected in the original clutch size. (That is, parents producing large clutches are likely to be in better "condition" than parents that produce small clutches.) Thus the third the last approach is the best. The importance of accounting for the original clutch size is

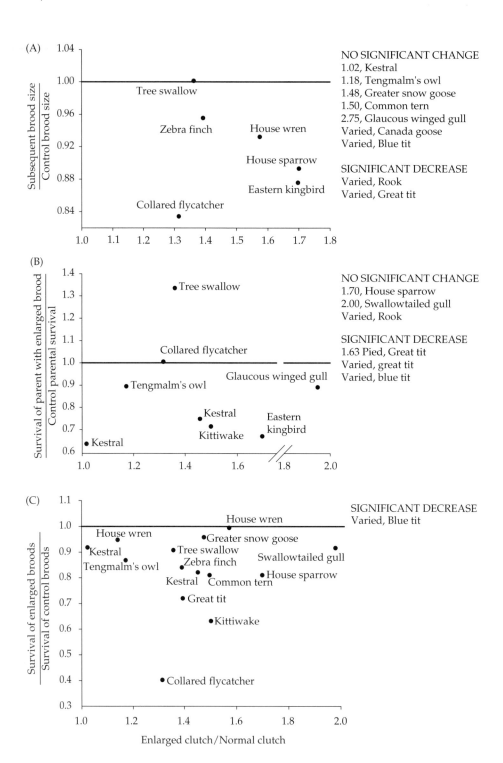

◀ **Figure 3.15** Experimental manipulation of brood size in birds to test for a trade-off between present reproduction and three other life history traits: future reproduction (A), future survival (B), and offspring survival (C). The lists at the side (brood manipulation ratio, species) show studies in which insufficient or inappropriate data were presented to construct a single ratio. Data sources from Wheelwright et al. (1991), Table 6.5 and 6.8 of Roff (1992a), Pettifor (1993a,b), Jacobsen et al. (1995), Daan et al. (1996), Deerenberg et al. (1996), Young (1996), Heaney and Monaghan (1995), Lepage et al. (1998), Maigret and Murphy (1997), Murphy (2000).

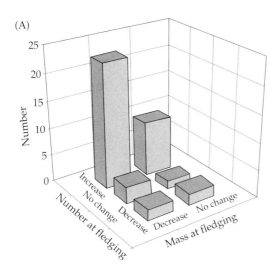

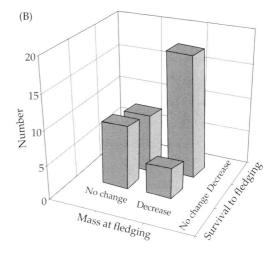

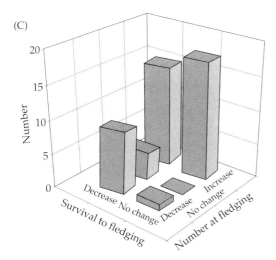

Figure 3.16 Comparisons among studies of the effect of brood enlargement on three life history traits. Data taken fom Table 9.2 of Roff (1992a).

shown by the study of blue tits by Pettifor (1993a). In this study there was both a significant effect due to the manipulation and due to the initial clutch size (Figure 3.17). Offspring from naturally large broods had a relatively higher survival than offspring from naturally small broods. Enlargement of broods reduced offspring survival. However, reduction of brood size also reduced offspring survival, as was also found in the Great tit (Pettifor et al. 1988). Pettifor (1993a, p. 136) suggested that "parents recognized the large decrease in the reproductive value of their nests and 'lost interest' in their broods when negative manipulations were particularly large." A reduction in reproductive effort with a reduction in the number of offspring is an adaptive response under some circumstances, with the extreme being desertion (see Chapter 4). Thus the trade-off that is observed as a consequence of experimental manipulation may also reflect phenotypic plasticity in other traits. Another example of this point is the change in behavior exhibited by gravid female common lizards, described previously.

A proximate factor in the trade-off between clutch size and reproductive success may be parasite load. Norris et al. (1994) and Richner et al. (1995) found a highly significant positive correlation between clutch size and the probability of infection by the haematozoan parasites in male but not female great tits. Enlargement of the clutch size increased the prevalence of parasites in males but not females. In contrast, Oppliger et al. (1997) measured a positive phenotypic correlation between the probability infection by the malarial parasite, *Plasmodium*,

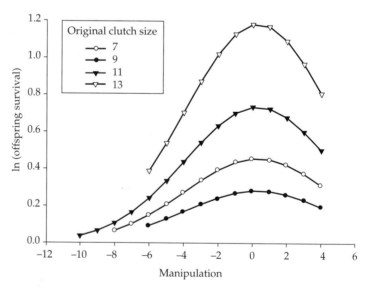

Figure 3.17 Predicted offspring survival in the Blue tit as a function of the original clutch size, C, and the manipulation, M. ln (offspring survival) $= -2.928 + 0.238C + 0.017M - 0.028M^2$. From Pettifor (1993a).

and clutch size in female great tits. Infected mothers had significantly larger clutches than uninfected females, which lead Opplinger et al. (1997) to hypothesize that females with larger clutches were weakened by the increased reproductive effort. An increase or decrease in parasite load following enlargement or reduction of clutch size supports this hypothesis (Ots and Horak 1996; Allander 1997). Manipulation of clutch size has also shown correlated responses in immune response consistent with effects of physiological stress in the zebra finch (Deerenberg et al. 1997), the collared flycatcher (Gustafsson et al. 1994; Nordling et al. 1998), the pied flycatcher (Siikamaki et al. 1997; Moreno et al. 1999) and the red jungle fowl (Johnsen and Zuk 1999). The detrimental effects of parasites on fledging success or other components of fitness have been demonstrated in a number of species by manipulation of parasite load (Moller 1990, 1993; Lope et al. 1993; Oppliger et al. 1994; Allander and Bennett 1995). Moller's study on the barn swallow is noteworthy for the simultaneous manipulation of parasite load and clutch size (Moller 1993). There is a very clear interaction between reproductive effort and parasite load on reproductive success (Figure 3.18). An additional effect of parasites is that a trade-off between clutch size and offspring quality may be negated by heavy parasite infestations of the offspring (Richner et al. 1993). A two-year study on Tengmalm's owl pointed to the deleterious effects of parasites being correlated to the food supply, females diverting energy to parasite resistance only when food was abundant (Korpimaki et al. 1993).

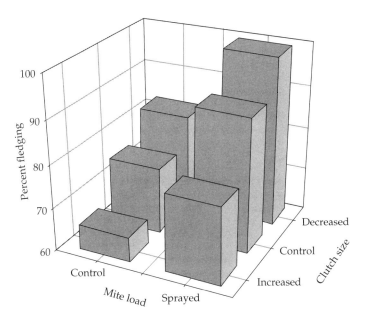

Figure 3.18 Effects of clutch size and mite load (control vs reduced by spraying) on percentage fledging in barn swallows. Data from Moller (1993).

Litter-size manipulations have been conducted on a few species of small mammals but the only variable consistently measured was weight at weaning, which showed a decline with litter size in all species studied (*Mus musculus*: Machin and Page 1973; Smith and McManus 1975. *Peromyscus polionotus*: Kaufman and Kaufman 1987. *Peromyscus leucopus*: Fleming and Rauscher 1978. *Clethrionomys glareolus*: Mappes et al. 1995; Koskela 1998. *Crocidura russula*: Genoud and Perrin 1994). These results are consistent with those from unmanipulated litters (see previous section). There are few data on survival rates, but in enclosures Mappes et al. (1995) observed a reduced nonsignificant survival rate of bank vole pups from enlarged litters (ratio = 0.85). They observed no significant effects on the parents, although the data does hint at effects on future reproduction. Given the sample size of 20, the failure to find significant effects cannot be taken very seriously. Litter-size manipulations in free ranging red squirrels showed that increased litter size decreased juvenile survival and growth rate (Humphries and Boutin 2000).

Male rock bass, *Ambloplites rupestris*, make nests and guard their eggs and early young. Sabat (1994) took advantage of this behavior and manipulated the brood size of newly hatched larvae, producing small, medium, and large size broods. There was a positive correlation between percent loss of weight and brood size (i.e., males with the largest broods lost the most weight). There was a negative correlation between survival probability and percent loss of weight (i.e., males losing the greatest weight had the lowest survival). These two relationships imply that males with the largest broods suffered the greatest mortality but, unfortunately, this trade-off was not directly tested.

Brood sizes have been manipulated in two species of insects that guard their eggs. Tallamy and Denno (1982) removed eggs from the lace bug, *Gargaphia solani*, as they were laid. This bug typically broods its eggs; loss of its brood causes it to lay further eggs. Females that were permitted to brood lived longer than those that continually produced eggs (Tallamy and Denno 1982). Thus in this species there is a trade-off between reproductive effort and survival. The heteropteran *Elasmucha grisea* defends its eggs and small nymphs against invertebrate predators. There is a positive correlation between female size and clutch size. Mappes and Kaitala (1994) hypothesized that small females could not defend larger clutches. To test this hypothesis they transplanted females so that large females were defending small clutches, LS, and small females were defending large clutches, SL. (Females apparently do not discriminate their own from another female's eggs.) Additionally, as a control, females were transplanted to clutches characteristic of their size (CL, CS). Small females defending large clutches lost significantly more eggs than other females (percent lost = 22%, SL; 3%, LS; 3%, CL; 11%, CS) and even more than the additional eggs she was given. Defending more eggs did not reduce the survival of the female parent.

Landwer (1994) and Sinervo and DeNardo (1996) experimentally altered clutch mass in free ranging populations of two species of lizard by surgical intervention. Landwer (1994), using the technique developed by Sinervo and Licht (1991a), reduced clutch masses carried by gravid female tree lizards, *Urosaurus ornatus*, by approximately 50%. There was a dramatic and highly significant difference in survival rates, with an annual survival rate of approximately 46% (depending on year and cohort) in the reduced-clutch females compared to 26% in the females with unaltered clutches. (Appropriate controls were used for effects of the anesthesia and the operation.) Using the aforementioned surgical technique, Sinervo and DeNardo (1996) decreased clutch mass in the side-blotched lizard *Uta stansburiana* and increased clutch mass by exogenous administration of either gonadotropin or corticosterone. As in the previous study, reduction of clutch mass increased survival in all comparisons. In one year the survival of females with enlarged clutches was reduced, whereas in a second year it was increased relative to the controls. Similar results were also observed using natural variation in clutch mass, but the ecological factor causing the variation in survival could not be determined. These data indicate both the importance of measuring trade-offs in the wild and also under several different circumstances.

Experimental manipulation of brood size in animals prior to birth or oviposition is difficult and possible in only a few species, such as the lizards described above. However, in plants it is relatively simple to alter reproductive effort by removal of flowers or artificial pollination. By such means, Montalvo and Ackerman (1987) were able to modulate the rate of fruit set in the orchid *Ionopsis utricularioides*. Both growth and the probability of not producing flowers in the following year were negatively related to fruit set. A similar treatment by Horvitz and Schemske (1988) using a perennial tropical herb, *Calathea ovandensis*, showed the same effects, but these were statistically nonsignificant. A power analysis showed that, for the given sample sizes, the differences in means required to produce significance ranged from 14% to 31%, while the observed differences ranged from 4% to 24%. Sample sizes required to obtain significance, given the observed differences, were two to three times those actually used. Interpretation of the results thus depends upon whether one considers the observed differences to be biologically important. Horvitz and Schemske concluded that they are not, but others may conclude otherwise. The trade-off between present and future reproduction may be cumulative as shown by the long-term studies of Ackerman and Montalvo (1990) and Primack and Hall (1990) on current and future reproduction in the orchids *Ionopsis utricularioid* and *Cypripedium acaule*. A similar result was obtained in the house wren *Troglodytes aedon*, where both the second brood in the same season and the first brood in the next season were reduced following enlargement of the experimental brood (Young 1996).

Examples of Trade-offs Between Life History Traits Determined from Pedigree Analysis

General Overview

The data to be considered consists of phenotypic and genetic correlations between life history traits as defined above. The traits were defined so that the correlation between two traits would be negative if a trade-off occurred. (Thus, for example, I considered the correlation between fecundity and survival not between fecundity and mortality.) Most of the data are for animals. There is no reason to expect all correlations between life history traits to be negative; indeed, there is no sound theoretical prediction except that some should be negative. Certainly, for any particular organism we would expect at least one negative genetic correlation or else there would be no constraints on selection.

For a full description of the data set, see Roff (1996b). There are 152 estimates of genetic correlations distributed among six species (which illustrates the relative paucity of data). The grand mean is 0.16 (SE = 0.051), which is similar to the mean phenotypic correlation ($\bar{r}_P$ = 0.15, SE = 0.031), and the distribution is roughly normal (Figure 3.19). In addition, 39% of the genetic correlations are negative compared to 34% of the phenotypic correlations. These proportions do not change significantly if species means or medians are used (see Table 2 in Roff, 1996b). Unfortunately, the 95% confidence regions of most of the estimates include zero and are sufficiently large that rather large negative or even positive correlations cannot frequently be excluded (Figure 3.20). As a consequence, it is very difficult to draw any general conclusions about the distribution of genetic correlations between life history traits. The problem of large standard errors plagues genetic analyses.

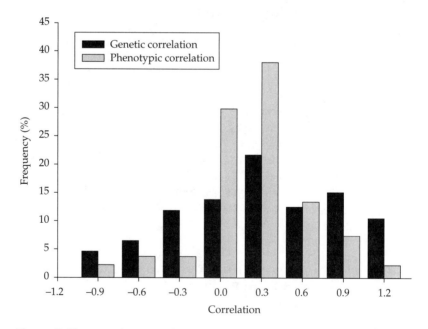

Figure 3.19 Distributions of the phenotypic and genetic correlations between life history traits. Data from Roff (1996).

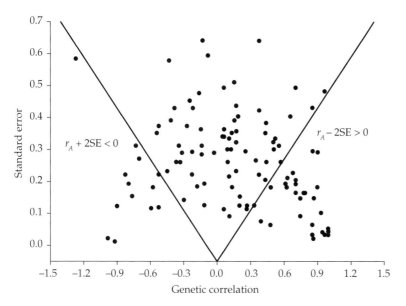

Figure 3.20 Plot of the genetic correlations between life history traits and their associated standard errors. Points lying below the two lines have 95% confidence ranges than exclude zero.

Examples from Plants

Solbrig and Simpson (1974) present data on clonal variation in the dandelion *Taraxacum officinale*, demonstrating the presence of genetic correlation between survival and fecundity that is dependent on growing conditions. From three sites in the Mathei Botanical Garden (University of Michigan), Solbrig and Simpson collected four clones of dandelions, designated A, B, C, and D. Clone A predominated in the two disturbed sites, comprising 73% and 53% of the total sample from each locality, while clone D predominated in the undisturbed site, comprising 64% of the sample. In contrast, clone D was virtually absent from the two disturbed sites (0% and 1%), and clone A made up only 17% of the sample from the undisturbed site. Plants from clone A and B were raised under two conditions: in pure culture and in a mixed culture of 50:50. In pure culture both clones fared about the same, although clone D tended to show a somewhat lower reproductive effort. In mixed culture clone A had a reduced survival and a lower dry weight than clone D, but clone A produced a significantly larger biomass of seeds. These data suggest a genetically based trade-off between competitive ability, reflected in survival rates, and fecundity.

Law (1979a) examined the present and future fecundity in the annual meadow grass *Poa annua*, basing his analysis on family means. Environmental effects were removed by growing the plants in a completely randomized design (Law et al. 1977), and hence differences in growth rate between families can be ascribed to genetic differences. The number of inflorescences per plant in the second season was negatively correlated with the number of inflorescences per plant in the first season, thus demonstrating a genetically based trade-off between present and future reproduction. There was also evidence of a negative genetic correlation between reproductive effort in the first season and plant size in the second season. This suggests that the causal reason for the negative correlation between present and future fecundity arose from a trade-off between present reproduction and future growth. Geber (1990) explored this trade-off in detail using *Polygonum arenastrum*. Growth and reproduction in higher plants depends on meristems, which can either differentiate into reproductive tissue, thereby eliminating further growth, or can continue to grow vegetatively. Because the two decisions are mutually exclusive, there is necessarily at the level of the individual meristem a correlation of –1 between growth and fecundity. Overall fecundity and growth is then a consequence of the combined action of all meristems. In *Polygonum arenastrum* the genetic correlation between early and late fecundity was –0.63 ($P < 0.05$) as qualitatively predicted by the hypothesized causal mechanism. In contrast to the two foregoing studies, Platenkamp and Shaw (1992) obtained both positive and negative genetic correlations between reproductive output across pairs of years in the grass *Anthoxanthum odoratum*. However, statistical problems in the estimations make these results suspect.

Examples from Animals

Table 3.4 shows the phenotypic and genetic correlations between life history traits in the milkweed bug *Oncopeltus fasciatus*. A trade-off between fecundity and age

Table 3.4 Genetic (lower triangle) and phenotypic (upper triangle) correlations between life history traits in the milkweed bug, *Oncopeltus fasciatus*.

	α	Clutch size 1	2	3	Percentage hatch in clutch # 1	2	3	ICI	D
		1	2	3	1	2	3		
	α	−0.09	−0.09	−0.12	−0.08	−0.08	−0.09	0.04	−0.09
1	−0.22	–	0.27	0.29	0.37	0.32	0.30	0.09	−0.10
2	−0.27	1.09[u]	–	0.29	0.20	0.26	0.23	0.20	−0.12
3	−0.98*	0.79*	1.04[u]	–	0.20	0.21	0.24	0.15	−0.12
%1	0.07	0.71*	1.06[u]	0.45	–	0.80	0.73	−0.01	−0.06
%2	0.11	0.62*	1.01[u]	0.53	1.02[u]	–	0.82	−0.06	−0.07
%3	0.18	0.63*	0.95*	0.56	1.01[u]	1.01[u]	–	−0.07	−0.06
ICI	0.07	−0.21	−0.08	−0.82*	−0.79*	−0.38	−0.54	–	−0.01
D	−0.13	−0.57	−0.76*	−0.58*	−0.52*	−0.33	−0.44	0.51	–

α = age at first reproduction, with day zero taken to be first day of adult life. ICI = interclutch interval. D = development time (egg to adult). * = Estimate ± 2SE does not include zero. u = SE undefined. Data from Hegmann and Dingle (1982).

at maturity would actually be evidenced by a positive correlation, because an increase in age at maturity decreases r while an increase in fecundity increases r. There is no evidence of a trade-off between age at maturity (taken from the first day of adult life) and clutch size ($r_A = -0.22, -0.27, -0.98$) or between development time and age at maturity ($r_A = -0.13$). There is slight evidence of a trade-off between age at maturity and percentage hatch ($r_A = 0.07, 0.11, 0.18$) but none of these correlations are statistically significant. The genetic correlations between clutch size at different ages, between percent hatch at different ages, and between the clutch size and percent hatch are all positive and several are statistically significant. Thus there is no evidence of a trade-off between these life history traits. Similarly, the genetic correlations between interclutch interval and the foregoing parameters do not indicate any trade-off. Therefore, within this entire correlation matrix there is no substantial evidence of any trade-offs!

A similar lack of trade-offs is observed in the age-specific fecundities of the bean beetle *Callosobruchus chinensis* (Table 3.5). The genetic correlation between components of fecundity are almost always positive, indicating that females that are genetically disposed to large fecundity at a particular age are also genetical-

---

Table 3.5 Genetic correlations (SE) between life history traits in the Azuki bean weevil, *Callosobruchus chinensis*.

Genetic correlations obtained in the analysis of Nomura and Yonezawa (1990).

	Fecundity for days 1–3	*Peak fecundity*	*Total fecundity*	*Egg-laying rate*
Peak fecundity	0.75 (0.14)			
Total fecundity	0.88 (0.14)	0.64 (0.21)		
Egg laying rate	0.86 (0.29)	0.26 (0.29)		
Fecundity for days 10–12	0.66 (0.40)	0.78 (0.16)	−0.12 (0.64)	−0.42 (0.57)
Fecundity for days 13–15	0.07 (0.36)	0.12 (0.22)	0.87 (0.43)	0.17 (0.36)
Fecundity for days 16–18	−0.24 (0.45)	−0.02 (0.29)	0.19 (0.43)	−0.16 (0.47)
Longevity	−0.34 (0.31)	−0.16 (0.18)	−0.89 (0.12)	−0.31 (0.29)
Number of days to last oviposition	−0.29 (0.14)	0.12 (0.09)	−0.10 (0.28)	−0.52 (0.11)

Genetic correlations obtained in the analysis of Tanaka (1993).

	F2	*F4*	*F5*	*Longevity*
F1	−0.11 (0.12)	1.00[u]	1.00[u]	−0.91 (0.01)
F2		0.14 (0.18)	0.87 (0.02)	0.28 (0.11)
F4			1.00[u]	0.21 (0.15)
F5				0.34 (0.12)

F1 = Age specific fecundity. The heritability of F3 was 0.06, and Tanaka (1993) provides no estimates of the genetic correlation between this and other traits. Given the low heritability, these correlations can be assumed to be close to zero. u = SE undefined.

ly disposed to high fecundities at other ages and overall. Nomura and Yonezawa (1990) found a trade-off between longevity and fecundity but Tanaka (1993) observed a trade-off only between early fecundity and longevity (Table 3.5). Working with a related species, *C. maculates*, Moller et al. (1989) found a highly significant positive genetic correlation between lifetime fecundity and longevity (0.91 ± 0.29) also indicating the lack of a trade-off. They did find a trade-off between development rate (1/egg-to-adult) and fecundity (–1.26 ± 0.58) and between development rate and adult longevity (–0.73 ± 0.31).

Disparate findings with respect to genetic correlations between age-specific fecundities have been reported for *Drosophila melanogaster*. Rose and Charlesworth (1981) reported negative genetic correlations (Table 3.6) but positive phenotypic correlations. In contrast, Engstrom et al. (1992) obtained highly significant positive genetic and phenotypic correlations (Figure 3.21). An interesting feature of this work is the observation that the strength of the correlation declines as the distance in time between the two traits increases (Figure 3.21). Such a pattern has also been observed in age-specific morphological measurements of mice (Atchley 1984) and maternally induced egg diapause in a cricket (Roff and Bradford 2000). As in the study of Engstrom's group, Tatar et al. (1996) found generally positive genetic correlations between fecundities at different ages but no very marked decline with the difference between ages.

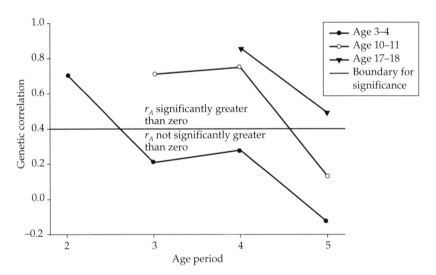

Figure 3.21 Genetic correlation in *D. melanogaster* between "adult offspring" (= number of adult offspring produced) at age *t* to *t* + 1 and "adult offspring" at a subsequent age (2 = 10 – 11d, 3 = 17 – 18d, 4 = 24 – 24d, 5 = 31 – 32d). Data from Engstrom et al. (1992).

Table 3.6   Genetic correlations between life history components in *Drosophila melanogaster.*

Trait 1	Trait 2	r_A (SE)[a]	Reference
Fecundity and Longevity			
Fecundity, lifetime	Longevity	0.76 (0.06)	1
Fecundity, days 5–10	Longevity	0.07 (0.11)	2
Fecundity, days 1–5	Longevity	**–1.431**[u]	3
Fecundity, days 6–10	Longevity	0.30[u]	3
Fecundity, days 11–15	Longevity	**–0.71**[u]	3
Fecundity, days 2–10	Longevity	**–1.76**[**]	4
Fecundity, days 20–26	Longevity	0.86[**]	4
Fecundity, day t	Age-specific survival	63, approx 50 pos[b]	5
Between Fecundity at Different Ages (see also Figure 3.21)			
Fecundity, days 1–5	Fecundity, days 6–10	**–0.16**[u]	3
	Fecundity, days 11–15	**–0.48**[u]	3
Fecundity, days 6–10	Fecundity, days 11–15	0.51[u]	3
Fecundity, days 2–10	Fecundity, days 20–26	**–0.96**[**]	4
Fecundity, day t	Fecundity at day $t+x$	25 pos, **7 neg**	5
Miscellaneous			
Egg-adult time	Pupal duration	0.40 (0.07)	6
Development time[c]	Fecundity, days 2–10	**0.21**[ns]	4
	Fecundity, days 20–26	–1.08[**]	4
Male virility	Development time, ♂	**0.33**[ns]	4
	Development time, ♀	–0.48[ns]	4
	Longevity, ♂	0.16	4
	Longevity, ♀	**–1.46**[**]	4
	Fecundity, days 2–10	0.27[ns]	4
	Fecundity, days 20–26	**–1.01**[*]	4
Last day of egg laying	Fecundity, days 1–5	**–0.13**[u]	3
	Fecundity, days 6–10	**–0.63**[u]	3
	Fecundity, days 11–15	**–0.50**[u]	3
	Longevity	**0.77**[u]	3
Survival at age t	Survival at age $t+x$	**3 neg**, 18 pos[c]	5

Reference: 1. Tantawy and Rakha (1964); 2. Tantawy and El-Helw (1966); 3. Rose and Charlesworth (1981); 4. Tucic et al. (1988); 5. Tatar et al. (1996); 6. Tantawy and El-Helw (1970); u = Unknown, ** $P < 0.01$, ns = not significant

[a]Genetic correlations in bold type are those in which the sign indicates a trade-off.

[b]Female data only presented (male mortality data similar). Tatar gives mortality but here I use survival for consistency with "longevity".

[c]A positive correlation between fecundity and development time denotes a trade-off because an increased development time, by itself, decreases the rate of increase.

Equally disparate results are reported for fecundity components and longevity in *D. melanogaster*, although the most statistically convincing data suggest a positive correlation (Table 3.6). Trade-offs are indicated by the signs of genetic correlations between other life history traits in *D. melanogaster*, but the large standard errors (or lack of standard errors) makes interpretation very dubious. Perhaps the most important message coming from these data is that the estimation of genetic parameters should be done with greater rigor than is generally the case (i.e., larger sample sizes).

Examples of Trade-offs Between Life History Traits Determined from Selection Experiments

Most selection experiments have focussed on morphological or behavioral traits. Experiments on two insects, *Tribolium* and *Drosophila*, in which individuals were selected for early or late fecundity and changes in survival monitored, give partial support to the hypothesis of a genetic trade-off between early and late fecundity (Table 3.7). However, the cause for this correlated change, when observed, may be a consequence not of antagonistic pleiotropy but of experimental artefacts or mutation accumulation (Hughes and Charlesworth 1994; Rose 1997). Selection experiments on chickens (Lerner 1958; Erasmus 1962; Nordskog and Festing 1962; Morris 1963), geese (Merritt 1962) and mice (Wallinga and Bakker 1978) support the hypothesis that early and late fecundity are negatively correlated.

Selection for a changed age-schedule of reproduction has produced correlated changes in other life history traits such as development time (Table 3.7), although direct selection on development time did not show correlated changes in adult lifespan (Zwaan et al. 1995). Females from lines selected for slow development did show relatively high early fecundity and low late fecundity. Nunney (1996) selected for fast larval growth and observed a significant decline in the fecundity of the selected lines. These changes are consistent with changes brought about by changes in adult body size (see Chapter 4).

The potential difficulty of interpreting correlated responses is illustrated by the attempts to select for early and late reproduction in the bean weevil, *Acanthoscelides obtectus*. Tucic et al. (1990) selected for early and late reproduction when adults were starved and when fed. Two control lines were maintained only under the starvation regime. The researchers observed no response when adults were fed but a significant change under the starvation regime. However, an inspection of the distribution of fecundities shows that, whereas the early fecundity increased relative to the controls in the "early fecundity" lines and decreased in the "late fecundity" lines, there was no increase in late fecundity of the "late fecundity" lines. The total fecundity of the "late fecundity" lines (21, 25 for the two replicate lines) was lower than either the control (38, 36) or the "early fecundity" (42, 48) lines. Further, there was no change in longevity. Later experiments suggested that there was inbreeding depression and the possibility of inadvertent selection for rapid larval development (Tucic et al. 1996). Because of

Table 3.7 Correlated responses to selection for age at maturity in *Tribolium casteneum* and *Drosophila melanogaster*. Selection regimes typically form pairs of early (E) and late (L) maturity selected lines.

Selection regime	Correlated responses	Reference
Tribolium casteneum		
E = Allowed to lay for 3 days	Increased development time	Sokal (1970)
L = Allowed to lay for 10–20 days	Increased early fecundity, higher mortality later	Mertz (1975)
Drosophila melanogaster		
E = Highest fecundity on days 1–5	No correlated response	Rose and Charlesworth (1981)
L = Highest fecundity on days 21–25	Early fecundity and mean egg-laying rate decreased	
L = Increasing delay in egg collection	Early fecundity decreased	Rose (1984)
E = Early reproduction, days 2–6	No control line to test against	Luckinbill et al. (1984)
L = Increasing delay in reproduction	Early fecundity decreased	
E = Early fecundity, days 14–17	Lines did not differ in early fecundity	Partridge and Fowler (1992)
L = Increasing delay in reproduction	Larval development time increased, lower larval survival rates	
E = Early reproduction	"Old" lines showed faster development time	Chippindale et al. (1994)
L = Increasing delay		

the deleterious effects of inbreeding on life history traits there could be substantial asymmetric responses to selection on fitness-related traits.

Summary

The concept of trade-offs is central to life history theory, but the focus of research has tended to be on study of assumed bivariate trade-offs, which can be misleading if the constraints involve more than two traits. The problem of "missing variables" in the analysis of trade-offs is a very important because the omission of a variable from the analysis can readily give the appearance of positive covariation between traits while in reality there is negative covariation when the third, "missing" variable is held constant. It is for this reason that phenotypic correlations estimated from unmanipulated conditions are unreliable unless a strong case can be advanced that all relevant variables are included in the analysis. A manipulation experiment is a very powerful means of demonstrating the presence of a trade-off and dissecting its causes. Whereas phenotypic correlations from unmanipulated populations have frequently failed to demonstrate a trade-off, manipulation experiments have consistently shown trade-offs. When selection acts on both traits involved in a trade-off, the response to selection depends not only on the genetic correlation but also the phenotypic correlation. The measurement of genetic correlations, either by pedigree analysis or selection, have not produced very satisfactory results, primarily because sample sizes or experimental design have lead to standard errors so large that the presence or absence of a trade-off cannot be resolved.

CHAPTER 4

Evolution in Constant Environments

Most of the analyses on the evolution of life history traits have assumed a constant environment and neither density-dependent nor frequency-dependent selection. Under such assumptions analysis is relatively simple from a theoretical perspective though the experimental examination is still formidable. Although we might expect that density-dependent selection occurs on occasion and that the world is not constant, we start with the simplest model and add complexity only as required by a failure of prediction to match observation. Indeed one of the major findings of the analyses presented below is that a considerable amount of variation in life histories can be explained without recourse to complex models.

In the first part of this chapter I address the general question, relevant both to this chapter and the following two chapters, of whether genetic variation might limit evolutionary change. In support of the proposition that there is indeed typically sufficient variation for evolutionary response, I present several examples in which very rapid evolutionary change in natural populations has been observed. The observation that there is an abundance of genetic variation raises the question of what factors might maintain this variation. In the remainder of the chapter I examine how selection on components of the characteristic equation generates variation in these components and how we can model and predict such variation. Finally, I expand the analysis to consider whether life history theory can explain larger-scale patterns such as allometric relationships frequently observed among species.

Is There Enough Genetic Variation for Evolution?

A prerequisite for evolution is the existence of genetic variation in traits. It is clear from the fact that evolution has and is occurring that additive genetic variation exists in natural populations. In most cases the estimation of genetic variation is done under laboratory conditions and hence the possibility exists that this variation is not representative of that in natural populations. In that case, laboratory estimates would be of very limited use.

Are Heritability Estimates from a Laboratory Environment Comparable to Field Estimates?

Conventional wisdom posits that organisms raised under the relatively constant conditions typical of the laboratory or greenhouse will exhibit lower levels of phenotypic variation than under the more variable field conditions (Mitchell-Olds and Rutledge 1986; Riska et al. 1989; Bull et al. 1982; Janzen 1992). The increased phenotypic variance in field populations is postulated to be a result of increased environmental variance, leading to reduced heritabilities (Bull et al. 1982; Coyne and Beecham 1987; Falconer 1989; Prout and Barker 1989; Riska et al. 1989; Schoen et al. 1994). If we make this assumption, then the ratio of the field to laboratory heritability can be reduced to give

$$h^2_{Field} = \frac{\sigma^2_A}{\sigma^2_{P, Field}}, \qquad h^2_{Lab} = \frac{\sigma^2_A}{\sigma^2_{P, Lab}}$$

$$\frac{h^2_{Lab}}{h^2_{Field}} = \left(\frac{\sigma^2_A}{\sigma^2_{P, Lab}}\right)\left(\frac{\sigma^2_{P, Field}}{\sigma^2_A}\right) = \frac{\sigma^2_{P, Field}}{\sigma^2_{P, Lab}} \qquad (4.1)$$

The ratio of the phenotypic variances is typically fairly modest lying between 1 and 2, though extreme values do occur most particularly in *Drosophila* species (Figure 4.1. Almost all traits are morphological traits). The median ratio for a variety of species is only 1.40 ($n = 24$, range 0.4–9.2, all data from Figure 4.1). At 2.51 the mean is considerably higher and results from extremely high ratios observed in the *Drosophila* species. When the genus *Drosophila* is excluded, the mean ratio drops to 1.22 and the median ratio is 1.30 ($n = 13$, range 0.4–1.7), with four estimates actually being less than 1. This suggests that field heritabilities will not be generally greatly reduced by a relative inflation of the environmental variance in the field (*Drosophila* appears to be an exception; see also Hoffman, 2000).

A comparison of heritability estimates of traits categorized as "morphological," "behavioral" and "life history" from the laboratory and field confirm this prediction, with the mean estimates of life history and morphological traits actually being larger from field studies than laboratory studies (Table 4.1). Pairwise comparisons using a Mann-Whitney test showed no differences between lab and field estimates (Weigensberg and Roff 1996). Eight studies have measured heritabilities simultaneously in the laboratory and the field, producing a total of 22 estimates (Figure 4.2). A paired t-test of these data indicates no significant difference (mean difference = 0.11, $t = 1.90$, $df = 21$, $P = 0.07$; note that the difference is significant if a one-tailed test is used). The confidence limits of the slope of the reduced major axis regression also include a slope of 1 (0.61–1.16). It is apparent

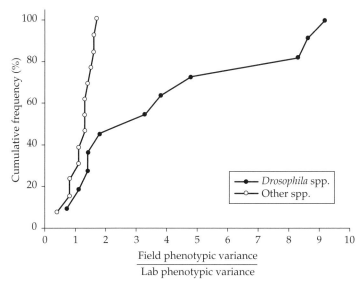

Figure 4.1 Cumulative frequencies of the ratio of phenotypic variances of the same trait measured in the field and the lab. Data from Table 2.11 of Roff (1997).

from Figure 4.2 that more estimates fall below the 1:1 line than above it (though the number is not significant). However, this is somewhat misleading since multiple estimates from the same species may not be independent. For example, morphological traits are generally highly correlated and thus the seven heritability estimates of morphological traits in *Gryllus pennsylvanicus* should perhaps be averaged. It is important to note that there is a strong correlation between laboratory and field estimates ($r = 0.63$, $n=22$, $P < 0.002$), and that the field estimates are by no means insignificant. I conclude that laboratory estimates of heritability are useful indicators of heritability in field populations.

Table 4.1 Mean heritabilities (sample size) estimated in the laboratory and field.

Trait category	Laboratory	Field	P (Mann-Whitney)
Life History	0.27 (75)	0.32 (12)	> 0.16
Morphology	0.50 (90)	0.56 (150)	> 0.31
Behavior	0.36 (24)	0.22 (3)	> 0.20

From Weigensberg and Roff (1996).

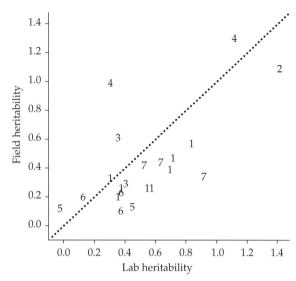

Figure 4.2   The distribution of heritabilities of traits measured in both the laboratory and the field. Each number represents a different species: 1. *Gryllus pennsylvanicus* (cricket; Simons and Roff, 1994); 2. *Poecilia reticulata* (guppy; Houde 1992); 3. *Partula taeniata* (snail; Murray and Clarke, 1967); 4. *P. suturalis* (Murray and Clarke, 1967); 5. *Drosophila montana* (fruitfly; Aspi and Hoikkala, 1993); 6. *Drosophila melanogaster, D. simulans* (fruitfly; Coyne and Beecham, 1987; Prout and Barker, 1989; Jenkins and Hoffman, 1994); 7. *Musca domestica* (housefly; Bryant, 1977).

Heritabilities of Different Types of Traits

AN HYPOTHESIS.   The next question to be addressed is whether the observed values of heritability are sufficient to permit reasonably rapid response to selection. In the above analysis I divided traits into three categories. The reason for this is that *a priori* it is reasonable to suppose that the strength of selection increases the more closely the trait is related to fitness. Consider some trait such as fecundity, which is closely connected to fitness; suppose that this trait is determined by the additive action of *n* loci at which there are two alleles 0 and 1, fecundity increasing as the sum increases. Obviously, over time, selection will favor loss of the 0-type alleles with a consequent decrease in additive genetic variance and thus heritability. The rate of fixation of alleles will depend upon the intensity of selection. Many morphological traits will approach fixation much more slowly than life history traits because they will be under relatively weaker selection. Therefore, if we begin with the same heritability, in general, life history traits will have lower heritabilities over their evolutionary trajectories than morphological traits. There will ultimately be an equilibrium established in which the erosion of variation is matched by the addition of variation by mutation. Providing that the rates of mutation are the same, the stronger selection on life history traits will give rise to smaller heritabilities. This mutation-selection balance prediction is potentially changed if genetic variation is maintained in part by antagonistic pleiotropy or frequency-dependent selection. The rankings of heritabilities of traits in other categories (behavioral, physiological) are more difficult to predict and in earlier analyses no prediction was made (Roff and Mousseau 1987; Mousseau and Roff 1987).

OBSERVATIONS. Gustafsson (1986) measured the heritabilities of a large number of traits in the collared flycatcher (*Ficedula albicollis*) and over a 5-year period estimated individual lifetime reproductive success. Likewise, Kruuk et al. (2000) estimated success in the red deer (*Cervus elaphus*) over an approximately 30-year period and estimated heritabilities in a wide range of traits, from fecundity to jaw length. As predicted, in both cases, there is a highly significant negative correlation between the heritability of the trait and its proportional contribution to the variance in lifetime reproductive success (Figure 4.3). For instance, in the collared flycatcher those traits such as lifespan and number of fledged young have low heritabilities (for females, –0.0160, –0.0161, respectively and for males –0.0001, –0.0052, respectively, all estimates being not significantly different from zero) but those traits make relatively large contributions to variance in lifetime reproductive success (18.9%, 7.4%, 26.9%, and 15.2%, respectively). The heritability of fitness itself (lifetime reproductive success) is very low (–0.0142 for females, 0.0083 for males) and not significantly different from zero.

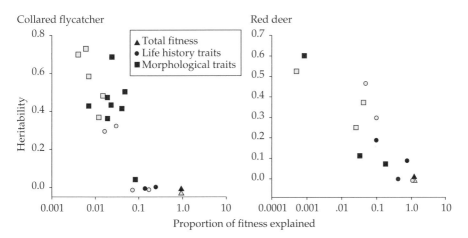

Figure 4.3 Relationship between heritability of a trait and the proportion of fitness explained by the trait in two animal species. Black = males, Gray = females. Excluding the total fitness measures (because they are composite traits), there are highly significant correlations between heritability and proportion of fitness explained. For the collared flycatcher, $r = 0.85$, $n = 19$, $P < 0.0001$, and for the red deer, $r = 0.85$, $n = 12$, $P < 0.0001$. Data from Gustafsson (1986) and Kruuk et al. (2000).

To test the hypothesis that traits most closely connected to fitness will have the lowest heritabilities, Roff and Mousseau (1987) surveyed estimates for a single genus, *Drosophila*, while Mousseau and Roff (1987) examined estimates for animals in general (75 species). In addition to the categories of morphology and life history, we defined two further classes: behavioral traits, which comprise such traits as activity level, alarm reaction and courtship behavior, and physiological traits such as oxygen consumption, resistance to heat stress, and body temperature. Because of potential statistical biases due to nonindependence of estimates, we analyzed both the entire data set and then, using only the median of the range in estimates, reported for each character of each species. Life history traits consistently have significantly lower heritabilities than morphological traits, with behavioral and physiological traits lying between (Table 4.2). A third method of comparing life history and morphological traits is to do a paired comparison using data from individual species (Figure 4.4). Of the 16 different species for which such data are available there is only one case in which the median heritability of life history traits exceeds that of the morphological traits, a distribution which is significantly different from the null hypothesis of 50:50 ($\chi^2 = 12$, $df = 1$, $P < 0.001$).

VARIATION OF HERITABILITY AMONG TRAIT CATEGORIES: AN ALTERNATIVE EXPLANATION. Price and Schluter (1991) suggested an alternative hypothesis for the difference in heritabilities between life history and morphological traits. Their hypothesis is summarized in Figure 4.5. They note that many, if not all, life history traits are directly connected to morphology—examples being a positive relationship between fecundity and body size or longevity and body size. Now let the additive genetic variance in the morphological trait be σ^2_{AM} and the environmental variance be σ^2_{EM}. Further, assume that the life history trait is determined

Table 4.2   Means of heritability estimates for non-domestic animals.

Comparison	Life history	Behavior	Physiology	Morphology
Drosophila	0.12	0.18	ns	0.32
All animals	0.26	0.30	0.33	0.46
Medians	0.26	0.37	0.31	0.51

Data are from Roff and Mousseau (1987) and Mousseau and Roff (1987).

Notes: The heritability estimates for *Drosophila* were estimated using from each study the median value of each character. "All animals" does not include the *Drosophila* data and consists of 1,120 estimates. "Medians" is based on the median heritability for a given species and character and consists of 283 separate values. Kolmogorov-Smirnov tests of the "medians" indicate that the heritability of morphology is significantly different from the other three categories, but that the remaining three do not differ from each other (for details see Table 3 of Mousseau and Roff, 1987).

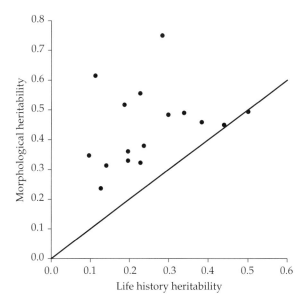

Figure 4.4 A comparison of the heritabilities of morphological and life history trait in 16 species—the 15 species listed in Figure 2.12 of Roff (1997) plus data for red deer from Kruuk et al. (2000).

solely by the morphological trait plus an additional environmental component E with variance σ_E^2. For simplicity, assume that the value of the life history variable is a simple linear function of the morphological trait and the second environmental factor

$$Y = a + bX + E \qquad (4.2)$$

where Y is the life history trait, X is the morphological trait and a, b are constants. For clarity, and without loss of generality, rescale the above equation so that $a = 0$ and $b = 1$. The heritability of the morphological trait is, by definition,

$$h_M^2 = \frac{\sigma_{AM}^2}{\sigma_{AM}^2 + \sigma_{EM}^2} \qquad (4.3)$$

The additive genetic variance of the life history trait is simply equal to the additive genetic variance of the morphological trait, but its phenotypic variance is increased by the variance of the second environmental factor:

$$h_{LH}^2 = \frac{\sigma_{AM}^2}{\sigma_{AM}^2 + \sigma_{EM}^2 + \sigma_E^2} \qquad (4.4)$$

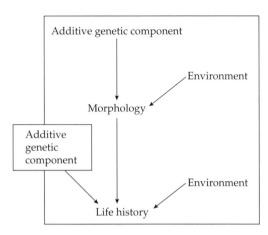

Figure 4.5 Schematic representation of the model postulated by Price and Schluter (1991) to account for the observation that heritabilities of life history traits are typically lower than those of morphological traits. The additive genetic variance component in the smaller box was not considered as a possible component by Price and Schluter.

and hence the heritability of the life history trait is reduced, giving the observed pattern. A different and more reasonable scenario is that the life history trait is determined in part by the morphological trait and in part by other genes that contribute an additive genetic variance (Figure 4.5). The two heritabilities are now

$$h_M^2 = \frac{\sigma_{AM}^2}{\sigma_{AM}^2 + \sigma_{EM}^2}, \quad h_{LH}^2 = \frac{\sigma_{AM}^2 + \sigma_{AO}^2}{\sigma_{AM}^2 + \sigma_{EM}^2 + \sigma_E^2} \tag{4.5}$$

If σ_{AO}^2 is much greater than σ_{AM}^2, then the heritability of the life history trait will be more dependent upon the former additive genetic variance than the latter. Selection on the life history trait will proceed by reducing σ_{AO}^2 at a faster rate than σ_{AM}^2, again producing the observed pattern, but because of the effect of selection, not the "downstream" nature of the relationship between the life history and morphological traits.

There is no theoretical way to resolve these two hypotheses, and indeed both may be playing roles. What is required is a greater understanding of the genetical and "nongenetical" (e.g., physiological, mechanical, ecological) architecture underlying life history and morphological variation. This is a more holistic approach than simply examining the genetic architecture. Its importance is illustrated by the traits body size and development time in insects. In a wide variety of insect species, metamorphosis into the adult form is triggered by the attainment of a critical size (*Rhodnius prolixus*, Wigglesworth 1934: *Manduca sexta*, Nijhout and Williams 1974a,b; Nijhout 1975: *Oncopeltus* spp., Blakley and Goodner 1978; Nijhout 1979: *Acheta domestica*, Woodring 1983). As a consequence, variation in development time may be primarily a result of genetic variation in the critical size for metamorphosis and environmental variation in the rate of growth. A similar situation also appears to occur in the plant *Cynoglossum officinale* in which there is an inherited threshold size for flowering (Wesselingh et al. 1993; Wesselingh and de Jong 1995). A size threshold for maturity has been suggested for other plant species (Werner 1975; Harper 1977; Kachi and Hirose 1985; Yokoi 1989) and mammals (Riska et al. 1985; Skogland 1989; Childs 1991). This situation is the scenario suggested by Price and Schluter (1991), the heritability of the life history trait (development time) being inflated by environmental factors and thus is larger than the morphological trait (body size). In fact, the situation is likely to be much more complicated than this, with genetic variation in rate of growth and possibly an interaction between development time and body size (discussed later). It is only with a detailed analysis of the genetic basis of the underlying traits in conjunction with a study of the mechanisms underlying the determination of metamorphic events that the various influences can be disentangled.

Are Heritabilities Large Enough to Permit Rapid Response to Selection?

Regardless of the ultimate cause, there is significant variation in heritabilities among trait categories. Heritabilities of life history traits have a central tendency of approximately 0.25: Is this sufficient to permit evolution within relatively short time limits? The single-trait response equation is $R = ih^2\sigma_P$. The response to selection depends not only upon the heritability of the trait but the selection differential ($S = i\sigma_P$). The total response over t generations of selection at the same selection intensity is simply Rt, and thus the same response can be achieved either by strong selection over a short period or weak selection over a longer one. If the top 99% of the population is selected each generation (i.e., if there is a 1% cull) then $R/\sigma_P = (0.0269)(0.25) = 0.00673$ and roughly 150 generations are required to change the mean trait value by one phenotypic standard deviation, which is a very short period of time. Very small selection pressures operating over a (geologically speaking) short period of time can produce very large phenotypic changes.

From an extensive review of estimates of directional selection in natural populations (see Figure 4.6), Endler (1986, p. 210) noted that the range of selection intensities extensively overlaps the values found in animal and artificial selection experiments. For example, the i values found in Falconer (1989) and in the papers cited by Robertson (1980) range from 0.15 to 1.39, with a geometric mean of 0.71. For comparison, the geometric mean of significant i in Figure 4.6 is 0.59. This suggests that natural selection is as often as strong as artificial selection. In summarizing his review of all the data Endler (1986, p. 220) concluded that "from the observed distribution of S, i . . . one cannot say either that selection is weak or strong in natural populations, but rather that it can take any value, up to and including values found in artificial-selection experiments and in animal and plant breeding. The observed distribution of significant values is roughly uniform, with a deficiency in very small values because they require very large sample size for detection. . . . It is not reasonable to draw any conclusion from these distributions other than that strong selection is not rare and may even be common . . . The frequent statement that selection is usually weak in natural populations is without merit."

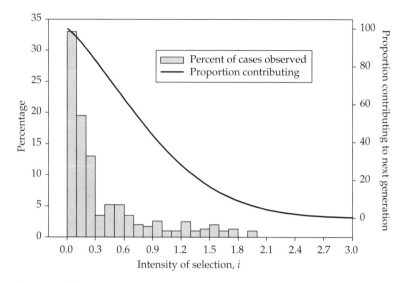

Figure 4.6   The observed distribution of directional selection intensities for 25 species measured in undisturbed habitats. Data from Endler (1986). The solid line shows the proportion of the population that contributes to the next generation as a function of selection intensity.

Estimates in natural populations are typically made over only one to a few generations and it is not likely that levels comparable with artificial selection are generally maintained over tens or hundreds of generations. But, as the example above shows, directional selection that is one or two orders of magnitude lower than used in artificial selection experiments can make large changes in a relatively few generations.

I conclude from the above that there is typically adequate genetic variation for selection to quickly drive trait combinations to their optima. A caveat that must be attached to this is that genetic correlations between traits may act as severe retardants to the attainment of the optima. At the present time we lack sufficient information to say if this is likely to be frequent. (For an example in which genetic correlations between the sexes may severely retard the attainment of equilibrium in sexual size dimorphism, see Reeve and Fairbairn 2001.)

Examples of Rapid Microevolutionary Change

Overview

Figure 4.7 shows the results of estimates of rates of microevolutionary change in various traits in 16 different species. Rates of change are presented here as **Haldanes**, which is the rate measured as number of standard deviations per generation. An alternate measure is the **Darwin**, which is the natural log of the change per time unit in the original units (i.e., 1 Darwin = ln $[x_t/x_1]/t$, where x_t is the trait value in the *t*th year or generation). In the present context, I prefer the Haldane because it permits a more meaningful comparison among species differing in scale and units of measurement, though the two measures are highly correlated (Hendry and Kinnison 1999). In the illustrative example given above, the change in standard deviation units per generation, and hence the number of Haldanes, was 0.00673, which leads to a change of one Haldane in approximately 150 generations. Observed rates in microevolutionary studies range from values comparable to this to rates approaching one Haldane per generation (Figure 4.7). A striking feature of the results is a highly significant negative correlation between the rate of change and the number of generations over which the change was measured. (Nonindependence among traits within studies inflates the degrees of freedom and hence the actual value of the probability should not be taken too seriously; Figure 4.7.) A possible explanation for this decline is that, in general, directional selection moves trait values to the new optimum faster than the time period over which it is measured, and hence the rate is artificially depressed by inclusion of subsequent generations in which no directional change takes place. This would suggest that rates of microevolutionary change are even faster than implied by the present data.

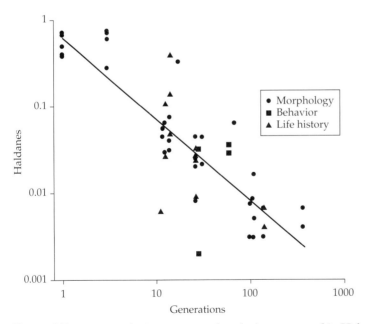

Figure 4.7   Microevolutionary rates of evolution measured in Haldanes (SD per generation) as a function of the number of measured generations and trait type. Where ranges were given, I used the midpoint. The regression equation is $\log_{10}(\text{Haldanes}) = -0.21 - 0.94 \log_{10}(\text{Generations})$, $r = -0.86$, $P < 0.0001$. Data from Hendry and Kinnison (1999).

Below I present some examples illustrating the potential for rapid response of quantitative traits under directional selection in nature: the evolution of life history traits in Trinidadian guppies, the evolution of a novel migratory route in the blackcap, and morphological evolution in sparrows, Anolis lizards, soapberry bugs, and *Drosophila subobscura*.

The Evolution of Life History Traits in Trinidadian Guppies

The guppy *Poecilia reticulata* is native to mountain streams of northeastern South America. It has been introduced into many tropical islands, Trinidad being one. Guppies breed throughout the year, maturing 2–3 months after birth and producing a brood every 3–4 weeks. In Trinidad sites can be roughly dichotomized as high-predation sites and low-predation sites. The principal predator in the high-predation sites is the pike cichlid, *Crenicichla alta*, which preys upon large, mature fish, whereas in the low-predation sites, *C. alta* is absent and the main predator, if any, is *Rivulus hartii*, approximately 10% of whose diet is small, immature guppies. Laboratory studies have confirmed this size-selective pattern of predation (Mattingly and Butler 1994). Now consider a population of guppies moved from a low-predation site to a high-predation site. In the new site the expected lifespan of a guppy is reduced in a manner that is "stage-specific" in the sense that the pike cichlid preys upon large fish and hence susceptibility to predation can be roughly divided into two stages, none ("small" fish) and heavy ("large" fish). (This rough division can be used for purposes of the present intuitive argument.) Thus there will be selection for the guppies to compress their age-schedule of reproduction into the shorter available average duration. Consequently, relative to sites without pike cichlids, guppies in sites with pike cichlids will reproduce earlier and have a higher effort per breeding attempt. If there is reasonably low variation in growth trajectories, an earlier age at maturity will translate into a smaller size at maturity. Possible changes in the size of newborns will depend upon the nature of the mortality schedule and cannot be predicted from the foregoing verbal argument. (The theory underlying these predictions is presented later.) A comparison of sites with and without pike cichlids showed that guppies in the former sites did indeed mature earlier, matured at a smaller size, had an increased reproductive effort and larger young (Reznick and Endler 1982). Common garden rearings showed that these differences were genetically based (Reznick, D. 1982).

Although the above differences are consistent with theory, evolutionary change itself was not observed. To test the prediction that these changes were a result of selection acting through predation pressure, Endler, Reznick, and colleagues did three experiments, in all cases rearing guppies in the lab under common conditions to ensure that differences were indeed due to genetic causes. In the first experiment guppies were taken from 18 localities among 11 streams covering all predator combinations and distributed among artificial streams, one group with *C. alta* (high predation) and the other with *R. hartii* (low predation, Endler 1980). These populations were sampled after 2.5 years, which corresponds to approximately four generations. The populations differed in the same manner as observed in the natural populations (Figure 4.8; age and size at maturity were not measured in this experiment), providing evidence that this variation was a result of predation pressure. Reproductive allocation (= relative clutch mass) differed between sites by approximately 0.9 standard deviations, whereas embryo weight differed by only 0.16 standard deviations. (The results for the separate high-predation treatments were rather disparate.)

In the second experiment Reznick and Endler transplanted guppies from a site containing *C. alta* (high predation) to a site without *C. alta* but with *R. hartii* (low predation) and no natural guppy population. After 2 years (8–10 generations) the populations were sampled: the results were comparable with those from the artificial streams (Figure 4.8). Similar to the selection experiment, reproductive allocation had diverged approximately 1.1 standard deviation units. But embryo weight had diverged by an enormous 5.2 standard deviation units!

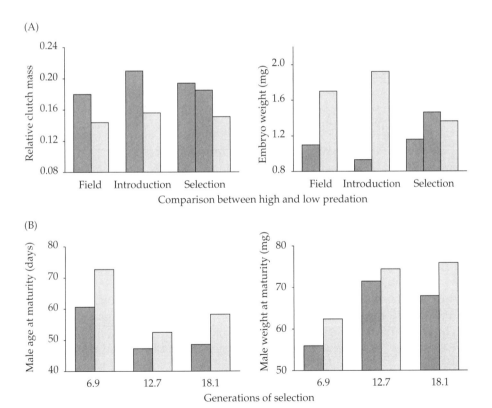

Figure 4.8   Evidence for microevolutionary change in life history and morphology in the Trinidadian guppy. Relative clutch mass = (dry wt of embryos)/(total dry wt). The dark bars indicate values for stream sites of high predation (*Crenichla*), whereas the light grey bars indicate sites of low predation (*Rivulus*). (A, data from Reznick and Endler, 1982 [each experiment in artificial streams—called selection— was run with two gravel types, but the *Rivulus* samples were accidently combined]; B, data from Reznick et al., 1997.)

In the third experiment guppies from *C. alta* (high predation) sites were transplanted upstream above waterfalls into two sites into which *R. hartii* but not *C. alta* could reach (Reznick and Bryga 1987; Reznick et al. 1990, 1997). In one site (El Cedro) the fish were sampled after 4 and 7.5 years and in the other site (Aripo) after 11 years. Fish were reared in the laboratory for two generations to ensure that differences between sites could be attributed to genetic variation rather than environmental or maternal effects. There were highly significant divergences in age and size at maturity in the males (Figure 4.8). Estimates of heritability were extremely high for the two traits (estimates with El Cedro site first are: age at maturity: 0.59 ± 0.02, 0.89 ± 0.02; size at maturity: 0.998 ± 0.01, 0.88 ± 0.02) as were the genetic correlations between the two traits (0.52 ± 0.08, 0.91 ± 0.02). Similar differences in age and size at maturity of female guppies was observed but only after 10 years (i.e., not in the earliest sample). These results are consistent with very low heritabilities for the female traits. (Only one estimated heritability was significantly different from zero, although the significant evolutionary response indicates that there was additive genetic variance and covariance for these traits.) The rates of response indicate selection intensities that are comparable with levels achieved in artificial selection experiments.

An important point to note is that the trait values for the "control" sites (high-predation sites in the second and third experiments) did not remain constant. This indicates the importance of matching control and treatment in "space" (i.e., the same environmental conditions except for the treatment effect) and time.

The Evolution of a Novel Migration Route in the Blackcap

The blackcap *Sylvia atricapilla* is a widely distributed warbler, ranging across much of Europe. Most populations migrate to overwintering grounds in Africa, though a few show either partial migration or no migration (Figure 4.9). Until the 1950s the species rarely overwintered in Britain (Stafford 1956) but the number of recorded overwintering individuals has risen dramatically since the 1970s (1978–79, 2000 recorded; 1981–84, 3000 estimated to be overwintering; Berthold 1995). These birds are not individuals that have taken up residence in Britain but are birds from Germany and Austria that are migrating in an entirely novel direction and for a shorter distance than is typical of the southwestern flying migrants from these countries (Langslow 1979; Berthold 1995). Hand-rearing of birds has shown that migratory orientation is not learned (Helbig, 1992), and hence the change in migration pattern represents a remarkably fast evolutionary change.

The advantages of overwintering in Britain are (Berthold 1995): (1) lower intraspecific competition, (2) shorter migration distance, and (3) earlier gonadal development and return migration. (This latter factor is a physiological response due to differences in photoperiod: since we can assume that this response is under genetic control it cannot be assumed to be adaptive unless conditions in Europe now favor such a change.) Other advantages are (4) earlier occupation of territories in spring, and (5) physiological acclimatization due to potentially harsh conditions initially experienced on the breeding grounds. These advan-

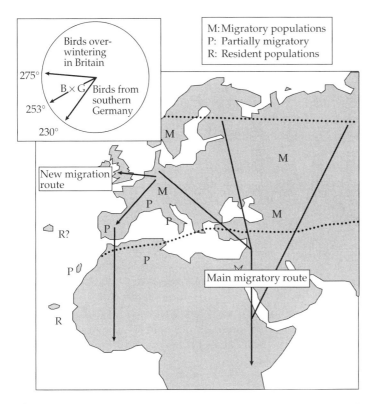

Figure 4.9   The migration pattern of the blackcap *Sylvia atricapilla*. The dotted lines demark the area over which the blackcap spends the summer but migrates south or southwest to overwinter. In southwest Europe there is partial migration (P), and on several islands the birds are resident year round (R). In recent years a new migration route northwest has evolved. Evidence for a genetic basis to orientation comes from crosses between birds from different zones (inset figure). Modified from Berthold (1988, 1995) and Berthold et al. (1992).

tages were presumably offset by the costs before 1960; the apparent reason why blackcaps can now overwinter successfully in Britain is climatic amelioration and the presence of a large number of bird feeders, on which the blackcaps rely almost entirely (Berthold 1995). Droughts in the Sahel zone may have also selected against birds overwintering in Africa (Sutherland 1988).

The rapid evolution of the new migratory pattern requires high additive genetic variance for migratory orientation and migration distance. The offspring of birds collected on the overwintering grounds in Britain orient westwards, the offspring of birds collected in Germany orient southwestwards, while the offspring of crosses between "British" and "German" birds orient in a direction almost midway between these two directions (Figure 4.9). Although the "British" birds probably originate in Germany or regions thereabouts, the population of birds that migrate to Britain is very small relative to that in which birds follow the traditional route. These data suggest a high heritability for orientation, but direct estimation through pedigree analysis is clearly required. The evidence for high additive genetic variation for migration distance is very strong. First, comparison among *Sylvia* species indicates that migratory restlessness as measured in a laboratory assay coincides with migratory distance (Berthold 1973). The presence of varying degrees of migration among blackcap populations indicates that migratory propensity can evolve, and this is supported by intraspecific crosses between partially migrant and fully migrant populations (Berthold and Querner 1981). More definitive evidence comes from selection for migration propensity (Berthold 1988) and offspring-parent regression (Berthold and Pulido 1994), which gives a heritability of 0.45 (see Figure 1.2). A similar heritability (0.52) was obtained for the European robin (Biebach, 1983). With such high heritabilities, migratory activity can be halved in approximately 10 generations with an 80% selection rate (Berthold and Pulido 1994). Further, there is a high genetic correlation between migratory activity and migratory incidence (Pulido et al. 1996). The experimental data are thus in accord with the rapid evolutionary response. Actually predicting the time course of the change will prove very difficult, given the number of selective factors that must be measured (that is, the relative success of the different migration patterns). Nevertheless, the work of Berthold and his colleagues has demonstrated that the rapid evolutionary change is explicable on the basis of quantitative genetic variation.

Rapid Evolution in Morphological Traits: House Sparrows, Anolis *Lizards, Soapberry Bugs, and* Drosophila subobscura

HOUSE SPARROWS. The house sparrow *Passer domesticus* was introduced into North America in Greenwood Cemetery in 1853 and a few years later in Portland, Boston, New York, Philadelphia, and Quebec and by 1879 they were introduced into Hawaii (Phillips 1928). They have expanded throughout North America and are common residents of human cities and towns. Given the wide geographic spread of this species, we might expect to find evolutionary adaptation to local environmental conditions. By 1947 there had occurred significant temporal changes in body size and there existed geographic variation (Calhoun 1947). This was confirmed more rigorously by Johnston and Selander (1964; see also Packard 1967) who showed that populations differed in both color and body size, the latter correlated with climatic conditions. The observed range in body weight of males covered approximately two standard deviations, with a mean ratio of largest to smallest of about 1.1 (means from Johnston and Selander, 1964, and standard deviations estimated from data in Packard 1967). All measurements were based on birds taken in the wild, and hence the degree to which this variation reflects environmental versus genetic effects cannot be judged.

ANOLES. *Anolis* lizards are aboreal reptiles widely distributed in mainland Central and South America and in the West Indies. Comparisons among and within species has shown that hind limb length covarys positively with perch height. It is hypothesized that such variation is a consequence of biomechanical factors (Losos 1994; Losos and Irschick 1996; Irschick et al. 1997). To test the general hypothesis that morphological variation was an evolutionary response to vegetation characteristics, Losos et al. (1997) released lizards onto 11 islands in 1977 and onto another 3 islands in 1981. The introduced lizards were all from a single source island and none of the new islands possessed anoles. Loso et al. (1997) speculated that this lack was a result of occasional catastrophic hurricanes (as described in Spiller et al.1998). In 1991, 14 years after the first introductions, lizards were sampled on all islands including the source island. As predicted, there was significant variation in morphology, which was a function of how different the new island was from the source island. Further, there was a highly significant correlation between hind limb length and perch height. The divergence between source and the new islands was approximately one-half standard deviation (Hendry and Kinnison 1999). As with the house sparrows, all measurements were made on individuals from the natal location; hence environmental and genetic factors cannot be separated.

SOAPBERRY BUGS. The soapberry bug *Jadera haematoloma* is an hemipteran that feeds exclusively on the mature and nearly mature seeds of plants in the family Sapindaceae. The bug is equipped with a long beak that enables it to penetrate the fruit walls to the seeds (Figure 4.10). The insect is found throughout the range of its three native hosts in North America. The native host varies across the range. In the southcentral region the native host is the *drummondii* variety of the soapberry tree (*Sapindus saponaria*), in southern Texas it is the serjania vine (*Serjania brachycarpa*), and in southern Florida it is the perennial balloon vine (*Cardiospermum corindum*). Three plant species belonging to the family Sapindaceae have been introduced into the United States and now serve as host plants for the soapberry bug (Carroll and Boyd 1992). The golden rain tree (*Koelreuteria paniculata*) has been planted as an ornamental throughout the range of the soapberry bug, while in central and northern peninsular Florida the species *K. elegans* has also been introduced as an ornamental. The third introduced host species is the heartseed vine (*Cardiospermum halicacabum*), which is a weed in Louisiana and adjacent parts of Mississippi. The host species differ in the morphology of the fruit in that the distance between the seed and the outer fruit coat varies enormously (Figure 4.10). The fruits of the soapberry tree are ovoid and approximately 1.5 cm by 1.3 cm, whereas the fruits of the balloon vine are spherical and approximately 3 cm in diameter (Carroll and Loye 1987). A consequence of this variation in size is that if there are costs to having a beak that is overly long, the optimal size of beak will differ between regions in which the two host species occur. There is a highly significant difference in beak length, bugs from Florida (ballon vine host) having beaks almost as long as their body (9.78 ± 0.79mm) while bugs from Oklahoma (soapberry tree host) have shorter beaks (7.12 ± 0.42). Florida bugs are also larger than those from Oklahoma, but the allometric relationship between beak length and body length differs, with Florida bugs having longer beaks for a given body length (Carroll and Loye 1987).

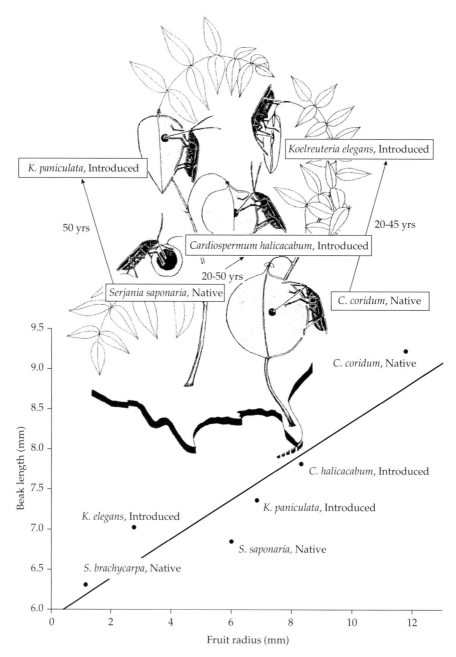

Figure 4.10 The relationship between fruit radius and beak length in the soapberry bug, *Jadera haematoloma*. The drawing above the graph shows the relative size and shape of the native and introduced fruits upon which the soapberry bug feeds. The arrows show the direction of evolutionary change in beak length and the approximate time since introduction of the new plant species in Florida (right) and the south-central United States (left). Data from Carroll and Boyd (1992). Drawing of bugs and seeds modified from Carroll and Boyd (1992).

If the introduced plants represent a significant addition to the food spectrum of the soapberry bug and/or there is little migration among host plants, then we would expect selection to have favored a divergence in beak lengths among populations living on different plants. The soapberry bug is flight dimorphic (i.e., some individuals have wings and are capable of flight whereas others either lack wings and/or the associated structures such as wing muscles and cannot fly; Carroll et al. 1997). Hence some individuals are capable of colonizing new plants, but the actual migratory behavior of the species is unknown. To test the prediction of population divergence, five populations were sampled over the peninsular region of Florida (Figure 4.10), two populations collected on the native host and three collected on the introduced host *K. elegans*, which was introduced mainly since 1960 (Carroll et al. 1998). The fruits of *K. elegans* are smaller than those of the balloon vine. Hence the prediction is that the beak length of bugs on the native host will be longer than those on the introduced host. This prediction was supported both for field-collected and lab-reared bugs (Carroll et al. 1997, 1998), demonstrating that the change was not simply environmentally induced. (Maternal effects are unlikely to account for the difference.) Significant differences among populations also indicate that if migration is occurring it is insufficient to prevent local adaptation.

In a wider set of comparisons, Carroll and Boyd (1992) found a highly significant correlation between beak length and fruit radius (Figure 4.10). Approximately 347 generations have occurred since the introduction of the alternate hosts, and beak length has shown a rate of change of between 0.003 and 0.009 Haldanes, which means a total change from 1.1 to 3.1 standard deviations (Hendry and Kinnison 1999).

DROSOPHILA SUBOBSCURA. *Drosophila subobscura* is a Palearctic species of the Old World. Like many species of both ectotherms and endotherms, it displays clinal variation in body size with size increasing with latitude (Prevosti 1955; McFarquhar and Robertson 1963; Misra and Reeve 1964; Misra 1966). It also shows clinal variation in chromosomal patterns (Larruga et al. 1993). Laboratory studies have shown that there is additive genetic variation for morphological traits (Sondhi 1961; Orengo and Prevosti 1999; Matos et al. 2000). Although it has been demonstrated in *Drosophila* that there is a phenotypic correlation between body size and temperature (Stanley 1941) and that laboratory populations maintained at different temperatures show genetic changes in body size that qualitatively match the clinal variation (Anderson 1966, 1973; Partridge et al. 1994), the reasons for this are still not understood (Partridge and Coyne 1997). Field collections of *D. subobscura* show an inverse but slightly curvilinear relationship with temperature, as do *D. melanogaster* and *D. simulans* (Kari and Huey 2000).

In 1978 a population of *D. subobscura* was discovered in Pueto Montt in southern Chile (South America) and in 1982 another population was found in Port Townsend in Washington State (Pascual et al. 1993). The evidence indicates that both populations are probably derived from the same founder stock and were most likely established around 1978 (Prevosti et al. 1988; Ayala et al. 1989). Since their introduction, the South American flies have spread northwards while the North American flies have spread southwards. Huey et al. (2000) collected flies across their European and North American range and raised them under common garden conditions. Remarkably, the flies from Europe and North America exhibit the same clinal variation in body size, though the components of the wing that contribute to the total length do not show congruence. The evolution of the cline in North America has taken place over one to two decades and the rate of divergence has been approximately 0.22 Haldanes, corresponding to a total change of 2–20 standard deviations.

The above examples demonstrate that in natural populations there is the potential for very rapid evolutionary change. From the perspective of life history analysis this is important because it means that it is not unreasonable to assume that organisms will evolve rapidly to the combination of trait values that maximizes fitness and that analyses can be made assuming that fitness has been maximized. However, it does raise the immediate question of what maintains genetic variation in populations, a topic to which we now turn.

What Processes Can Maintain Additive Genetic Variation In Populations Living In Constant Environments?

Directional selection can quickly change trait values. Observation, however, tells us that trait values are not constantly changing, at least not radically. Trait values appear to typically fluctuate about some long-term average value. This implies either that there is directional selection that changes direction in a haphazard manner over time and/or selection is typically stabilizing or frequency-dependent. A third possibility is disruptive selection but this is theoretically highly unlikely to maintain genetic variation (Bulmer 1971a) and does not appear in many analyses. Heterozygous advantage is a theoretically possible mechanism for maintaining genetic variation, but it is not clear that it would lead to a particular combination of phenotypic trait values. For a discussion of its possible role in maintaining genetic variation see Roff (1997). Frequency-dependent selection is clearly a restrictive circumstance and will be dealt with after considering what possible factors can maintain genetic variation when selection is frequency-independent. Three candidate processes are stabilizing selection, antagonistic pleiotropy and mutation-selection balance. The second in this list has been discussed in detail in the previous chapter.

Can Stabilizing Selection Maintain Genetic Variation in a Population?

DEFINITION AND DETECTION. **Stabilizing selection** is defined as selection in which an intermediate optimum is favored. This optimum can arise because selection acts directly on the focal trait, or because there is balancing selection in which selection on several traits or at different episodes oppose each other to generate apparent stabilizing selection on the focal trait. An example of this type of selection acting on body size in the water strider *Aquarius remigis* is shown in

Figure 2.24. Directional selection on daily fecundity favors an increased size, whereas directional selection on longevity favors a decreased size, the result being that an intermediate optimum body size maximizes lifetime fecundity.

Stabilizing selection occurs when there is selection against extreme phenotypes, which can occur through two processes (Waddington 1957). First, there may be the elimination of genotype combinations at the extremes, and secondly there may be selection for genes that canalize development such that the phenotypic range is reduced. The first process Waddington called **normalizing selection** and the second **canalizing selection**. Most of the theoretical analyses of stabilizing selection have focused on the former process.

Stabilizing selection can be detected by plotting fitness or more typically some component of fitness, against the trait of interest. A simpler technique is to compare the means and variances before and after selection. This method has to be applied carefully since directional selection will also lead to a reduction in variance (Endler 1986). Stabilizing selection has been observed in a wide range of organisms under both domestic and field conditions. Estimates of the percentage change in variance before and after selection shows that in more than 20% of reported cases there is greater than a 35% reduction in variance, a change that indicates strong stabilizing selection (Figure 4.11).

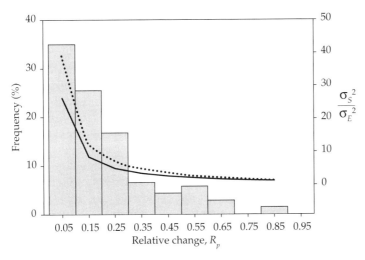

Figure 4.11 The distribution of percent reduction in phenotypic variance resulting from stabilizing selection in nonfossil organisms. Data from Table 7.1 of Endler (1986). Lines show estimates of the ratio σ_S^2 to σ_E^2 (where $\sigma_S^2 = \sigma_E^2 + \gamma$) for $h^2 = 0.25$ (solid line) and $h^2 = 0.50$ (dotted line).

THEORY: INFINITESIMAL MODEL WITHOUT EPISTASIS. Starting with Haldane (1954), stabilizing selection has typically been modeled as a **Gaussian type curve**

$$W(X) = \exp\left[-\frac{(X-\theta)^2}{2\gamma}\right] \tag{4.6}$$

where $W(x)$ is the relative fitness of trait X, θ is the optimal value, and γ is a parameter that measures the strength of the stabilizing selection. (As γ increases the strength of stabilizing selection decreases.) This model is sometimes referred to as **nor-optimal selection**. If we assume that before selection the trait is normally distributed, then from considerations of symmetry it is obvious that the population mean will converge to θ. If the phenotypic distribution is not symmetric, then at equilibrium there will persist a difference between the optimum and θ (Zhivotovsky and Feldman 1992).

For the infinitesimal model, in which there are an infinite number of alleles, the additive genetic variance declines initially, following the recursive equation (Bulmer 1985a, p. 152)

$$\sigma_A^2(t+1) = \tfrac{1}{2}\sigma_A^2(0) + \tfrac{1}{2}\sigma_A^2(t)\left[\frac{\gamma + \sigma_D^2 + \sigma_E^2}{\sigma_A^2(t) + \gamma + \sigma_D^2 + \sigma_E^2}\right] \tag{4.7}$$

and eventually converges to

$$\sigma_A^2 = \tfrac{1}{4}\left\{\sigma_A^2(0) - C + \sqrt{\left[C + \sigma_A^2(0)\right]^2 + 8C\sigma_A^2(0)}\right\} \tag{4.8}$$

where $C = \gamma + \sigma_D^2 + \sigma_E^2$, σ_D^2 is the dominance variance, and σ_E^2 is the environmental variance. The change in additive genetic variance is brought about entirely by the linkage disequilibrium induced by selection. Hence, if selection is relaxed the variance returns to its initial value (Bulmer 1971a; Tallis 1987). In the presence of physical linkage between loci, the change in the variance will be larger, although no exact solutions can be obtained (Bulmer 1985a, pp. 158–160).

THEORY: FINITE MODEL WITHOUT EPISTASIS. In reality, the number of loci will be finite. Hence, while the above model may be a reasonable predictor of changes in variance in the short term, it is unlikely to be adequate in the long term. First consider the simple situation of two alleles per locus, n unlinked loci that contribute equally to the phenotype, no dominance or epistasis. Under these conditions, equilibrium allelic frequencies are unstable and genetic variance will be lost (Bulmer 1985a, p. 167). A critical assumption of this model appears to be that each locus contributes equally to the phenotype; violation of this assumption may lead to stable equilibria (Gale and Kearsey 1968; Kearsey and Gale 1968). Suppose, as shown in Figure 4.12, there are two loci with the contributions at each locus being strictly additive both within loci and between loci but in which the contribution of one

(A)

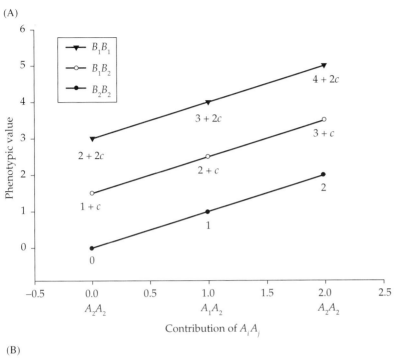

(B)

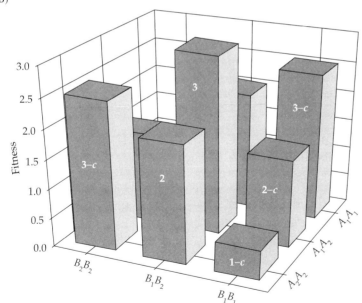

Figure 4.12 (A) Phenotypic values for the two-locus additive model analyzed by Gale and Kearsey (1968). (B) Fitness values of the above genotypes. For display purposes $c = 0.5$. Note the symmetry in fitness values along the different axes.

locus is greater than the other. (In the present case, locus *B* contributes more than locus *A*, Figure 4.12A). Further, assume that fitness is maximal at the double heterozygote and decreases by 1 for a unit change of value on either side of the optimum (Figure 4.12B). In this model, stable equilibria are possible, contingent on the value of recombination; as the recombination rate approaches 0.5 (no linkage), the disparity between loci must increase for equilibrium to be possible. Although it may be argued that this model is simplistic and unrealistic—an argument that might be applied to the majority of theoretical genetic models—it makes the point that the details of the model may be critical for its dynamics and that we must take great care in extrapolating from any particular model to the general case.

Bulmer (1971b) considered the case in which there is dominance, so that $A_1A_1 = a$, $A_1A_2 = da$, $A_2A_2 = -a$. Equilibria at intermediate frequencies are possible only if there is overdominance. Even though the region of stability increases with the value of d, over most of the parameter space no intermediate equilibrium is possible. This conclusion conflicts with that arrived at by Kojima (1959) who analyzed a two-locus model with a quadratic fitness function. (In other words, fitness decreases as the square of the distance from the optimum phenotype.) From his analysis Kojima concluded that equilibria at intermediate allele frequencies are possible even with partial dominance. But, as noted by Lewontin (1964, p. 761), "to maintain successively larger numbers of loci in stable equilibrium, the optimum phenotype must be successively a larger and larger proportion of the extreme phenotype." Extending the analysis of Kojima, Lewontin (1964, p. 764) concluded "that selection based upon squared deviations from an optimum cannot maintain much variance for a character although it may maintain large number of loci segregating. However, when large numbers of loci are segregating each is maintained so close to fixation that random events are sure to reduce the number of segregating loci to a very few where the net selection per locus becomes more substantial." This phenomenon is illustrated by the simulation model of Mani et al. (1990), which is discussed in the section on frequency-dependent selection. Introduction of linkage into this model produces a quasi-stable equilibrium of allele frequencies, which may last for a long period, during which linkage disequilibrium is generated (Lewontin 1964), and further, the linkage increases the region of stability (Singh and Lewontin 1966).

THEORY: FINITE MODEL WITH EPISTASIS. Genetic determination involving only additive or dominance components does not appear to lead to the maintenance of additive genetic variance. Fraser (1960) introduced epistasis and found that, although no stable equilibria appeared possible, the erosion of additive genetic variance was considerably slowed by epistatic interactions. Gimelfarb (1989) has shown that equibria are possible with a very simple model that includes epistasis. His model supposes that there are n loci, each with two alleles, with the genotypic value, X, of an individual being given by

$$X = (1-c)\sum_{i=1}^{n} x_i + c\prod_{i=1}^{n} x_i \tag{4.9}$$

where x_i is the effect of the genotypic value of the ith locus and c is a value between zero and 1. The first term describes a strictly additive component, whereas the second term is strictly multiplicative. If $c = 1$, the first term disappears, and we are left with a multiplicative model that would be additive on a logarithmic scale. A quadratic fitness function was used with the range being set such that fitness was never negative. Analysis of this model showed that, over a wide range of parameter values stable equilibria are possible. Furthermore, when stabilizing selection is relaxed most of the genetic variation appears as additive genetic. (This is also evident from the "heritability" estimated from offspring-on-parent regression in the population undergoing stabilizing selection, although in this case the value is inflated slightly because of nonlinearity introduced by the epistasis.) The appearance of most of the genetic variation as additive genetic variation in this epistatic model is not in itself surprising (see Chapter 2), but what is surprising is that equilibria are so readily obtained. Particularly disconcerting is that the usual genetic analyses would not pick up the epistatic effects and one would conclude that an additive model is appropriate, leading to the paradoxical situation perhaps of observing the maintenance of additive genetic variance in an additive model in the face of stabilizing selection. Such an observation may lead one to postulate some other cause such as mutation being responsible for the variation.

This analysis indicates quite clearly the distinction that should be made between the statistical and functional description of additive and epistatic models. As Gimelfarb (1986, p. 218) notes "It is important, however, to recognize the difference between epistasis as a nonadditive action of loci controlling a character and 'epistasis' measured by the component of the total genotypic variance attributable to the nonadditive action of the loci. The 'epistatic component of variance' depends on a particular distribution of genotypes, and, hence, 'epistasis' measured by such a component does not represent a property of a quantitative character, but rather a characteristic of a particular population. Therefore, the fact that very little variation is attributable to non-additivity in the action of loci does not necessarily mean that the loci do indeed act additively."

OBSERVATIONS AND CONCLUSION. Stabilizing selection experiments are almost universal in showing a reduction in phenotypic variance and a reduction in heritability (see Table 9.5 in Roff, 1997). There has been no satisfactory statistical demonstration of a decline in the environmental variance as predicted by Waddington's model of canalization (Thoday 1959; Prout 1962; Kaufman et al. 1977), but the genetic variance has been shown to decrease in three studies (Prout 1962; Gibson and Bradley 1974; Kaufman et al. 1977), and the general trend for h^2 to decrease indicates that additive genetic variance is eroded (Roff 1997). These results are somewhat confounded because there is also typically a component of directional selection, which itself reduces additive genetic variance.

The theoretical and empirical evidence suggests that both directional and stabilizing selection will generally be an erosive factor. Therefore, what other factors will maintain genetic variation in the face of frequency-independent selection in a constant environment? One phenomenon is antagonistic pleiotropy, considered in detail in the previous chapter. The jury is still out on the importance of this process but the theory and quantitative genetic analyses of dominance variance in life history traits make it a viable hypothesis. The remaining possibility is that of **mutation-selection balance**, namely that the observed genetic variance is a consequence of the balance between selection, which is continually eroding variance, and mutation, which is adding variance back.

Can Mutation-Selection Balance Account for the Observed Additive Genetic Variance in Populations?

One of the greatest difficulties in resolving the question of mutation-selection balance is that the answer depends upon the type of mathematical approximation used (Lande 1975; Turelli 1984), and it is only relatively recently that simulation modeling has been extensively used to address the problem (e.g., Burger et al. 1989; Houle 1989; Foley 1992; Burger and Lande 1994).

THE CONTINUUM-OF-ALLELES MODEL. The **continuum-of-alleles model** assumes that there are n loci, at which mutation can produce an infinite series of alleles, the average effect being zero (i.e., there is symmetry of effects). There is no exact solution for the continuum-of-alleles model, but it can be assumed that the variance due to a new mutation is much smaller than the existing genetic variance at the locus

$$\sigma_A^2 = \sqrt{2n\sigma_m^2\left(\gamma + \sigma_E^2\right)} = \sigma_m \sigma_S \sqrt{2n} \tag{4.10}$$

where σ_m is the average amount of new genetic variance introduced per zygote per generation by mutation, σ_E is the environmental variance, γ is the parameter of the Gaussian stabilizing selection function, and $\sigma_S^2 = \gamma + \sigma_E^2$. The above equation is based on the assumption that parameter values are the same at all loci; if this is not the case, then n must be replaced by n_e, the effective number of loci (see Bulmer 1989, p. 764).

THE HOUSE-OF-CARDS MODEL. Turelli (1984) argued that the variance associated with new mutations should generally overwhelm existing genetic variation at a locus (the **house-of-cards model**). As a consequence, each locus will comprise one common "wild-type" allele in high frequency and rare mutant alleles with large effect (Lande 1995). Correcting the continuum-of-alleles model for this assumption leads to the same prediction as the diallelic model

$$\sigma_A^2 = 4n\mu\sigma_S^2 \tag{4.11}$$

where μ is the mutation rate per locus and with there being effectively two alleles per locus. Burger et al. (1989) extended the analysis to include the effect of finite population size,

$$\sigma_A^2 = \frac{4n\mu\sigma_S^2}{1+\sigma_S^2/N_e\alpha^2} \tag{4.12}$$

where α^2 is the variance of the mutational effect. The factor $n\mu$ is the per trait gametic mutation rate, designated μ_g, and the variance introduced by mutation, σ_m^2, is equal to $2\mu_g\alpha^2$. Equation (4.12) is called the stochastic house-of-cards approximation. Simulation modeling demonstrated that this model is an excellent predictor of the genetic variance when the assumptions of the model are upheld (Burger et al. 1989; Burger and Lande 1994). There are, however, several assumptions that require further investigation. The first is that dominance and epistasis are assumed to be absent. From the data previously presented, absence is not likely to be generally correct for life history traits. Burger et al. (1989) argue that the difference will not be greater than a factor of 2 or 3. The work of Lynch and Hill (1986), earlier theoretical analyses of mutation-selection balance models, and the simulation modeling of Burger (1989; see this paper for references to previous analyses) is used by Burger et al. (1989) to argue than linkage will not dramatically alter their predictions. A far more difficult issue is that of pleiotropic effects. From his analysis of the effect of pleiotropy, Turelli (1985, p. 188–189) concluded, "no simple message emerges from my analysis of mutation-selection balance with hidden pleiotropic effects . . . The difficulties imposed by pleiotropy may well preclude accurate predictions concerning mutation-selection balance for polygenic traits." A similar conclusion was reached using slightly different approaches by Wagner (1989), Barton (1990), Keightley and Hill (1990), and Slatkin and Frank (1990). Nevertheless, it would seem that, despite the analytical difficulties, the effect of pleiotropy will be to decrease the additive genetic variance, though by an amount that cannot be readily determined (Turelli 1985).

From Equation (4.12) the heritability at mutation-selection balance is

$$h^2 = \frac{2\left(\dfrac{\sigma_m^2}{\sigma_E^2}\right)\left(\dfrac{\sigma_S^2}{\sigma_E^2}\right)}{N_e\left(\dfrac{\sigma_m^2}{\sigma_E^2}\right)\left(2\mu_g\right)^{-1}+\left(\dfrac{\sigma_S^2}{\sigma_E^2}\right)+2N_e\left(\dfrac{\sigma_m^2}{\sigma_E^2}\right)\left(\dfrac{\sigma_S^2}{\sigma_E^2}\right)} \tag{4.13}$$

The ratio σ_m^2/σ_E^2 is termed the **mutational heritability** (denoted as h_m^2), because it is approximately the expected heritability of an initially homozygous population following one generation of mutation (Lynch 1988). From a survey of the scant literature available, Turelli (1984) suggested that a typical set of parameter values are $\mu_g = 0.01$, $\sigma_S^2/\sigma_E^2 = 20$, and $h_m^2 = 0.001$, with possible lower values of $\mu_g = 0.002$ and $\sigma_S^2/\sigma_E^2 = 5$. Further analysis by Lynch (1988) and others has shown that the mutational heritability can vary enormously, the range of mean values alone being 0.00002–0.0161, with an observed upper value of 0.0716 (see Table 9.7 in Roff, 1997, and Table 12.1 in Lynch and Walsh, 1998). Data compiled by

Endler (1986) can be used to crudely estimate σ_S^2/σ_E^2 (Roff 1997). The change in phenotypic variance before, σ_P^2, and after, σ_P^{2*}, selection, predicted by the infinitesimal model, is (Bulmer 1985a, p. 151):

$$\sigma_P^{2*} = \sigma_P^2 - \frac{\sigma_P^4}{\sigma_P^2 + \gamma} \tag{4.14}$$

Letting $\sigma_P^2 = c\sigma_E^2$, (hence $c = 1/[1 - h^2]$), and rearranging the above, we obtain

$$\frac{\sigma_S^2}{\sigma_E^2} = 1 + \frac{c}{R_P} = 1 + \frac{1}{R_P\left(1 - h^2\right)} \tag{4.15}$$

where $R_P = |\sigma_P^2 - \sigma_P^{2*}|/\sigma_P^2$ (i.e., the observed relative change in phenotypic variance). The value of σ_P^2/σ_E^2 is influenced by the assumed heritability and R_P (Figure 4.11), and the possible range of values is larger than given by Turelli (1984). The equilibrium heritability maintained at mutation-selection balance is decreased, with a decrease in the value of σ_P^2/σ_E^2 and in this regard it is notable that 14.6% of observed values of R_P suggest a value of σ_P^2/σ_E^2 less than 5 (Figure 4.11). I estimated the equilibrium heritabilities for four combinations of μ_g and σ_P^2/σ_E^2 encompassing the range given by Turelli (1984). For a low value of μ_g (0.002) very little additive genetic variance is maintained at mutation-selection balance, regardless of the value of the other parameters (Figure 4.13). The maximum value of $h^2 = 0.14$ is below that typically observed for morphological traits, though not uncommonly found for life history traits. Heritabilities in the range of 0.4, which is typical for morphological traits, require $\mu_g = 0.01$, $\sigma_P^2/\sigma_E^2 = 20$, $h_m^2 > 0.02$, and $N_e > 100$. Whether such values are typical of organisms in natural populations can only be decided by the gathering of much more data than presently available. Endler's compilation of selection values for natural populations does suggest that σ_P^2/σ_E^2 may be frequently considerably less than 20 (Figure 4.11; Barton, 1990). The gametic mutation rate also appears high if one accepts the estimates of per locus mutation rate of $\mu < 10^{-4}$. In fact, estimates of the gametic mutation rate have been obtained that are an order of magnitude higher than used here (Lynch and Walsh 1998). Three resolutions to this problem have been proposed (reviewed by Turelli, 1984; Lynch and Walsh, 1998): (1) The number of loci controlling a trait is in the range 10^3–10^4, rather than 10–10^2 as indicated by mapping techniques (see Chapter 1 in Roff, 1997 for evidence that present techniques may be underestimating the number of loci); (2) Mutational effects are not uniform, there being many more that produce only minor changes. For evidence of this in *D. melanogaster,* see Lopez and Lopez-Fanjul (1993) and Mackey et al. (1992), and for a general review, (Crow 1993); (3) The observed mutations are the product of unusual genetic events and are not representative of most mutations; (4) The high estimates for the gametic mutation rates are artifacts of the experimental design, and the actual rates are much lower.

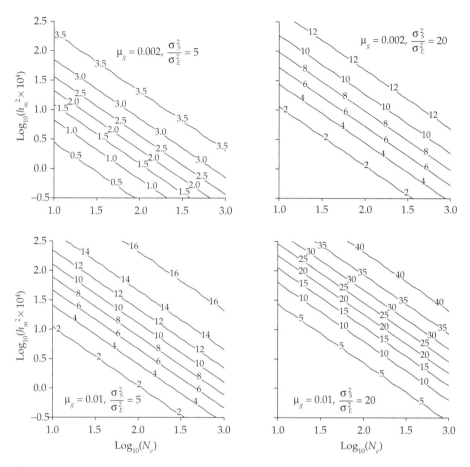

Figure 4.13 Heritabilities ($\times 100$) maintained at mutation-selection balance for a realistic range of parameter values.

CONCLUSION. The question of the amount of observed additive genetic variation that is being maintained by mutation-selection balance remains open, but it is plausible mechanism for life history traits that are under relatively weak selection. Needed are both better estimates on the genetic parameters and also the strength of selection on life history traits in natural populations.

Frequency-Dependent Selection and the Maintenance of Additive Genetic Variance

Frequency-dependent selection is defined as that selection which causes the fitness of a phenotype or genotype to vary with the phenotypic or genotypic composition of the population (Ayala and Campbell 1974; Gromko 1977; DeBenedictis 1978). This type of selection can readily maintain genetic variation at a single locus (Wright 1948; Haldane and Jayakar 1963b; Clarke and O'Donald 1964;

Clarke, 1964; Anderson, 1969). The case for multiple loci is not so clear. If there is neither dominance nor epistasis, and n unlinked diallelic loci, a stable polymorphism is possible (Bulmer, 1985a, p. 169) provided the change in phenotypic variance due to selection is positive and that the frequency-dependent selection for rare phenotypes must be sufficiently strong to counter the tendency of disruptive selection to destabilize the mean (Box 4.1). Slatkin (1979) extended Bulmer's analysis showing that the qualitative results do not depend upon the underlying genetic model, provided that the mean and variance are not constrained.

Using a simulation model, Mani et al. (1990) explored the combined effect of mutation, stabilizing-selection, and frequency-dependent selection on a genetic system in which there are n (≤ 12) loci, each with up to 32 alleles that act additively, the ith allele contributing an amount i to the genotypic value. The envi-

---

Box 4.1
A model in which frequency-dependent selection maintains additive genetic variance

The general theory of frequency-dependent selection can be illustrated by a model of competition between individuals (Bulmer 1974, 1985). The fitness of an individual with a phenotypic value of X is given by

$$W(X) = \left[c_1 - c_2 f(X)\right] e^{-\left[\frac{(X-\theta)^2}{2\gamma}\right]} \tag{4.16}$$

where c_1 and c_2 are positive constants and $f(X)$ is a function describing the effect of competition between individuals; as $f(X)$ increases, the fitness of an individual with phenotypic value X decreases. Bulmer assumed that the effect of competition between two individuals of phenotypes X and X' to be a Gaussian function and that phenotypic values were distributed normally with mean m and variance σ_P^2, giving

$$f(X) = N \left(\frac{\alpha}{\alpha + \sigma_P^2}\right)^{1/2} e^{-\left[\frac{(X-m)^2}{2(\alpha+\sigma_P^2)}\right]} \tag{4.17}$$

The term α determines the strength of the interaction, larger values leading to reduced competition, and N is population size. The second term in Equation (4.15) represents stabilizing selection tending to push the mean value to θ. At equilibrium, $m = \theta$ and $W = 1$. In the above model, selection always forces the population to the mean value, but the genetic variance is not necessarily preserved. Specifically, if stabilizing selection (which reduces variance) is stronger than the effect of competition—a condition that occurs when $\alpha/\gamma > c_1 - 1$—then all of the variance will be eliminated. However, the condition $\alpha/\gamma < c_1 - 1$ does not guarantee the preservation of variance. The actual numerical conditions under which variation is preserved need not be specified here (see Bulmer 1985a, p. 172); we may simply note that they are not excessively restrictive.

ronmental variance was assumed to be zero, and linkage between loci was allowed. As in the Bulmer model, the fitness of an individual with value X was set equal to the product of a frequency-dependent function and a stabilizing-selection function, $W(X) = W_F(f_X)W_S(X)$, where W_S is stabilizing selection, W_F is frequency-dependent selection, and f_X is the frequency of X. The stabilizing selection component was assumed to be Gaussian, and the frequency-dependent function chosen such that changes in parameters altered the shape of the curve from linear, to a concave curve, to a convex curve. The results did not differ significantly for the linked and unlinked case. With the exception of no frequency-dependence ($a = 1$), the number of alleles maintained at equilibrium was independent of the strength of the frequency-dependent selection and only weakly related to the strength of stabilizing selection. Despite beginning with 32 alleles per locus, at equilibrium the number of alleles per locus was only 1.5–3.0. Nevertheless, these simulations show very clearly that frequency-dependent selection can play a major role in the maintenance of genetic variation.

Selection On the Components of the Characteristic Equation

Components of the Characteristic Equation and their Heritabilities

HERITABILITY OF THE AGE OF FIRST REPRODUCTION (α). Recall that in a constant environment in which there is neither density-dependent selection or density-dependent selection does not operate on the trait of interest or in which there is no frequency-dependent selection, then **fitness** can be operationally defined as the rate of increase, r, obtained from the characteristic equation $\int_{\alpha}^{\infty} e^{-rx}l(x)m(x)dx = 1$, where α is the age at first reproduction, $l(x)$ is the age-schedule of mortality, and $m(x)$ is the age-schedule of female births, which I shall simply refer to as the **age-schedule of fecundity**. The age at first reproduction, α, is clearly the simplest component as it consists of a single parameter. The observation of variation in α within and between populations of the same species is evidence that the age of first reproduction can evolve. Estimates of the heritability of the α are relatively uncommon, but components of this trait, such as development time from hatching to maturity or time to metamorphosis, have been studied and provide a reasonable index of the magnitude of genetic variation in α. From the literature I obtained such data for 27 animal species. By using all the data (i.e., not correcting for multiple estimates per species) the mean heritability is 0.36 (SE = 0.03, $n = 112$) and by using species means the mean heritability is 0.38 (SE = 0.04, $n = 27$). The mean heritability is comparable to that of morphological traits (= 0.46, Table 4.2) and indicates that the age at first reproduction can respond readily to selection. The presence of such a high heritability may appear paradoxical if, as we might suppose, the age at first reproduction is likely to be under strong selection. As discussed above, there are mechanisms, most particularly antagonistic pleiotropy that will maintain genetic variation even in the face of strong selection.

HERITABILITY OF AGE-SPECIFIC FECUNDITY PARAMETERS. The age-schedule of fecundity can generally be placed into one of three categories; **decelerating,** **accelerating,** and **triangular.** In the first category I include the limiting case in which fecundity remains constant. This pattern is typical of animals such as mammals and birds and the decelerating curve seems in most cases to approach an asymptote. Examples of species with an asymptotic fecundity curve are some populations of the grass *Poa annua* (Law et al. 1977), lizards (e.g., Tinkle and Ballinger 1972; Tinkle 1973; Vinegar 1975), the Iceland scallop, *Chlamys islandica* (Vahl 1981), and most species of fish. A general heuristic equation describing this type of curve is

$$m(x) = M\left[1 - e^{-a(x-x_0)}\right]^b \tag{4.18}$$

where M is the potential maximum fecundity, a and b are constants, and x_0 is a constant that translates the curve left or right along the age axis. This curve can in many cases be viewed as consisting of two components; a growth curve in which length (generally) is proportional to $[1 - e^{-a(x-x_0)}]$ and in which fecundity is an allometric function of length, $m(x) \propto (length)^b$. The growth curve is generally called the von Bertalanffy function and has been used extensively in fish (Pauley 1978, 1980), reptiles (Shine and Charnov 1992; Shine and Iverson 1995) and mammals (Van Jaarsveld et al. 1995). The parameters of the von Bertalanffy function should not be given biological interpretation (Knight 1968; Roff 1980a) because they are composites of an allocation process between growth and reproduction (Roff 1992a; Day and Taylor 1997). Given that the von Bertalanffy function describes growth trajectories in many species, it is not surprising to find that many models that explicitly include an allocation procedure produce a curve that is of this form (Roff 1983a; Kozlowski 1996). Alternate growth function models, which should equally be viewed heuristically, are the Gompertz, logistic, and Richards (Zach 1988). Accelerating curves are in principle possible but I know of none and consider them unlikely since these models can lead to the optimal age at first reproduction to be delayed forever. However, accelerating curves for $m(\alpha)$ versus α are plausible and, as shown later, can have significant impacts on the evolution of α. Triangular age-specific fecundity functions are common in insects: for example, the beetle, *Pterostichus coerulescens* (van Dijk 1979), milkweed beetles, *Oncopeltus* (Landahl and Root 1969), the cabbage butterfly *Pieris rapae* (Jones et al. 1982) and crickets (Roff 1984a). Other taxa in which a triangular $m(x)$ function is found are nematodes (Woombs and Laybourn-Parry 1984), *Artemia* (Browne 1982; Browne et al. 1984), rotifers (Jennings and Lynch 1928; Bell 1984), small mammals (Krohne 1981) and many species of plants (Harper and White 1974). A useful descriptor of this curve is (McMillan et al. 1970a,b):

$$m(x) = F_{max}\left[1 - e^{-a(x-x_0)}\right]e^{-bx} \tag{4.19}$$

where b is a constant. Notice that the first component of the function is simply the von Bertalanffy function. Three examples of this curve are shown in Figure

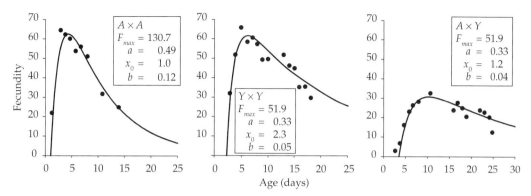

Figure 4.14　Examples of triangular $m(x)$ functions for two different strains of *D. melanogaster* ($A \times A$, $Y \times Y$) and their hybrids ($A \times Y$). Dots show the data; the line shows the fitted function $m(x) = F_{max}(1 - e^{-a(x-\bar{x}_0)})e^{-bx}$. Data from McMillan et al. (1970b).

4.14: In this example there is evidently genetic variation in at least some of the parameters, the hybrid females showing a dramatically different fecundity curve. McMillan et al. (1970b) estimated the parameters for three strains of *D. melanogaster*, their hybrids and a couple of backcrosses giving 9 sets of parameters in all. Simple pairwise correlation coefficient are significant for F_{max} and b ($r = 0.864$, $P = 0.003$, $P = 0.016$ after Bonferroni correction) and for a and x_0 ($r = -0.759$, $P = 0.018$, $P = 0.107$ after Bonferroni correction). Unfortunately, although there is abundant evidence that fecundity is typically heritable, there are no studies of which I am aware that have estimated the heritabilities of the parameters of the $m(x)$ function. There are a number of studies that have measured the heritabilities of parts of the $m(x)$ function, such as "early" or "late" fecundity; the mean heritability of these fecundity components is surprisingly large (using all estimates $\bar{h}^2 = 0.38 \pm 0.05$, $n = 53$; using species means, $\bar{h}^2 = 0.41 \pm 0.06$) and comparable with the age at first reproduction and morphological traits. Such genetic variation suggests that the $m(x)$ function should be readily altered by selection. The extent to which selection can change the shape of the function will depend on the genetic correlation between parameters. The data of McMillan et al. (1970b) and those presented in the previous chapter indicate that such correlations are unlikely to be long-term constraints on the evolution of the age-specific fecundity function.

HERITABILITY OF SURVIVAL, *l*(*x*).　The age-specific survival function, $l(x)$, can be written in a general form as $l(x) = e^{-M(x)}$, where $M(x)$ is some function of age. If mortality is constant with age, $M(x) = $ a constant (say, M), which can be estimated by plotting the logarithm of the numbers alive at time x, $N(x)$, against x. In other words, $N(x) = N(0)e^{-M(x)}$, and hence $\ln[N(t)] = \ln[N(0)] - Mx$. Two alternate extreme scenarios are that mortality is low throughout most of the life, increasing sharply at the older ages, and secondly that there is extremely high early mortality followed by a relatively high survival rate thereafter. Pearl and Miner (1935) termed these Type 1 (concave), Type 2 (linear) and Type 3 (convex) curves. Examples are presented in Figure 4.15. Typical examples of these three types of curves are man (Pearl and Doering 1923), birds (Deevey 1947), and marine invertebrates (Brousseau and Baglivo 1988). However, the classification into three types does not adequately describe the diversity of survivorship curves. To remove this deficiency, Pearl (1940) later added two additional models (low-high-low survivorship, and high-low-high), though there is a general tendency to retain the simple threefold classification. Caughley (1966) argued that, at least for mammals, we know too little about mortality rates to attempt any method of classification, though his analysis and those of Spinage (1972) on African ungulates, Clutton-Brock et al. (1983) on red deer, and Gage and Dyke (1988) on Old World Monkeys, show that, in general, mammals follow the low-high-low pattern. The early stage of any life history is likely to be a period of relatively high mortality, the young being small, underdeveloped, and inexperienced; thus, a low-high survivorship component of the $l(x)$ curve can be generally expected.

   Siler (1979) combined the three types of age schedules of mortality described by Pearl and Miner (1935) into a single general function relating the instantaneous rate of mortality to age,

$$M(x) = a_1 e^{b_1 x} + a_2 + a_3 e^{-b_3 x} \qquad (4.20)$$

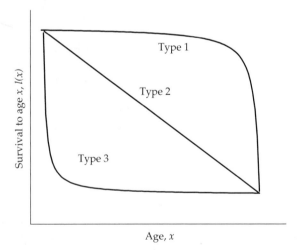

Figure 4.15 A schematic illustration of the threefold classification proposed by Pearl and Miner (1935) for survival curves.

where $M(x)$ is mortality rate, and the three terms, individually produce the three types of survivorship curves. The first term corresponds to an increasing hazard occurring as a result of senescence (Type 1 curve), the second term to a constant hazard, to which the organism cannot adjust (Type 2 curve), and the third term to a hazard that decreases as a result of the organism adapting with age to its environment (Type 3 curve, Siler 1979). There is no particular biological significance to the functions that comprise this survival function—they are simply mathematically convenient. Siler obtained an excellent fit of the model to data in all species examined (6 mammals, 1 bird, and 1 fish species), though frequently the lifespan is too short for the first term (senescence effect) to have an impact. Given the flexibility of the model with respect to variation in shape and five parameters, the satisfactory fit to data is not surprising. It is a very satisfactory descriptor of the survival function, but its utility in life history analysis is limited unless the values of the parameters of the model can be made functions of particular life history events such as the age of maturation. Chen and Watanabe (1989) took a similar approach to Siler (1979) in assuming in fish that the mortality rate is composed of three components (which they termed, "initial," "stable death," and "death by senescence"), but developed the functions on a hypothesized relationship between survival and growth. To the extent that such models allow for trade-offs between traits, they are to be preferred. Other types of curves that have been fit to survival data are the Gompertz curve (Easton 1997; Wilmoth 1997; Haybittle 1998; Ricklefs 1998) and the Weibull function (Fukui et al. 1993; Wilson, D. 1994; Manton 1998; Vanfleteren et al. 1998).

As with the fecundity function, the genetic basis of the components of the survival function have not been specifically investigated. However, heritabilities of survival rates over specified periods, longevity, and other measures of the $l(x)$ have been estimated in a few species and these show that, on average, the heritability of survival components is low (for all species, $\bar{h}^2 = 0.21 \pm 0.03$, $n = 56$; for species means, $\bar{h}^2 = 0.17 \pm 0.05$, $n = 12$). It is the very low heritabilities of survival that account for the overall trend for the heritabilities of life history traits to be lower than those of morphological traits (Roff 2000). I compared the three categories of traits using both all the data and species means alone. The results are qualitatively the same and so I shall discuss only the analysis of the species means. Visually, the distributions of heritabilities of fecundity and age at maturity components are very similar whereas, there is a marked difference in the heritabilities of survival components (Figure 4.16). A oneway ANOVA followed by a Tukey test confirmed this observation ($F_{2,56} = 4.45$, $P = 0.016$; Tukey test, $P = 0.03$ for age vs survival, $P = 0.02$ for fecundity vs survival, $P = 0.92$ for age vs fecundity. A Kruskall-Wallis test gave the same result as the ANOVA).

The Relative Importance of the Age at First Reproduction versus the Effective Age-Schedule of Fecundity

From a mathematical perspective, the product $l(x)m(x)$ can be considered as a single function, which I shall call the **effective age-specific fecundity** function,

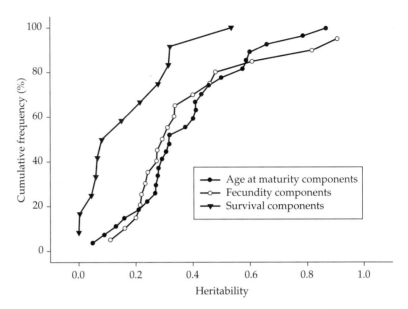

Figure 4.16   Cumulative frequency distributions of species-mean heritabilities for three categories of traits. Data sources: Robertson (1957), Sang and Clayton (1957), Prout (1962), Tantawy and Rakha (1964), Tantaway and El-Helw (1966, 1970), Englert and Bell (1970), Perrins and Jones (1974), Dawson (1975), McLaren (1976), Naedval et al. (1976), Orozco (1976), McLaren and Corkett (1978), Derr (1980), Fairfull et al. (1981), Rose and Charlesworth (1981), Flux and Flux (1982), Hegmann and Dingle (1982), Berven and Gill (1983), Busack and Gall (1983), Findlay and Cooke (1983), Gjerde and Gjedrem (1984), Rinder et al. (1983), Saxton et al. (1984), Via (1984a), Palmer and Dingle (1986), Berven (1987), Travis et al. (1987), Henrich and Travis (1988), Newman (1988), Pashley (1988), Tucic et al. (1988, 1991), Gjerde and Schaeffer (1989), Moller et al. (1989), Tanaka (1989, 1991a,b), Snyder (1991), Blouin (1992), Kasule (1992), Hard et al. (1993b), Miyatake and Yamagishi (1993), Simons and Roff (1994), Ueno (1994), Gu and Barker (1995), Dupont-Nivet et al. (1997), Klingenberg and Spence (1997), Reznick et al. (1997), Roff et al. (1997c), Simons et al. (1998), Kruuk et al. (2000).

denoted as $F(x)$. This function will always be roughly triangular in shape. For example, suppose $m(x)$ is asymptotic; we then have

$$F(x) = F_{\max}\left(1 - e^{-a(x-x_0)}\right)e^{-M(x)} \qquad (4.21)$$

Now since $e^{-M(x)}$ is an exponentially declining function, we have the same shape as the triangular fecundity function shown in Figure 4.14; (if $M(x) = b$, they are identical). Which is more important in determining fitness, the age at first reproduction or the effective age-specific fecundity? This question was considered by Cole (1954), who correctly noted that the effect of changes in parameter values are dependent upon their initial values, but a later analysis by Lewontin (1965) unfortunately generated the conventional wisdom that selection will operate most strongly on the age at first reproduction. Lewontin's frame of reference was specifically a colonizing episode, which, as we shall see, is an important distinction. Lewontin assumed that the $l(x)m(x)$ function can be represented as a triangle and asked: "What changes in a single parameter are required to increase r by some specified amount?" Suppose, for example, r is increased from 0.30 to 0.33; for the particular life history parameters chosen by Lewontin, the age at first reproduction must be decreased by only 10% whereas the total expected fecundity must be increased by almost 100%. From this and other numerical examples, Lewontin concluded that selection should act more strongly on the age at first reproduction than on fecundity. Lewontin's conclusion is valid when the effective fecundity is high, as might be expected in a colonizing organism, but is not true when fecundity is low (MacArthur and Wilson 1967; Meats 1971; Green and Painter 1975; Snell 1978; Caswell and Hastings 1980; Caswell 1982b). This distinction can be most easily seen by considering the approximation for r (Andrewartha and Birch 1954)

$$r = \frac{\ln\left[\int F(x)dx\right]}{T} = \frac{\ln R_0}{T} \tag{4.22}$$

where T is generation time. For the present purposes, I shall equate generation time with age at first reproduction, which would be exactly true if the organism were semelparous. Suppose we double R_0 (effective fecundity) or halve T (age at maturity). r for the two cases is, respectively,

$$r_{2R_0} = \frac{\ln(2R_0)}{T} = \frac{\ln 2 + \ln R_0}{T}$$
$$r_{T/2} = \frac{\ln R_0}{T/2} = \frac{\ln R_0 + \ln R_0}{T} \tag{4.23}$$

The effect of doubling the effective fecundity will be greater than a one-half reduction in the age at first reproduction whenever the effective fecundity is less than 2. This can be generalized to the following statement: "If R_0 is multiplied by a factor c and T divided by an equivalent amount $(1/c)$, the effect of a change in fecundity on r will be greater whenever $R_0 < c$." As the rate of increase approaches zero an increase in fecundity become increasingly more important in its relative effect on r. Thus for a bird such as the Californian condor *Gymnogyps californicus*, which lays only one egg every second year and has an annual rate of increase of only 5%, an increase in fecundity or survival (which would increase the effective fecundity) would be far more significant than an equivalent change in the age at maturity (Mertz 1971).

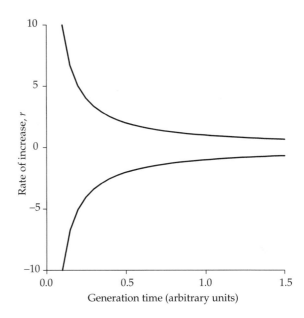

Figure 4.17   The general relationship between the rate of increase, *r*, and generation time.

Mertz (1971) and later Caswell (1982b), in a more rigorous fashion demonstrated that in a declining population the effects of selection will be reversed, favoring a delay in the age at first reproduction. This result can be readily demonstrated by the simple model given above. The relationship between *r* and generation time is hyperbolic (Figure 4.17). When *r* is positive an increase in *T* decreases *r*, but when *r* is less than zero a delay in the age at maturity will increase *r* (Figure 4.17). A continuous decline leads to extinction, and hence selection in a declining population will be of significance only if the decline is temporary.

Iteroparity versus Semelparity

PRELIMINARY OBSERVATIONS.   The extreme form of a life history is semelparity, in which the organism dies immediately after breeding. Examples of this type of life history occur across a wide range of taxa; mayflies (Borror and DeLong 1964), Pacific salmon (Croot and Margolis 1991), European and American eels (Wheeler 1969), octopods (Wodinsky 1977), males of the marsupial genus *Antechinus* (Eisenberg 1988), bamboo (Keeley and Bond 1999), agaves (Howell and Roth 1981), and leeks (Boscher 1981). In at least one species, the colonial ascidian *Botryllus schlosseri*, there are both semelparous and iteroparous morphs in the population (Grosberg 1988). The characteristic equation for the semelparous life history reduces to

$$e^{-r\alpha}l(\alpha)m(\alpha) = 1 \tag{4.24}$$

which after log transformation and rearrangement gives

$$r = \frac{\ln[l(\alpha)m(\alpha)]}{\alpha} \tag{4.25}$$

Semelparity may arise because the amount that must be allocated to reproduction so drains the organism's resources that its death is certain. In some cases the organism may have little choice in the matter. For example, semelparity and iteroparity among various anadromous fish species may, in part, be a consequence of difficulties of migration. This may account for Pacific salmon, *Oncorhynchus* spp, being semelparous, while Atlantic salmon, *Salmo salar*, are iteroparous, though Schaffer (1979b) disputes this reason, suggesting that differences in physiology or phylogenetic constraints are responsible. Evidence in favor of the cost-of-migration hypothesis is that Chinook salmon, *O. tshawytscha*, grown in hatcheries in New Zealand show an incidence of repeat spawners from 5–12% (Unwin et al. 1999). Difficulties of migration may determine not only whether a fish is semelparous or iteroparous but also the degree of iteroparity. For example, the proportion of repeat spawners in sea-run migrant trout, *Salmo trutta*, increases with passable river length and water discharge, which l'Abée-Lund et al. (1989) attributed in part to more effective passive expulsion of spent adults in larger rivers. Further, Leggett and Carscadden (1978) hypothesized that the relative allocation of energy to migration may account for a latitudinal gradient in the proportion of repeat spawners in American shad, *Alosa sapidissima*. The important point to draw from these examples is that semelparity or a significant reduction in potential lifespan may be an unavoidable consequence of reproduction. If an individual female has little or no chance of surviving after the breeding episode, she has no reason to conserve resources. On the other hand, death might be avoidable if the organism devoted less energy to reproduction. Reproduction is likely to increase the likelihood of dying either because the act of reproduction places the organism directly at risk, or because it increases its susceptibility to death in the period subsequent to reproduction (Chapter 3). These arguments suggest that, in general, iteroparous species will allocate fewer resources to reproduction than semelparous species.

COLE'S PARADOX. From his analysis Cole (1954, p. 118) concluded that selection would favor a switch from iteroparity to semelparity when the difference in reproduction was extraordinarily small: "For an annual species, the absolute gain in intrinsic population growth that can be achieved by changing to the perennial reproductive habit would be exactly equivalent to adding one more individual to the average litter size." Thus, an annual (semelparous) species with a clutch size of 101 would increase in numbers as fast as a perennial (iteroparous) that produces 100 young every year forever. There is obviously something amiss with this result, for perennials are common. Although there is good evidence that survival and reproduction are negatively correlated (see previous chapter) it seems highly unlikely that perennial species are committing so much energy into reproduction that they cannot produce one more offspring. The error in Cole's analysis stems from the assumption that there was no age-specific mortality. The importance of mortality was recognized by Gadgil and Bossert (1970) who, however, mistakenly attributed Cole's result to the absence of mortality per se. This error was noted by Bryant (1971) who derived Cole's result with mortality present. A correct solution to the paradox was given by Charnov and Schaffer (1973).

COLE'S PARADOX RESOLVED. Consider two species, one annual and one perennial, both breeding at age 1 and producing m_a and m_p female offspring per year, respectively. The proportion surviving the first year (juvenile survival) is s, and the annual adult survival of the perennial is S. The annual population therefore grows according to

$$N(t+1) = m_a s N(t) \tag{4.26}$$

whereas the perennial increases as

$$N(t+1) = m_p s N(t) + S N(t) \tag{4.27}$$

where the first term represents the number of recruits to the population and the second the surviving adult population. To find the annual rates of increase, we divide throughout by $N(t)$, $(\lambda = N(t+1)/N(t))$. Letting the two rates for annual and perennial be λ_a and λ_p, respectively, we have

$$\begin{aligned} \lambda_a &= m_a s \\ \lambda_p &= m_p s + S \end{aligned} \tag{4.28}$$

The two patterns of reproduction confer equal fitnesses when $\lambda_a = \lambda_p$, which is equivalent to $m_a s = m_p s + S$, from which we obtain

$$m_a = m_p + \frac{S}{s} \tag{4.29}$$

Cole considered the case in which $s = S = 1$, and Bryant, that in which $s = S < 1$. In both cases the result is $m_a = m_p + 1$, Cole's paradox. Survival rates of juveniles are generally lower than those of adults and hence in general S/s will be greater than one. Thus Cole's result applies only to a very special circumstance.

Charlesworth (1980; see also Young 1981) extended the model of Charnov and Schaffer to the more general case of semelparous and iteroparous organisms with an arbitary age at first reproduction. Assuming the same model formulation as above but an age of first reproduction of α, the characteristic equation becomes

$$\sum_{i=0}^{\infty} l(\alpha)m_p e^{-r_p(\alpha+i)} S^i = 1 \tag{4.30}$$

The product $l(\alpha)m_p$ is a constant and the summation $\sum e^{-r_p(\alpha+i)} S^i$ is a geometric series, from which the solution is obtained as

$$\frac{l(\alpha)m_p e^{-r_p\alpha}}{1-Se^{-r_p}} = 1 \tag{4.31}$$

The characteristic equation for the semelparous life history with the same age at maturity and survival to maturity is

$$l(\alpha)m_a e^{-r_a\alpha} = 1 \tag{4.32}$$

The two life histories have equal fitness when $r_a = r_p = r$, which is when

$$\frac{m_p}{m_a} = 1 - Se^{-r} = 1 - \frac{S}{\lambda} \tag{4.33}$$

The ratio of iteroparous to semelparous reproduction decreases with decreasing λ and adult survival (Figure 4.18). High adult survival and a low rate of increase will tend to shift the selective advantage to iteroparity since the reproductive allocation of a semelparous organism must be large to equal that of the iteroparous. For example, when adult survival is 0.1 and the per generation rate of increase is 2, a semelparous organism must produce only 1.05, (1/0.95), times the number of progeny of an iteroparous species with the same population parameters; but when adult survival is 0.9 and the population is stationary ($\lambda = 1$), the semelparous life history requires a productivity 10 times that of the iteroparous to have equal fitness.

If there is density-dependent regulation of population size, then Equation (4.33) has to be modified (Charlesworth 1994, pp. 202–203). If the density dependence acts through fecundy or juvenile survival then the condition becomes

$$\frac{m_{pI}}{m_{aI}} = 1 - S_I \tag{4.34}$$

where I refers to the density-independent component of the parameters. Note that this is equivalent to Equation (4.33) with $\lambda = 1$. If density-dependence acts through age-independent adult survival, the two characteristic equations are

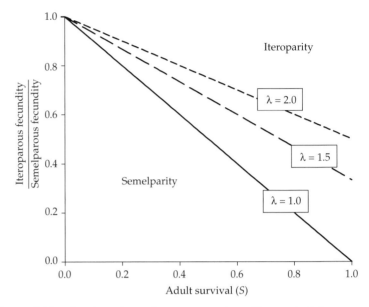

Figure 4.18   Combinations of parameter values (S vs m_p/m_a) that separate the region in which semelparity is the most fit life history trait from the region in which iteroparity is the most fit life history trait.

$$\frac{l(\alpha)_I\, S_D\!\left(\hat{N}\right)^{\alpha} m_p}{1 - S_I S_D\!\left(\hat{N}\right)} = 1$$

$$l(\alpha)_I\, S_D\!\left(\hat{N}\right)^{\alpha} m_a = 1 \tag{4.35}$$

where I signifies the density-independent components as before, D signifies the density-dependent components ans $S_D(\hat{N})$ is the survival over the critical age when density-dependence operates and $\hat{N}$ is the number in the critical age group. The equilibrium conditions between semelparity and iteroparity are then

$$\frac{m_p}{m_a} = 1 - S_I S_D\!\left(\hat{N}\right) \tag{4.36}$$

The qualitative results with density-dependence are the same as with density independence.

EMPIRICAL STUDIES. Females of the spider *Stegodyphus lineatus* almost always die after producing only one brood, the female being consumed by the babies, on average, 14 days after they hatch. If removed from the first brood on day 10 or day 13 post-hatch, the female is capable of producing a second brood, provided she survives (Schneider and Lubin 1997). Two factors select against this iteroparous behavior. First, the young from broods with incomplete maternal care are smaller than those in which the mother is eaten (approximately 3.0, 3.3, 3.7 mm for 10d, 13d, and controls, respectively) and, second, the probability of surviving to the time of dispersal of the babies of the second brood is extremely low (5%). Survival rates were estimated in the field and most likely represent predation on the adults (known in 75% of the deaths). Schneider and Lubin (1997) estimated the reproductive output for the three life histories using total reproductive mass rather than the number of dispersing young: $R = (n_1 m_1 + n_2 m_2 S)/3.7$, where n_i is the number of dispersing spiders in the ith brood, m_i is the average length of the dispersing spider, S is the probability of producing the second brood, and 3.7 is the length of the "standard" young (Table 4.3). This index is reasonable on the argument that the increased size of the young probably lead to increased survival and possibly increased adult size. "Desertion" after 10 days produced the lowest expected reproductive output (9.1 standard young), whereas 13-day-desertion produced 14.4 standard young and the control produced 15 standard young. Schneider and Lubin (1997) used the observed values for the various components, which lead to the mean number of surviving young in the second brood being higher than in the first (Table 4.3). As this is based on few females, it is suspect. While it is possible that survival increased over the season, a perhaps more reasonable assumption, would be that brood size would be the same or less in the second brood. Under this assumption the reproductive output of the iteroparity option would be even less than that of semelparity. These data support the hypothesis that semelparity is favored in this spider by virtue of the low adult survival, but the data are not in a form that can readily be plotted on the surface of Figure 4.18.

From species data it is possible to get a rough idea using Equation (4.34) of the increase in fecundity required for semelparity to be favored. The ratio $1/(1 - S)$, which I shall refer to as R_{min}, is approximately equal to the ratio by which fecundity must be increased in order for the switch from iteroparity to semelparity to be selected. In birds this minimum appears to be at least a twofold increase in

Table 4.3 **Data used by Schneider and Lubin (1997) to compute the expected reproductive output of female *Stegodyphus lineatus* under three life history scenarios.**

	Control	13d removal	10d removal
Number of young in brood1, n_1	15	15	10
Size of young in brood 1, m_1	3.7	3.3	3.0
Number of young in brood 2, n_2	0	19.6	19.6
Size of young in brood 2, m_2	0	3.9	3.9
Reproductive output, $R = (n_1 m_1 + n_2 m_2 0.05)/3.7$	15	14.4	9.1

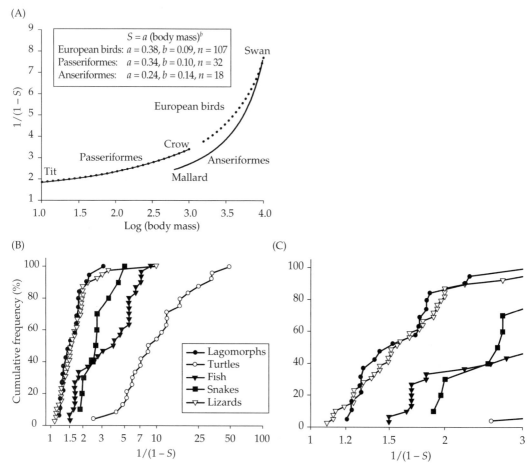

Figure 4.19 Illustrations of the minimum increase in annual fecundity required to favor the evolution of semelparity over iteroparity. The approximate minimum require-ment for the two life histories to be equal is $1/(1-S)$. (A) The minimum ratio as a func-tion of body size in European birds (all species and two separate taxa) using the equa-tions relating body mass and S of Saether (1989). (B) The cumulative frequency plots for lagomorphs (Swihart 1984), lizards and snakes (Shine and Schwarzkopf 1992), turtles (Shine and Iverson 1995) and fish (Roff 1984b). (C) An expansion of the first portion of the x axis in (B).

fecundity (Figure 4.19). It is highly unlikely that a bird could increase its brood size to this extent and still maintain the same fledging success in terms of both fledging weight and survival (see previous chapter). The cumulative distribu-tions of R_{min} for lizards and lagomorphs are very similar, and there are a number of species for which relatively small increases in fecundity (<20%) would favor a switch from iteroparity to semelparity. The minimum ratio for fish is 1.5 and some of these species, such as *Gasterosteus aculeatus* (three-spined stickleback), do approach semelparity in that few survive to breed a second time (Wootton 1984; Poizat et al. 1999). A few snakes but no turtle species have a ratio less than 2 and some turtles have extremely large ratios (Figure 4.19). If these six data sets are representative of animals in general (or at least vertebrates), it is not surprising that semelparity is such a rare life history trait in vertebrates.

The above results provide an indirect test of the evolution of semelparity in vertebrates. Another indirect test is the requirement that the fecundity of semelparous forms be greater than that of iteroparous forms. Again, this can be approached only by using a comparative among-species analysis. There is indeed a highly significant difference between the reproductive efforts (= mass of reproductive parts or seeds/total mass) of annual and perennial plants (Figure 4.20. Because of differences in variances, the two groups were tested using the Mann-Whitney test. For the seed weight data, $U_{33,15} = 16$ $P < 0.0005$. For the data using all reproductive parts, $U_{12,6} = 2.5$, $P = 0.002$).

Calow (1978) obtained support for the hypothesis using semelparous and iteroparous gastropods (Figure 4.20). Calow defined an indirect index of effort, IIE, as the ratio (number of eggs per breeding season × egg volume)/(volume of parent). Though the sample sizes are small (5 semelparous species, 3 iteroparous species), the differences are statistically significant (Mann Whitney U test, U = 15, P < 0.025, one-tailed test). Calow and Woollhead (1977) estimated reproductive effort over a 70-day period for two iteroparous and one semelparous species of freshwater triclads (flatworms). They computed four reproductive effort measures, based on number and energy content of young relative to the parent (Figure 4.20) and provided sufficient data to construct a gonadosomatic index for the three species. All measures of reproductive effort agree that the semelparous species, *Dendrocoelum lacteum*, has a higher reproductive effort than the two iteroparous species. But there is no consistent pattern between the two iteroparous species, perhaps due to the two species having very similar efforts. Because of the small number of species used, the results, though consistent with the hypothesis, must be viewed with caution.

CONCLUSION. The sum of evidence presented above does suggest that iteroparous organisms expend less energy per period of reproduction than semelparous species. An important general assumption of the theoretical analyses presented above is that there is no trade-off between adult survival and fecundity. What generalities can be made given such a trade-off? The most robust statement is that iteroparity will be favored only if the trade-off is concave down (Figure 4.21; Gadgil and Bossert 1970; Schaffer 1974a; Takada 1995). This feature can be shown using the above general equation $\lambda = ms + S$. Replacing adult survival, S, with its trade-off function with m, we have $\lambda = ms + f(m)$. It is reasonable to assume that the trade-off function will generally be linear, convex (concave up) or concave down, as shown in Figure 4.21. On this plane can be plotted the contours of λ. These increase in value from the lower-left to upper-right corners,

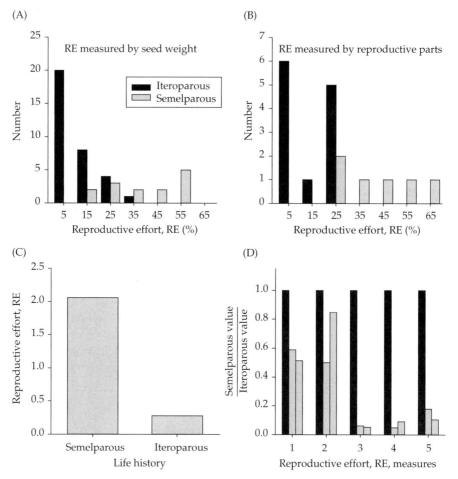

Figure 4.20 Comparisons of reproductive effort (RE) in semelparous and iteroparous species. (A,B) The frequency distributions of reproductive effort in different plants species. Data sources: Harper and Ogden (1970), Hickman (1975), Snell and Burch (1975), Werner (1976), Pitelka (1977), Caswell and Werner (1978), Bostock and Benton (1979), Boscher (1981), Howell and Roth (1981), Pinero et al. (1982), Wilson and Thompson (1989). (C) A comparison of the mean (±1 SE) reproductive effort of five gastropod species with that of 3 iteroparous species. (Where there were several values for a species I used the median.) Data from Calow (1978). (D) Five reproductive effort measures for three species of triclad, one an annual species (*Dendrocoelum lacteum*, black bar) and two perennial species (*Dendrocoelum lugubris*, *Polycelis tenuis*, grey bars). Data from Calow and Woollhead (1977). Reproductive measures: 1, Number of young per parent; 2. Number of young per unit weight; 3. Energy (J) in young per parent; 4. Energy (J) in young per unit weight; 5. weight of young per unit weight.

which can be shown by simply writing λ as a function of the dependent variable S, $S = f(m) = \lambda - ms$. Possible values of λ are those that intersect the trade-off curve. It is visually clear that if the trade-offs are linear or convex the optimal fecundity is at $S = 0$ or $m = 0$. On the other hand, if the curve is concave down,

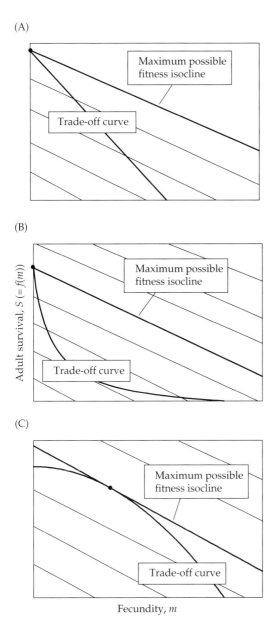

(A)

(B)

(C)

Adult survival, $S\ (=f(m))$

Fecundity, m

Maximum possible fitness isocline

Trade-off curve

Figure 4.21   The optimal combination between fecundity and adult survival for three plausible trade-off functions. (A,B) Selection drives the life history to extreme values (such as semelparity), only one possible combination being shown, whereas iteroparity will generally be the most fit result (C). For the linear trade-off, the direction of selection depends on the slope of the trade-off function. Let this function be $F(m) = a - bm$. Now $\lambda = ms + a - bm = m(s - b) + a$. There is no maximum value $(\partial\lambda/\partial m = s - b)$. When $s - b < 0$, then λ decreases with m, whereas when $s - b > 0$, then λ increases with m. For the convex trade-off there are also two possible end points depending on the points of intersection on the axes.

the optimal fecundity is that at which the fitness contour is tangent to the trade-off curve. (It is possible for this point to be when $S = 0$, but this requires a very specific set of parameter values.) This result applies to both the density-independent and density-dependent case (Takada 1995). Despite the importance of establishing the shape of the trade-off function, I know of no studies that have successfully done so.

The Evolution of the Age and Size At First Reproduction

We have seen from the foregoing analyses that the age of first reproduction can contribute very significantly to fitness. It has thus been a major focus of study in life history analyses. Very frequently size at first reproduction is studied in lieu of age at maturity. There are two reasons for this: first, it is often easier to measure size than age, and second, it is generally assumed that size, age and reproductive success (fecundity or mating success) are related.

Reproductive Success and Size

FECUNDITY AND SIZE. Among ectotherms and plants a common finding is that fecundity increases with (body) size: for example, plants (Primack 1979; Watkinson and White 1985); crustacea (Green 1954; Jensen 1958; Daborn 1975; Bertness 1981b,c; Corey 1981; Rhodes and Holdich 1982); mollusca (Glynn 1970; Green and Hobson 1970; Hughes 1971; Spight and Emlen 1976); annelida (Creaser 1973); echinodermata (Gonor 1972; Rutherford 1973; Menge 1974); arichnida (Peterson 1950; Kessler 1971); insecta (Chiang and Hodson 1950; Tyndale-Biscoe and Hughes 1968; Clifford and Boerger 1974; Waage and Ng 1984; Cook 1988; Marshall 1990); tunicata (Wyatt 1973); fish (Bagenal 1966; Wootton 1979); amphibia (Tilley 1968, 1972; Salthe 1969; Wilbur 1977) and reptilia (Tinkle 1967; Lemen and Voris 1981; Gibbons et al. 1982; Ford and Seigel 1989a,b).

The relationship between clutch size and body size is frequently plotted on an arithmetic scale. Although such a regression may be statistically appropriate if the relationship is linear within the observed range, a better method is to use a log/log transformation. The reason for this is that when the range in body size is large, it is generally found that the relationship is allometric, that is, is best expressed in the form: *Fecundity* $= a(size)^b$, where a and b are constants. This relationship is linearized by taking logarithms, $\log(fecundity) = \log(a) + b\log(size)$. If fecundity varies allometrically, but the range in length (or some other body dimension) is small, there may be no statistical difference between the linear/linear and log/log regressions. However, for purposes of comparison the latter regression is to be preferred as a general rule. Is there any biological reason to expect fecundity to vary allometrically? The simplest explanation is that the amount of space available for eggs increases with size. On such a scenario the value of b should be about 3, the amount of space increasing as the cube of length. This prediction depends on size dimensions increasing isometrically, that is, that there is no change in shape with size, and that egg size remains constant.

The only extensive analysis of the distribution of the exponent b is that on fish by Wootton (1979). The range in values of b is enormous, from about 1 to 7, though this reflects both real and sampling variation, and the true range might be much smaller. The modal class is 3.25–3.75, suggesting that fecundity increases slightly faster than expected according to the geometric argument; but, given the variability in the data it would be premature to draw any definite conclusions. An exponent of 3 need not necessarily imply a limited amount of space for eggs. In many fish species, energy for egg production is stored in organs such as the liver, and the size of this organ might constrain fecundity, rather than actual space for the maturation of the eggs. The rate at which resources can be gathered might be proportional to weight, which would also produce an exponent of 3.

Across mammalian species there is a positive allometric relationship between litter size and body size in endothermic vertebrates (Millar 1981), but within species no such relationship is generally found. Why is this so? The obvious answer is that the range in body size is small, the relative size of offspring large and clutch size small. Large mammals, such as deer, goats, and marine mammals tend to have only one or two young. Considering the size at birth, it would take a relatively enormous increase in body size to accommodate another offspring. Small mammals are frequently quite fecund, but their range in body size and fecundity doesn't match that of ectothermic vertebrates. Nevertheless, a positive correlation between litter size and the mother's body weight has been demonstrated in various cricetine rodents: *Peromyscus longicaudus* (Hamilton 1962), *P. leucopus* (Svendsen 1964; Lackey 1978), *P. maniculatus* (Svendsen 1964; Myers and Master 1983), *P. yucatanicus* (Lackey 1976), *P. polinotus* (Kaufman and Kaufman 1987), *Neotoma floridana* (McClure 1981), *Sigmodon hispidus* (McClenaghan and Gaines 1978), and in the microtine rodents, *Microtus ochrogaster* (Fitch 1957; Keller and Krebs 1970), *M. pennsylvanicus* (Keller and Krebs 1970), *M. townsendii* (Anderson and Boonstra 1979), and *Clethrionomys rufocanus* (Kalela 1957). A positive correlation between body size and litter size is also found in mountain hares, *Lepus timidus* (Iason 1990) and the white-toothed shrew, *Crocidura russula* (Genoud and Perrin 1994). Selection experiments on growth in rats and mice have typically shown that selection for an increased/decreased weight at a given age (or weight gain) leads to a correlated increase/decrease in size at maturity and increase/decrease in litter size but a decreased/increased age at maturity (Table 4.4). These results indicate that selection on weight at a given age is increasing both the threshold size for maturation and the growth rate, an issue that is taken up below.

MATING SUCCESS AND SIZE IN MALES. Reproductive success in females can be measured in part by their fecundity: the equivalent trait in males is their success at inseminating females. As with fecundity, in animals at least there is frequently a positive correlation between size and mating success in males. Though a large size advantage has been observed in mammals and birds (Bryant 1979; Clutton-Brock et al. 1983; Price 1984; Sauer and Slade 1987), much of the detailed investi-

Table 4.4   **Correlated responses in females to selection on weight at a given age or weight gain in rats and mice.**

Weight at maturity	Age[a] at maturity	Litter size	Longevity	Number of studies
Mice, selected for increased weight or weight gain				
Increased	Decreased	Increased	—	1
Increased	—	Increased	—	5
Increased	Decreased	—	—	3
Increased	Decreased	—	Decreased	1
Increased	—	No change	—	1
Rats, selected for increased weight or weight gain				
Increased	Increased	Increased	—	1
Increased	Decreased	ncreased	—	1
Mice, selected for decreased weight or weight gain				
Decreased	Increased	—	—	2
Decreased	—	Decreased	—	2
Decreased	—	—	Increased	1
Rats, selected for decreased weight or weight gain				
Decreased	No change	Decreased	—	1

Data from review by Millar and Hickling (1991).

[a]Included are studies showing an increased weight at an age later than the point of selection but less than the age at maturity.

Blank cells indicate that the trait was not measured.

gations have dealt with ectotherms, where numerous studies have shown that large males are able to secure more matings. (See Table 5.2 in Roff 1992a for a list of over 60 examples.) A small male advantage has been detected in two species of pyralid moths (Marshall 1988), but this appears to be unusual. However, a common finding is that mating is assortative with respect to body size (reviewed in Ridley 1983; Fairbairn 1988; Crespi 1989). Given the greater fecundity of large females, size-assortative mating should differentially increase the fitness of large males. The advantage of large size in males does not enter into the characteristic equation and hence this approach to analysis is not appropriate unless one can equate matings with fecundity. In many if not most cases, male size-selective advantages will be frequency-dependent and hence the optimal body size can be found using the game theoretic approach (Maynard Smith and Brown 1986).

THE COSTS OF GETTING BIGGER.   The above observations indicate that increased body size is favored because it increases reproductive success. For an organism to achieve a larger body size, the assumption most commonly made is that it must delay maturity and continue to grow, which will itself reduce r. Thus, under this scenario, although reproductive success increases with the age at first reproduction, there may not necessarily be a concomitant increase in fitness. The interaction of these factors will produce an optimal age/size at first reproduction.

An alternative assumption to increased development time is that the organism grows faster. An increased body size can be achieved either by an increased development time or an increased rate of growth. In the latter case there may be no resulting correlation between development time and adult body size and hence no positive correlation between reproductive success and age at first reproduction. An increased growth rate could be selected against because of a trade-off between growth rate and mortality rate (Roff 1992a; Arendt 1997). For example, an organism that grows fast is likely to have to spend more time foraging, and this may increase the risk of predation. Alternatively a fast growing organism may require a larger maintenance ration and hence be susceptible to starvation. For any given individual, only two of the three traits, adult size, age at first reproduction, α (which I shall equate with development time) and growth rate, are independent. Suppose, for example, that growth is linear on some scale,

$$A = c + G\alpha \qquad (4.37)$$

where G is growth rate, A is size at maturity (= adult size) and c is the initial size. I shall assume c is very small relative to adult size and hence can be ignored (i.e., $c = 0$). In a variety of animal and plant species adult size appears to be determined by a size threshold for maturation, in which case development time would be the dependent variable and the growth relationship is best represented by the equation

$$\alpha = \frac{A}{G} \qquad (4.38)$$

By taking logs, this equation can be converted into the simple additive model $\ln(\alpha) = \ln(A) - \ln(G)$. For simplicity I shall consider the transformation to be understood and write this equation as $\alpha = A - G$. Assuming that these traits are normally distributed on the log scale, then the variance of α, $\sigma^2_{(\alpha)}$, can be written as a function of the variances of growth rate and adult size and their covariance:

$$\sigma^2_{(\alpha)} = \sigma^2_{(A)} + \sigma^2_{(G)} - 2\sigma_{(AG)} \qquad (4.39)$$

As discussed above, a typical assumption in life history analysis is that an increased body size will require an increased development time. The argument for this rests on the implicit assumption that individuals are all genetically programmed to follow the same growth trajectory. It follows from this that the genetic correlation between adult size and α is 1. Because of data limitations, in testing this prediction I used primarily development time, assuming that this is a reasonable surrogate for age at first reproduction (Roff 2000). There are 15 estimated genetic correlations that are significantly different from zero; these represent 60% of the cases listed in Table 4.5, indicating that a genetic correlation between development time and adult size is a common phenomenon. There are 17 positive and 6 negative genetic correlations (the two zeros are ignored), which is significantly different from the null hypothesis of a 1:1 ratio ($\chi^2 = 5.26$, df = 1, P = 0.02). Thus we can conclude that there is typically a positive genetic correlation between development time and size at maturity. The negative genetic correlations cannot be dismissed as due to small sample size, because in three cases they are significantly different from zero (Table 4.5). In general, the genetic correlation is significantly less than 1 (11 significantly less than 1 and 5 with 95% confidence limits that include 1; Table 4.5). These results indicate that the assumption of a single growth trajectory can be rejected and that, in general, there will be genetic variation in growth rate. Given genetic variation in adult size and growth rate, the heritabilities and genetic correlations between traits can be derived as shown in Box 4.2.

The analysis outlined in Box 4.2 shows that the genetic correlation between α and adult size can be either positive or negative, depending upon the size of the component variances. How will this impact upon the genetic correlation between fecundity and α? In Box 4.3 I show that even with a very simple model the genetic correlation between fecundity and α can be either negative or positive even if both the genetic correlation between adult size and fecundity and the genetic correlation between α and adult size are positive.

The equations derived in Box 4.3 show that it is possible to have a positive genetic correlation between fecundity and adult size, a positive genetic correlation between adult size and age of first reproduction, but a negative genetic correlation between fecundity and age of first reproduction, or any other combination for the last two. Further, it is possible to have a negative genetic correlation between adult size and age of first reproduction but a positive genetic correlation between fecundity and age at first reproduction. Predictions of the sign of the genetic correlation between age of first reproduction and fecundity (and hence the existence of the generally presumed trade-off) cannot be made on the basis of the signs of the genetic correlation between these and other traits, such as adult body size.

Despite the importance of the genetic correlation between age at first reproduction and fecundity, I could find data for only five species and the standard errors on the estimates were so large that no conclusions could be drawn (Roff 2000). The important point to make is that a positive relationship between age at first reproduction and fecundity cannot be assumed even when the regressions with body size suggest that it is so. Nevertheless, the relative simplicity of analyses under the assumption of a fixed growth trajectory make it an attractive first model. Although obtaining a satisfactory fit between predicted and observed age or size at first reproduction does not demonstrate that the assumption is correct it at least points in that direction.

Table 4.5   Heritabilities of development time (D_{time}), taken as an index of age to first reproduction, and size at maturity (A_{size}) and the genetic correlation between them in various species, arranged in ascending order of the genetic correlation.

Species	Heritability		Correlations			Source(s)
	D_{time}	A_{size}	r_p	r_a	$SE(r_a)$	
Harmonia axyridis	0.31	0.45	−0.34	−0.52	ng	Ueno (1994)
Gerris buenoi	0.29	0.31	−0.29	−0.46	0.14	Klingenberg and Spence (1997)
Sitophilus oryzae				−0.24	0.35	Holloway et al. (1990)
Gasterosteus aculeatus	0.46	0.31		−0.22	0.41	Snyder (1991)
Tribolium casteneum[a]			−0.05	−0.17	0.04	Englert and Bell (1969)
Gryllus pennsylvanicus	0.31	0.60	−0.08	−0.14	Sig	Simons and Roff (1994, 1996)
Helix aspersa	0.40	0.48	0.05	0.003	0.07	Dupont-Nivet et al. (1997)
Salmo gairdneri	0.21	0.29	0.21	0.13	0.35	Gjerde and Gjedrem (1984)
Scathophaga stercoraria			0.06	0.20	0.07	Blanckenhorn (1998)
Oncopeltus fasciatus	0.28	0.61	−0.19	0.24	0.27	Palmer and Dingle (1986)
Tribolium casteneum[b]			0.03	0.28	0.13	Englert and Bell (1969)
Liriomyza sativae	0.16	0.45	0.15	0.37	Sig?	Via (1984a,b)
Salmo salar	0.39	0.35	0.31	0.40	0.15	Gjerde and Gjedrem (1984)
Drosophila melanogaster	0.25	0.30	0.25	0.45	0.06	Tantawy and El-Helw (1970); Robertson, F.W. (1960a,b,1963); Nunney (1996)[c]
Pieris rapae	0.29	0.90	−0.18	0.51	Sig	Tanaka (1989, 1991a)
Brassica campestris	0.68	0.33	0.38	0.61	Sig	Dorn and Mitchell-Olds (1991)
Spodoptera frugiperda	0.81	0.36	0.51	0.68	Sig	Pashley (1988)
Drosophila aldrichi	0.37	0.09	0.24	0.80	0.14	Gu and Barker (1995)
Drosophila buzzatii	0.01	0.52	0.20	0.90	0.34	Gu and Barker (1995); Betran et al. (1998)[d]
Callosobruchus maculatus	0.23	0.55	0.02	1.01	0.22	Moller et al. (1989)
Poecilia reticulata	0.26	0.09		1.11	ns	Reznick et al. (1997)
Results based on correlated response to selection						
Wyeomyia smithii				0		Bradshaw and Holzapfel (1996)
Gryllus firmus	0.36	0.39		0		Roff (1990b, 1998c)
Aedes aegypti				>0		Koella and Offenberg (1999)
Acanthoscelides obtectus	0.09			>0		Tucic et al. (1998)

[a]Realized genetic correlation from selection on development time.
[b]Realized genetic correlation from selection on pupal weight (adult size).
[c]see also Hillesheim and Stearns (1991), Partridge and Fowler (1992), Zwann et al. (1995).
[d]$r_a > 0$ from pleiotropic effect of second-chromosome inversions.

Relationship between Survival and Age and Survival and Size

An increase in the age of first reproduction increases fitness by increasing reproductive success but also has a detrimental effect in an increased α directly

Box 4.2
Heritabilities and genetic correlations between growth parameters

An environmental covariance between growth rate and adult size, $\sigma^2_{E(AG)}$, could arise and be either positive or negative, but it would most likely be positive. Suppose, for simplicity of exposition, that there were no genetic variation in any of the components and that the process of maturation commenced when the organism exceeded some size threshold. If the period of maturation were constant, then fast-growing individuals, if they continued to grow relatively fast after the initiation of maturation, would be relatively large. Therefore, in this case the covariance between growth rate and adult size would be positive. A positive covariance would also ensue if maturation were initiated after a fixed development time. An apparent covariance could arise if the laboratory environment is not entirely homogeneous, because, in general, there will be selection for a reaction norm between growth rate and adult size. The evolution of such a reaction norm is beyond the scope of the present chapter and I shall assume, unless otherwise stated, that $\sigma^2_{E(AG)} > 0$.

The heritability of adult size is

$$h^2_A = \frac{\sigma^2_{A(A)}}{\sigma^2_{P(A)}} \tag{4.40}$$

and from the growth equation, $\alpha = A - G$, the heritability of age at first reproduction is

$$\sigma^2_{A(\alpha)} = \sigma^2_{A(A)} + \sigma^2_{A(G)} - 2\sigma_{A(AG)}$$

$$\sigma^2_{P(\alpha)} = \sigma^2_{P(A)} + \sigma^2_{P(G)} - 2\sigma_{P(AG)}$$

$$h^2_\alpha = \frac{\sigma^2_{A(A)} + \sigma^2_{A(G)} - 2\sigma_{A(AG)}}{\sigma^2_{P(A)} + \sigma^2_{P(G)} - 2\sigma_{P(AG)}} \tag{4.41}$$

There is no general relationship between the two heritabilities, the relative sizes depending on the variance in growth rate and the covariance between growth rate and adult size. The two heritabilities are very similar in magnitude (Roff 2000), which would occur if there were little variation in growth rate. This is at least some slight evidence that most of the variation in α is associated with adult size.

To obtain the genetic covariance between adult size and α we proceed as follows:

$$\sigma^2_{A(G)} = \sigma^2_{A(A)} + \sigma^2_{A(\alpha)} - 2\sigma_{A(A\alpha)} \tag{4.42}$$

Rearranging and substituting the right hand side of Equation (4.39), we get

$$\sigma_{A(A\alpha)} = \sigma^2_{A(A)} - \sigma_{A(AG)} \tag{4.43}$$

The genetic correlation will be negative if $\sigma_{A(AG)} > \sigma^2_{A(G)}$. In other words, under this condition an increased adult size will be accompanied by a decrease in age at first reproduction! Selection will not, however, necessarily drive adult size up because such selection will also increase growth rate ($\sigma_{A(AG)}$ must be positive to be greater than the adult size variance). This growth rate may be directly selected against (such as through increased predation risk because of the required increased foraging). The covariance between growth rate and α, using the above approach, is $\sigma_{A(G\alpha)} = \sigma_{A(GA)} - \sigma_{A(G)}^2$.

decreases r and indirectly decreases r, and R_0 by decreasing the probability of surviving to reproduce. Remember that $l(\alpha)$ must decrease with α, unless the process of maturing itself increases the mortality rate. This latter is feasible if, for example, large females can escape gape-limited predators, whereas small females cannot. Mortality rate may decrease with age because individuals become too large for the majority of predators or it could increase with age because individuals increase their reproductive effort and hence put themselves at greater risk both from predators, disease, and abiotic factors (such as hypothermia in small mammals breeding early in the year). The actual mortality rate is likely to be due to a combination of these factors.

A constant mortality rate implies a linear relationship between $l(x)$ and x. Birds come the closest to generally fitting this pattern (Nice 1937; Deevey 1947; Lack 1943a,b; Davis 1951; Hickey 1952; Gibb 1961; Slobodkin 1966; Bulmer and Perrins 1973) but more recent analyses have shown that there is an age-specific decline in survival, though given typical sample sizes it is difficult to detect (Botkin and Miller 1974; Ricklefs 1998). An age-specific decline in survival is taken by some as an appropriate definition of **senescence** (e.g., Promislow 1991; Tatar 2001). But others regard **senescence** and aging as "a persistent decline in the age-specific fitness components of an organism due to internal physiological deterioration" (Rose 1991, p. 20; see also Kirkwood et al. 1999, p. 219; Partridge 1997). The first definition of senescence does not necessarily include physiological deterioration—an organism may show no physiological decline in function but may suffer and increased age-specific mortality rate because of an increased reproductive effort—whereas it is pivotal in the second. I regard the last as the most appropriate definition and will use senescence in this sense. Under this definition, there are probably few organisms that show senescence in wild populations (Austad 1997). Semelparous organisms are an extreme example of very rapid senescence. In wild populations of iteroparous species, senescence has been reported in elephants, primates, some cetaceans (Carey and Gruenfelder 1997; Austad 1997) and the Virginia opossum (Austad 1993).

SURVIVAL AND SIZE. There is abundant evidence that mortality rates are-age specific, but it is unlikely that in most cases age per se is the important factor; it is far more likely that experience, developmental stage, size, or reproductive status are the factors that are functionally related to survival. Experience may play a role in the survival of endothermic vertebrates and perhaps some ectothermic

Box 4.3

Deriving the correlation between fecundity and α and between fecundity and size

In general, fecundity can be related to body size by the allometric relationship $F = aA^b$. The actual fecundity function will often be more complex than this in that it will have an age component (e.g., be triangular). For the present case I shall assume that this component does not change with adult size, so, for example, the full fecundity function might be $F = aA^b f(x)$, where $f(x)$ is an age-dependent but not size-dependent function. Genetic variation in fecundity implies only genetic variation in adult size, A, although there may also be variation in c and/or b. Analysis becomes quite complex if b is variable, because we have the product of two random variables. To keep the analysis simple I shall assume that b is a constant equal to 1. (This does not qualitatively change the results.) I also assume that there is genetic variation in A and a. Taking logs, we have $\log F = \log a + \log A$, which, as before, for simplicity shall be written as $F = a + A$. As before, to fulfill the quantitative genetic assumptions, I shall assume that the traits are normally distributed on the log scale. Hence $\sigma^2_{A(F)} = \sigma^2_{A(A)} + \sigma^2_{A(a)} + 2\sigma^2_{A(Aa)}$ and the heritability of fecundity is

$$h_F^2 = \frac{\sigma^2_{A(a)} + \sigma^2_{A(A)} + 2\sigma_{A(Aa)}}{\sigma^2_{P(a)} + \sigma^2_{P(A)} + 2\sigma_{P(Aa)}} \tag{4.44}$$

A positive environmental covariance might be expected, given that conditions that promote large phenotypic size might also be expected to promote high fecundity. Under constant laboratory conditions, the covariance should be zero. As in the previous model, there are no simple general qualitative relationships between the relative sizes of h_F^2 and h_A^2.

There are abundant phenotypic data showing that fecundity typically increases with body size (see above). Genetic correlations between fecundity and body size range from –0.16 to 1.00 (Table 4.6), although the low correlations are associated with relatively large standard errors. There are 10 positive genetic correlations and 1 negative, which is significantly different from the null hypothesis of no association ($P = 0.006$, one-sided binomial test). Thus, overall, there is strong evidence for a positive genetic correlation between body size and fecundity. Consequently, I shall assume that fecundity and adult size are positively genetically correlated. There seems no *a priori* reason to suppose that a and F will be genetically correlated, and so I shall assume that $\sigma^2_{A(aF)} = 0$.

From the above, $A = \alpha - G$ and thus $F = a + \alpha - G$. Rearranging gives $G = a + \alpha - F$. Variance relationships are then

$$\sigma^2_{A(G)} = \sigma^2_{A(a)} + \sigma^2_{A(\alpha)} - 2\sigma_{A(F\alpha)} \tag{4.45}$$

If there is no additive genetic variance in growth rate ($\sigma^2_{A(G)} = 0$) then from the previous equation we have

$$\sigma_{A(F\alpha)} = \tfrac{1}{2}\left(\sigma^2_{A(a)} + \sigma^2_{A(A)} + \sigma^2_{A(F)}\right) \tag{4.46}$$

Table 4.6   Heritabilities of adult size (A_{size}) and fecundity (F_{eggs}) and the genetic correlation between them in 11 species, arranged in ascending order of the genetic correlation.

Species	F_{eggs}	A_{size}	r_p	r_a	$SE(r_a)$	Source(s)
			Correlations			
Pieris rapae	1.27	0.90	0.02	–0.16	0.32	Tanaka (1991a)
Drosophila simulans	0.11	0.23	0.24	0.01		Tantawy and Rakha (1964)
Drosophila melanogaster	0.07	0.26	0.35	0.05	0.11	Robertson (1957); Martin and Bell (1960); Tantawy and Rakha (1964); Tantawy and El-Helw (1966)
Oncopeltus fasciatus	0.50	0.52	0.15	0.24	0.22	Palmer and Dingle (1986)
Gryllus firmus	0.45	0.36	0.35	0.40	Sig	Roff (1995)
Gryllus pennsylvanicus	0.55	0.65	0.24	0.52	Sig	Simons and Roff (1994, 1996)
Salmo gairdneri	0.31	0.31	0.54	0.54		Gjerde and Schaeffer (1989)
Callosobruchus maculatus	0.10	0.55	0.61	0.68	0.26	Moller et al. (1989)
Dysdercus fasciatus	0.82	0.35	0.35	0.75		Kasule (1992)
Geranium carolinianum				0.91		Roach (1986)
Gasterosteus aculeatus	0.22	0.31		1.00		Snyder (1991)

that is, we have a positive genetic correlation between fecundity and age at first reproduction. However, if there is additive genetic variance in growth rate, then

$$\sigma_{A(F\alpha)} = \tfrac{1}{2}\left(\sigma^2_{A(a)} + \sigma^2_{A(F)} + \sigma^2_{A(A)} - 2\sigma_{A(AG)}\right)$$

$$= \tfrac{1}{2}\left(\sigma^2_{A(a)} + \sigma^2_{A(F)} - \sigma^2_{A(A)} - 2\sigma_{A(A\alpha)}\right) \tag{4.47}$$

Now the covariance between fecundity and age at first reproduction could be negative. This can occur even if the covariance between adult size and age at first reproduction is positive (recall from previously that $\sigma^2_{A(A\alpha)} = \sigma^2_{A(A)} - \sigma^2_{A(AG)}$).

vertebrates, but for most organisms (invertebrates and plants) experience will be of no consequence. Most animals are generally undeveloped at birth and hence relatively unable to flee from predators or move to alternate habitats if necessary. This ontogenetic factor undoubtedly is an important component of the high mortality rates experienced by most organisms during the early phase of their lives. Another factor is size; most predators take prey only within a particular size range and hence the growing organism must pass through a gauntlet of predators. An increase in mortality associated with reproduction may also be important.

Age-related changes in survival may be a consequence of increased experience and development, but size itself can be a significant component of mortality rates. Inter-specifically, there is a strong correlation between body size and survival in mammals (Sacher 1959; Millar and Zammuto 1983; Western 1979; Ohsumi 1979; Jones 1985; Promislow and Harvey 1990; Promislow 1991), birds (Lindstedt and Calder 1976, 1981; Calder, 1983; Saether 1989; Dobson 1990), reptiles (Calder 1976) and fish (Beverton and Holt 1959; Beverton 1963; Adams 1980; Pauly 1980), though the mechanism(s) generating these relationships are far from clear. A common feature of all the fitted relationships is that they are allometric, taking the form $y = ax^b$, where y is survival rate or an index of it, x is body size, and a, b are fitted constants. Assuming that predation is the primary mortality factor and that the weight of prey is a constant fraction of the weight of the predator, Peterson and Wroblewski (1984) developed a model of this form for the mortality of fish within the pelagic ecosystem. McGurk (1986) later empirically showed this model to be applicable to invertebrates and marine mammals (whales). The constant of proportionality, a, depends upon a number of metabolic parameters, and the exponent, b, is equal to the exponent in the allometric relationship between growth and metabolic rate. Empirical estimates of these parameters for pelagic fish lead to the equation $y = 1.92x^{-0.25}$, where y is the instantaneous mortality rate, the time scale is per year and weight is measured in grams. Over 16 orders of magnitude the equation of Peterson and Wroblewski predicts the observed mortality rates with reasonable accuracy (Figure 4.22), although the scatter covers two orders of magnitude and hence only a general pattern of size-related mortality can really be inferred. Allometric relationships between survival and body size are recorded for mammals, birds, and reptiles, the exponent ranging from –0.15 to 0.26, with a mean of 0.16. Whether there exists a general allometric relationship between size and survival among terrestrial animals comparable to that found by Peterson and Wroblewski remains to be determined, though the data are not encouraging. It should be noted that the relationships for terrestrial animals cover a much smaller range in values and this certainly accounts in part for the variety of regressions.

EXAMPLES OF SIZE-SPECIFIC SURVIVAL RATES WITHIN SPECIES. The foregoing analyses all show that inter-specific mortality rates decline with size. Studies within species also support the hypothesis that large size can be a protection against predators and environmental stress. Within the subtidal region, large size can be a refuge from predators (Connell 1970; Menge 1973; Vermeij 1974; Miller and Carefoot 1989; Palmer 1990). If sufficiently large, barnacles and mussels can withstand attacks from their principal predators, starfish (*Pisaster*) and gastropods (*Thais*), (Connell 1970, 1972; Dayton 1971; Paine 1976). Species within the bivalve genera, *Venus, Cerastoderma,* and *Modiolus* can also achieve a size refuge (Ansell 1960; Seed and Browne 1978). Because it swallows its prey whole, the opisthobranch, *Navanax inermis,* has limited prey size; large size in potential prey can make them invulnerable to this predator (Paine 1965). Similarly, shell-crushing crabs cannot crush the shells of larger hermit crabs (Bertness 1981a).

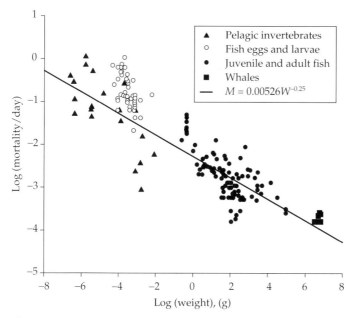

Figure 4.22 Mortality, *M*, versus dry weight, *W*, for marine organisms. The solid line shows the relationship predicted by the model of Peterson and Wroblewski (1984). Data from McGurk (1986).

Size selective predation and a size refuge have been demonstrated in a number of studies of larval amphibia. Invertebrate predation is a major source of mortality in larval amphibia (Travis 1983a; Werner 1986) and may frequently be size-selective (Pritchard 1965; Travis 1983b). Size-selective predation has been demonstrated in dragonfly nymphs (Calef 1973; Caldwell et al. 1980; Brodie and Formanowicz 1983), the backswimmer, *Notonecta undulata* (Licht 1974), and larvae of the predacious diving beetle *Dytiscus verticalis* (Brodie and Formanowicz 1983). Vertebrate predators may also be important, and in these, limitations of gape can restrict the size of prey. A size refuge from salamander predation has been demonstrated for the larvae of *Rana aurora* (Calef 1973), *Hyla gratiosa* (Caldwell et al. 1980), *Rana clamitans, Rana catesbiana* (Brodie and Formanowicz 1983), and *Ambystoma maculatum* (Stenhouse et al. 1983). The importance of size in the survival of amphibia can extend beyond the larval period, postmetamorphic survival in the wood frog, *Rana sylvatica* being dependent on size but not age (Berven and Gill 1983).

The spines of sticklebacks are generally acknowledged to be morphological structures that have a defensive role, increasing manipulation time, and hence the probability of escape (Werner 1974; Zaret 1980; Hoyle and Keast 1987). An analysis of scars resulting from unsuccessful attacks on sticklebacks showed that injuries were absent among juvenile fish (<50 mm), rare among subadult fish

(50–70 mm), and common among adults, the frequency of scars increasing with body size (Reimchen 1988). From the pattern of scarring, Reimchen concluded that at least one-third of the unsuccessful attacks were from avian piscivores. Further, since sticklebacks of all sizes are preyed upon, the pattern of scarring reflects an increasing likelihood of surviving an attack with body size (Reimchen 1988).

Perhaps one of the most well-known (certainly most reanalyzed) cases of size selective mortality is that of Bumpus's sparrows. Bumpus (1899) collected the bodies of house sparrows, *Passer domesticus*, killed after a severe winter storm and compared these with the size of birds that survived. Reanalysis of his data and the analysis of further samples showed that males were subject to directional selection, the larger individuals surviving, while the surviving females were both smaller and less variable (Johnston et al. 1972; Fleischer and Johnston 1984). The actual selective factors have not yet been isolated, but large size is thought to be favored because of increased fasting ability (Murphy 1985). This cannot explain the decreases in size and variability of the females, which remain enigmatic. A recent analysis of massive weather-caused mortality in cliff swallows also showed strong size-selective mortality, with the larger individuals (as indicated by some measures but not others) being more likely to survive (Price et al. 2000).

A severe drought on the Galápagos Island of Daphne Major greatly reduced the population of Darwin's medium ground finch *Geospiza fortis*. The size of birds surviving this population crash was larger than the size of birds prior to the drought (Boag and Grant 1981). Concomitant with the drought and population decline, the seeds upon which the finch feeds, decreased in abundance and increased in size. Boag and Grant (1981) hypothesized that the large birds survived best because they were able to crack the large and hard seeds that predominated in the drought. Although large adult size may be favored, further studies suggest that selection may favor small juveniles, though the mechanism generating size-specific mortality is unknown (Price and Grant 1984).

The importance of predation as a factor in size selective mortality has been demonstrated in the freshwater zooplankters, in which it is the *larger* forms that are differentially removed by vertebrate predators (Galbraith 1967; Brooks 1968; Hall et al. 1976; Lynch 1977, 1980a; Mittelbach 1981). Lakes with fish have smaller sized zooplankters than those without (Hrbáek and Hrbáková-Esslová 1960; Brooks and Dodson 1965), while the introduction of fish or increase in fish abundance also leads to a shift to smaller zooplankters (Brooks and Dodson 1965; Wells 1970; Warshaw 1972). The effect of fish predation occurs not only across species but also within a population. For example, the mean size of *Daphnia lumholtzii* in the stomachs of the fish *Alestes baremose* was 1.28 mm (SE = 0.026), compared to 1.06mm (SE = 0.024) in the plankton (Green 1967). Increased visibility as a consequence of increased size may be important (O'Brien et al. 1976, 1979), but in at least one species, *Bosmina longirostris*, it is an increased eye size rather than total size that appears to be critical (Zaret and Kerfoot 1975; Hessen 1985).

Age and Size at First Reproduction in Females

OUTLINING THE GENERAL APPROACH. When one attempts to predict the evolution of age and size at first reproduction, a general strategy is to assume all life history components other than those that are functions of α are fixed and then write the characteristic equation as a function α and the fitness measure, say, r:

$$\int_{x=\alpha}^{\infty} l(x,\alpha)m(x,\alpha)dx = 1 \tag{4.48}$$

The next step is to solve the integral in terms of r and α and separate the two to give r as a function of α, say, $r = f(\alpha)$. The optimal age at first reproduction is then the value at which $f(\alpha)$ has a maximum, which is when

$$\frac{\partial f(\alpha)}{\partial \alpha} = 0 \tag{4.49}$$

and

$$\frac{\partial^2 f(\alpha)}{\partial \alpha^2} < 0 \tag{4.50}$$

Because it is a good policy to plot fitness versus the trait of interest (α), the latter condition can be checked visually by simply noting that the turning point is a maximum. To illustrate this principle, consider the case of a semelparous life history in which the product of the two functions is $l(\alpha)m(\alpha) = e^{-M\alpha}a\alpha^b$. One interpretation of this function is that survival is constant per unit time (e^{-M}) and female births is an allometric function of body size ($a\alpha^b$), which increases with age. The coefficient a may also include any mortalities, such as hatching success, that act in such a short period that they can be considered point events. Thus there are potentially a number of different biological life histories that are encompassed by the product $l(x)m(x)$. This is an important principle to keep in mind, because it may mean that a particular model is actually more general than implied by the initial biological considerations from which it was derived.

If, for the above model, we take r as the appropriate fitness measure we have

$$r = \frac{\ln\left(a\alpha^b e^{-M\alpha}\right)}{\alpha} = \frac{b\ln\alpha + \ln a}{\alpha} - M \tag{4.51}$$

Differentiating, we have

$$\frac{\partial r}{\partial \alpha} = \frac{1}{\alpha^2}(b - b\ln\alpha - \ln a) \tag{4.52}$$

And $\dfrac{\partial r}{\partial \alpha} = 0$ when

$$\ln(\alpha) = \frac{b - \ln a}{b} \tag{4.53}$$

If R_0 is taken as the fitness measure, we have

$$R_0 = a\alpha^b e^{-M\alpha} \tag{4.54}$$

Proceeding as above

$$\frac{\partial R_0}{\partial \alpha} = e^{-M\alpha} a\alpha^{b-1}(b - M\alpha)$$

$$\frac{\partial R_0}{\partial \alpha} = 0 \quad \text{when} \quad \alpha = \frac{b}{M} \tag{4.55}$$

Note that the optimal age of reproduction depends upon the particular choice fitness measure (Figure 4.23). Different answers will not necessarily follow from this choice, but it is important to recognize that the most fundamental assumptions can be very important.

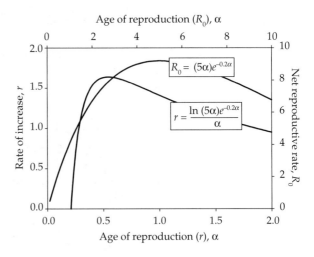

Figure 4.23 The relationship between two measures of fitness and the age at reproduction (r = bottom axis, R_0 = top axis) for a semelparous life history in which $l(\alpha)m(\alpha) = e^{-M\alpha}a\alpha^b$ with $M = 0.2$, $a = 5$, $b = 1$. For the case in which r is maximized, the maximum is at $e^{\frac{1-\ln 5}{1}} = 0.54$, and when R_0 is maximized, the maximum occurs at $1/0.2 = 5$.

APPLICATION TO *DROSOPHILA MELANOGASTER*. The above model can be extended to the iteroparous life history in a manner that is relevant to the evolution of age and size at maturity in *Drosophila melanogaster* (Roff 1981a). Pterygote insects, which comprise more than 99.9% of all insects, have three characteristics important in the modeling of their life histories: first, growth ceases at maturity (Ephemeroptera undergo a moult as adults but this does not change body mass); second, the age schedule of births is typically triangular, though fecundity typically increases with size; third, at least in holometabolous insects, the larval and adult forms may inhabit completely different habitats and hence be subject to independent mortality regimes.

In *D. melanogaster,* body size increases with development time and fecundity increases allometrically with body size. Although there are convincing data demonstrating that both these traits are heritable and that body size and development time and fecundity and development time are genetically correlated, the estimates on the genetic correlation between development time and fecundity are relatively scarce (Roff 2000). In the absence of strong contrary evidence it is reasonable to assume, as a working hypothesis, that there is a single growth trajectory, or at least that there is little genetic variation in growth rate. Development time, $d(L)$, is an allometric function of adult thorax length, L:

$$d(L) = a_d L^{b_d} + c_d \qquad (4.56)$$

The constant a_d represents the time required for the eggs to hatch and the development within the pupa, both of these components being independent of size. The age-specific fecundity curve of *D. melanogaster* is triangular in shape (Figure 4.14) and, measured on a single day or over some fixed time period, varies allometrically with thorax length. There are no data for the relationship between the specific parameters of the $m(x)$ curve and body size but, given the observed allometry, a simple model that captures this is

$$m(x,L) = \tfrac{1}{2} a_f L^{b_f} \left(1 - e^{-c_f(x-x_0)}\right) e^{-c_{ff}x} \qquad (4.57)$$

Thorax length, L, scales the age schedule of reproduction, larger females producing more eggs, but does not change its position. The constant a_f is the product of two constants: the coefficient of proportionality within the allometric relationship between fecundity and length, and the proportion of eggs that fail to hatch.

Because information on mortality rates are so poorly known, I assumed that instantaneous rates remain constant in the adult and larval phases at M_a and M_l, respectively. The probability of surviving the larval period for a given adult thorax length, $l(L)$, is thus,

$$l(L) = e^{-M_l d(L)} \qquad (4.58)$$

and the probability of surviving to some adult age x, where x is taken from the day of emergence into the adult form is

$$l(L,x) = l(L)e^{-M_a x} \tag{4.59}$$

Fecundity increases with adult size, but this will be opposed by the increased mortality entailed by the increased development time and the increased generation interval (Figure 4.24). These three opposing forces will produce an intermediate optimum adult size/age at maturity, the particular value being dependent upon the fitness measure used. *Drosophila melanogaster* is a colonizing species and hence the appropriate measure of fitness is r. Combining the above relationships, we obtain the characteristic equation

$$\sum_{x=x_0+1}^{\infty} m(x,L)l(L,x)e^{-r(d(L)+x)} = 1 \tag{4.60}$$

Though this equation is tediously long, its solution presents no great difficulty. Briefly, the method is first to evaluate the series making use of the fact that it is a

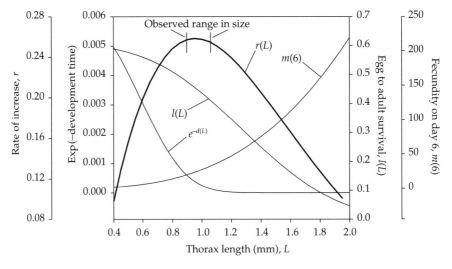

Figure 4.24   The relationship between r and thorax length, L, in *Drosophila melanogaster*, and the three components that contribute to the function. The characteristic equation is $\sum_{x=x_0+1}^{\infty} m(x,L)l(L,x)e^{-r(L)(d(L)+x)} = 1$, where r is written as a function of L, $r(L)$.

Development time increases generation length and thereby decreases r, the two being related in the characteristic equation as $e^{-r(L)d(L)}$. Fecundity increases r because $m(x,L)x \propto L^{bf}$. Survival from egg to adult decreases r, because $l(L) = e^{-M_f d(L)}$. Parameter values used to generate the curves are as given in Table 1 of Roff (1981b).

geometric progression; and second, to differentiate implicitly to obtain the optimal length at maturity (full details are given in Roff 1981a). The optimum thorax length depends upon all components of the characteristic equation, none dropping out as in the previous semelparous model. Thus adult mortality, larval mortality, egg hatching rate, and the components of the fecundity function all play a role in the evolution of the age and size at reproduction in *D. melanogaster*. Figure 4.24 shows the relationship between *r* and thorax length using values obtained from laboratory stocks and estimates from wild populations. The predicted maximum, 0.95 mm, falls very nicely within the observed range in thorax length of 0.90 to 1.15 mm. The rates of increase predicted in the above analysis may be unrealistically high (Ricklefs 1982). Restricting the parameter space to only low values of *r* does not change the conclusions (Roff 1983b).

Given the possible error in the estimation of parameter values, the close fit between prediction and observation bears closer scrutiny. *A priori*, the close fit suggests that either the right values were fortuitously chosen, or the optimum is relatively insensitive to variation in parameter values. To test the second hypothesis, I varied two parameters at a time and computed the change in the optimum thorax length. Wide variation in the three pairs of components that are likely to be poorly estimated produces relatively little variation in the optimum thorax length, the range in the optimum thorax length being only 0.80–1.10 mm (Roff 1981a). An estimate of the total variation possible can be obtained by selecting the set of extreme values that favor either an increase or decrease in body size. From this, the smallest thorax length is 0.60 mm and the largest, 1.38 mm. Although this range is larger than that observed (0.90–1.15) it is still surprisingly small, given the range in parameter values. The maximum range obtained covers almost the full range of body size of North American drosphilids (0.7–1.5 mm), demonstrating that this set of species could easily have evolved from relatively minor changes in parameter values.

EXTENDING THE ANALYSIS TO OTHER DROSOPHILIDS. Hawaiian drosophilids may be two or three times as large as *D. melanogaster*. Can such extreme variation be predicted from the above model and, if so, do the corresponding changes match what is observed in the Hawaiian species? According to the body size-development time trade-off function, the large sizes of the Hawaiian species should be accomplished by a very extended development time. This is, in fact what is observed (Figure 4.25). Analysis of Equation (4.60) shows that large adult size will be favored by high survival rates and a very low adult fecundity. In *D. melanogaster*, fecundity is correlated with the number of ovarioles and hence can be used as an approximate index of fecundity in other *Drosophila* species. For the Hawaiian Drosophillidae (*Drosophila, Scaptomyza, Antopercus, Ateledrosophila*) there is a highly significant regression between ovariole number and thorax length [log(ovariole number) = 0.73 + 1.9 log(thorax length), $r = 0.68$, $P < 0.00001$], with the number of ovarioles being substantially less than found in *D. melanogaster* (Figure 4.25). This is indirect evidence for the low predicted fecundity of the Hawaiian species. This conclusion is reinforced by fecundity data for the

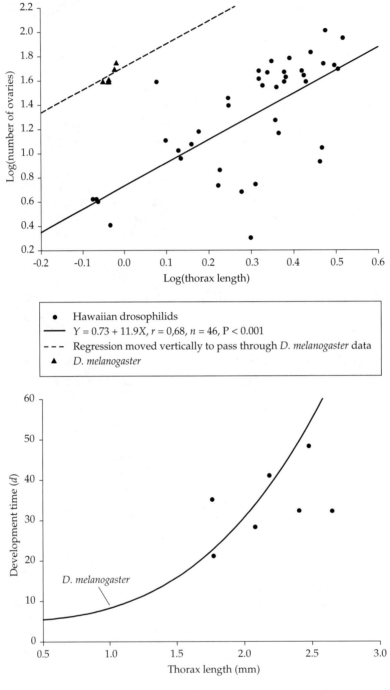

Figure 4.25   Development time and ovariole number vs thorax length in the Hawaiian drosophilids compared to *D. melanogaster*. Data sources: Carson et al. (1970); Kambysellis and Heed (1971); Capy et al. (1983).

Hawaiian species, *D. disticha* which indicates a fecundity in the laboratory of less than 0.5 eggs per day (Robertson et al., 1968) and an ovariole number typical of Hawaiian species and far below that of *D. melanogaster*.

CHANGING THE FITNESS MEASURE. Although the assumption that R_0 is the appropriate fitness measure is unlikely to be correct it is instructive to see how the predictions are altered. The solution is actually much simpler than the previous, and the optimal thorax length, L_{opt} is

$$L_{opt} = \left[\left(\frac{b_f}{b_d} \right) \left(\frac{1}{M_l} \right) \left(\frac{1}{a_d} \right) \right]^{\frac{1}{b_d}} \qquad (4.61)$$

and the optimal development time ($= \alpha - x_0$) is

$$d\left(L_{opt} \right) = c_d + \left(\frac{b_f}{b_d} \right) \left(\frac{1}{M_l} \right) \qquad (4.62)$$

Thus, both the optimal development time and body size depend on the rate at which fecundity increases with body size versus the rate at which development time is increased (b_f/b_d) and is inversely related to the larval mortality rate ($1/M_l$). Adult mortality rate (M_a) and the fecundity/egg viability constant (a_f) no longer influence adult body size or α. This result is consistent with the earlier analysis of the semelparous model, the larval mortality rate corresponding to the mortality rate prior to breeding. The optimal thorax length using the same values as in Figure 4.25 is 1.46, substantially larger than the observed. However, the optimum is very sensitive to the estimate of larval mortality, which, as noted above, is one of the parameters for which we have the least amount of data. The value used was 0.1, corresponding to a daily mortality rate of 10%: increasing the rate to 0.31, corresponding to a daily mortality rate of 27%, gives an optimum thorax length of 1 mm. Under natural conditions, the higher figure is not unrealistic. (In the sensitivity analysis, the range in larval rate considered was 0.0–0.5).

MAKING A MORE GENERAL MODEL—THE FISH MODEL. The *Drosophila* model was tailored specifically for a single species, although it was extended to account for variation among drosophiliids in general. Similar species-specific models have been constructed for the chaetognath *Sagitta elegans* (McLaren 1966), a salamander, *Hynobius nebulosus* (Kusano 1982), the semelparous plant *Oenothera glazioviana* (Kachi and Hirose 1985), and northern elephant seals, *Mirounga angustirostris* (Reiter and Le Boeuf 1991). An alternate approach is to construct a more general model with fewer parameters. Such a model may not be as accurate as the species-specific models, but it may account for variation across a broader range of taxa. Such a model was produced by Roff (1984b) to predict the age/size at first reproduction in fish, with possible extension to other groups such as reptiles and amphibians. I shall refer to this model as the **Fish model**.

Growth in fish can be described by the von Bertalanffy function

$$L(x) = L_\infty\left(1 - e^{-k(x-x_0)}\right)$$ (4.63)

where L_∞ is the asymptotic length, k is a parameter that dictates the rate of approach to L_∞, and x_0 is a parameter that shifts the curve along the age axis. Typically, x_0 is very close to zero, and I shall assume this in the further development of the model. The above growth curve describes the growth trajectory but does not explicitly include a trade-off between growth and reproduction. Models that incorporate such a trade-off do generate this type of curve (Roff 1983b, Kozlowski 1996). The curve is used in the present context to describe the growth until the age of maturity.

Fecundity is generally a power function of length, and since growth does not cease at maturity, the $m(x)$ curve can be written as

$$m(x) = aL(x)^b$$ (4.64)

The parameter a includes both the fecundity coefficient, egg hatching success and any early larval mortality that cannot be accounted by a constant mortality rate, M, (so $l(x) = e^{-Mx}$). Thus, written out in full, the product of survival and fecundity at the age of first reproduction, α, is

$$l(\alpha)m(\alpha) = e^{-M\alpha}\left[aL_\infty\left(1 - e^{-k\alpha}\right)\right]^b$$ (4.65)

To incorporate a cost of reproduction without introducing more parameters, we can write the $l(x)m(x)$ function as

$$l(x)m(x) = e^{-M(x-\alpha)}l(\alpha)m(\alpha)$$ (4.66)

This model assumes that the cost of reproduction is such that increases in fecundity due to increases in size with increasing age are offset by increases in mortality. This is likely to be at least approximately true if there is a survival cost associated with reproduction. The optimal age at first reproduction depends upon the fitness measure chosen but not the assumption of iteroparity or semelparity (Roff 1984b). For most fish, populations are not frequently undergoing colonizing episodes, and hence the most suitable measure of fitness is R_0, from which we obtain the optimal age at first reproduction as

$$\alpha = \frac{1}{k}\ln\left(\frac{bk}{M} + 1\right)$$ (4.67)

and the optimal length at first reproduction as

$$L(\alpha) = L_\infty\left(\frac{bk}{bk + M}\right)$$ (4.68)

The above equations will apply not only to fish but other taxa, such as reptiles, in which growth can be described by a Von Bertalanffy function and fecundity increases allometrically with body size. For fish, b is generally around 3 (Wootton 1979) but there has been no large-scale survey of reptiles. In the absence of such a survey, I have used three. (This is consistent with the widespread use of relative clutch mass as an index of reproductive effort in lizards and snakes, because such an index assumes that clutch mass is proportional to body mass.) Absolute deviations in predicted values can be expected to increase with the value of the prediction. Therefore a log-log plot is more appropriate than a linear-linear plot of observed and predicted values (Figure 4.26). Overall, there is an

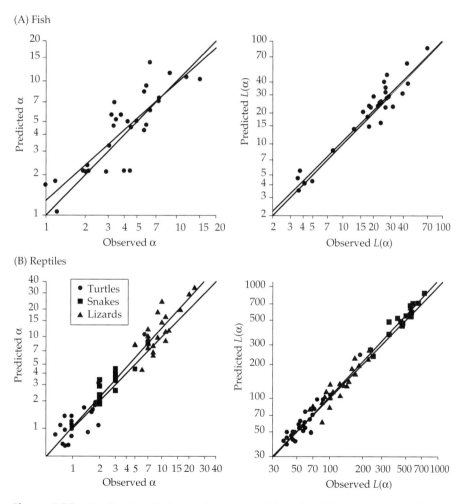

Figure 4.26 Predicted and observed ages, α, and lengths, $L(\alpha)$, at maturity in fish (A) and reptiles (B). Data from Roff (1984b), Shine and Charnov (1992), Shine and Iverson (1995). Lines shown are the 1:1 line and the fitted regression line of predicted on observed.

excellent fit between observed and predicted values of both α and $L(α)$. Significant deviations occur for the age at maturity in turtles and snakes, the predicted values being too high. The most likely explanation for this is that fecundity does not scale as the cube of length but some lower value. (For turtles, see Gibbons et al. 1982 and for snakes, Ford and Siegel 1989a.)

The *Drosophila* and Fish models share the feature that $m(α)$ is a decelerating function of α. (For other examples, see Stearns and Koella 1986; Berrigan and Koella 1994; Figure 4.27). In other models, such as that described by Equation (4.54), $m(α)$ can be an accelerating function. (For examples, see Cohen 1971; Roff 1983a; Kozlowski and Wiegert 1987; Perrin and Rubin 1990; Hutchings 1993; Gemmil et al. 1999; Figure 4.27). A realistic example applies to the growth of American plaice *Hippoglossoides platessoides,* in which the length of immatures increases linearly with age (Roff 1983a), and so the $m(α)$ function is

$$m(α) = a(c + dα)^b \tag{4.69}$$

Assuming the same conditions as for the Fish model, we have $l(x)m(x) = e^{-M(x-α)}l(α)m(α)$ and the optimal age at maturity (the same for both iteroparous and semelparous life histories) is

$$α = \frac{b}{M} - \frac{d}{c} \tag{4.70}$$

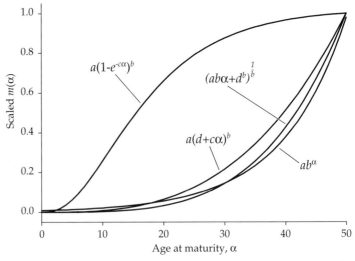

Figure 4.27   Examples that show decelerating fecundity curve, Roff (1983a), $a(1 - e^{-cα})^b$, and accelerating fecundity curves with age at maturity: Cohen (1971), $ab^α$; Kozlowski and Wiegert (1987), $\left(abα + d^b\right)^{\frac{1}{b}}$; Perrin and Rubin (1990), $a(d + cα)^b$. The curves have been arbitrarily scaled to a value of 1 at age 50.

Intuitively, we might suppose that an increased growth rate would favor a delay in the age at first reproduction, because this gives an increased marginal value of fecundity relative to survival. This hypothesis is supported by the above model—increased values of c decreasing d/c, and hence increasing α. However, an increase in the growth rate, k, of the Fish model leads to a decrease in α. The reason for this is that because growth is asymptotic in the Fish model, the age at which marginal increases in fecundity no longer favor continued growth is reached earlier with an increased k. The importance of the growth function depends upon its shape. Growth curves that are accelerating will produce delayed reproduction with an increased growth parameter, whereas growth curves that are decelerating may favor earlier reproduction, contingent upon the parameter altered. (For example, in the above model, an increase in c will produce an earlier reproduction, but an increase in b, which could be a parameter composed of a growth and fecundity components, will favor a delay in the age at maturity.)

THE IMPORTANCE OF CORRECTLY DEFINING THE *l(x)* FUNCTION.   The models considered thus far assume either that mortality rate is constant throughout life or that it is stage-specific in that it may change its value when the organism matures (as in metamorphosis in the *Drosophila* model). In the Fish model the mortality rate can be considered to be composed of three components: a "point" mortality event (e.g., egg and early larval mortality) that is part of the coefficient a, a constant rate, M, and an increase at maturity that contributes to maintaining the "realized" fecundity constant despite an increased body size with age. An alternative type of mortality regime is one that is dependent on age not stage. This difference is of crucial importance in determining the evolutionarily optimal age/size at maturity. Suppose mortality is age/size-dependent; an increase in adult mortality will select for a decreased age at first reproduction, whereas an increase in juvenile mortality will select for an increased age at first reproduction (Gadgil and Bossert 1970; Law 1979b; Michod 1979; Taylor and Gabriel 1992). This is exactly the opposite to what is predicted by the *Drosophila* model. In this model a female incurs the mortality associated with the adult phase only by metamorphosing into the adult form. If adult mortality is high, selection favors females that delay entry into the adult stage, remain as larvae, and grow larger. Likewise, a high larval mortality favors rapid exit from this stage into the adult. The animal thus has the option of varying how long it is subjected to either mortality regime. In contrast, because in the age/size-survival models mortality rate depends upon age not developmental stage, a female cannot escape the source of mortality by delaying reproduction; therefore, if mortality in later age groups increases, selection will favor females that reproduce before they are subjected to the increased mortality. If the cost of reproduction in terms of mortality declines with age, selection will act to delay maturation. By appropriate changes in the relationship between age, reproductive effort, and survival, a variety of more complex responses can be obtained (Schaffer and Rosenzweig 1977; Law 1979b).

Because the two types of mortality regimes make very different predictions, care should be taken in distinguishing between the two in any tests. Hutchings

(1993) and Bertschy and Fox (1999) used the ratio of adult to juvenile survival, assuming that mortality was age/size dependent rather than stage-dependent. In both studies the authors predicted that selection would favor an increased age at maturity with an increased adult/juvenile survival ratio. However, two populations could have the same ratio but different mortality curves, and in this case they could have different life histories. Thus the survival ratio is a relatively crude metric, although it can predict the direction of change. Hutchings (1993) studied three populations of brook trout while Bertschy and Fox (1999) examined five populations of pumkinseed sunfish. Both studies showed that the ages at maturity changed in the appropriate direction. Belk (1995) assayed for genetic variation in life history parameter in two populations of bluegill sunfish using a common garden design and found no evidence of genetic variation. In a subsequent experiment, Belk (1998) raised fish from three populations in the visual (but otherwise noninteracting) presence and absence of a predator and obtained evidence that age at maturity was altered simply by the presence of the predator. Evidence for genetically based changes in age and size at maturity as a consequence of size-specific mortality rates is provided by the study of life history variation in guppies discussed earlier in this chapter.

Edley and Law (1988) experimentally tested the prediction that growth rate will evolve in response to size-selective predation by using two culling regimes on laboratory populations of *Daphnia magna*. In one regime only small individuals were removed, while in the second only large individuals were sieved from the population. After 150 days of culling, clonal differences in growth rate were evident. In the clones subjected to predation on small size, growth was rapid during the early life stages, thereby pushing the females rapidly through the vulnerable period. Maturity in these clones was delayed, as expected, if there is a trade-off between growth and reproduction. In contrast, populations subjected to predation on large size showed a reduced growth rate and early maturation, a phenology that reduces the impact of predation. Unfortunately, the culling regimes also changed densities differentially, and hence it is possible, though Edley and Law think unlikely, that the observed responses were a result of density-dependent selection.

Age and Size at First Reproduction in Males

THE IMPORTANCE OF SIZE. In many ectotherms and endotherms, competition for mates is frequently contest competition in which the victor is generally the larger. To grow large, an animal must forage, an activity that increases its susceptibility to predators. Therefore, when size is not a critical factor in mating success, the optimal pattern of growth is to grow to the minimal size necessary to sustain the investment in gonads and success in the scramble for mates. Consequently, the optimal size and age at maturity of males relative to the female should be smaller in species with scramble-mate competition than in species with contest-mate competition.

Bell (1980) tested this hypothesis on a large scale, comparing freshwater North American fish with birds and mammals. As predicted, whereas the vast majority of male fish mature earlier than female, the reverse is true in birds and mammals.

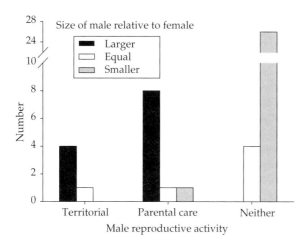

Figure 4.28 The distribution of the size of male teleost fishes relative to females as a function of the male reproductive behavior. Data from Roff (1983a).

Potential problems with differences ascribable to phylogeny can be resolved by considering a single taxon. To this end I collated the relative size at maturity of North American fish species into three groups: those in which the male is territorial but show no parental care, those in which the male exhibits parental care, and those in which the male shows neither of the preceding behaviors. Males in the last group are predicted to be smaller at maturity than the female, and in the first two categories the male will be larger. Both predictions are supported (Figure 4.28). Factors other than contest for mates may also shape the relative age and size schedules of maturity; for example, if one sex carries the other for a prolonged period, the "carrying sex" may have to be larger than the "carried" for purely mechanical reasons (Fairbairn 1990). An important factor for a male is the presence of receptive females. Emerging after all females have been mated is clearly maladaptive. The selective advantage of emerging first may favor the evolution of protandry (Bulmer 1983; Wiklund et al. 1991, 1992; Wiklund and Forsberg 1991; Kleckner et al. 1995; Baughman 1991; Bradshaw et al. 1997).

DIRECTIONAL SELECTION FOR INCREASED SIZE WITHOUT AN EVOLUTIONARY RESPONSE. Female fitness typically increases with body size because of increased fecundity. In males the probability of obtaining a mate is frequently dependent on relative body size, larger males winning contests against other males or being preferred by females. Under these circumstances we might expect selection to favor continuously increasing size. Increased body size will be selected against, as in the case of females, by the detrimental effects of increased development time (= decreased survival), the possibility of not meeting maintenance ration, and so on. A model proposed by Cooke et al. (1990) originally for the evolution of clutch size can be used to illustrate the interaction between genotype and phenotype in determining the equilibrium body size for animals that defend territories. Suppose that obtaining a mate depends on body size, which itself depends on the amount of nutrients the male can sequester and which is a function of territory size and/or quality. Selection may favor an increased size, but this can only be achieved by increasing the size/quality of the terrritory. Thus selection acts upon both body size and some behavioral component that relates to the ability to defend a territory. Selection for, say, aggressiveness, will result in a better territory only if a limited number of individuals in the population change their behavior. If, as is likely to happen, the general level of aggressiveness rises, no individual will be "better off," and hence each bird may, on average, be able only to defend the same territory as before; hence there can be no response in body size since the nutritional status of the bird remains the same (Box 4.4).

MODELING THE OPTIMAL SIZE IN MALES: A GENERAL APPROACH. Given that male fitness depends on relative body size, what approaches are appropriate for its prediction? In the case of continuous variation in body size, the optimal size under the assumption of a stationary population can be calculated from the equation

$$\mu = \int_{-\infty}^{+\infty}\int_{-\infty}^{+\infty} f(Y,X)w(X)g(X,\mu)dXdY \tag{4.71}$$

where μ is the mean body size of males in the equilibrium population, $f(Y,X)$ is the probability of a male of size Y winning a contest against a male of size X, $w(X)$ is the fitness of a male of size X that is independent of its mating function (e.g., the survival of the male to reproductive age), and $g(X,\mu)$ is the frequency of males of size X in the population, which is obviously a function of μ. If body size and age are related, then changes in $f(\bullet)$ will lead to changes in the mean body size and hence the optimal age at first reproduction. Although $f(\bullet)$ cannot be exactly specified, several properties that it must satisfy can be stated. First, when $Y = X$, the probability of success must be one-half. Second as $Y-X$ increases, $f(\bullet)$ will typically approach 1, whereas as $Y-X$ decreases, $f(\bullet)$ will typically approach 0.

MODELING THE OPTIMAL SIZE IN MALES: DIMORPHIC VARIATION. In a wide range of invertebrates and a number of fish species there occurs two male morphs, one small and one large (Roff 1996a; Table 4.7). The larger male invariably has an advantage in contests over territories or females but may pay a cost in that it must take longer to grow to the larger size. The smaller male frequently takes up a **satellite strategy**. This strategy is also found in species in which size is continuous, again the smaller males in the population adopting the strategy (see references in Table 7.7 of Roff 1992a). The large morph defends a territory, which may comprise simply the immediate area around a receptive female. The smaller morph acts as a satellite and attempts to obtain copulations either by intercepting the incoming female or by sneaking between the male and female at the moment the eggs are released. Variations on this basic pattern are numerous. Within a population several types of satellite males may occur (Taborsky et al. 1987); differences in morphology between satellites and territorial males may form a continuum rather than distinct classes; and the satellite behavior may consist of sim-

---

Box 4.4
Selection on size without response: Mathematical details

The process suggested by Cooke et al. (1990) can be described mathematically as follows: Suppose that body size, x, is the product of three factors, $x = RTe$, where R is the ration per unit area determined by the foraging ability of the animal; T, territory size; e, a random environmental component. Territory size is determined by a large number of factors, which Cooke and colleagues lump together under the general term, $A =$ "aggressiveness." The amount of territory a particular bird can sequester is dependent upon the aggressiveness of this bird relative to the mean aggressiveness ($\hat{A}$), and the number of birds, N, that have territories. Letting the total area available be a, the value of T is given by

$$T = \left(\frac{a}{N}\right)\left(\frac{A}{\hat{A}}\right) \tag{4.72}$$

The two component traits, R (the actual trait would be foraging ability) and A, can be decomposed into their genetic and environmental components, $R = R_G R_E$ and $T = A = A_G A_E$, and hence $x = (R_G A_G)(R_E A_E)(ea)/(N\hat{A})$. In a constant environment, the change in mean phenotype in response to selection is $h^2 S$, where $h^2 S$ is the heritability of the trait and S is the selection differential (the difference between the means of the selected and unselected populations). If the environment is changing with time, this equation is modified to $Response = h^2 S + \Delta\hat{E}$, where $\Delta\hat{E}$ is the change in the mean environmental component over one generation. Now, even if the environmental components R_E and A_E do not change, the overall value of E will change if $\hat{E}$ changes, as might occur if there is a change in one of the behavioral components. If $\hat{A}$ increases, $\hat{E}$ will decrease, and hence $\Delta\hat{E}$ will be negative, and may negate any change in body size resulting from a positive selection differential, S. The difficulty with this hypothesis is that it requires that the two components be fairly well balanced to prevent any significant response. The most important message is that responses may depend upon how the environment, both biotic and abiotic, is changing in concert with selection for increasing body size.

---

ply remaining in the vicinity of the territorial males, taking over the territory after the territory is vacated (Wells 1977b).

Satellite behavior may represent an alternate behavior that has an equal fitness with territorial behavior, or it may represent the "best of a bad lot" situation (Eberhard 1982; Jonsson and Hinder 1982; Dominey 1984). Suppose that for some reason (such as a local scarcity of food) a male cannot achieve a size at which it can be successful as a territorial male. In this case the male should adopt the satellite behavior, because even though its lifetime fitness will be less than that of a large territorial male, it will be greater than that of a small territorial male. For this scenario, there is no assumption of equal fitnesses; the behavior is maintained in the population because it is adaptive, given unavoidable variation in size at maturity. The behavioral response is an adaptive norm of reaction and

Table 4.7   **A review of costs and benefits of size-related dimorphisms in male animals. Adapted from Roff (1996a).**

Species	Associated dimorphism	Advantage to larger morph	"Cost" to larger morph	References
Trachyderes mandibularis[a]	Mandibles	Wins fights Mating probability greater when sap ooze sites are rare	Mating probability the same on plants with many sap ooze sites	Goldsmith (1985); Goldsmith and Alcock (1993)
Podischnus agenor[a]	Horns	Wins fights	Unknown	Eberhard (1980b, 1982)
Onthophagus binodis[a]	Horns	Wins fights Female fecundity larger than with similar-sized hornless male	Assists females in constructing dung nests	Cook (1988)
Hoplothrips karnyi[b]	Large front forelegs	Shorter development time More matings	Wingless, cannot migrate	Crespi (1988a,b)
Myrmica ruginodis[c]	Overall size	More matings	unknown	Elmes (1991)
Rhizoglyphus robinii[d]	Large third legs No difference in overall size	More matings	unknown	Radwan (1995)
Caloglyphus berlesei[d]	Large third legs No difference in overall size	More matings in small colonies	Fewer matings in large colonies	Radwan (1993)
Euterpina acutifrons[e]	Overall size	Female fecundity larger	1) Longer development time 2) Fewer copulations	Haq (1972); Stancyck and Moreira (1988)
Pachypgus gibber[e]	Overall size	Produces more spermatophores	Later age at reproduction	Hipeau-Jacquotte (1984)
Fish	Territorial (small morph is satellite)	Wins fights, attracts more females (*Oncorynchus kisutch, O. keta, Salmo salar, Salvelinus malma, Lepomis macrochirus, Porichthys notatus*)	1) Small males better at sneaking (*O. kisutch, S. salar, L. macrochirus, P. notatus*) 2) Delayed age at maturity (*O. kisutch, S. salar, L. macrochirus, P. notatus*)	Gross and Charnov (1980); Gross (1985); Maekawa and Onozato (1986); Myers (1986); Hutchings and Myers (1987, 1994); Brantley et al. (1993)

[a]Beetle, [b]thrips, [c]ant, [d]mite, [e]copepod

implies that there exists or existed genetic variation for the behavioral repertoire. (For a more detailed discussion of reaction norms, see Chapter 6.)

The second situation is that in which the two behaviors (= size morphs) are maintained because they have equal fitnesses. It is possible to construct a circumstance in which the fitnesses would be equal at all frequencies of the two morphs; but such cases must be exceedingly rare because they require a "knife-edge" balance that will be upset with the slightest variation in parameter values. A more usual circumstance will be that in which the relative fitnesses are frequency-dependent, in which case stability is assured. Age has been demonstrated to be an important component in the fitness of alternate male behaviors in several species of teleost. Species in which smaller, younger males mature at an early age and become satellites, while other males mature at a later age, are bigger and become territorial comprise various species within the subfamily Salmoninae (Scott and Crossman 1973; Gross 1985; Maekawa and Onozato 1986; Hutchings and Myers 1987, 1988) and the swordtails, *Xiphoporus* spp (Zimmerer and Kallman 1989). Satellite and territorial behavior has been observed in the bluegill sunfish *Lepomis macrochirus*, but it is not known if this represents two entirely different patterns of maturation or an ontogenetic shift from satellite to territorial, though the former seems most likely (Dominey 1980; Gross 1982, 1991). In other indeterminate growers such as amphibians, the shift between tactics might represent an ontogenetic shift, though there appears to be more flexibility in these cases, with individuals shifting between behavioral modes opportunistically.

The phenomenon of **precocial maturation** or "**jacking**" is well known within various salmon species. Although some progress towards understanding the factors favoring this phenomenon is possible using a frequency-independent approach (e.g., Caswell et al. 1984), a game-theoretic framework as outlined above will likely prove the more successful (Myers 1986). Precocial maturation is clearly related to environmental conditions (Lundqvist and Fridberg 1982), but there is unambiguous evidence for a genetic basis (Heath et al. 1994). Although there are two morphs, genetic determination is polygenic and can be understood using the threshold model of quantitative genetics (Myers and Hutchings 1986; Hazel et al. 1990; Hutchings and Myers 1994; Roff 1996a). As previously described, under this model there is a normally distributed underlying trait known as the liability and a threshold of sensitivity. Individuals above the threshold develop into one morph, whereas individuals below the threshold develop into the alternate morph. (See Roff 1997 for a full discussion of estimation of genetic parameters.) Using this model Heath et al. (1994) estimated the heritability of jacking to be approximately 0.4. Analysis of a wide range of other threshold traits has shown that under a fixed environment, the heritability of the trait is typically in the region of 0.5, but that these traits show considerable phenotypic plasticity (Roff 1996a). Under directional selection, phenotypic variation is eroded, but a considerable amount of genetic variance is maintained by mutation-selection balance (Roff 1998a). With frequency-dependent selection, both genetic and phenotypic variances can be maintained at levels consistent with those observed (Roff 1998b). Under the threshold model, the equilibrium frequency of the two morphs is exactly the same as that obtained using a strictly phenotypic model (Roff 1998b), which is analytically much more tractable.

As discussed above, precocial maturation is most likely to be determined by the action of many genes with environmental interaction. However, in at least one species of *Xiphophorus*, size and age at maturity are determined by a sex-linked gene (Kallman et al. 1973; Kallman and Borkoski 1977; Kallman 1983; Zimmerer and Kallman 1989). Males of the species *X. nigrensis* from the river Río Coy (Mexico) comprise four size classes, and these differ in their Y-linked *P* alleles (*s, I, II,* or *L*). All X chromosomes carry the *s* allele, which when homozygous, produce small, early maturing males; the *L* allele produces large, late-maturing males, and alleles *I, II* produce intermediates (Zimmerer and Kallman 1989). Large males display exclusively and are significantly more successful in obtaining copulations; small males switch from display to sneaking behavior according to the size of competing males. Controlling size by using genotypes carrying the *I* allele, Zimmerer and Kallman were able to show that only small males homozygous for the *s* allele display the sneaking behavior. How the three alleles are maintained in the population has not been determined.

MODELING DIMORPHIC VARIATION: TWO EXAMPLES. Gross (1985) examined the relative fitnesses of precocial ("jack") and late-maturing ("hooknose") coho salmon, *Oncorhynchus kisutch*, in a single stream (Deer Creek Junior, Washington State). For both forms to be maintained in the population, the relative lifetime fitnesses must be equal. Lifetime fitness is the product of the ratios of jack to hooknose survivorship to maturity, breeding lifespan, and mating success. Survivorship was estimated as 13% for jacks and 6% for hooknose (no standard errors). Breeding lifespan was estimated as 8.4 ± 2.3 days ($n = 7$) for jacks and 12.7 ± 1.2 days ($n = 35$) for hooknose males. Finally, Gross assumed that mating success was proportional to the proximity to the females, which was 124.6 ± 15.5 cm for jacks and 93.0 ± 6.1 cm for hooknose males. Jack success was then estimated as $1 - (124.6 - 93.0)/93.0 = 0.66$. The relative lifetime fitness is then

$$\frac{w_{jack}}{w_{hooknose}} = \left(\frac{0.13}{0.06}\right)\left(\frac{8.4}{12.7}\right)\left(\frac{0.66}{1.00}\right) = 0.95 \tag{4.73}$$

where w_{jack} is the fitness of the jack, $w_{hooknose}$ is the fitness of the hooknose, and the figures in the three parentheses refer to the three fitness components. The predicted value of 0.95 is remarkably close to the expected value of 1, but, because the estimate lacks confidence intervals, it is difficult to know if the result might not be due to chance. To approximately estimate confidence limits, I assumed that each parameter was independent and normally distributed with the observed mean and standard deviation (= standard error of the estimate) and then generated 1000 random combinations and estimated the relative fitness. The estimates were more or less normally distributed (there was a slight skew towards larger values) with a mean of 0.95 and a median of 0.91. The 95% confidence region (= ± twice the standard deviation) was 0.15 to 1.754. This region is very large and indicates that the results must be viewed very cautiously. The proportion of precocial males varies widely in Atlantic salmon, *Salmo salar* (Myers and Hutchings 1986), but though Myers (1986) has produced a game-theoretic model, the quality of the data is too poor for a meaningful test. Hutchings and Myers (1994) produced a model based on a comparison of rates of increase of the two morphs, but although the model does provide an important conceptual framework for further analysis it could not be effectively tested with the available data. Further experiments based on this model are warranted.

Three morphs are found in the bluegill sunfish *Lepomis macrochirus* (Gross 1982), although these are generally compressed into two categories, territorial males and satellites (= sneakers). Let the $l(x)$, $m(x)$ functions for the two morphs be designated by the subscripts s (= satellite) and t (= territorial). Satellites mature at 2 yr and territorial males at 8 yr. Assuming a stationary population and the fitnesses of the two morphs to be equal, we have (Gross and Charnov 1980)

$$\int_2^\infty l_s(x)m_s(x)dx = \int_8^\infty l_t(x)m_t(x)dx \tag{4.74}$$

Letting the proportion of males that become satellites be p and the fraction of eggs fertilized by a satellite male be h, we have, by definition

$$\frac{p\int_2^\infty l_s(x)m_s(x)dx}{(1-p)\int_8^\infty l_t(x)m_t(x)dx} = \frac{h}{1-h} \tag{4.75}$$

By Equation (4.74), the above equation is equal to

$$\frac{p}{1-p} = \frac{h}{1-h} \tag{4.76}$$

That is, the proportion of males becoming satellites should be equal to the fraction of eggs they manage to fertilize. For the fish of Lake Opinicon, 21% (= p) of all males become satellites and fertilize 14% (= h) of the eggs. Binomial confidence limits (95%) for the former statistic are 11–31%. The hypothesis of equal fitnesses cannot, therefore, be rejected.

Maekawa and Hino (1987) extended the model of Gross and Charnov to include cannibalism by satellite males, a phenomenon observed in the Miyabe char *Salvelinus malma miyabei*. The addition of cannibalism has very little effect on the predicted value of p. The estimated value of p does fall within the observed range (23.1–37.5%) and is not too far removed from the observed mean (31%).

The Evolution of the *l(x)m(x)* Function

The previous examination of the evolution of the age at first reproduction did not consider how the $l(x)m(x)$ function will evolve, but saw it as bearing a fixed relationship with α. In this section I reverse the viewpoint and consider how the $l(x)m(x)$ function evolves, keeping α either fixed or with a fixed relationship to the $l(x)m(x)$ function. This issue has already been broached when we examined the conditions under which semelparity would evolve. In that section it was shown that iteroparity would evolve if there is the trade-off between adult survival and fecundity that is concave down (Figure 4.21). There may be circumstances in which the trade-off is not between fecundity and adult survival, but between adult and juvenile survival. For example, the survival of offspring might be increased by parental defence, though this may decrease the survival probability of the parent. Mathematically, this is the same as the trade-off between fecundity and adult survival. For the model previously explored, the two equations would be $\lambda = ms + f(m)$, and $\lambda = ms + g(s)$, where f and g express the relevant trade-off. Another possible scenario is for juvenile survival and fecundity to be related, as might occur if offspring survival depended on initial size, which for any given total clutch mass would lead to a reduced fecundity. In this case we have $\lambda = mh(m) + S$ and iteroparity will be favored if the trade-off between juvenile survival and fecundity is concave down. Such a restriction also applies to the conditions favoring an increase in reproductive effort with age, a phenomenon discussed in detail below.

l(x)m(x) and Age: The evolution of Reproductive Effort

THEORY. If both fecundity and survival covary, what is the optimal age-schedule of survival and fecundity? The analysis of this problem has typically focused on the fecundity component and such models are termed **reproductive effort** models because it is assumed that increasing effort put into reproduction will increase the effective fecundity of an organism (in other words, the number that survive either because more are produced or because they are better provisioned). Following the same rationale as used to predict the shape of the trade-off function between fecundity and survival necessary to favor iteroparity, it can be shown that an intermediate reproductive effort will be favored when the trade-off between reproductive effort and fecundity is concave down (Schaffer 1974a; Leon 1976).

Intuitively, we might suppose—because the effect of a given $m(x)$ on r declines with age—that evolution would favor an increased effort with age (Williams 1966). This hypothesis has received support from particular models (Gadgil and Bossert 1970; Cohen 1971; Kitahara et al. 1987; Kozlowski and Uchmanski 1987), but the lack of complete generality of this hypothesis has been demonstrated by several counter-examples (Fagen 1972; Charlesworth and Leon 1976; Charlesworth 1990; Roff 1992a; a model by Dixon et al. 1993 does not provide a counter-example as they assumed a triangular fecundity function and hence "built in" the decline as a function of age). A fairly general treatment of this model was given by Charlesworth and Leon (1976: see also Taylor 1991). Details of the model are given in Box 4.5. Briefly, they assumed three trade-offs: one between growth and reproductive effort, one between survival and reproductive effort, and a third between fecundity and reproductive effort. The first two were combined into a single concave trade-off function with reproductive effort .

Charlesworth and León (1976) showed that, at an evolutionary equilibrium, a *sufficient* condition for reproductive effort to increase from age $x-1$ to x is $\ln(P_{max})$ $\geq r^*$, where P_{max} is the value of $P(E,x)$ when $R(E,x) = 0$ (i.e., at zero reproductive effort), and the rate of increase is at its global maximum r^*. Recall that $P(E,x)$ is the product of the survival and increment in growth; thus an increase in reproductive effort will be favored if r is not too large and growth and/or survival are high at low levels of reproductive effort. As r approaches zero (a stable population size), the above inequality becomes $P_{max} \geq 1$ and the likelihood of reproductive effort increasing with age increases. In a population controlled by density-dependent factors, reproductive effort is likely to decrease with age if a population is subjected to a high density-dependent mortality or reduction of individual growth rate (Charlesworth and León, 1976, p. 455).

As the above inequality is a *sufficient* condition only, optimal reproductive effort might still increase with age even if the inequality is not satisfied. The *necessary* condition for reproductive effort to increase with age is

$$\ln\left(\tfrac{1}{2}W(0)\left|\frac{\partial P}{\partial R}\right| \right) \geq r^* \tag{4.77}$$

where $\dfrac{\partial P}{\partial R}$ is the derivative at $P(\bullet) = 0$ (i.e., the largest value of $R(\bullet) = 0$). The trade-off curve between $P(\bullet)$ and $R(\bullet)$ has not been estimated to my knowledge. If we can make the assumption that a particular population is stable, then it is possible to test the theory using the requirement that reproductive effort will increase whenever P_{max} exceeds unity.

TESTS OF THE THEORY. Ware (1980) estimated surplus energy and allocation to reproduction in the southern Gulf of St. Lawrence cod stock. Annual survival of fish from this stock is estimated at 0.86 (Myers and Doyle 1983). Using these two pieces of information, we can construct the relationships between P_{max} and age, and between reproductive effort and age. P_{max} declines with age but is always greater than 1, and hence reproductive effort should increase with age, which indeed it does (Figure 4.29).

For most animal populations detailed information is not at hand as it is for the cod stock above, but it is possible to predict at least the response for the first years after maturation by considering the growth rate of immature individuals.

Box 4.5
Reproductive effort and age

Charlesworth and Leon (1976) assumed that $m(x)$ is a product of size, $W(x)$, and a function of reproductive effort, $R(E,x)$, so

$$m(x) = R(E,x)W(x) \qquad (4.78)$$

Effort, E, is scaled such that it varies between 0 and 1, the latter meaning that all surplus energy is diverted into reproduction. In reality, the assumption that fecundity is a function of body size is not explicitly required. What is required is that fecundity increases with age. For simplicity I shall continue the assumption that growth continues after maturation. This makes description easier without altering the mathematics. There is a trade-off between growth and reproduction such that size at age x is the product of size at age x–1 and a function of the reproductive effort at that age

$$W(x) = g(E,x-1)W(x-1) = g(E,x-1)g(E,x-2)g(E,x-3)\cdots W(0) = \prod_{i=0}^{x-1} g(E,i)W(0) \quad (4.79)$$

Because growth rate decreases with reproductive effort, $g(E,x)$ must be a decreasing function of E. Survival is also a decreasing function of reproductive effort, $S(E, x)$. Because both component functions are decreasing functions of E, the product $g(E,x)S(E, x)$ must itself be a decreasing function. Therefore, we can combine these into a single function $P(E,x) = S(E, x)g(E,x)$, (the product of survival and growth rate). This is essentially what was done in the Fish model (see above). The characteristic equation is

$$\sum_{x=1}^{\infty} e^{-r(x+1)}W(0)R(E,x)\prod_{i=0}^{x-1} P(E,x) = 1 \qquad (4.80)$$

As discussed previously, iteroparity (an assumption of this model) requires a concave-down relationship between $P(E,x)$ and $R(E,x)$.

In the year prior to maturation, energy is being diverted into gonad production, and hence the appropriate year in which to estimate the growth rate is the year two years prior to maturation. If we assume Von Bertalanffy growth, the growth increment in weight (i.e., $g(\bullet)$) is approximately

$$\left[\left(1 - e^{-k(\alpha-1)}\right)\middle/\left(1 - e^{-k(\alpha-2)}\right)\right]^3 \qquad (4.81)$$

For species in which maturation occurred in the second year of life, I used the year prior to maturation, which should be an underestimate. I estimated the

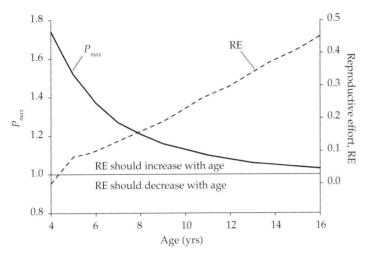

Figure 4.29 Estimated values of P_{max} and reproductive effort, RE (= proportion of surplus energy devoted to reproduction) for the Southern Gulf of St. Lawrence cod. Data from Ware (1980).

product of survival times growth rate for a wide range of fish and reptile species (Figure 4.30). In none of the turtle species did the ratio fall below unity, though most were close, while for snakes and lizards 4.8% (1/21) were less than 1 and in fish 15.4% (4/26) were less than 1. These data suggest that, in general, in vertebrate ectotherms reproductive effort should increase with age. For one species in this data set, *Pleuronectes platessa*, the allocation of energy into eggs and soma has been directly measured and, as predicted, it increases substantially with age (Rijnsdorp et al. 1983). Analysis in other species, both fish and other taxa, must rely upon indirect measures of reproductive effort.

Reproductive effort is typically measured in terms of annual productivity or, more frequently, clutch weight relative to total body weight (variously called the **gonadosomatic index**, GSI, or **relative clutch mass,** RCM). How do we interpret age-related changes in the gonadosomatic index? For fish the energy available for somatic growth is known to be an allometric function of body weight (Parker and Larkin 1959; Ware 1975a, 1978), Surplus energy = aW^b, where a and b are constants characteristic of a species and population. Many other animals (e.g., crustacea, molluscs, and lizards) show the same form of growth function as fish, and thus this equation may be a generally applicable description of the relationship between surplus energy and size in indeterminate growers. From the foregoing allometric equation we can infer that if an organism allocates a constant fraction of its surplus energy to reproduction, clutch mass should vary allometrically with body weight, with an exponent of b. In nine species of fish, the value of b ranges from 0.46 to 1.1, with a mean of 0.81, and only two greater than 1

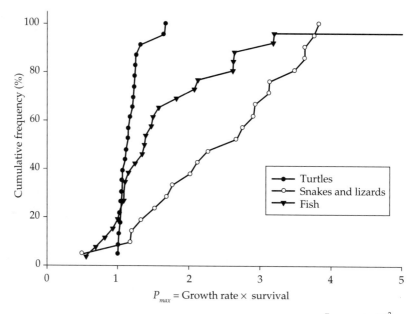

Figure 4.30   Cumulative frequency plots of P_{max} estimated as $P_{max} \approx e^{-M}\left[\dfrac{1-e^{-k(\alpha-1)}}{1-e^{-k(\alpha-2)}}\right]^3$.

Data from Roff (1984b), Shine and Schwarzkopf (1992), Shine and Iverson (1995).

(Myers and Doyle 1983). Inter-species regressions of log(clutch mass) on log(body mass) give similar exponents (spiders 0.84; hoverflies 0.95; mammals 0.80–0.84; birds 0.74; reptiles 0.88; frogs 0.90; salamanders 0.64; aquatic poikilotherms 0.92; data from appendix VIIa, Peters 1983, and Table 4, Reiss 1985) but because such regressions typically underestimate the within-species slopes, these results are inconclusive.

If an organism allocates a constant fraction of its surplus energy to reproduction, clutch mass should increase with weight, but with an exponent less than 1; the gonadosomatic index should therefore decrease with size. If we assume that size and age are closely connected, then an increase in reproductive effort with age will be indicated by either an increase in the gonadosomatic index with age *or an index that remains constant with age*. Given the uncertainty in the value of the exponent and statistical problems of determining slopes, good evidence of an increasing reproductive effort is provided only by those cases in which the index increases significantly. (For a discussion of tests of reproductive effort. RE versus size, see Klinkhamer et al. 1990). In some instances the relationship between gonadosomatic index and age has been estimated by examining the allometric relationship between clutch mass and body mass. An exponent greater than 1 (where body weight is the independent variable) indicates an increased reproductive effort with size, and hence age.

Whether one uses data on energy budgets or relative clutch mass, among vertebrates reproductive effort generally increases with age. Increases have been observed in 17 species, no change detected in 2 species, and in only one case, the harp seal, is there sufficient evidence to suggest that reproductive effort may decline (data from Table 8.4 of Roff 1992a plus reindeer data from Kojola 1991). Harp seals feed their pups for about 13 days, the pups growing at the rate of 2 kg per day (Kovacs and Lavigne 1985). During this period the females do not feed, and all lactation costs are met from stored reserves (Kovacs et al. 1990). Young female harp seals used 36% of their stored energy for milk production, middle-aged females use 29%, and old females 25% (Stewart 1986). The actual amount transferred remains the same, with younger females beginning with smaller reserves.

In molluscs an increase in reproductive effort has been detected in all 11 species studied (Table 8.4 in Roff 1992a). However, a disturbing feature of the reproductive effort estimates for the two molluscs, *Viviparus georgianus*, and *Corbicula manilensis*, is that while the reproductive efforts estimated from the energy budgets increase with age, the reproductive efforts calculated from the ratio of the amount of carbon channelled into reproduction versus the amount contained within the average adult female shows a decline with age (see Table 2 in Browne and Russell-Hunter 1978). Similarly, reproductive effort for various gastropod species shows an increase with age when estimated from energy budgets but no change or a decrease when calculated on the basis of reproductive production per time unit over the somatic energy content at the beginning of the time unit over which reproduction is measured (see Figures 2 and 4 in Hughes and Roberts 1980). The second of the two measures in each of these two studies is roughly equivalent to the gonadosomatic index, and in the first case the two measures are *positively* correlated when comparisons are made across species (Roff 1992a). These results suggest some caution be used in the interpretation of indirect measures of reproductive effort that do not vary substantially.

An increasing reproductive effort (RE) with age was observed in the insect *Gargaphia solani* (Tallamy 1982), but in two other species, the willow-carrot aphid and the vetch aphid, GSI declines with age (Dixon et al. 1993). An increase in GSI is evidence of increasing effort, but a decline in GSI is more difficult to interpret because input of energy may be declining with age but structural components may remain more or less constant. Again, the message is that measurement of reproductive effort should be most preferably based on direct rather than indirect measures.

The probability that reproductive effort will increase with age increases as r approaches zero (see above). The tropical palm *Astrocaryum mexicanum* is very long lived (> 100 yrs) and shows a consistent increase in reproductive effort with age (Piñero et al. 1982). Though rates of increase cannot be estimated, they are undoubtedly very low. Most of the molluscs for which RE data are available are also relatively long-lived, the Iceland scallop reaching 26 years (Vahl 1981), winkles up to 16 years (Hughes and Roberts 1980), and the mussel *Anodonta pisci-*

nalis 3–6 years (Haukioja and Hakala 1978). Though the early mortality of these molluscs may be very high—for example, the larval survival of Iceland scallop is estimated to be only 0.00005% (Vahl 1981)—the long lifespan of the adults suggests a generally low rate of increase. Among the vertebrate ectotherms for which RE data are avaliable, lifespans range widely, from the long-lived American plaice (20–30 yrs) to lizards with lifespans of about four years. Lifetables have been constructed for *Sceloporus jarrovi*, one of the lizard species included in the present data set, and two related species *Sceloporus undulatus* and *Sceloporus poinsetti*. Estimates of R_0 range from 0.61 to 1.23 (Tinkle and Ballinger 1972; Ballinger 1973), indicating a low rate of increase. Rates of increase of the endotherms are also probably quite low.

The crude estimate of P_{max} predicts that, for the majority of vertebrate ectotherms, reproductive effort will increase with age. The observed increase of reproductive effort with age is consistent with theory. Extension of this prediction to endotherms is not immediate because growth rates of endotherms after maturity are either low or nonexistent. Nevertheless, the data do indicate a trend for increasing effort with age. Data on invertebrates and plants are too few for us to come to any general conclusion. More accurate estimates of age schedules of reproductive effort and experimental manipulations of effort are required to adequately test the model.

Evolution of l(x)m(x) and Age: Conditions for Continued Growth after Maturation

There are at least two conditions under which growth will continue after maturity.

CONCAVE TRADE-OFF BETWEEN SURVIVAL AND REPRODUCTIVE EFFORT. This is actually the same situation as in the previous section, because if selection favors a reproductive effort that is less than 100% the remainder must, by the definition of the model, be shunted into growth. For formal analyses, see Schaffer (1974a), Taylor et al. (1974), Pianka and Parker (1975) and Leon (1976). An excellent illustration of the two phenomena is provided by the analysis of growth and reproduction in cod and herring presented by Kitahara et al. (1987: Box 4.6).

The model of Kitahara et al. (1987) is very similar to that of Myers and Doyle (1983), who also used the same fish stocks in their analysis. The analyses of Kitahara et al. (1987) and Myers and Doyle (1983) are important because they demonstrate that the theoretical analyses of León and others are consistent with the real world. What we require is an experimental analysis of the relationship between reproductive effort and survival that can be applied to a field situation; at present no such analyses exist.

FECUNDITY INCREASES WITH SIZE WHILE SURVIVAL DECREASES. A second condition under which both growth and reproduction are favored occurs when fecun-

Box 4.6
Details of the model of Kitahara et al. (1987)

Instead of using r as the measure of fitness, Kitahara et al. chose R_0, the expected lifetime reproductive success, justification for which has already been presented. The components of the model were essentially the same as described above:

$$m(x) = ap(x)g(x)$$
$$g(x) = bW(x)^c \qquad (4.82)$$
$$W(x) = W(x-1) + [1 - p(x)]g(x)$$

where $p(x)$ is the proportion of surplus energy allocated to reproduction, $g(x)$ is the surplus energy assumed to be an allometric function of weight, W, which increases each year by the amount of surplus energy not devoted to reproduction, and a, b, c are constants.

The mortality schedule was complicated by splitting the year x to $x + 1$ into two parts, the first comprising the period from the spawning season to the beginning of gonad development, and the second, the remainder. Survival in the first period was assumed to be a consequence of post-spawning stress in year x, and in the second part due to pre-spawning stress in year $x + 1$. This model leads to some difficulties in defining $l(x)$ because the effect in year $x + 1$ is now split between an effect due to year x and an effect due to year $x + 1$. For this reason Kitahara et al. (1987) defined the survival function in terms of a new function $L(x)$. In their application of the model Kitahara and colleagues make some simplifying assumptions that boil down to assuming survival in year x, in terms of this new function $L(x)$, to be a function of reproductive effort in the previous 'year'.

To ensure a graded pattern of allocation to growth and reproduction, Kitahara et al. adopted the simplest arbitrary concave function, a parabola, relating survival and reproductive effort:

$$L(x) = \left[A - Bp(x)^2\right]L(x-1) \qquad (4.83)$$

where A and B are constants. The expected lifetime reproductive success is

$$R_0 = \sum l(x)m(x) = \sum \left[A - Bp(x)^2\right]L(x-1)acW(x)^b \qquad (4.84)$$

The two constants a and c were estimated independently, but the remaining parameters were obtained by finding the set that gave the best fit to the observed growth and all allocation patterns. This negates the independence between prediction and observation, and hence any fit obtained must be viewed with caution. Thus the apparent excellent fit between prediction and observation for growth and reproductive effort for the two species analyzed should not be taken as evidence that the model is correct, only that it can be fitted to data.

dity increases with size but survival decreases with size (Perrin et al. 1993; Heino and Kaitala 1996). On the surface this seems to be a quite different circumstance to the first, but the two are closely linked in the following manner. Since fecundity increases with size, W, while survival decreases, it follows that survival decreases with fecundity, which is the same trade-off as originally considered. (See Box 4.5: the actual relationship also involves the growth increment, which is the same here.) We can see this mathematically by writing the two sets of equations side by side (the Charlesworth and Leon model on the left):

$$m(x) = R(E,x)W(x) \qquad\qquad m(x) = R(E,x)W(x)$$
$$W(x) = \prod g(E,i)W(0) \qquad\qquad W(x) = \prod g(E,i)W(0)$$
$$\text{Survival} = S(E,x) \qquad\qquad \text{Survival} = h(W) \qquad (4.85)$$
$$P(E,x) = g(E,x)S(E,x) \qquad\qquad P(E,x) = g(E,x)h(W)$$

Survival decreases with body size, $h(W)$, but the compound function $P(E,x)$ is still a decreasing function of reproductive effort. (Note also that W is itself a function of reproductive effort and hence $h(W)$ is also a function of previous reproductive efforts.) It is not clear, however, if the same mathematical conditions hold, namely a concave relationship between P and R.

Effects of Changes in the Trade-off between l(x) and m(x)—Between Survival and Fecundity

THEORY. Consider the concave trade-off function shown in Figure 4.31. Now suppose the organism adopts a new strategy that decreases its survival for a given m. The trade-off function will shift leftward most probably as shown in Figure 4.31. This shift will favor a decrease in the optimal clutch size, $= m(x)$. To illustrate this relationship I shall use the previously described three-parameter model $\lambda = ms + S = ms + f(m)$. The optimal value of m is

$$\frac{\partial\lambda}{\partial m} = s + \frac{\partial f(m)}{\partial m} = 0 \quad \text{when} \quad \frac{\partial f(m)}{\partial m} = -s \qquad (4.86)$$

The optimal m occurs when the slope of the trade-off function equals $-s$. For any given value of $f(m)$ (=S, adult survival), the slope of curve A is less than that of curve B: therefore to satisfy Equation (4.86) the optimal m on curve B must lie above that on curve A, which implies a lower m. To provide a concrete example let $f(m) = 1 - am^2$. The derivative of this function is $2am$ and hence the optimal combination of S and m is thus $m = s/(2a)$ and $S = 1 - a(s/(2a))^2 = 1 - s^2/(4a)$. If $a = 0.278$, the optimal combination is $m = 0.9$, $S = 0.774$, whereas if $a = 2.5$ the optimal combination is 0.975 (Figure 4.31).

TESTS: INTRODUCTION. An ecological circumstance that meets this scenario is one in which there are two possible categories of lifestyles, producing the two types

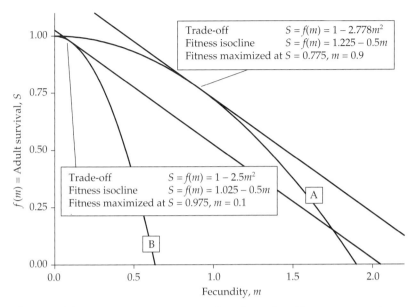

Figure 4.31 and caption text:

Trade-off $\quad\quad\quad S = f(m) = 1 - 2.778m^2$
Fitness isocline $\quad\quad S = f(m) = 1.225 - 0.5m$
Fitness maximized at $S = 0.775$, $m = 0.9$

Trade-off $\quad\quad\quad S = f(m) = 1 - 2.5m^2$
Fitness isocline $\quad\quad S = f(m) = 1.025 - 0.5m$
Fitness maximized at $S = 0.975$, $m = 0.1$

$f(m)$ = Adult survival, S

Fecundity, m

A

B

Figure 4.31 Two hypothetical examples of a trade-off between m and adult survival. Note that a leftward shift of the trade-off function from A to B changes the optimal combination of S and m, provided that they have the same maximum and functional form.

of curves described above. Two examples that may meet this criterion are foraging mode and reproductive mode. Ideally, one would like to test this hypothesis within a species that possessed two distinct foraging morphs or modes of reproduction. Such data are lacking and the next best alternative is to compare sets of different species, preferably with controls for phylogenetic history. In examining different species it is typically insufficient to consider only m because propagule size is likely to vary among species. Assuming that variation in propagule size changes early survival, we can incorporate this variation into m and use total clutch mass as a relative index. Five data sets for which the above requirements can at least be partially satisfied are available for lizards, spiders (foraging mode) and snakes, copepods, and fish (reproductive mode).

The prediction for the above dichotomous circumstance can be expanded to predict that there will be a negative correlation between adult survival and relative clutch mass, all other things being equal. Of course, all other things are rarely equal, and hence a negative finding may indicate that other factors in addition to foraging or reproductive mode are contributing to survival. For example, survival may differ with body size, which would potentially confound the predicted relationship. Thus it may be possible to demonstrate a broad scale relationship (i.e., differences between groups) without necessarily being able to demonstrate a more refined correlation.

TESTS: FORAGING MODE VERSUS RELATIVE CLUTCH MASS IN LIZARDS. Vitt and Congdon (1978) classified lizards into two foraging groups, "sit and wait" predators versus "widely foraging" predators, and hypothesized that lizards in the former category will experience a reduced "loading" constraint and hence will have larger relative clutch masses. Reference has already been made to the decreased running speed in lizards and hence likely increased susceptibility to predators (see Chapter 3). Unfortunately, there is no clear field evidence for a relationship between loading and vulnerability to predators. Data on the stomach contents of field-collected horned adders (*Bitis caudalis*) suggest that widely foraging lacertids are more vulnerable than sedentary species (Huey and Pianka 1981), but this does not demonstrate that loading per se changes vulnerability. (Lee et al. 1996 did show that in starlings escape performance is degraded in gravid females.) McLaughlin (1989) examined the concept of a dichotomy in search patterns in both birds and lizards and found that species can be classified as "mobile" or "sedentary" on the basis of the frequency of moves, the former type moving approximately 7–10 times as often as the latter.

A detailed field study of the relationship between foraging type and relative clutch mass in four species of lizard was undertaken by Magnusson et al. (1985). Three of these species, *Cnemidophorus lemniscatus*, *Ameiva ameiva*, and *Kentrophyx striatus*, are widely foraging species, the last, *Anolis auratus*, a sit-and-wait predator. The first three exhibited different rates of activity, as measured by movements per hour, mean speed of movement, and area used, demonstrating that within categories there is a continuum of rates of activity. As expected, *A. auratus* showed the lowest level of activity. The relative clutch masses of the three actively foraging species varied in the opposite direction to foraging rate, as predicted by the loading constraints hypothesis. The anole has a very low clutch mass, but this is typical of anoles and, as discussed in Chapter 2, may be a consequence of an arboreal habit.

The prediction that the relative clutch mass of "sit-and-wait" lizards will be greater than that of "widely foraging" species is supported by data on relative clutch mass in 130 species of lizards (Figure 4.32). Unfortunately, with the exception of one family (for which there are 11 data points), the two categories of lizards belong to different taxonomic families (Figure 4.32), and thus phylogeny cannot be excluded as a confounding factor. I attempted a more detailed analysis by testing for a correlation between annual survival and relative clutch mass (data from Dunham et al. 1988; Shine and Schwarzkopf 1992), but the correlation, though negative, was not significant ($r = -0.09$, $n = 39$, $P = 0.59$). Incorporation of size did not increase the significance. As there were only four "widely foraging" species, the difference between groups in survival rates could not be tested, although there did not appear to be even a trend towards a difference.

TESTS: FORAGING MODE VERSUS RELATIVE CLUTCH MASS IN SPIDERS. Spiders capture their prey either by active hunting (cursorial mode) or by sitting and waiting in a web (web builders). As with lizards, I shall assume that the cursorial mode is a more risky existence, and hence I predict that the relative clutch mass of curso-

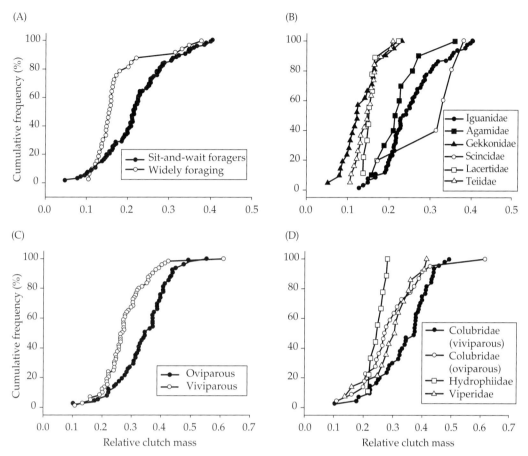

Figure 4.32 (A,B) Distributions of relative clutch mass of lizards classified according to foraging mode (A; *n* = 33, 97 for widely foraging and sit-and wait, respectively) and taxonomic family (B; only families with five or more data points shown). Data sources: Huey and Pianka (1981), Ananeva and Shammakov (1986). (C,D) Distributions of relative clutch mass of snakes classified according to reproductive mode (C; *n* = 52, 60 for oviparous and viviparous, respectively) and taxonomic family (D; only families with nine or more data points are shown). Data sources: Seigel and Fitch (1984), Seigel et al. (1986).

rial spiders will be smaller than that of web builders. Simpson (1995) correlated egg and clutch size with body size in the two groups but did not regress total reproductive effort (= egg size × clutch size) on body size. Combining his separate regressions gives log(clutch size × egg size, cursorial) = −0.12 + 2.06 log(body size), log(clutch size × egg size, web builders) = 0.07 + 2.06 log(body size). The two regressions have the same slope, but the intercept for web-building spiders is greater than that of cursorial spiders, meaning that for a given body size web-building spiders have a larger relative clutch mass, as predicted. These results are based on genus means and hence should not be overly influenced by phylogenetic effects.

ESCAPE MODE VERSUS REPRODUCTIVE MODE IN SNAKES. If clutch mass impedes the movement of lizards, it might also be expected to have a similar effect on locomotor performance in snakes. Gravid female garter snakes, *Thamnophis marcianus*, show a reduced locomotor performance relative to nongravid females, and this effect increases with relative clutch mass (Seigel et al. 1987). Another factor that may be important is whether reproduction is viviparous or oviparous. The risk of being preyed upon is likely to be a function of the relative clutch mass and the period over which this is maintained, and hence viviparous snakes may be at greater risk than oviparous snakes (Tinkle and Gibbons 1977). Seigel and Fitch (1984) examined the relationship between relative clutch mass, reproductive mode, method of predator escape ("flee" vs "crypsis"), prey location ("active foraging" vs "sit-and-wait") and prey capture ("pursue" vs "ambush"). They reported a significant effect of reproductive mode (Figure 4.32), but not foraging mode or escape behavior. The lack of statistical significance may be a consequence of misclassification of the behavior of snakes, since these are more difficult to observe than lizards, or that snakes do not fit easily into dichotomous groups (Seigel and Fitch 1984). Further field studies are required to assess these possibilities. As in the case of the lizard data, the analysis is somewhat confounded by taxonomic family, but the one family for which there is a relatively large data set for both reproductive modes (Colubridae) does show the same separation as found for the combined data set (Figure 4.32).

TESTS: SPAWNING TYPE VERSUS RELATIVE CLUTCH MASS IN ZOOPLANKTERS. Evidence has been presented earlier (Chapter 3) that reproduction for zooplankters is costly in terms of survival, because they become more visible and probably less mobile. Marine planktonic copepods either carry their eggs in sacs or broadcast their eggs into the water (Kiorboe and Sabatini 1995). From the laboratory experiments with copepods it is reasonable (though unproven) that sac spawners are more likely to be preyed upon than broadcast spawners. Thus the relative clutch mass of sac spawners should be less than that of broadcast spawners. Indeed, the weight-specific fecundity (= number of eggs per day × egg wt/female wt) is significantly larger in broadcast spawners than sac spawners (0.211 ± 0.02 vs 0.097 ± 0.02, $P = 0.0005$, one-tailed Mann-Whitney test).

Although most fish lay eggs, a few do retain their young. There is a highly significant difference between oviparous and viviparous species, the latter having a mean GSI (gonadosomatic index = wt of ovaries/somatic wt. This is typically the same definition as relative clutch mass) of 0.069 (SE = 0.01, $n = 7$) for oviparous species compared to 0.183 (SE = 0.03, $n = 21$) for viviparous species ($P = 0.0015$, one-tailed Mann-Whitney test because of different variances). Similarly, as predicted, the mortality rate of viviparous fish is less than that of oviparous ($M = 0.164 \pm 0.05$ vs 0.340 ± 0.06 [where annual survival = e^{-M}] $P = 0.021$,

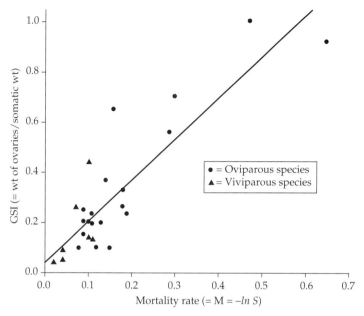

Figure 4.33   Plot of GSI on mortality rate for a wide variety of fish species. Data from Gunderson (1997).

one-tailed Mann-Whitney test). In accord with prediction, there is a highly significant positive regression between GSI and mortality rate, and there is no independent effect due to mode of reproduction (covariance analysis, Figure 4.33).

The above results support the predictions from an assumption that there is a trade-off between $l(x)$ and $m(x)$ (= adult survival and reproductive effort). In the next section this trade-off will be explored further, considering how the optimal allocation to each reproductive episode differs when there is a survival cost to reproduction.

The Trade-off between l(x) and m(x) continued: Predicting the Optimal Clutch Size

LACK'S HYPOTHESIS.   Throughout this book I emphasize that a study of the evolution of life history traits should focus upon the $l(x)m(x)$ function and not more narrowly upon, say, the $m(x)$ function. Lack's hypothesis is a good example of how far astray one may go in ignoring components of the life history. Lack (1947) laid out the problem and his hypothesis thus:

> "[I]f clutch-size is inherited *and if other things are equal,* (my italics) those individuals laying larger clutches will come to predominate in the population over those laying smaller clutches. It is easy to see why a species

which normally lays four eggs and raises four young should not, instead, lay only three eggs. The difficult problem is to discover why such a species should not normally lay five eggs. I believe that, *in nidicolous* [= altricial] *species, the average clutch-size is ultimately determined by the average maximum number of young which the parents can successfully raise in the region and season in question* (Lack's italics), i.e. that natural selection eliminates a disproportionately large number of young in those clutches which are higher than the average, through the inability of the parents to get enough food for their young, so that some or all of the brood die before or soon after fledging, with the result that few or no descendants are left with their parent's propensity to lay a larger clutch "(Lack 1947, p. 319).

Lack later applied this hypothesis to birds in general, with possible exception of some gulls where the number of eggs may be constrained by the size of the brood patch (Lack 1947, p. 326). In a later paper (Lack 1948) he extended the hypothesis to mammals. The emphasis of Lack's hypothesis is on the survival of offspring not parents; this is made clear in his book, *The Natural Regulation of Animal Numbers,* first published in 1954, where he states (p. 22 of 1967 edition): "[I]n most birds clutch-size has been evolved through natural selection to correspond with the largest number of young for which the parents can on average find enough food."

Lack's hypothesis assumes that the only important interactions are negative density-dependent interactions between siblings within a clutch. It is not a hypothesis restricted to birds and mammals and does not require that parental care be given after the eggs are laid. It will be strictly true for a semelparous life history, for here it is only the interactions between siblings and nonsiblings that is important.

Lack's hypothesis predicts that the most productive clutch should also be the most frequent clutch observed. Klomp (1970) in his review of the hypothesis noted that the most productive brood-size was similar to the most frequent in 10 species but larger in 11. The "simple" version of Lack's hypothesis, namely that the average number of observed chicks fledged represents the best the parents can do has been tested by experimental enlargement of clutches. According to Lack's hypothesis such enlarged clutches should produce fewer fledged young. One might argue that the hypothesis refers to survival to reproduction, but most manipulation experiments are unable to follow birds to this age. Roff (1992a) surveyed 55 studies and found that in 41 cases enlarged clutches produced more fledged offspring than control clutches, whereas in 10 cases fewer offspring were produced, and in 4 there was no effect of enlargement. Comparing only the positive and negative cases, we find that there is a significant deviation from 50:50 ($\chi^2 = 18.74$, $df = 1$, $P < 0.0001$). VanderWerf (1992) used a meta-analysis in which **effect size** was defined as $(\bar{n}_E - \bar{n}_C)/SD$, where $\bar{n}_E$ is the mean number from the enlarged brood in a given study, $\bar{n}_C$ is the mean number from the control, and SD is the pooled standard deviation. Thus the result of each experiment is standardized into standard deviation units. The null hypothesis of no effect of manipulation was tested with a two-tailed *t*-test. The mean effect size of 0.55 (SE = 0.22) was highly significantly different from zero ($t = 2.52$, $n = 42$, $P = 0.016$). There is a positive but nonsignificant correlation between effect size and the relative size of the enlargement, and a significant negative correlation with study length. VanderWerf (1992) suggested that the latter finding supported Boyce and Perrin's (1987) hypothesis that large clutches suffer significant losses in the infrequent "bad" years. However, it is possible that the finding is a statistical artifact due to VanderWerf's selection criterion; given multiple years, VanderWerf either pooled the data or if this were not possible selected the data from the *smallest* enlarged broods. This procedure is conservative with respect to the null hypothesis, which is good, but it will tend to make effects from multiple years on average smaller than single years.

Thus Lack's hypothesis can be rejected for number at fledging. But fledgling mass is lower in enlarged broods (Figure 3.16), which might lead to reduced survival between fledging. But even if the survival of chicks is not affected there remains a second problem with Lack's hypothesis. The problem, as pointed out by Williams (1966) and Klomp (1970), is that Lack's hypothesis ignores adult survival. Lack was not unaware of the possible effect of adult survival in relation to clutch size but considered it unimportant, at least with respect to one species, the Great tit (Lack 1966). Data on more species (see Chapter 3) indicate that adult survival is reduced when clutch size is enlarged. Thus we need to consider the $l(x)m(x)$ function of the parents, not simply their $m(x)$ function.

PREDICTING THE OPTIMAL CLUTCH SIZE. The effect of a trade-off between clutch size and adult survival can be examined using the model of Charnov and Krebs (1973)

$$\lambda = m(X)s(X) + S(X) \tag{4.87}$$

This model assumes that all parameters are functions of clutch size, X, that reproduction occurs after one year, and that the three functions do not change with age. The last two assumptions are not critical to the qualitative results. Parental care, as expressed by the relation between clutch size and adult survival, can equally refer to the post-hatching care provided by birds and mammals or the provisioning of eggs or seeds. The issue is thus far more general than simply an issue of bird life history.

The most productive clutch size, Y, is $Y = Xs(X)$. The value of X at which Y is maximized, X^*, is obtained by differentiating Y with respect to X, $\partial Y/\partial X$ and setting the result equal to zero. The clutch size that maximizes λ (= fitness) is obtained in the same manner,

$$\frac{\partial \lambda}{\partial X} = \frac{\partial Y}{\partial X} + \frac{\partial S(X)}{\partial X} = 0, \text{ when } \frac{\partial S(X)}{\partial X} = -\frac{\partial Y}{\partial X} \tag{4.88}$$

Now note that (1) $\partial S(X)/\partial X$ is negative for all X (lower-left panel, Figure 4.34), and (2) $\partial Y/\partial X$ is positive when X is less than X^* and negative when X is greater than X^* (upper-right panel, Figure 4.34). Thus Equation (4.88) will be satisfied

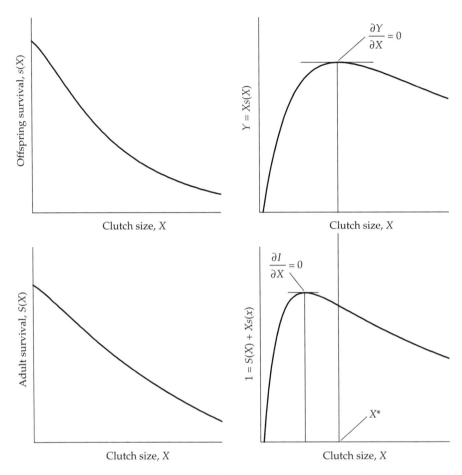

Figure 4.34   Graphical demonstration of the mathematical argument that if parental survival or future fecundity is a function of clutch size, the optimal clutch size will be less than the most productive clutch size.

only when X is less than X^*. A negative relationship between clutch size and parental survival and/or future fecundity favors an optimal clutch size below that which produces the most offspring. The Lack value will be the optimal clutch size when the only important factors are negative density-dependent interactions within the clutch. The incorporation of additional factors in the life history will almost certainly reduce the optimal clutch size. Two such factors are decreased adult survival and fecundity.

In this chapter I have assumed that the environment is constant. This does not, however, mean that there is no variance in life history traits. As discussed in the first part of the chapter there are a number of mechanisms that preserve additive genetic variance even in a constant environment. The high heritability of fecundity components, such as clutch size, has already been noted. As a consequence, there will be a distribution of clutch sizes within the population. As pointed out by Mountford (1968), this itself will generally lead to an optimal clutch size that is less than the Lack value (most productive clutch size). Mountford illustrated his proposition with data on guinea pigs. Litter size in guinea pigs is lognormally distributed, with a mean of approximately ln(5) (Figure 4.35). Survival to weaning declines rapidly with litter size (Figure 4.35). The Lack litter size is equal to the product of litter size and survival to weaning, and it

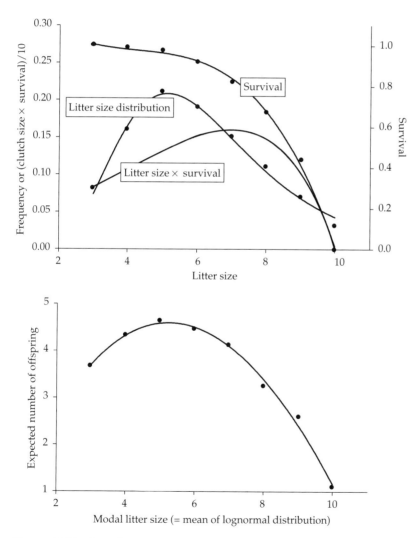

Figure 4.35 Data from Mountford (1968) on litter-size variation in guinea pigs. Solid lines show fitted curves (lognormal for the litter-size distribution and cubic for survival and expected number of offspring).

occurs at approximately 7. The expected number of offspring from a given female is equal to $\sum is_i f_i$, where i is litter size, s_i is survival to weaning, and f_i is the frequency of this clutch size. If we assume that the distribution function applies to individuals, the expected number of offspring for individuals with the observed mean is approximately 4.6, which corresponds to the observed mean litter size not the most productive. Now consider females for which the mean litter size is more or less than 5. Assuming that the frequency distribution retains the same shape and only the mean (of the lognormal) changes, we can calculate the expected number of offspring across a range of litter sizes (Figure 4.35). Females with means greater or less than 5 produce fewer offspring. The reason for this is that below 5 there is a loss due to small litter sizes, whereas above 5 there is an increased loss due to the increased mortality. The latter has a greater influence than the increased number of young initially produced.

The above phenomenon is quite general but not universal. Suppose, for example, clutch size is symmetrically distributed and survival is a linear function of clutch size across the entire range (i.e., it must not reach 1 or 0). Under these conditions the mean clutch size is also the most productive clutch size. Any skew in the clutch size distribution or nonlinearity in the survival function will make the Lack value suboptimal.

Effects of a Three-way Trade-off Involving Immature Survival, Adult survival and Fecundity

The likelihood of a trade-off not only between adult survival and brood size but also between brood size and early survival was suggested in the previous section. There is abundant evidence that enlarged broods in both birds and mammals give rise to smaller offspring, which are not likely to survive as well as better-nourished offspring (Chapter 3). The same phenomenon applies to organisms that show no extended parental care—it only requires that parental investment in the offspring affects their survival or that the offspring compete with each other for resources. Thus, for example, plants can increase the survival of their offspring by increasing seed size (Roff 1992a) and by dispersing them over a sufficiently large area that the seedlings do not compete against each other.

When there is a trade-off between adult survival and reproductive effort there is a severe constraint on the shape of the trade-off function. To produce intermediate values of reproductive effort the curve must be concave down. Such a restriction does not apply for the trade-off between immature survival, s, and clutch size, m. This is readily seen by again considering the relationship $\lambda = ms + S$. Now suppose that s is a function of m, as might occur if there is density-dependent interaction among siblings (as could occur simply because of close proximity to each other). We can write the equation as $\lambda = mf(m) + S$, where $f(m)$ is a decreasing function of m. The isoclines of λ with respect to $f(m)$ are $f(m) = (\lambda - S)/m$, and hence are hyperbolic rather than linear. Thus the largest isocline can intersect at a single point on the trade-off function even if it is not concave.

Whereas birds and mammals typically produce broods sequentially and show parental care to this single brood until the time of offspring independence, other organisms, such as many amphibians and insects, deposit their eggs over a number of sites, parental care being exercised in the placement and provisioning of the eggs. The female must select both the appropriate site and the appropriate clumping of eggs to ensure the maximum number of offspring. This decision involves a number of considerations: (1) the number of females per site, (2) the type of density-dependence, (3) the limiting "resource." This limiting resource could be the number of eggs the female has, the number of sites available, a nonzero adult mortality, limited time, limited reserves, site quality, or a combination of the above. I shall consider only the effects of single limiting factors; no qualitative differences emerging in models with multiple factors (Wilson and Lessells 1994).

ONE FEMALE PER SITE, NEGATIVE DENSITY-DEPENDENCE, LIMITED NUMBER OF EGGS. In this case it is assumed that any increase in the density of offspring at a site will decrease survival or adult size (and hence fecundity), that the number of sites exceeds the number of eggs, and there is no cost in terms of time or survival to the female of visiting sites. The optimal number of eggs per site is obviously 1.

ONE FEMALE PER SITE, NEGATIVE DENSITY-DEPENDENCE, NUMBER OF SITES LIMITED. To evaluate the optimal number of eggs per site we must consider the number of eggs, X, the survival of offspring, $s(X)$, and any decrements in fecundity due to density-dependent effects, $m(X)$. The expected contribution to fitness from a single patch, $W(X)$, is then $W(X) = Xs(X)m(X)$. Typically, increased density leads to decreased body size, which leads to decreased fecundity, and the product of the three components is dome-shaped (Figure 4.36; Godfray 1987; Vet et al. 1994; Wilson 1994; Nakamura 1995). The most productive clutch size per patch is where $\partial W(X)/\partial X = 0$, which is simply the local maximum (Figure 4.36). This clutch size is referred to as the **Lack clutch size** (Charnov and Skinner 1984; Godfray et al. 1991). Because it is assumed that mortality of the female is insignificant over the oviposition period, in the present case the Lack clutch size has the highest fitness.

ONE FEMALE PER SITE, NEGATIVE DENSITY-DEPENDENCE, NONZERO MORTALITY RATE. If adult mortality rate were an important factor, we would expect that, all other things being equal, that females laying only one egg at a time would be less likely to lay as close to their maximum than females that lay in batches. Courtney (1984) tested this hypothesis by comparing the realized fecundity of insects that lay their eggs singly with insects that lay their eggs in batches. As predicted, batch-layers realize more of their potential fecundity than single-layers.

What effect does mortality rate have on the optimal clutch size? The qualitative results do not depend upon the fitness measure used. Here I maximize R_0. Letting the adult mortality rate be M we have

$$R_0 = \int_0^{F(X)} e^{-Mt} W(X) dt = \frac{W(X)}{M}\left[1 - e^{-MF(X)}\right] \qquad (4.89)$$

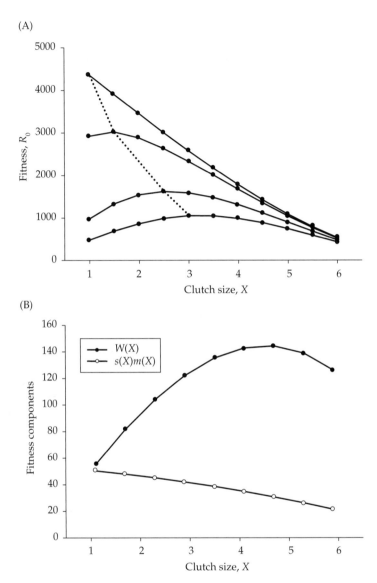

Figure 4.36 The optimal clutch size when there is mortality and negative density-dependence at a site. The data are based on that given by Weis et al. (1983). (A) Fitness as a function of clutch size and survival probability (from top to bottom e^{-M} = 1.00, 0.99, 0.95, 0.90). The dotted line shows the optimal clutch size. (B) The functional relationships are $m(c) = 98 - 12c$, $s(c) = 0.55 + 0.04c$.

where α is the period from egg to first reproduction and $F(X) = m(X)/X$ is the maximum number of clutches that can be laid. To find the optimum X, we differentiate R_0 with respect to X and set this equal to zero. Unfortunately, this does not give a readily interpretable result. We can, however, readily show that the optimum must be less than the Lack clutch size by showing that when $F(X) = \infty$, the optimum is the Lack value. The derivative of R_0 with respect to X is

$$\frac{\partial W(X)}{\partial X}\frac{1}{M} \tag{4.90}$$

and hence the optimum clutch size, given $F(X) = \infty$, is the Lack clutch size. Thus for finite values of $F(X)$, the optimal is less than the Lack value. This result is illustrated in Figure 4.36 using relationships based on the data for the gall maker *Asteronyia carbonifera* (Weis et al. 1983). As expected, in *A. carbonaria* fecundity decreases with clutch size, but mortality rate due to parasitism actually goes down with clutch size, which Weis et al. ascribe to gall size not increasing much with the number of larvae and hence being no more likely to be found (the **convoy hypothesis**—see Roff 1992a). The optimum number per gall when there are no time or survival constraints is 1 larvae per gall, and the Lack value is approximately 4.5 larvae per gall (Figure 4.36). As survival between sites declines from 1 to 0.9, the optimal clutch size increases from 1 to 3 (Figure 4.36).

ONE FEMALE PER SITE, NEGATIVE DENSITY-DEPENDENCE, LIMITED TIME OR RESERVES. In the case of limited time, the evolutionary problem is to incorporate the time it takes to locate suitable sites. Limited reserves can be converted into time units and hence the same arguments can be used for both scenarios (Smith and Lessells 1985). The approach taken by Parker and Courtney (1984) and Skinner (1985) was to maximize productivity per unit time using the marginal value theorem (Charnov and Skinner 1984, 1985). To apply the marginal value theorem we first convert clutch size into oviposition time.That is, oviposition time equals clutch size X time to produce and lay an egg. For convenience, time is scaled such that 1 time unit is equivalent to the time required to produce and lay one egg, in which case oviposition time and clutch size are the same magnitude and we can simply refer to clutch size rather than oviposition time. The optimal clutch size is that at which a line drawn from the point $-t$, where t is search time, is tangential to the curve of per clutch productivity on clutch size (Figure 4.37). It is immediately apparent that this value will be less than that which maximizes productivity per clutch (i.e., the Lack clutch size). This is formally demonstrated in Box 4.7.

As already shown from a consideration of mortality, as search time increases, the optimal clutch size increases (Figure 4.37), and in the limit ($t = 4$) the optimal and most productive clutch sizes converge. This makes intuitive sense; if search

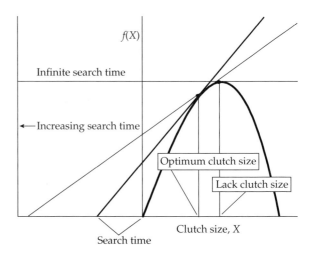

Figure 4.37   Graphical calculation of the optimal clutch size when a female must search for several sites. $f(X)$ is the productivity for a clutch size X. Unless search time is infinite (in which case only one site is visited), the optimum clutch size is less than the Lack clutch size and increases towards it as search time increases.

time is infinite, the female has only a single chance of laying, and hence she should lay the Lack clutch size. A decrease in site density should, in general, lead to an increase in search time and hence and increase in clutch size. Similarly, an increase in female density will increase the search time for unoccupied sites and hence also lead to an increase in clutch size. These predictions are supported by studies on two parasitoid wasps (Feijen and Schulten 1981; Pak and Oatman 1982; for analysis, see Roff 1992a), and on the bean beetle *Callosobruchus maculatus* (Mitchell 1975).

ONE FEMALE PER SITE, NEGATIVE DENSITY-DEPENDENCE, BETTER SITE QUALITY. Skinner (1985) suggested that, for a constant search time, clutch size will increase with host quality. This prediction is not universally true but depends upon the particular manner in which host quality is entered into the per clutch production function, $W(X)$. Skinner based his argument on the marginal value approach outlined in the previous section. It can be shown using this approach that Skinner's hypothesis is not necessarily true (Roff 1992a). A simple counter-example is a model in which site quality varies in proportion to some reference function: $W_0(X)$; so $W(X) = aW_0(X)$. Because the optimal clutch size is

$$X = \frac{W(X)}{\partial W(X)/\partial X} - t \tag{4.91}$$

Box 4.7
Proof that the optimal clutch size is less than the Lack value, given limited time

Fitness, w, is defined as the maximal productivity of a female per unit time,

$$w = \frac{W(X)}{X + t} \qquad (4.92)$$

Differentiating w with respect to X gives

$$\frac{\partial w}{\partial X} = \left(\frac{\partial W(X)}{\partial X} \right) \left(\frac{1}{X + t} \right) - \frac{W(X)}{(X + t)^2}$$

$$= 0 \qquad \text{when} \quad \frac{\partial W(X)}{\partial X} = \frac{W(X)}{X + t} \qquad (4.93)$$

All the terms on the right-hand side are positive and hence the optimal clutch size must occur when the derivative is positive. This is to the left of the clutch size at which the derivative is zero, in other words, at a lower clutch size.

(see Equation [4.93]), the constant of proportionality cancels and the optimum does not change. Without specification of $W(X)$ as a function of site quality it is not possible to predict how the optimal clutch size will change with site quality.

Increases in clutch size with site quality have been observed in *Trichogramma embryophagum* (insect egg parasite, Klomp and Teerink 1962, 1967), *Callosobruchus maculatus* (beans, Mitchell 1975), *Aphytis nerii* and *A. lingnanensis* (scale parasites, Luck et al. 1982), *Pegomya nigritarsus* (leaf miner, Godfray 1986), *Goniozus nephantidis* (ectoparasitoid of lepidopteran larvae, Hardy et al. 1992) and *Aphaereta minuta* (larval parasitoid, Vet et al. 1993). The wide diversity of taxa over which increased clutch size has been found suggests that a positive relationship is general. More work is needed in this area to clarify why increases are common. Specifically, what realistic trade-off functions generate an increased clutch size with site quality? In one study, that of Klomp and Teerink (1967) enough information is presented to test the hypothesis that observed changes in clutch size are optimal.

Trichogramma embryophagum is an egg parasite of insect eggs, mainly lepidopteran. Klomp and Teerink (1967) measured the changes in survival, body size, and fecundity as a function of clutch size (= number of eggs introduced onto the host egg) and the species of lepidopteran. Data were presented for three species, two with very similar sized eggs (*Bupalus* and *Ellopia* with egg sizes of 0.26 mg and 0.25 mg, respectively) and one with a very much smaller egg (*Ana-*

gasta, 0.028 mg). Individual fecundity data were not presented, but their analysis showed that fecundity was a linear function of adult head width, independent of the host upon which the insect was raised (see their Figure 5). Therefore, instead of fecundity we can use head width (which they did present). The three "productivity" curves are shown in Figure 4.38. As expected from the size of the host, the productivity on *Anagasta* eggs is much lower than on the other two. Further, there is virtually no difference between *Bupalus* and *Ellopia* (the former data are actually hidden in the graph by the latter data), demonstrating that site quality is largely due to size. I fitted a cubic polynomial to the *Bupalus* and *Anagasta* data to the given fitted curves. (I chopped the *Anagasta* curve at a clutch size of 11 as the cubic gave an increasing function after this point). The observed clutch sizes for *Bupalus* and *Angasta* were 7.3 and 1.05 eggs, respectively. Using the two polynomials I calculated the tangent at these two points and then extended the lines back to the value on the clutch-size axis at which productivity equaled zero. If *T. embryophagum* is optimizing its clutch size, these two values should be the same. The value for *Bupalus* was –2.47 and for *Anagasta* it was –1.98 (Figure 4.38). Given the large differences in the productivity curves and spread of clutch sizes,

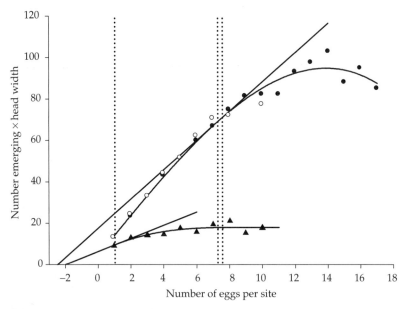

Figure 4.38 The productivity curves for *Trichogramma embryophagum* laying eggs on three lepidopteran hosts, *Anagasta* (triangle), *Bupalus* (solid circle) and *Ellopia* (open circle). Dotted lines are drawn at the observed clutch sizes (1.05, 7.3, and 7.6, respectively). The cubic equation fitted to the *Anagasta* data was $y = 5.8209 + 4.0621x - 0.4505x^2 + 0.0164x^3$ and for *Bupalis* it was $y = 4.1759 + 9.8461x - 0.0080x^2 - 0.0166x^3$. The equations for the two tangents were, respectively, $y = 6.28 + 3.17x$ and $y = 17.52 + 7.075x$, where x is clutch size and y is productivity. Data from Klomp and Teerink (1967).

these two values are very close and provide at least initial support for the hypothesis that *T. embryophagum* is optimizing its clutch size to site quality.

MULTIPLE FEMALES PER SITE, NEGATIVE DENSITY-DEPENDENCE. Suppose that each female lays only once at each site and that n females lay at each site. With a single female, the productivity function is $W(X) = Xs(X)m(X)$; it is now necessary to take into account the eggs laid by the other females, and the productivity function is $W(X) = Xs(nX)m(nX)$. Two points can immediately be made: First, the optimal clutch size may increase, remain the same, or decrease as the number of females per site changes (Ives 1989), and second, the optimal clutch size for any single female is highly unlikely to be the Lack clutch size, but will generally be smaller. The foregoing results are based on the assumption that larvae cannot disperse from patches. The situation becomes quite complex when spatial effects are included.

The models thus far have all assumed negative density-dependence, which is probably the most general case. A general conclusion from these models is that the optimum clutch size will be less than the Lack clutch size. There has been no general survey in insects as in birds but an analysis of 10 species of parasitic wasps showed that in all cases the observed clutch size was less than the Lack clutch (Takagi 1985; Dijkstra 1986; Godfray 1987; Taylor 1988a; Hardy et al. 1992; Vet et al. 1994). The mean clutch size in the gall maker *Asteromyia carbonifera* is also less than the Lack value (Weis et al. 1983, Figure 4.36).

SINGLE FEMALE PER SITE, POSITIVE DENSITY-DEPENDENCE (TO A POINT). With negative density-dependence the optimal clutch size, given no searching costs and an unlimited resource, will be a single individual per site. In some insects the survival of larvae is enhanced by aggregation, a phenomenon known as the **Allee effect**. The Allee effect can be a result of several factors (Godfray 1987): Insects in a group may be better able to exploit their resource, to defend themselves from a predator or parasite, attract mutualists (e.g., ants), or increase the probability of mating. Regardless of how such an effect is generated it will clearly favor clutch sizes in excess of 1. However, at some density there will be negative effects and hence the $W(X)$ will be dome-shaped. (With only negative density-dependence the curve can be monotonically declining.) The foregoing argument thus still in large measure applies and the usual clutch size is predicted to be less than the Lack clutch size.

Godfray (1986) examined the Allee effect in the leaf-mining fly *Pegomya nigritarsis*. This species lays its eggs on the leaves of various species of *Rumex*, into which the newly hatched larvae mine; they remain until they drop to the ground and pupate in the litter layer or soil. Godfray (1986) studied the life history in *Rumex obtusifolius*. Mortality can be divided into two categories (Godfray 1986): "larval death" and mortality from endoparasitic attack. The first category, "larval death" includes death due to a number of causes that cannot be separated: starvation in early instars, invertebrate predation, parasitoid host feeding, and interference from other larvae. The proportion of larvae surviving this category decreases monotonically with clutch size.

Attacks by endoparasites generally occur when the larva is in the first instar. The parasite egg hatches, but the larva remains as a first instar until the fly pupates, at which time the parasite completes development, killing the fly. Only one parasite emerges per host. The proportion of larvae surviving attacks by endoparasites is a concave function of clutch size, but the mechanism generating this relationship is unknown. The overall mortality rate is also a concave function of clutch size.

The adult size of *P. nigritarsus* is independent of clutch size and thus there is probably no fecundity cost to increased larval density (at least within the range of 1–9 larvae per leaf). Therefore, the optimal clutch size will be greater than 1 even if there are no costs associated with searching and egg production, or even if resources are not limiting. Indeed, for a leaf miner the amount of resources available are likely to be very large, and the time to move from one suitable oviposition site to another are probably very small. Under these conditions the optimal clutch size is that at which larval survival is greatest (since fecundity does not vary with larval density), which is also the Lack clutch size. Recall that, at the other extreme, if resources are scarce or survival between ovipositions is very low, the optimal clutch size is also the Lack value. For *P. nigritarsis* this value is 3 eggs per clutch, which is most frequently observed in the leaves of *R. obtusifolius*.

The Evolution of the *m(x)* Components: Clutch and Offspring Size

Thus far we have examined how trade-offs influence the evolution of the product function $l(x)m(x)$ or $m(x)$. In the latter case it is still not $m(x)$ per se that is under selection but again the combined effects of a trade-off between $m(x)$ and either early survival or future fecundity. In general, it is therefore sufficient to consider not $m(x)$ itself but some index of reproductive output such as total offspring biomass, as indeed was done in considering the effects of age, and optimal clutch size in organisms that disperse their clutches among sites. A female can modify the survival or growth of her offspring by varying the number of offspring as in the previous section or by varying the quality of her offspring, which would either lead to a decreased fecundity or an increased reproductive effort by the parents and hence a reduction in their future fecundity and/or survival. Because selection acts upon individuals it may be easier to evaluate the evolution of the $l(x)m(x)$ function by considering how selection acts upon its components. In the present case this means considering the consequences of varying offspring quality on $l(x)$ and $m(x)$.

Empirical Evidence of a Trade-off between the Number and Size of Offspring

The most obvious way in which offspring quality might be enhanced is by increasing their size. A survey of the effects of phenotypic variation in offspring size produced very strong evidence that increased offspring size increased early survival (germination success, hatching success) and increased early growth rate. In plants increased offspring size increased the seedling competitive ability (Figure 4.39). Because in many cases size at maturity or metamorphosis is cued by a size threshold there is no a priori reason to suppose that initial size will determine adult size, though in fact there is such a correlation, both among and within species. The most likely manner in which final size could be impacted by initial size is if the initial size significantly changes the development time. Such a trade-off would favor large offspring size. What factor would select against a large offspring size? If the amount of material available to convert into offspring is limited, then there will be a trade-off between the size and number of offspring. A study of teleost fish (Elgar 1990) and another of copepods (Poulin 1995) using correction for phyologenetic effects indicated a trade-off across a very broad taxonomic range. Ten studies of the trade-off using species as the basic datum all showed a trade-off (see Table 10.4 in Roff 1992a). The negative case for birds reported by Rohwer (1988) was later refuted by Blackburn (1991) and Olsen et al. (1994), as did four studies using population within species as the unit of comparison (Roff 1992a).

The most important test is that within populations, and here the results also support the hypothesis of a trade-off, with the exception of birds for which the data are very mixed (Table 4.8). Unfortunately, estimates of the genetic correlation between size and number of offspring are very scarce. The only one that I have found is that for the three-spined stickleback for which $r_A = -0.98 \pm 0.02$ (Snyder 1991; the phenotypic correlation is not given). In the side-blotched lizard *Uta stansburiana,* Sinervo and Doughty (1996) obtained a significant negative regression between clutch size in daughters regressed on egg size in mothers, indicating a negative genetic correlation between these two traits. For the same species Ferguson and Fox (1984) measured a phenotypic correlation of -0.54. Estimates for other reptiles are comparable with this value, whereas three estimates for insects are slightly smaller (Table 4.9).

The two component traits have high heritabilities (for fecundity, or at least fecundity components, $\bar{h}^2 = 0.41 \pm 0.06$, (discussed earlier in the chapter) and for propagule size, $\bar{h}^2 = 0.49 \pm 0.06$, $n = 23$; data from Table 7.7, Roff (1997) with four additional estimates from Schwaegerle and Levin (1990, 1991); Sinervo and Doughty (1996); Weigensberg et al. (1998); Schwarzkopf et al. (1999). Recall that

$$r_P = r_A \sqrt{h_x^2 h_y^2} + r_E \sqrt{\left(1 - h_x^2\right)\left(1 - h_y^2\right)} \tag{4.94}$$

which upon substitution of the mean heritabilities gives $r_P = 0.45 r_A + 0.55 r_E$. If the environmental correlation took its smallest value of -1 then, if the genetic correlation were zero, the phenotypic correlation would be -0.55. This value is reasonable, given the published estimates. On the other hand, if $r_E = -0.5$, and $r_P = -0.5$, then $r_A = -0.5$, which is also not unreasonable. The point is that, given

(A) Plants

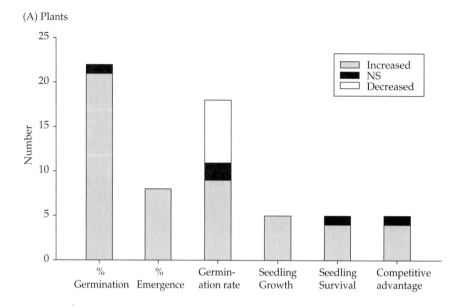

(B) Animals

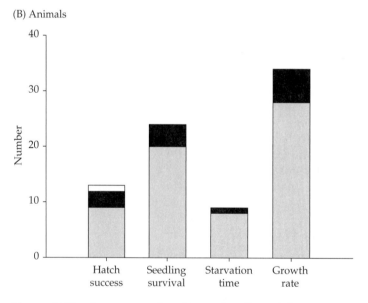

Figure 4.39 A summary of studies on the effect of phenotypic variation in offspring size on fitness traits. (A) For plants, traits measured, in sequence as shown, were % germination, % emergence, germination rate, seedling growth, seedling survival, and competitive advantage. (B) For animals, the traits measured were hatching success (lepidoptera, fish, birds), survival (insects, fish, herps, birds, mammals), starvation time (invertebrates, fish) and growth rate (insects, fish, birds, mammals). Data from Tables 10.1 and 10.2 of Roff (1992a), with additional material from Sibly and Monk (1987), Fox (1993a, 1994), Hart (1995) and Roosenburg and Kelley (1996).

Table 4.8   A summary of studies that have tested for a trade-off between offspring size and clutch number.

Group	Trade-off	NS	Correlation +
Plants	3	0	0
Invertebrates	11	0	0
Ectothermic vertebrates	12	2	0
Mammals	11	7	0
Birds	2	11	2

Data from Tables 10.4 and 9.5 of Roff (1992a) with additional data from Clarke (1993), Leprince and Foil (1993), Genoud and Perrin (1994), Carriere and Roff (1995b, 3 spp tested, trade-off observed with a multivariate analysis but not significant within individual species), Booth (1998), den Bosch and Bout (1998), Abell (1999).

the present data it is not possible to achieve even a crude estimate of the genetic correlation between propagule size and fecundity. The high heritabilities of egg size and fecundity components indicate that both should respond to selection, but this does not tell us anything about the corresponding response of the other. Experimental reduction of the number of ovulated eggs in the side-blotched lizard resulted in increased egg size of the subsequent reduced clutch, suggesting severe constraints on egg size with clutch size (Sinervo and Licht 1991b).

Table 4.9   Phenotypic correlations between egg mass and clutch size in some animals.

Species	Group	Phenotypic correlation	Reference
Uta stansburiana	lizard	−0.54	Ferguson and Fox (1984)
Sceloporus occidentalis	lizard	−0.32 to −0.35[a]	Sinervo (1990)
Algyroides fitzingeri	lizard	−0.33	den Bosch and Bout (1998)
Algyroides moreoticus	lizard	−0.72	den Bosch and Bout (1998)
Podarcis milensis	lizard	−0.67	den Bosch and Bout (1998)
Sceloporus virgatus	lizard	−0.44 to −0.52[b]	Abell (1999)
Chelodina expansa	turtle	−0.60	Booth (1998)
Gryllus firmus	cricket	−0.28	Carriere and Roff (1995b)
Gryllus veletis	cricket	−0.12	Carriere and Roff (1995b)
Gryllus pennsylvanicus	cricket	−0.17	Carriere and Roff (1995b)
Tabonus fuscicostatus	fly	−0.33	Leprince and Foil (1993)
Aquarius remigis	waterstrider	−0.53	Preziosi et al. (1996)

[a]Path coefficient after correcting for effects of size, lay date, and condition (four estimates).

[b]Two sets of estimates (different years), one using offspring mass and another using egg mass.

Such constraints would almost certainly be manifested as a negative genetic correlation, but such data remain to be collected.

Selection experiments on egg size in *D. melanogaster* (Parsons 1964; Schwarzkopf et al. 1999) and the spruce budworm *Choristoneura funiferana,* (Harvey 1983, 1985) demonstrated genetic correlations between egg size and other life history traits, but only the study by Schwarzkopf et al. (1999) tested for a correlated response in fecundity. In this study the results were confusing, to say the least. Selection for increased and decreased egg length was successful and while, in all lines, including the controls, there was a negative phenotypic correlation between egg size and fecundity (but statistically significant only in the large egg size line) there was no statistically significant change in mean fecundity indicative of a trade-off (mean 10 day fecundities were 305 ± 8.5, 330 ± 9.4, 314 ± 10.3, for large, control, and small egg-size lines, respectively). Because of possible differences in body size, Schwarzkopf et al. (1999) did all tests using residuals from two allometric regressions, one for egg size on body size and another for fecundity on body size. The problem with this approach is that the trade-off arises because total egg volume is constrained. Therefore, although it is necessary to correct egg number for body size, it is wrong to also correct egg size. The appropriate regression model is $ZY = aX^b$, where Z is fecundity, Y is egg size, X is body size, and a, b are constants. Taking logs and rearranging gives $\ln(Z) = \ln(a) + b \ln(X) - \ln(Y)$. If we use the mean values for each line, the product of fecundity and egg volume in order of mean body size (in parentheses) was 6.04 (control, 1.577), 5.24 (small eggs, 1.575), and 5.83 (large eggs, 1.565), which is still not in agreement with prediction. One possibility is that selection for small eggs has selected for physiologically inferior individuals. In support of this hypothesis is the observation of decreased viability in the small-egg line. In nature, selection for small eggs would select jointly for small, viable eggs. Whether this is or is not the reason, this example points out the difficulty of interpreting artificial selection experiments. Considering only the control and large-egg line, I conclude that there is evidence for a genetic correlation between egg size and fecundity in *D. melanogaster*, though clearly this needs further investigation.

Several selection experiments on plants have suggested a negative genetic correlation between seed size and clutch size. Selection for large seed size in *Sinapis alba, Brassica napus* and *B. campestris* resulted in a decreased number per pod (Olsson 1960) indicating a negative genetic correlation between size and number. Tedin (1925) obtained similar results for selection on seed size in an annual species of *Camelina*: A decline in seed number per pod accompanied selection for increased seed size, but selection for both large seed and high number could be achieved by simultaneous selection for both seed size and pod size. In an attempt to select for high yield in winter barley, Nickell and Grafius (1969) selected plants on the basis of yield, kernel weight, and seed number per unit area. Instead of an increase in yield there was a significant drop from 4271 kg/ha to 3267 kg/ha. This decline was attributed to negative genetic correlations between the number of heads, the number of kernels per head, and kernel weight. Even though the number of heads per unit area increased from 100.9 to 199.5, the number of kernels per head decreased from 43.8 to 20.2, and kernel weight decreased from 37.4 mg to 28.5 mg. As no control line was grown, these results must be interpreted with caution.

It is reasonable to suppose, at least as a working hypothesis, that propagule size and clutch size are negatively correlated both phenotypically and genetically. In addition to a trade-off between egg size and fecundity there may also be stabilizing selection on propagule size; for example, stabilizing selection has been demonstrated in hatching rate in chickens (Lerner and Gunns 1952), human birth weight (Van Valen and Mellin 1967), hatching success in *D. melanogaster* (Curtsinger 1976a,b), calf survival (Martinez et al. 1983), and survival in hatchling lizards (Sinervo et al. 1992, discussed below).

Modeling the Evolution of Maternally Affected Traits such as Clutch-size–Litter-size Combinations

Initial offspring size is most probably a maternal trait, but future size is a consequence of the effects due to the maternal input plus the effect of the offspring genes. For example, egg size in the sand cricket is entirely maternally determined (possibly with indirect effects of the father). By using a color mutant, Weigensberg et al. (1998) were able to mate two males to a single female and hence separate effects due to the maternal environment from effects due to the genetic constitution of the offspring. After oviposition, egg-size changes as a consequnce of absorption of water and growth of the embryo. It normally takes 14 days for the eggs to hatch; by day 10 post-oviposition egg size is due to both initial size (maternal effect) and genetic constitution of the offspring (shown by a significant "paternal" effect). If there is a genetic component to a trait in addition to the maternal effect, over time we would expect the ratio of the maternal variance to the direct additive genetic variance of the trait to decline; this has been observed in body weight of mice, the ratio declining from approximately 3 at two weeks of age to less than 0.5 by 10 weeks (Riska et al. 1984). In many cases maternal effects continue to have an impact throughout the life of the offspring (Roach and Wulff 1987; Mousseau and Dingle 1991; Mousseau and Fox 1998).

There are two models for the inclusion of maternal effects into a quantitative genetic framework. The first was first introduced by Dickerson (1947) and later developed by Willham (1963, 1972) and the model is generally referred to as the **Wilham model**. Kirkpatrick and Lande (1989) proposed an alternative formulation based on a conceptual framework suggested by Falconer (1965). The latter approach is more general and is illustrated here by consideration of three cases (Figure 4.40).

CASE 1: A SINGLE CHARACTER, *X*, MATERNALLY AFFECTING ITSELF. An example of this type of character is early offspring size, which might itself affect size at maturity; the size of hatchlings, seedlings, or neonates is determined in part by the

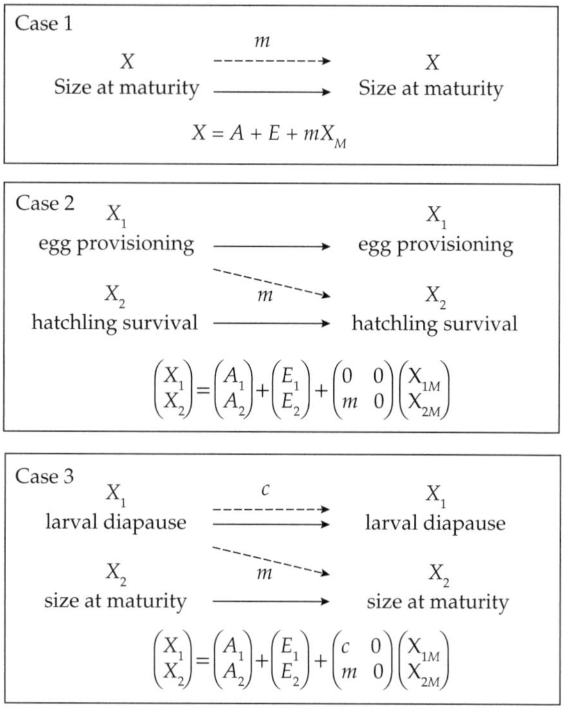

Figure 4.40 Three possible maternal-effects models using the theoretical framework of Kirkpatrick and Lande (1989). Modified from Roff (1997). Case 1: the trait, such as offspring size, which may determine size at maturity, is determined by the genes inherited by the offspring and by the mother's phenotype, which itself is genetically determined. For example, size at maturity may be determined both by the genes of the offspring and the environment in which the offspring develop. Large mothers have high fecundity, which causes density-dependent reduction in growth, leading to small offspring which then have low fecundities and hence large offspring both because of the environment (low density) and the genetic constitution of the offspring. Note that the effect functionally involves more than one trait. It is difficult to erect a reasonable scenario in which the effect is functionally direct, although it can be written mathematically in such a fashion. Case 2: Two traits, one maternally affecting the other. Egg provisioning affects the survival of hatchlings but does not itself directly affect the amount of egg provisioning by the offspring. Case 3: Two traits, one of which maternally affects both itself and the other trait. In the example shown, larval diapause is determined by the environmental conditions experienced both by the mother and the offspring. An offspring that enters diapause may show a different growth pattern upon emergence from the diapause than a larva that undergoes direct development, leading to different sizes at maturity. Thus size at maturity is a function both of the genes controlling body size, the genes controlling larval diapause, and the maternal effect that comes through the diapause pathway.

genes inherited from the mother and, in part, by the phenotype of the mother. The phenotypic value of the trait in generation $t + 1$, $X(t + 1)$ is given by $X(t + 1) = A(t + 1) + E(t + 1) + C_M X_M(t)$, where $A(t + 1)$ is the additive genetic component, $E(t + 1)$ is the environmental component (including nonadditive effects such as dominance and epistasis), C_M is the maternal-effect coefficient, and $X_M(t)$ is the phenotypic value of the mother. The foregoing equation is a linear regression relationship between offspring and parent; it reduces to the usual offspring on parent regression when $C_M = 0$. The coefficient C_M measures the effect of the phenotypic value of the mother; it is defined as the partial regression of the offspring's phenotype on its mother's phenotype, holding genetic sources of variation constant. The maternal effect can be either positive or negative. In the latter case, large mothers may produce genetically large offspring (i.e., large A) but their phenotypic effect (C_M) is to reduce phenotypic size, so that if C_M is sufficiently large in magnitude, small offspring are produced by large mothers. This is equivalent to the negative covariance in the Willham model. Negative covariances are commonly observed in domestic animals (mice, swine, sheep and cattle, see Table 7.5 in Roff 1997) and also in body weight in *Drosophila melanogaster* (DeFries and Touchberry 1961.) This effect comes about because large mothers lay large clutches, which lead to small adult offspring, because of negative density-dependence). Negative covariances are also observed in pupal weight and family size in *Tribolium* (Bondari et al. 1978), and body size in springtails (Janssen et al. 1988).

The response to selection, $R(t)$, is the difference between the population mean at generation $t + 1$ and generation t, $R(t) = \overline{X}(t + 1) - \overline{X}(t)$ is

$$R(t) = \left(\sigma_{AX} + C_M \sigma_P^2\right)\beta(t) + C_M R(t-1) - C_M \sigma_P^2 \beta(t-1) \qquad (4.95)$$

where σ_{AX} is the covariance between the additive genetic value and the phenotypic value. In the absence of any maternal effects, σ_{AX} is equal to $\frac{1}{2}\sigma_A^2$, where σ_A^2 is the additive genetic variance of the trait. In the presence of maternal effects, σ_{AX} is equal to $\sigma_A^2/(2 - C_M)$. The response to selection is the sum of three components: (1) the change in the genetic component (σ_{AX}) plus the maternal effect (C_M), (2) the change resulting from the maternal effect as determined by the change in the phenotype in the previous generation, and (3) the loss in response due to purely phenotypic effects resulting from maternal influence in the previous generation. If we assume the population to be initially at equilibrium, the response to one generation of selection is approximately $R(1) \approx (2\sigma_{AX} + C_M\sigma_P^2)\,\beta(1)$. Maternal effects introduce a time lag in the evolutionary response. Two consequences follow from this: First, if $\sigma_{AX} < -C_M\sigma_P^2$, there will initially be a reversed response to selection, and second, the response to selection changes each generation even under constant directional selection, only asymptotically reaching a constant value, which in the present case is $R(\infty) \approx (2\sigma_{AX}\beta)/(1 - C_M)$. In contrast to the situation without maternal effects, the response to selection does not cease immediately upon cessation of selection but declines exponentially, this continuation being called the **evolutionary momentum** by Kirkpatrick and Lande (1989).

Important for the present analyses is that maternal effects do not prevent the population attaining an optimal value, only the time and trajectory to this point (Naylor 1964; Kirkpatrick and Lande 1989).

CASE 2: A MATERNAL TRAIT, X_1, THAT AFFECTS ANOTHER TRAIT, X_2. Possible candidates for this case are litter size, milk production or egg provisioning as the maternal trait, and body size or offspring survival as the offspring trait. (Egg provisioning phenotypically affects offspring survival but does not affect the egg provisioning of the offspring.) This is the model most frequently considered. Letting 1 denote the maternal trait and 2 the offspring trait, there are two equations describing this case. First, for trait X_1, which affects the second trait but is not itself affected, we have the usual regression $X_1(t + 1) = A_1(t + 1) + E_1(t + 1)$. For the second trait we have to include the maternal component due to the first trait, $X_2(t + 1) = A_2(t + 1) + E_2(t + 1) + C_M X_{M1}$, where X_{M1} is the phenotypic value of the maternal trait, and $|C_M| < 1$. These two equations can be conveniently represented by the set of matrices shown in Figure 4.40. The response to selection on the maternal trait (X_1) follows the usual formulation, since it is not itself maternally affected. For the second (offspring) trait (X_2), the response in the first generation and at equilibrium (see Appendix 1 in Kirkpatrick and Lande, 1989), assuming that there is no environmental correlation between traits 1 and 2, is approximately

$$R_2(1) \approx \left(\sigma_{A22}^2 + \frac{3C_M}{2} \sigma_{A12} + \frac{C_M^2}{2} \sigma_{A11}^2 \right) \beta_2(1) \qquad (4.96)$$

where the subscripts denote the position in the additive genetic variance matrix; hence, σ_{A22}^2 is the additive genetic variance of trait 2 (offspring trait), σ_{A12}^2 is the additive genetic covariance between the two traits, and σ_{A11}^2 is the additive genetic variance of trait 1 (maternal trait). The term $\beta_2(1)$ represents the selection gradient on trait 2 in the first generation of selection.

CASE 3: A TRAIT, X_1, THAT MATERNALLY AFFECTS ITSELF AND ANOTHER TRAIT, X_2. The two types of traits discussed above can be combined as indicated in Figure 4.40. For example, a trait such as litter size may both affect the body size of the offspring and also their own future litter size. This phenomenon can easily be accommodated by addition of another coefficient, c, into the maternal-effects matrix, as shown in Figure 4.40. This process can be extended to any sort of pathway of maternal effects. The equations can also be modified to include common family environment, and maternal and paternal effects separately (Kirkpatrick and Lande 1989; Lande and Kirkpatrick 1990). The phenotypic covariances become quite complex, as also does the complete description of response to selection, and the reader is referred to Appendix 1 of Kirkpatrick and Lande (1989) for further details.

The impact of maternal effects can cause the population to evolve in the opposite direction for an indefinitely long time (Cheverud 1984; Kirkpatrick and Lande 1989), although we lack sufficient empirical data to say how important such effects might be in natural populations.

Although no modeling of the evolutionary trajectory of fecundity and offspring size combinations can ignore maternal effects, a reasonable case can be made for ignoring them when considering equilibrium conditions. Although it is possible for maternal effects to prevent the attainment of equilibrium, in most cases maternal effects probably cause only an increase in the time taken to achieve the equilibrium combination or cause the combination to fluctuate about the optimal combination.

Effects of a Trade-off between the Number and Quality of Offspring

THE SMITH-FRETWELL MODEL. The general starting point for this question is the **model of Smith and Fretwell** (1974). They made three assumptions: (1) that R_0 is the appropriate measure of fitness, and (2) that there is no trade-off between clutch size and future fecundity or survival of the parent. (The model actually assumes a fixed investment and assumes that dividing this among different clutch sizes does not affect adult survival or future fecundity.) Another assumption is (3) that there is a concave-down relationship between the expected fecundity of offspring and offspring size (Figure 4.41). Assumption 1 is not a necessary requirement, as can be shown by the model $\lambda = sm + S$, where the parameters are as previously defined. This model does not assume a stationary population. (It does assume a fixed age at maturity and so, if the cost of a decreased propagule size is increased α, then assumption 1 is required.) Given assumption (2), we can find the optimal offspring-clutch size combination by maximizing sm, which is $c(X)f(X)$. Here $c(X)$ is clutch size for offspring of size (quality) X and $f(X)$ is the functional relationship between expected fecundity of offspring ($= s(X)m(X)$) and offspring size, X. Graphically, the optimal propagule size occurs where a line drawn through the origin is tangent to $f(X)$, the fitness function (Figure 4.41). This procedure is justified as follows: Let fitness be $W(X)$; we then have

$$W(X) = c(X)f(X) = \frac{I}{X}f(X) \tag{4.97}$$

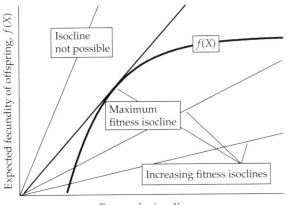

Expected fecundity of offspring, $f(X)$

Isocline not possible

$f(X)$

Maximum fitness isocline

Increasing fitness isoclines

Propagule size, X

Figure 4.41 A hypothetical example of the asymptotic function proposed by Smith and Fretwell (1974) relating the expected fecundity of offspring, $f(X) = s(X)m(X)$, to propagule size (or quality), X. For a given fitness, $W(X)$, $f(X)$ is also a straight line passing through the origin according to the equation $f(X) = XW(X)/I$, where I is the total amount of resource to be divided among the propagules. Only lines that intersect the curve are permissible, and hence the optimum propagule size is that value of X at which an isocline is tangent to the curve.

where I is the total invested into offspring and assumed to distributed equally among the offspring. Since $f(X) = XW(X)/I$, for a given value of fitness, $W(X)$ and a fixed investment I, $f(X)$ is a linear function of x passing through the origin (Figure 4.41). Each line passing through the origin represents a fitness isocline (i.e., a particular value of $W(X)$), the value increasing with the angle subtended at the origin. The investment per offspring, X, that generates a fitness value $W(X)$ is the value of x at the intersection of the fitness isocline and the mean fecundity of offspring function, $f(X)$, (Figure 4.41). From the above argument it is obvious that the maximal value of $W(X)$ is that point at which the fitness isocline is tangent to $f(X)$. The actual value of propagule size, X, is obtained as

$$\frac{\partial W(X)}{X} = \left(\frac{\partial f(X)}{\partial X}\right)\left(\frac{I}{X}\right) - \left(\frac{f(X)}{X}\right)\left(\frac{I}{X}\right) = 0 \quad \text{when } X = \left(\frac{\partial f(X)}{\partial X}\right)\left(\frac{1}{f(X)}\right) \tag{4.98}$$

The model of Smith and Fretwell establishes that, given a trade-off between propagule size and propagule number, there will exist a particular combination that maximizes fitness.

TESTS OF THE SMITH-FRETWELL MODEL. Females in the cricket genus *Gryllus* bury their eggs in the soil, the depth of burial depending in part upon the length of the ovipositor. In three species Carriere and Roff (1995a) examined the trade-off between egg size and fecundity and between survival of eggs and egg size. The eggs were buried just prior to hatching, and hence the survival measured was probably the ability of the newly hatched nymph to climb to the surface. As described above, a trade-off between egg size and fecundity was found in all three species. In two species, *G. firmus* and *G. veletis*, the data permitted the estimation of the emergence success as a function of egg size. We fitted these functions using the cubic spline method, which minimizes the assumption about the shape of the curve (Schluter 1988). In both species the curve was concave and following the algorithm given above. (We assumed that survival and fecundity were the only components varying with egg size.) The optimal egg sizes were calculated to be 33 and 27.5 g × 10^{-5} for *G. firmus* and *G. veletis*, respectively. These predicted values are close to the observed values of 34.5 and 27.0.

A field test of the hypothesis that selection will optimize egg size as a consequence of the interaction between survival and the clutch size-egg size trade-off was done by Sinervo et al. (1992) in the side-blotched lizard. They increased the range in size of hatchlings using surgical techniques, which do not appear to affect the hatchlings except by virtue of their changed size. Removal of yolk from newly laid eggs produced miniaturized hatchlings, whereas removal of yolk from a female's follicles produced enlarged babies. The experiment was replicated both in time (two years) and space (two sites). Survival was measured for the first month after release, with hatchlings that were released initially being kept separate from later releases. The argument for doing this was that the latter had an additional source of competition. Linear regression was used to estimate the relationship between fecundity selection and egg size, but because of differing shapes, the cubic spline technique was used to estimate the survival functions. In all cases selection on fecundity was linear and directional, fecundity selection favoring, not unexpectedly, small eggs. Survival in 7 of 8 cases either increased monotonically with egg size or was dome-shaped (i.e., stabilizing selection), and in all of these 7 cases the $l(x)m(x)$ function was dome-shaped indicating an optimal egg size. Sinervo et al. (1992) did not attempt a quantitative assessment of the fit between observed and estimated optimal egg sizes, although overall the correspondence seemed quite good.

In the above examples a graphical estimation was necessary because of the cubic spline fit of the data. When a specific equation can be fitted to the data, the optimal offspring size can be obtained using an analytical method such as calculus. This is illustrated with the analysis by Lawlor (1976) of the optimal offspring size in the isopod *Amadillidium vulgare*. The size at maturity of offspring is linearly related to the size-specific fecundity (F = young per mg) of the mother, *Size* ∝ (2.587–0.317F). Fecundity is proportional to weight and hence the total number of offspring is FW_F, where W_F is the weight of the mother. The total biomass of offspring is proportional to (2.587–0.317F)FW_F. To obtain the optimal size-specific fecundity we differentiate the equation with respect to F to get (2.587–0.634F)W_F, and find the value at which this is zero (the turning point). It is 2.587/0.634 = 4.08 young per mg of female. The optimal size of young is 1/4.08 = 0.245 mg, which compares favorably with the observed value of 0.287 mg (from an observed size-specific fecundity of 3.48 mg).

OFFSPRING SIZE AND TIME TO MATURITY. In the cricket and lizard examples, the third factor in the three-way trade-off was survival, whereas in the isopod example, it was a change in adult size. Such a trade-off assumes a fixed age at maturity (= development time). It might be argued that because initial size is typically a very small fraction of adult size (certainly true for plants, most invertebrates and most ectothermic vertebrates), variation in propagule size will not change development time to any significant extent. Placing doubt on this are numerous observations on a trade-off between propagule size and growth as measured by development time or size achieved after a fixed period (Figure 4.39; see also Sibly and Monk 1987; Fox 1993a, 1994; Hart 1995; Roosenburg and Kelley 1996). The general importance of a delay in growth has yet to be rigorously tested. To illustrate, the potential impact suppose that growth is exponential $X(t) = X_0 e^{bt}$, where $X(t)$ is size at age t and X_0 is the initial size. If the initial size is increased by some fraction f, the ratio of development times for the same adult size is (Roff 1992a)

$$R = 1 - \frac{\ln f}{\ln X_{adult} - \ln X_0} \tag{4.99}$$

Numerous studies have shown that among species propagule size varies allometrically with adult size. Using the upper and lower range for each group I calculated R for the taxa given in Visman et al. (1996), which included plants, crustaceans, and all the vertebrate major taxa (metatherians, eutherians, birds, reptiles, amphibians, fish). A 10% increase in propagule size would, assuming exponential growth, reduce development time by a maximum of 10% and more typically by 1–4%. Exponential growth is probably at the upper limit of growth rates and hence these rates are probably greater than general. The impact of a 4% decrease in growth versus a 10% decrease in fecundity depends, as shown earlier in this chapter, on the relative value of the gross fecundity.

SITE QUALITY AND THE SMITH-FRETWELL MODEL. It was previously shown that clutch size increased with site quality although there is no general theoretical prediction that this should be so. An increase in site quality might reasonably be expected to increase the rate of growth of an immature organism (plant or animal). From the perspective of the Smith-Fretwell model do we expect a change in site quality that affects growth rate to affect the optimal propagule size? This question has been explored by Sibly and Calow (1983, 1985), Parker and Begon (1986) and Sibly et al. (1988). Here I shall present a simple verbal argument based on the model of Sibly et al. (1988). Suppose the environmental conditions are relatively poor and the female has either a fixed development time or a fixed period as an adult to gather the resources to be transferred to the offspring. Under this environmental condition, the curve relating resource invested per offspring as a function of offspring size will be fairly shallow, as shown in Figure 4.42. Under better environmental conditions the asymptotic limit of the investment curve will be raised, and the optimal propagule size will be shifted to a smaller size, as shown (Figure 4.42). A simple mathematical example is given in Box 4.8.

Another conceptual model that addresses the question of optimal propagule size and female quality, taken in the broad sense, was proposed by Parker and

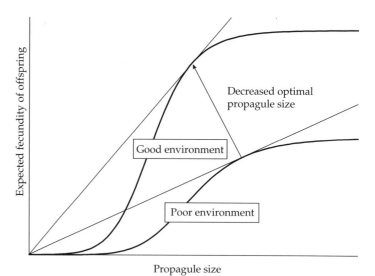

Figure 4.42 A schematic illustration of the effect of environmental quality on the optimal propagule size.

Box 4.8
Site quality and the optimal propagule–clutch-size combination: A simple mathematical example

We begin with the model previously introduced, $\lambda = sm + S$. Two trade-offs are proposed: (1) immature survival increases linearly with propagule mass, $s(X) = -a_1 + b_1 X$, (2) propagule mass decreases linearly with clutch size, $c(X)$, and is increased by site or female quality, Q, $X = a_2 + Q - b_2 c(X)$. Writing the fitness function in terms of propagule mass (from step 2 we have $c(X) = (a_2 + Q - X)/b_2$),

$$\lambda = \frac{(-a_1 + b_1 X)(a_2 + Q - X)}{b_2} + S \qquad (4.100)$$

Differentiating and setting equal to zero, we arrive at the optimal propagule mass:

$$X^* = \tfrac{1}{2}\left[a_2 + \left(a_1/b_1\right) + Q\right] \qquad (4.101)$$

The optimal propagule mass increases as female or site quality increases. There is also an optimal clutch size, which can be obtained simply by substituting in the above equation to give

$$c\left(X^*\right) = \frac{1}{2b_2}\left(a_2 - \frac{a_1}{b_1} + Q\right) \qquad (4.102)$$

Thus an increase in female and/or site quality favors an increase in both propagule size and clutch size.

Begon (1986). They specifically considered the case of how ration would select for propagule size, assuming that selection maximizes a female's productivity per unit time (Box 4.9). Because of the complexity of the analytical result (Box 4.9, Equation [4.104]), Parker and Begon inserted specific functions and for these they found that, as resources increased, so did the optimal propagule and clutch size. The Parker and Begon model posits negative density-dependence, while a model proposed by McGinley (1989) assumes positive density-dependence and still generates the prediction that propagule size should increase with resource quality. Similar effects also occur if offspring survival is a size-specific function of site quality (Hutchings 1991, 1997).

The diversity of models that predict an increasing propagule size with some index of quality (e.g., ration, female size, site quality) suggests that this relationship might be quite general, although the actual mechanism may vary from species to species. The majority of studies of animals do show that propagule size increases with ration and female size, though there are a large number in which either no relationship was found or not a simple monotonic function (Figure 4.43; Fox 2000). It would be very useful to explore specific cases in detail to understand the biological factors driving the responses. The cases displayed in Figure 4.43 represent phenotypic responses within a population, and hence are

---

Box 4.9
The model of Parker and Begon (1986)

The female spends time t foraging for resources and then lays a clutch of eggs. Productivity per unit time is

$$W(X) = \frac{g(t)}{X} h(C) \frac{f(X)}{t + t_{min}} \qquad (4.103)$$

where $g(t)/X = C$ is clutch size, which, as indicated, is itself a function of the resources gathered. A higher-quality site or female will result in more resources and hence $g(t)$ is an increasing function of time and female/habitat quality. The second function $h(C)$ specifies the effect of competition between immatures, and $f(X)$ designates density-independent effects. The constant t_{min} is the minimum time required between clutches (e.g., time taken from feeding site to oviposition site). Differentiating the above with respect to propagule size and setting the differential to zero, the optimal propagule size is the value of X that satisfies the relationship

$$\frac{\partial h(C)}{\partial X} f(X) - \frac{h(C) f(X)}{X} + h(C) \frac{\partial f(X)}{\partial X} = 0 \qquad (4.104)$$

If there is no density-dependence, then $h(C) = 1$ and the above reduces to the condition

$$\frac{\partial f(X)}{\partial X} - \frac{f(X)}{X} = 0 \qquad (4.105)$$

which does not contain the resource function, $g(t)$. Hence, in the absence of clutch-size effects, the optimal propagule size is independent of female or site quality for this model. With such an effect, the clutch-size function does not cancel out and hence the optimal propagule and clutch size are functions of ration, site quality, and so on.

---

norms of reaction. In the present chapter we are assuming a constant environment; therefore the specific evolutionary response modeled is appropriate for environments that are constant but vary in ration, size, and other factors. The case of variable environments is discussed in the next two chapters. In the present context it is sufficient to note that we can use the phenotypic response of individuals as a test of the theory because the response is optimal in both a constant environment and in an environment in which the offspring are subject to a particular set of constant conditions, even if those conditions might vary from generation to generation. (A caveat must be attached to this statement with respect to bet-hedging strategies, and so it may be better to state that the response will be approximately optimal.)

THE OPTIMAL NUMBER AND SIZE OF PROPAGULES IN RELATION TO FEMALE AGE. It was previously shown that, under fairly general conditions, reproductive effort should increase with age. This implies that either propagule number or size

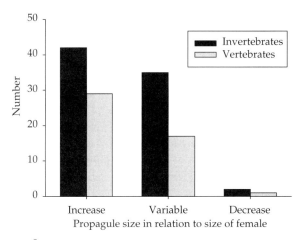

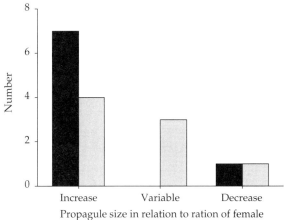

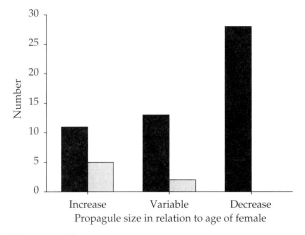

Figure 4.43 Relationships between propagule size and three factors (size, ration, age). The category "variable" includes those cases in which no relationship was observed or in which it was not monotonic. Data from Roff (1992) and Fox and Czesak (2000).

Box 4.10
Propagule size and female age

This model (Begon and Parker 1986) was constructed specifically to match the life history pattern of an insect and is based on the following five assumptions; (1) the adult female accumulates a total reproductive reserve of I, to be divided among all eggs to be produced during the lifetime of the female, (2) the probability that a female will survive to lay the ith clutch is S_i, (3) the size of eggs in the ith clutch is x_i, (4) clutch size is fixed at C, and the female produces a maximum of N clutches in her lifetime, (5) the expected fecundity of offspring (survival x fecundity) is an asymptotic function of egg size, $f(X)$ (Figure 4.44), (6) population size is stationary, making expected lifetime

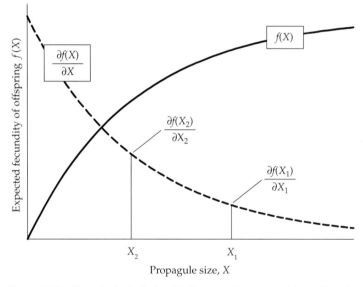

Figure 4.44 Hypothetical relationship between the expected fecundity of offspring and propagule size, (solid line) and its derivative (dashed line). For a female laying two clutches of eggs that are optimal in size the following inequality holds (see text): $\dfrac{\partial f(X_1)}{\partial X_1} < \dfrac{\partial f(X_2)}{\partial X_2}$ where X_i is the size of propagules in the ith clutch. Thus, as can be seen from the above figure, $X_2 < X_1$.

fecundity an appropriate measure of fitness. This model does not fit a vertebrate or plant life history, because resources are gathered throughout life in these taxa. There are also many insects in which fecundity depends on resources gathered after maturation.

    Given the model assumptions, the optimal propagule size is that which maximizes the expected lifetime fecundity

$$R_0 = \sum_{i=1}^{N} S_i f(X_i) \tag{4.106}$$

subject to the constraint of a fixed reserve,

$$I = \sum_{i=1}^{N} CX_i \qquad (4.107)$$

To solve the foregoing equation, Begon and Parker (1986) used Lagrange's method of optimization subject to constraint. A simpler method of analysis is as follows. First we shall consider an insect that produces only two clutches, for which the fitness function is $R_0 = S_1 Cf(X_1) + S_2 Cf(X_2)$. Since $I = C(X_1 + X_2)$, R_0 can be rewritten as

$$R_0 = CS_1 f(X_1) + CS_2 f\left(\frac{I}{C} - X_1\right) \qquad (4.108)$$

We differentiate and set the differential to zero (see Roff 1992a for details). The maximum fitness occurs when

$$S_1 \frac{\partial f(X_1)}{\partial X_1} = S_2 \frac{\partial f(X_2)}{\partial X_2} \qquad (4.109)$$

Now since the probability of surviving to lay the second clutch (S_2) must be less than the probability of surviving to lay the first clutch (S_1), then

$$\frac{\partial f(X_1)}{\partial X_1} < \frac{\partial f(X_2)}{\partial X_2} \qquad (4.110)$$

which requires that X_1 be larger than X_2 (Figure 4.44). In other words, eggs laid in the second clutch will be smaller than eggs laid in the first. We can extend this analysis to an arbitrary number of clutches by considering pairwise comparisons. In each case the preceding egg size will be greater, and hence egg size will decrease with age, as observed in a variety of insects (Figure 4.43).

should increase with age. From the above models we have also seen that, under particular but not very restrictive conditions, propagule size and number should covary positively. In invertebrates the overwhelming observation is that propagule size decreases with age, whereas in vertebrates propagule size either increases with age or shows a more complex pattern (Figure 4.43). A more detailed breakdown of the invertebrate group suggests that the overall pattern may be illusory. Only three noninsect species are included in the data set, and in these species (all crustacea), there is an increase in egg size with age (Fox and Czesak 2000). Even within the insects there is considerable diversity among the Orthoptera and Heteroptera, and it is primarily the Lepidoptera that generate the overall pattern of a negative trend (0 increase, 20 decrease, and 5 variable, Fox and Czesak 2000). It is very unlikely that we are dealing with a single phenomenon, and the relationship between propagule size and age is a result of an interaction between female quality varying with age and changes in residual reproductive value. The latter consideration can be illustrated by the model of Begon and Parker (1986; Box 4.10).

A weakness of the above Begon and Parker model is that it assumes a fixed clutch size. If offspring survival depends on egg size but not clutch number, then the analyses given above would suggest that a fixed clutch size is optimal. However, if the is a density-dependent interaction between immatures, then a fixed clutch size may not be optimal.

EFFECTS OF CHANGES IN CLUTCH SIZE AND NUMBER ON THE OPTIMAL AGE AT MATURITY.   The foregoing analyses have focussed on the joint evolution of clutch size and propagule size. Evolutionary changes in these traits will also affect other life history traits, most particularly the age at maturity. This can be demonstrated using the semelparous model discussed earlier

$$r = \frac{\ln\left[e^{-M(X)\alpha}m(X,\alpha)\right]}{\alpha} \tag{4.111}$$

Here the mortality rate, $M(X)$, is now set as a function of propagule size, X, and fecundity is a function of the age at maturity (generally an increasing function as a consequence of greater size) and propagule size (fecundity decreased with increased X). For initial simplicity we can combine these two functions into one, $f(X, \alpha) = e^{-M(X)\alpha}m(X, \alpha)$. Differentiating Equation (4.111) with respect to X gives

$$\frac{\partial r}{\partial X} = \left(\frac{1}{\alpha}\right)\left(\frac{1}{f(X,\alpha)}\right)\left(\frac{\partial f(X,\alpha)}{\partial X}\right) \tag{4.112}$$

which equals zero when $\partial f(X,\alpha)/\partial X = 0$, and is simply the maximum value of the expected fecundity (survival x fecundity). At first glance this may appear to be independent of the age at maturity but, as shown in Box 4.11, this is not necessarily the case, as α appears in $f(X,\alpha)$.

A trade-off between propagule size and survival will not necessarily lead to a change in the age at maturity. Suppose, for example, that the effect of propagule size on survival is only exhibited in the immediate period following hatching and this period is very short compared to the time to maturity. In this case we might reasonably regard the survival consequent on propagule size to be a point event and write the survival function as $s(X)e^{-M\alpha}$, where $s(X)$ is the survival fraction that is dependent on propagule size. For the Fish model this requires modification of the product $l(\alpha)m(\alpha)$:

$$l(\alpha)m(\alpha) = e^{-M\alpha}\left[aL_\infty\left(1-e^{-k\alpha}\right)\right]^b \tag{4.113}$$

by replacing a with $a = s(X)g(X)$, where $g(X)$ is the reduction in fecundity due to propagule size. The constant a plays no role in the optimal age at maturity and hence the optimal propagule size evolves independently of the optimal age at maturity (unless genetically correlated for some other reason).

Box 4.11
Propagule size and age at maturity are not independent

To demonstrate that propagule size and age at maturity are not independent, consider the specific example

$$f(X,\alpha) = \left(1 - e^{-bX}\right)^{\alpha} c\frac{\alpha}{X} \tag{4.114}$$

where b and c are constants. As propagule size increases, the probability of surviving per unit time interval increases while the fecundity decreases. Increasing the age at maturity increases fecundity but decreases the probability of surviving to reproduce. Now

$$\frac{\partial f(X,\alpha)}{\partial X} = \frac{c\alpha\left(1 - e^{-bX}\right)^{\alpha-1}}{X}\left(\alpha b e^{-bx} - \frac{1 - e^{-bX}}{X}\right)$$

$$= 0 \quad \text{when } e^{-bX}(1 + \alpha bX) = 1 \tag{4.115}$$

The optimal propagule size depends upon the age at maturity, which is obtained by setting $\partial r/\partial\alpha$ equal to zero,

$$\alpha - \alpha\ln\left(1 - e^{-bX}\right) - \ln\left(\frac{c\alpha}{X}\right) = 0 \tag{4.116}$$

The solution for the above two simultaneous equations can be found numerically for given values of c and b: if $c = 1$, $b = 1$, $X = 0.268$, $\alpha = 1.147$, whereas if $a = 2$, $b = 2$, $x = 0.181$, $\alpha = 2.410$.

Explaining Larger Scale Patterns

The preceding analyses have focused on the evolution of particular traits or trait combinations on the assumption that *within a species* there are trade-offs that limit the scope of variation. It is natural to ask if we can extend the analysis to variation among species. In other words is there *a priori* any reason to expect that we shall find patterns of variation at taxonomic levels above that of the species (i.e., among species, genera, families, and so forth)? There is indeed considerable empirical evidence that life history traits covary among species (Table 4.10, Figure 4.45). This variation may match that seen among populations of the same species (Figure 4.45); further study is required to tell how general the pattern may be.

Invariants

In their analysis of patterns of variation in growth and mortality of fish, Beverton and Holt (1959) and Beverton (1963, 1987, 1992) noted that relationships may exist

Table 4.10   **Some examples of allometric relationships among life history traits.**

Taxa	Traits[a] "Dependent"	Traits[a] "Independent"	Other factors	References
Copepods	Egg size	Body size, type of spawner (Sac vs broadcast)		Kiorboe and Sabatini (1995)
Nematodes	Fecundity α	Body size Mortality	Phylogeny	Morand (1996)
Intestinal nematodes	Prepatency, patency, Fecundity, body size, egg size		Phylogeny	Skorping et al. (1991)
Platyhelminthes	Adult size, fecundity, α, Longevity	Body size	Phylogeny Habitat	Trouve et al. (1998) Duarte and Alcaraz (1989)
Fish	Clutch wt, fecundity			
Fish	α, mortality, Ratio asymptotic length to length at maturity	Growth rate (k)		Roff (1984b)
Fish	GSI, mortality, longevity, α, asymptotic length			Beverton and Holt (1959); Gunderson (1980)
Salmonidae	Numerous LH and PCs	Anadromy, body size		Hutchings and Morris (1985)
Teleosts	Growth parameter (k)	Lifespan		Kawasaki (1980)
Amphibians	Clutch size	Egg wt, body size		Kuramoto (1978)
Lizards	Clutch size	Body size		Tinkle et al. (1970)
Lizards	Annual fecundity	Adult mortality		Tinkle (1969)
Reptiles	α, Mode of reproduction, Broods per year		Phylogeny	Dunham and Miles (1985)
Reptiles	α, Growth (k)	Adult mortality		Shine and Charnov (1992); Shine and Iverson (1995)
Gamebirds	Clutch size, longevity, Body mass		Latitude	Arnold (1988)
Birds	Active and resting metabolic rate	Body size, prey capture, nest dispersion	Habitat	Bennett and Harvey (1987)
New world passerines	Clutch size	Nest type	Nest predation	Kulesza (1990)

Table 4.10 (continued)

Taxa	Traits[a]		Other factors	References
	"Dependent"	"Independent"		
Birds	Annual fecundity	Adult survival, migration style		Martin (1995)
	Clutch size	Number of broods		
Birds	Longevity	Body size		Lindstedt and Calder (1976)
Nonpasserines	Egg size	Body size, clutch size	Phylogeny	Olsen et al. (1994a,b)
Passerines	Eggs/year, incubation time, nestling period, growth rate	Body size (BMR adjusted), BMR (body size adjusted) NS		Padley (1985)
European birds	Survival rate, egg wt, clutch size, nestling period, α	Body size		Saether (1987, 1989)
European birds	Clutch size, α	Adult survival	Phylogeny	Saether (1988)
Mammals	Numerous LH	Body size	Phylogeny	Millar (1977, 1984); Harvey and Clutton-Brock (1985); Martin and MacLarnon (1985)
Mammals	Birth wt, litter wt, gestation time, fetal growth rate, body size, life expectancy, etc.			Millar, 1981; Millar and Zammuto (1983); Read and Harvey (1989)
Mammals	α, offspring size, litter size, gestation period	Mortality (body mass effects removed)		Promislow and Harvey (1990)
Mammals	R	Body mass, offspring type (altricial, precocial)		Hennemann (1983, 1984); Thompson (1987)
Placental mammals	Prenatal growth rate	Growth rate at birth, juvenile period	Phylogeny	Pontier et al., 1993
Mammals	Ratio of neonatal wt to litter size	Body size, mating system		Zeveloff and Boyce (1980)
Birds and Mammals	Mortality rate	Body size	Flight ability	Pomeroy, 1990

[a]These columns are so designated purely for convenience and do not imply causality. No entry in the "dependent" column indicates that correlations are found among all or several combinations in the "independent column."

[b]Defined as "the rate of increase in age-specific mortality with age."

[c]Both variables corrected for body size.

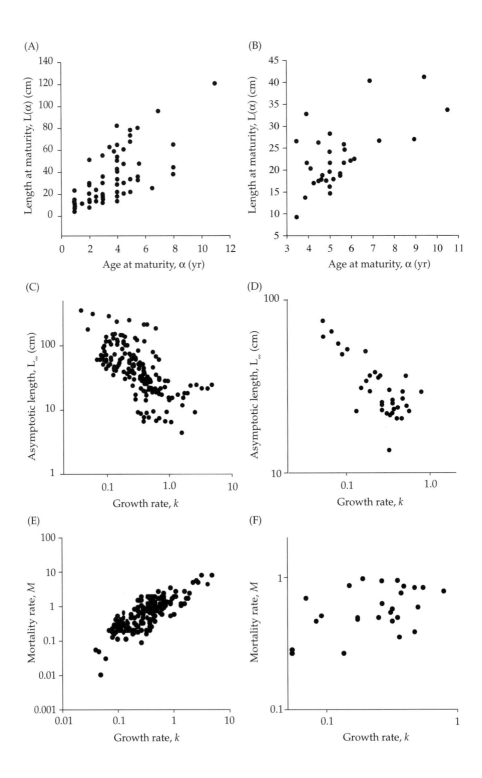

◀ **Figure 4.45**   Some examples of covariation in life history traits among a wide range of fish species (A, C) and among populations of Arctic char (B, D). The growth parameters are those from the Von Bertalanffy growth equation $L(x) = L_\infty(1 - e^{-kx})$, where $L(x)$ is length at age x. Data for (A) from Powles (1958), Wheeler (1969), Hart (1973), Oostuizen and Daan (1974), Ni (1978), Baltz (1984), Hutchings and Morris (1985), Roff (1981b). Data in the other leftside panels from Pauly (1980). Data for Arctic char from Vollestad and L'Abu-Lund (1994).

not only between particular pairs of traits but also aggregates of traits. They noted that these aggregates appeared to be characteristic of particular taxa. In a long series of papers, Charnov and his colleagues developed this idea further, calling such quantities **life history invariants** (Charnov 1989b, 1990, 1991a,b; Charnov and Berrigan 1990, 1991a,b: for a summary of the work see Charnov 1993). From an analysis of 27 populations (5 species) of pandalid shrimp, Charnov (1989b) obtained the equation $M = 2.7k$. Substituting this in the Fish model, he noted that the proportionality between M and k implied that the product αk was a constant equal to 0.75. Observed values for the shrimp are close to this number (0.78–0.83; Charnov, 1989b). Charnov and Berrigan (1991a) noted that if the ratio $L(\alpha) = L_\infty$ were constant within a group, then the von Bertalanffy equation, $L(x) = L_\infty(1 - e^{-kx})$, implies that the product αk will be constant. Other ratios and products put forward as invariants are k/M and αM (Charnov and Berrigan 1990, 1991b; Beverton 1992; Shine and Charnov 1992; Jensen 1996; Morand 1996; Gemmill et al. 1999) or as even more complex invariants (Mangel 1996). Considerable care must be taken in discerning if an aggregate is invariant, because it is easy to slip into the statistical error of plotting something like x versus $1/x$ (Mangel 1996). Another problem is that there is no objective method of deciding when the variability in an "invariant" is small enough to be called invariant. For example, Vollestad et al. (1993, 1994) concluded that the proposed invariants were highly variable for brown trout but not so for Arctic char. These comparisons were for variation within a species, whereas those presented in other papers are for variation among species, where one is more likely to accept a larger degree of variation. If an aggregate measure can be shown to be invariant, then we have a valuable objective for further study since we can then ask what processes would lead to such invariance. On the other hand, if an invariant is taken to exist when it does not do so, then a lot of wasted experimental and theoretical work could ensue.

Patterns and the Persistence Criterion

There are two reasons why some degree of patterning is expected among species. First a minimum requirement for persistence is that R_0 equals 1 (Roff 1984b; Sutherland et al. 1986), and, second the trade-offs constraining variation within a population of species are likely to be similar among closely related species, with the degree of difference in trade-off patterns increasing with phylogenetic distance.

Consider the simple semelparous model

$$R_0 = l(\alpha)m(\alpha) = e^{-M\alpha}c\alpha^b \tag{4.117}$$

where M is the mortality rate, c and b are constants. Given the condition that $R_0 = 1$, we have

$$
\begin{aligned}
1 &= e^{-M\alpha}c\alpha^b \\
e^{M\alpha} &= c\alpha^b \\
M &= \frac{\ln\left(c\alpha^b\right)}{\alpha} \\
\log(M) &= \log\left[\ln\left(c\alpha^b\right)\right] - \log(\alpha)
\end{aligned}
\tag{4.118}
$$

Therefore, if c and b are fixed, then there must inevitably be a negative correlation between mortality and age at maturity (Figure 4.46). If R_o is maximized, then $\alpha = b/M$ and, again, for a fixed b there will be a negative correlation between α and M, with approximately the same slope as previously (slope ≈ -1). In reality, the constants c and b are unlikely to fixed but vary among populations, species, genera, and so on. However, provided the variation is not too great, there will still remain a negative correlation between α and M.

The above model is obviously very simplistic. A more realistic model is the Fish model previously described:

$$R_0 = \frac{e^{-\alpha M}c\left(1-e^{-k\alpha}\right)^b}{1-e^{-M}} \tag{4.119}$$

where, for comparison with the previous model, a single constant c is used in place of the product $(cL_\infty)^b$. Proceeding as above,

$$
\begin{aligned}
1 &= \frac{e^{-\alpha M}c\left(1-e^{-k\alpha}\right)^b}{1-e^{-M}} \\
M &= \frac{\ln c + b\ln\left(1-e^{-k\alpha}\right) - \ln\left(1-e^{-M}\right)}{\alpha} \\
\ln(M) &= \ln\left[\ln c + b\ln\left(1-e^{-k\alpha}\right) - \ln\left(1-e^{-M}\right)\right] - \ln(\alpha)
\end{aligned}
\tag{4.120}
$$

In this case M occurs on both sides of the equation, and so it must be solved numerically. In the example shown, there is a negative correlation for all α

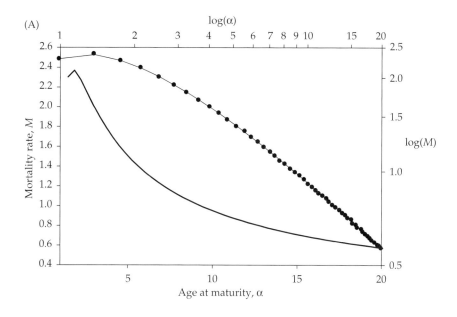

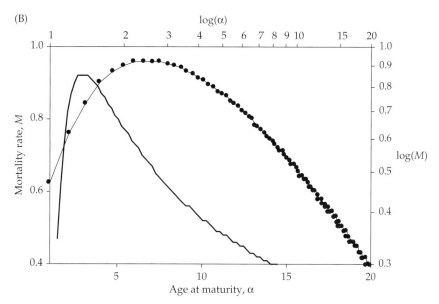

Figure 4.46   Functional relationships between the two life history variables, age at maturity (α) and mortality rate (M) for a simple semelparous life history (A) and the fish model (B). The dotted line indicates the relationship on a log-log scale. Parameter values: $b = 3$, $c = 10$ (A) and 100 (B), $k = 0.2$

greater than 3. The increasing portion of the curve is due to the particular value of c, and for $c = 1000$, the curve closely resembles that of the simpler model. The maximization of fitness gives the general relationship

$$M = \frac{bk}{e^{\alpha k} - 1} = \frac{bke^{-\alpha k}}{1 - e^{-\alpha k}}$$

$$\ln M = \ln(bk) - \ln(1 - e^{-\alpha k}) - \alpha k \tag{4.121}$$

When R_0 is maximized, the logarithm of mortality rate is predicted to be roughly a linear function of the product αk.

Both the simple persistence condition $R_0 = 1$ and the principle of optimization of fitness are capable of generating patterns of covariation. In fact, the latter can be true even if $R_0 = 1$, provided that the density-dependence acts on traits not correlated with those under examination (see Chapter 2). Although life history theory can predict the sign of the correlation between life history traits (e.g., Charnov 1991a) this does not provide sufficient evidence that this pattern has resulted from natural selection.

Distinguishing between the Persistence Criterion and Natural Selection

Can we distinguish between the two sources of constraint? The Fish model predicts that the age at maturity depends only upon b, k, and M and not c, whereas under the persistence criterion, α is a function of all four parameters plus any other variables that reduce the rate of increase to zero. We have seen that the Fish model makes excellent predictions on the age and length at maturity in both fish and reptiles (the two groups for which the necessary data exist). We can thus conclude that the predictions are consistent with observations. There are insufficient data to address the persistence criterion in this regard, but we can ask whether there is a good correlation between predicted and observed α ignoring c. That is, instead of

$$\alpha = \left[\ln c \left(1 - e^{-k\alpha}\right)^3 - \ln\left(1 - e^{-M}\right) \right] \Big/ M \tag{4.122}$$

we use

$$\alpha = \left[\ln\left(1 - e^{-k\alpha}\right)^3 - \ln\left(1 - e^{-M}\right) \right] \Big/ M \tag{4.123}$$

In this comparison we are considering only covariation between prediction and observation, not the difference between the two. (We are not concerned with whether the data fall close to the 1:1 line.) The correlation using the combined fish and reptile data is 0.52 compared to 0.88 using the Fish model. Thus, in this case, the predictions based on the maximization of fitness account for considerably more variance than predictions based on the persistence criterion. This does not provide sufficient grounds to conclude that persistence constraints will not produce covariation in other taxa, but it does provide support for the hypothesis that covariation arises from similarity of trade-offs.

Eliminating Body Size Does Not Remove Covariation

In the analysis of allometric covariation among traits, a frequent approach is to statistically eliminate the effect of body size. Since life history theory predicts that body size variation will be intimately tied to life history traits, I do not see the benefit of excluding it from the analyses. If body size is held constant, do we still expect covariation among life history traits? The answer is YES, and resort to r–K arguments (e.g., Promislow and Harvey 1990) is not required. Suppose we take body size at the age at maturity in the von Bertalanffy function: $L(\alpha) = L_\infty (1 - e^{-k\alpha})$. After algebraic manipulation we get two relationships relevant to Equations (4.120) and (4.121):

$$\left(1 - e^{-k\alpha}\right) = \frac{L(\alpha)}{L_\infty} \tag{4.124}$$

and

$$k\alpha = -\ln\left[1 - \frac{L(\alpha)}{L_\infty}\right] \tag{4.125}$$

Substituting these into the equations and rearranging we get

Persistence criterion $\quad \ln(M) = \ln\left[c + b\ln\left(\frac{L(\alpha)}{L_\infty}\right) - \ln\left(1 - e^{-M}\right)\right] - \ln(\alpha)$

Fish model $\qquad\qquad \ln(M) = \ln(3k) + \ln\left[\frac{L_\infty}{L(\alpha)} - 1\right]$

$$\tag{4.126}$$

Both approaches indicate that covariation among life history traits will remain after keeping body size constant. There are interesting differences between the two equations: The persistence criterion indicates that mortality rate will vary with the ratio of size at maturity to adult size and age at maturity, whereas the Fish model predicts variation with the same ratio and the growth parameter. In a stepwise regression model using the fish and reptile data with $\ln(M)$ as the dependent variable and $\ln(\alpha)$, $\ln[L(\alpha)/L_\infty]$, $\ln(k)$ as independent variables, only $\ln[L(\alpha)/L_\infty]$ and $\ln(k)$ remained in, as predicted by the Fish model ($F_{2,95} = 98.8$, $P < 10^{-6}$, $R^2 = 0.68$).

Several authors have attempted to account for allometric relationships by constraints on energy production imposed by body size (Table 4.11; Enquist et al. 1999; Charnov et al. 2000). Beginning with an allometric function that defines the rate of energy production as a function of size one can derive, using either a nat-

Table 4.11 **A list of models that attempt to predict allometric covariation among life history parameters based on an allometric relationship between growth rate and body size, X.**

Condition	Charnov (1991)	Kozlowski (1996)	Kozlowski and Weiner (1997)	Jensen (1998)
Growth	Determinate	Indeterminate	Determinate	Indeterminate
$\dfrac{\partial P(X)}{\partial X} =$	$AX^{0.75}$	aX^b	$aX^b - \eta X^\beta$	$C^*X^m - kX^n$
$\dfrac{\partial m}{\partial t} =$	$\dfrac{A}{C\delta}X_\alpha^{-0.25}$	aX^b if $Y = 1$	aX^b if $Y = 1$	$m = H\sum N_{i,t}X_{i,t}$
$M =$	$0.75\,AX_\alpha^{-0.25}$ from optimality	Constant, independent of X	γX^λ	$aX_{i,t}^b$
Fitness	Maximized	Maximized	Maximized	No, $R_0 = 1$
Parameters	Constant	Variable	Variable	Variable
L_∞, k		Negative		Negative
M, k		Positive		Positive
L_α, L_∞		Positive		Positive
L_α, α		Positive	Positive	
α, M	Negative	Negative	Negative	Negative
k, M		Positive		Positive

Unless specified, the parameter symbols correspond to those used in the original text and should not be considered necessarily the same from model to model. In the predictions I have designated body size with the length symbol L (L_∞ = asymptotic length, L_α = length at maturity).

Notes (by row):

1. Growth refers to whether growth ceases at maturity (Determinate) or continues (Indeterminate).

2. $\dfrac{\partial P(X)}{\partial X}$ is the rate of energy production per unit size.

In both the Charnov and Jensen models the rate of production per unit size, $\dfrac{\partial P(X)}{\partial X}$, is equated with rate of production per unit time $\dfrac{\partial X}{\partial t}$. In Kozlowski (1996) growth is indeterminate but seasonal. In Jensen, $C^* = (e_f - a_0)b_0(1 - Af)h\left(1 - e^{-s_f N_1}\right)$, where f is feeding level ($0 \leq f \leq 1$), h is a physiological coefficient of food utilization, $b_0(1 - Af)$ is the amount of food absorbed, $1 - e^{-s_f N_1}$ is Ivlev's food-abundance equation in which N_1 is prey density and s_f is a constant, and the remaining parameters are constants.

3. $\dfrac{\partial m}{\partial t}$ is the rate of offspring production per unit time.

In Charnov (1991) fecundity was derived assuming that each offspring takes a fixed amount of energy and $\delta = X_0/X_\alpha$, where X_0 is the initial size and X_α is the size at maturity. The ratio δ is assumed to be an invariant.

Table 4.11   (continued)

In the model of Kozlowski (1996), if the following condition is satisfied in any year then the organism switches from growth to reproduction and diverts all energy production into offspring production. Switch at time E_j, where E_j is the time remaining to the end of the season, if/when $Y = 1$, where

$$Y = \frac{\partial P(X_j)}{\partial X_j} E_j + \frac{P(X_{j+1})}{X_j} S ,$$ and S is annual survival, as defined in the above table.

A similar condition applies to determinate growth model of Kozlowski and Weiner (1997),

$$Y = \frac{\partial P(X)}{\partial X} E(X) + \frac{\partial E(X)}{\partial X} P(X),$$ where $E(X)$ is the life expectancy of an organism of size X.

$m = H \sum N_{i,t} X_{i,t}$, refers to annual fecundity with H a constant, $N_{i,t}$ the number of females in the ith age of year t, and $X_{i,t}$ is their individual weight.

ural selection argument or a persistence constraint, all of the generally observed bivariate relationships. However, this shows only that both assumptions are qualitatively consistent with the data. Although I do think that much of the variation is driven by natural selection, I do not think that focusing upon inter-specific variation is very illuminating unless a rigorous exclusion of the null hypothesis that the persistence criterion alone is responsible can be formulated and tested.

Summary

Evolutionary change requires additive genetic variance. Data from both laboratory and field studies indicate that, in general, there is sufficient genetic variation to permit rapid evolutionary change. Case studies of evolutionary change in natural populations of fish (guppies), birds (blackcap, house sparrow), lizards (anoles), and insects (soaperry bugs and *D. subobscura*) clearly show that rates of change can be as fast as found in artificial selection experiments. Directional and stabilizing selection generally erode additive genetic variance. Factors that can potentially preserve such variation are mutation-selection balance, frequency-dependence and antagonistic pleiotropy. At present there are still insufficient data to decide if any or all of these process are important.

Most analyses of evolutionary change in life history traits have used r or R_0 as the measure of fitness. In the former case, the relative importance of generation length versus fecundity depends upon their relative magnitudes. Semelparity is an extreme condition but is found in a wide range of organisms. Its evolution depends on the pattern of trade-offs between survival and fecundity (Figures 4.19 and 4.20). The prediction that semelparous organisms will show increased reproductive effort has received support from comparative analyses (Figure 4.21).

Reproductive success typically increases with size, because of increased fecundity (females) or increased mating success (males). Increased size can be achieved by either increased development time or increased rate of growth. The latter phenomenon can generate a nonsignificant correlation between adult size and development time. However, the assumption of a single growth trajectory with increased size being determined by increased development time appears reasonable for some species. To illustrate, I show how the size and age at maturity evolve in response to variation in life history traits using *D. melanogaster* and the Hawaiian drosophilids. It is important to distinguish between stage-specific mortality regimes from age or size-specific mortality regimes as the predictions are quite different.

The characteristic equation is composed of the product $l(x)m(x)$ and hence evolutionary change is contingent on this product rather than simply $l(x)$ or $m(x)$. From various theoretical and empirical analyses, the following generalizations can be made: (1) Reproductive effort typically increases with age. (2) Continued growth after maturity requires a concave trade-off between survival and reproductive effort. (3) Foraging mode (sit-and-wait, widely foraging) and reproductive mode (vivparous, oviparous), because they alter the survival schedule, are predicted to alter reproductive effort. Support for this statement comes from data on reptiles, fish, and copepods. (4) The Lack clutch size is generally not the predicted optimal clutch size and is rarely observed to be the modal or mean clutch size in natural populations. (5) The optimal clutch size for various oviposition scenarios (primarily insects) has been investigated and the most general message that emerges is that biological details are critical.

The single component, $m(x)$ has received a lot of attention from the perspective of the proposed trade-off between propagule size and number. There is fairly extensive evidence for a phenotypic trade-off between these two but few estimates of the genetic correlation between them. A potentially important confounding factor in the evolution of propagule size and related traits is the presence of maternal (or paternal) effects that can generate dynamics that override the selection response. Various models, starting from the Smith-Fretwell have predicted that propagule size will increase with the quality of the mother and decrease with age. There is considerable evidence for the former, but a wide variety of age-specific responses have been observed.

Finally, I consider how evolutionary models might be used to predict allometric patterns among species and higher taxonomic units. Attempts to produce life-history invariants have not been entirely successful, in no small measure from the lack of a quantitative definition of what is meant statistically by "invariant." A further problem with the use of evolutionary models is that they have not been contrasted with the appropriate null model of a constraint due to the requirement that persistence requires that $R_0 = 1$. Much remains to be done in this area.

CHAPTER 5

Evolution in Stochastic Environments

Thus far we have considered environments in which all parameter values remain constant. In this chapter I examine the alternate extreme, in which one or more parameter values vary but in a manner that cannot be predicted other than in terms of a probability distribution. In the chapter that follows I shall examine the intermediate and probably most realistic case—one in which the environment (and hence parameter values) vary, but there are cues that indicate the likelihood of the state of the environment. In such an environment we expect that reaction norms will evolve, whereas in the unpredictable environments discussed in this chapter we expect that selection may favor the production of diversity in offspring types.

Maintaining Genetic Variation in Stochastic Environments

Temporal Variation

TEMPORAL VARIATION IS UNLIKELY TO PRESERVE GENETIC VARIATION WHEN GENERATIONS ARE NONOVERLAPPING. The conditions necessary for temporal or spatial heterogeneity to generate stable single-locus, two-allele polymorphisms have been extensively investigated (see reviews by Felsenstein 1976; Hedrick et al. 1976; Hedrick 1986; Frank and Slatkin 1990). For temporally variable environments, the necessary condition is that the heterozygote has a higher geometric mean fitness than either homozygote. This result meshes nicely with the finding that, in phenotypic models with no age structure, fitness is maximized by maximizing the geometric mean of λ. However, this does not mean that these models also predict the maintenance of variation, phenotypic or genetic, by such a mechanism; in fact, in general we may expect to find only a single optimal trait combination. As far as I am aware, there has been no formal analysis of temporal variation and the maintenance of genetic variation in quantitative characters (Barton and Turelli 1989).

As discussed in the preceding chapter, a significant component of the selection regime is likely stabilizing selection. (This is the rationale for the presence of optimal trait combinations.) Therefore, we might ask if temporal heterogeneity acting on stabilizing selection might preserve genetic variation. It seems unlikely that temporal variation alone will preserve additive genetic variation. An intuitive argument suggesting this is as follows: In each generation there is selection against extremes at one tail of the distribution; the tail selected against varies over time, and hence there is a steady erosion of genetic variation. Alternatively, we may think of temporal variation as stabilizing selection in which the optimal value fluctuates: Since stabilizing selection erodes genetic variance, there is no reason to suppose that fluctuation in the optimal value will be itself sufficient to prevent this erosion.

TEMPORAL VARIATION CAN PRESERVE GENETIC VARIATION WHEN GENERATIONS OVERLAP. The conclusion that temporal variation will not preserve genetic variation is contingent on the assumption of nonoverlapping generations. With overlapping generations and a Gaussian (stabilizing) selection function as shown above, genetic variance is maintained provided $\eta\sigma_\theta^2/\sigma_\gamma^2>1$, where η is a measure of generation overlap. This refers to generation overlap when individuals are weighted by their reproductive value as calculated from the transition matrix for a monomorphic population: (Ellner and Hairston 1994; Ellner 1996; Hairston et al. 1996). In the case of non-Gaussian selection, increased generation overlap and stronger selection also favors maintenance of genetic variation (Ellner 1996; Ellner and Sasaki 1996). This condition for preservation of genetic variation is predicated on the assumption that there does not exist a single genotype that produces a range of phenotypes; in the presence of such a genotype, genetic variance will be eroded (Ellner and Hairston 1994; Hairston et al. 1996). By definition, $\eta \leq 1$, and hence the maintenance of genetic variation requires that the variance in the mean (σ_θ^2) must be greater than the variance in the strength of selection (σ_γ^2). Such a condition implies relatively frequent, strong selection, which has been assumed to be unlikely (e.g., Slatkin and Lande 1976). However, the survey of selection estimates in the wild by Endler (1986, see Figure 4.6) makes this conclusion questionable.

TEMPORAL VARIATION CAN PRESERVE PHENOTYPIC VARIATION. Although fluctuating selection may not by itself favor the maintenance of genetic variation, it can favor the maintenance of phenotypic variance. This issue is discussed in detail in other sections of this chapter. However, it is appropriate here to describe an analysis by Bull (1987) that specifically addresses the issue using a genetic model. Bull assumed nonoverlapping generations with stabilizing selection with a fluctuating mean,

$$W(X,\theta_t) \propto \exp\left[-\frac{(X-\theta_t)^2}{2\gamma}\right] \tag{5.1}$$

The total phenotypic variance, assuming independent loci (no epistasis) producing n genotypes is

$$\sigma_P^2 = \sum_{i=1}^{n} p_i \sigma_{E,i}^2 + \sum_{i=1}^{n} p_i (\bar{\mu} - \sigma_i)^2$$

$$\bar{\mu} = \sum_{i=1}^{n} p_i \mu_i \tag{5.2}$$

where p_i is the frequency of the the ith genotype, $\sigma_{E,i}^2$ is the environmental variance contributed by the ith genotype, and μ_i is the mean contributed by the ith genotype. The first term represents the average environmental variance, while the second term is the variance in means. We now focus upon a single locus with one allele, say a, and ask if a mutant allele, say A, can invade the population. Because A must be initially rare, the invasibility of A requires consideration only of the fitness of the heterozygote, Aa. The allele A cannot invaded if (Gillespie 1973; Karlin and Liberman 1974, 1975; Chapter 2)

$$E\big\{\ln\big[W(aa,\theta_t)\big]\big\} > E\big\{\ln\big[W(Aa,\theta_t)\big]\big\} \tag{5.3}$$

This equation essentially states that the geometric mean fitness of aa exceeds that of Aa. Therefore, when $E\{\ln[W(i,\theta_t)]\}$ is maximized with respect to μ_i and $\sigma_{E,i}^2$, a homozygous genotype with these two values is resistant to invasion by any other genotype. Incorporating Equation (5.2) into the fitness function and then maximizing with respect to the mean and variance gives the result, that at equilibrium,

$$\hat{\sigma}_E^2 = \sigma_\theta^2 - \tfrac{1}{2}\sqrt{\gamma} \tag{5.4}$$

where σ_θ^2 is the variance in the optimal phenotype, θ_t. Note that genetic variance is eroded but environmental variance can be preserved. The reason for this is that, in a variable environment, the production of only a single phenotype may have a lower geometric mean than one that produces variation. This point is taken up in greater detail in the following sections. The foregoing analysis implies that environmental variance might itself evolve. A simple manner in which this could occur is if an allele existed that caused developmental instability (Simons and Johnston 1997), which can be equated with the coin-tossing phenomenon suggested by Kaplan and Cooper (1984). If there is a single allele that will generate the appropriate environmental variance, this allele will become fixed in the population. It is not clear what will happen if no such allele exists. Would selection maintain genetic variation for the induction of environmental variation? In practice it is unlikely that there exist no cues to future conditions, and in the presence of cues we can expect reaction norms to evolve, though the question of preservation of genetic variance in these has been little studied.

Spatial Variation

Genetic variation can also be preserved by spatial variation. As before, an intuitive argument can be advanced: Genetic variation lost in one patch may be restored by migration of individuals from another patch in which a different selection regime operates. This intuitive argument also applies to the case of overlapping generations, "patches" being replaced by "generations."

The first analysis of spatial variation was that of Levene (1953), who assumed a single, diallelic locus model in which the fitness of the homozygote is 1 in each patch, and that of the two homozygotes in the ith patch is W_i and V_i. Selection occurs after migration, and both mating and patch selection are random. Selection is "soft," with each patch contributing a fixed fraction of the next generation regardless of how many individuals get eliminated from the patch. A sufficient, but not necessary, condition for a stable equilibrium is that the weighted harmonic means of the fitnesses are less than 1, that is,

$$\left(\frac{c_1}{V_1} + \frac{c_2}{V_2}\right)^{-1} < 1 \tag{5.5}$$

and

$$\left(\frac{c_1}{W_1} + \frac{c_2}{W_2}\right)^{-1} < 1 \tag{5.6}$$

where c_i is the fraction of the population occurring in the ith patch. If the foregoing conditions are not satisfied, it is still possible for there to be an equilibrium solution. But the solution is unstable in that any perturbation from the equilibrium leads to fixation of one of the alleles (see Figure 9.12 in Roff 1997).

Numerous variants of Levene's model have been analyzed, the two most intensively studied being models allowing for variation in migration rates and habitat choice (e.g., Deakin 1966; Maynard Smith 1966, 1970; Christiensen 1974; Gillespie 1975; Taylor 1976; Namkoong and Gregorius 1985; Hoekstra et al. 1985), and models with different rules for the number of parents from each patch. (For example, for hard and soft selection, see Karlin and Campbell 1981; Arnold and Anderson 1983; Walsh 1984; Christiansen 1985.) The results of these investigations have verified the qualitative conditions for a polymorphism to be maintained in a spatial environment. These conditions are formally equivalent to those required in frequency-dependent selection (Bryant 1976). Three important models, fundamentally different from that of Levene, are those of Bulmer (1971b), Gillespie (1976, 1977a, 1978), and Gillespie and Turelli (1989).

Bulmer (1971b, 1985a) investigated the stability of an additive genetic model in a two-patch universe. Within each patch there is stabilizing selection with the optimal value θ differing between patches and a migration rate m between patches. Additive genetic variance will be maintained provided

$$\left(\theta_1 - \theta_2\right)^2 > \frac{4m\left(\sigma_E^2 + \gamma\right)}{1 - m} \tag{5.7}$$

This condition bears a striking resemblance to that for overlapping generations. The term $(\theta_1 - \theta_2)^2$ is equivalent to σ_θ^2, $\sigma_E^2 + \gamma$ is equivalent to γ, and $(1 - m)/(4m)$ is equivalent to η. Significant genetic variation can be maintained by migration, but the phenotypic variance must be significantly smaller than the difference between the optima and the stabilizing selection coefficient (γ). Clinal models also suggest that large amounts of additive genetic variance can be maintained by spatial heterogeneity (Felsenstein 1977; Slatkin 1978).

A critical assumption of the above analyses is that an organism cannot jointly satisfy both optima. However, under the infinitesimal model, unless the genetic correlation between environments is exactly ±1, a reaction norm can evolve that will lead to the character value being θ_1 in environment 1 and θ_2 in environment 2. In this case there would be subsequent total erosion of additive genetic variance. The same argument applies to the case of overlapping generations. One way to circumvent this problem is to assume that there are no or imperfect cues that permit the evolution of a reaction norm. Two models based on this premise were developed and analyzed by Gillespie and Turelli.

Gillespie (1976, 1978) introduced a model in which enzyme activity is an additive function of alleles, and fitness is a concave function of enzyme activity. As a consequence, the heterozygote is intermediate in fitness between the two homozygotes. The fitnesses of the homozygotes are assumed to vary across patches. In the two-patch case, one homozygote, say, *Aa*, is the most fit in the first habitat, while the second homozygote, *aa*, is dominant in the second habitat. This reversal in dominance is a critical element of the model (reminiscent of the conditions necessary for antagonistic pleiotropy to maintain additive genetic variance). Because the function relating fitness to enzyme activity is concave, the heterozygote is fitter than the arithmetic mean of the two homozygotes. The parameter space over which this model produces a stable polymorphism far exceeds that of Levene's model (Maynard Smith and Hoekstra 1980).

Gillespie and Turelli (1989) proposed an alternative model in which there cannot be a single genotype that is most fit in all environments. The phenotype of an individual is the sum of three components, G, E, and Z, where G is the average phenotype produced by a given genotype averaged over all environments, E is an environmental effect that is independent of the genotype (assumed to be normally distributed with mean 0), and Z is a genotype × environment effect. There are n loci with a finite but unspecified number of alleles. If we assume neither dominance nor epistasis, $G + Z$ is simply the sum of the individual contributions of the alleles. A consequence of this particular formulation is that the variance of the average phenotype produced by a given genotype across all environments is a decreasing function of the number of heterozygous loci. Gillespie and Turreli then assumed that there is a single phenotype that is optimal in all environments (i.e., there is a single stabilizing selection function). With this assumption it can be shown that the mean fitness of a genotype is an increasing function of the number of heterozygous loci. Because of the overdominance averaged across all environments, selection will tend to preserve genetic variation. The important assumption of this model is that increasing heterozygosity "buffers" the organism against environmental perturbations.

An Empirical Test

Although environmental heterogeneity is now an acknowledged "major player" in the evolution of life history traits, there has been a surprising lack of empirical research on the relationship between environmental heterogeneity and the maintenance of genetic variation. Most of the research has focused on phenotypic variation (see below). To my knowledge, there has been only one experimental investigation of the effect of environmental heterogeneity on quantitative genetic variation. Mackay (1980, 1981) attempted to test the hypothesis that spatial and temporal variation would maintain genetic variance in three morphological traits (body weight, sternopleural bristles, abdominal bristles) in *D. melanogaster*. The experiment consisted of four treatments, with two replicates per treatment, to population cages: (1) Control—Weekly addition of two bottles of control medium (C); (2) Spatial variation—Weekly addition of one bottle of C medium and one bottle of medium to which 15% alcohol (medium A) had been added; (3) Short-term temporal variation—Alternately, two bottles of C and two bottles of A; (4) Long-term temporal variation—Alternation of A and C media every four weeks. The experiment was run for one year (approximately 20–25 generations). Unfortunately, the results are rather difficult to interpret. Heritability and additive genetic variance of sternopleural bristle number and body weight increased in all environmental treatments, but abdominal bristle number showed no response. Because it is not possible to isolate the forces of selection, it is not possible to suggest with any confidence why such disparate results were obtained. The fact that temporal variation appeared to maintain even higher levels of additive genetic variance than spatial variation suggests that heterozygote advantage might be important (Mackay 1981). However, there is a possible trivial reason for the results obtained. From later analysis Mackay concluded that the difference in genetic variances was a result of a decline in the genetic variance within the control lines rather than an increase in the treated lines; therefore, it is conceivable that there was simply less directional selection imposed by the A medium. Instead of two replicates of C medium for the control lines, a better control would have been four lines, two with C medium and two with A medium.

Evolution in Temporally Stochastic Environments

General Theory

The simplest type of stochastic variation we might imagine is one in which there are no cues to future conditions. In the absence of density dependence, the appropriate fitness measure for populations lacking age structure is the geometric mean rate of increase (Chapter 2). With age structure, the appropriate measure is somewhat more complicated but can be at least approximated by the geometric average rate of increase. In any event, analyses of the evolution of life histories in temporally stochastic environments typically assume either no age structure or have used the approach of considering the geometric rate of increase of clones adopting particular strategies.

There are two components that need to be considered: First, how does temporal, stochastic variation affect the optimal trait value(s), and second, does temporal variation favor the production of variable offspring? From the observation that the geometric mean is a nonlinear function of the arithmetic mean (Chapter 2), it can be concluded that, in general, the trait values that maximize the geometric mean will not be the same as those that maximize the arithmetic mean, though obviously the solutions must converge as the variance declines to zero. Consider two traits, X and Y, for which increases in either trait increase fitness when all other trait values are held constant. Now suppose that there is a trade-off between X and Y, Y, $Y = f(X, c_i)$, where c_i is a parameter of the trade-off function that varies temporally. Let the finite rate of increase be $\lambda_i = g(X, f(X), c_i)$. The geometric mean, $\bar{\lambda}_{GM}$, is then

$$\bar{\lambda}_{GM} = \prod g\big(X, f(X), c_i\big)^{P_i} \tag{5.8}$$

where P_i is the probability of c_i. Taking logs and assuming that the probability density is continuous (e.g., normal) we have

$$W(X) = \ln \bar{\lambda}_{GM} = \int P(c) \ln g\big(X, f(X), c\big)dc \tag{5.9}$$

The integral is evaluated and the optimal trait value obtained in the usual manner (i.e., trait value at which $\partial W(X)/\partial X = 0$). An example demonstrating that this optimum will not generally be the same as obtained using the arithmetic average is presented in Box 5.1.

There is a curious multiple history to the idea that, in a temporally variable world, selection may favor variation in offspring phenotypes. This idea appears to have been suggested independently at least three times since 1966. The idea is implicit in Cohen's 1966 analysis of optimal germination rate in a randomly varying environment, was explicitly put forward as a verbal argument by den Boer in 1968 who termed the phenomenon "**spreading the risk**," and finally developed by Gillespie (1974, 1977b) in the context of variation in offspring number. In reviewing Gillespie's analysis, Slatkin (1974) labeled it as "**bet-hedging**," a term that has stuck, although den Boer's term actually has priority. Den Boer (1968, p. 166) postulated that "the effects of 'factors' that influence the chance to survive and reproduce are spread over individuals differing in various respects, . . . often favourably influences the chance of survival of the population as a whole." As stated, the postulate smacks of group selection, but I believe that this is not the case; rather, den Boer was implicitly using the "clonal" model of

Box 5.1
The optimum obtained using the geometric mean is not the same as that obtained using the arithmetic mean

Suppose that there are just two values of c, c_1 and c_2, occurring with probability P and $(1 - P)$, respectively. For notational simplicity I shall denote the finite rates of increase as $g(1)$ and $g(2)$, respectively. Using the recipe above we have

$$W(X) = P \ln g(1) + (1 - P) \ln g(2)$$

$$\frac{\partial W(X)}{\partial X} = P \frac{g'(1)}{g(1)} + (1 - P) \frac{g'(2)}{g(2)}$$

$$= 0 \quad \text{when} \quad \frac{g'(1)}{g'(2)} = \left(\frac{P - 1}{P} \right) \left(\frac{g(1)}{g(2)} \right) \tag{5.10}$$

where $g'(\bullet)$ denotes the derivative of g with respect to trait X. Using the arithmetic average of the finite rates of increase gives

$$W_A(X) = P g(1) + (1 - P) g(2)$$

$$\frac{\partial W_A(X)}{\partial X} = P g'(1) + (1 - P) g'(2)$$

$$= 0 \quad \text{when} \quad \frac{g'(1)}{g'(2)} = \frac{P - 1}{P} \tag{5.11}$$

In general, there will be no single optimal trait value that will simultaneously satisfy Equations (5.4) and (5.5). The optimal trait value in a temporally variable environment will generally be more "conservative" than that expected by simple averaging, because the geometric mean is more strongly influenced by low values. Specific examples are discussed below.

fitness. We now know that the risk of extinction is not the appropriate measure of fitness, but the general tenure of den Boer's argument still holds. Cooper and Kaplan (1982) used extinction probability as an index of fitness and suggested that environmental variation would favor random variation in phenotypic values, calling this phenomenon "**adaptive coin-flipping**." Again, this idea is really implicit in those of Cohen, den Boer, and Gillespie, nonrandom variation falling under the umbrella of **phenotypic plasticity**, to be discussed later in this chapter. Cohen's original analysis is still one of the best illustrations of the general nature of bet-hedging in a life history trait.

The Optimal Germination Rate in a Temporally Variable Environment

Cohen considered a very particular situation, although his model can be expanded to cover a wider variety of circumstances. The specific life history on which

Cohen focused was that of an annual plant that has a single period of reproduction, producing seeds that germinate with some probability G each year. Conceptually, we have a model in which there is a fixed period of growth and development but a variable timing of emergence, which could be germination or emergence from some dormancy state of any life-cycle stage. For example, the chestnut weevil *Curculio elephas* shows variable lengths of dormancy consistent with the notion of bet-hedging (Menu 1993; Menu and Debouzie 1993). To emphasize the generality of Cohen's model I shall use the term emerge rather than germinate, but still use the symbol G. Failure to emerge is not without cost, some fraction D dieing each year. The recursion equation for population (clone) size is

$$N_{t+1} = F(t)GN_t + (1-G)(1-D)N_t \tag{5.12}$$

The first term on the right represents the contribution from the organisms that emerge in the present year and the second term is the "**seed bank**" (= proportion not emerging/germinating times the proportion not dieing). For any given year, the fitness is measured by the per-year rate of increase

$$W = N_{t+1}/N_t = F(t)G + (1-G)(1-D) \tag{5.13}$$

Variation in the rate of increase is generated by temporal variation in the expected fecundity, $F(t)$, of each emergent. For initial simplicity, let us assume that fecundity can take two values, 0 and F, with the probability of F being P. Taking the geometric mean to be the correct measure of fitness, the fitness, W_G for emergence fraction, G, is

$$W_G = \left[G0 + (1-G)(1-D)\right]^{(1-P)}\left[GF + (1-G)(1-D)\right]^P \tag{5.14}$$

To find the optimum emergence fraction, G_{opt}, we first take logs and then differentiate with respect to G:

$$\frac{\partial \ln W_G}{\partial G} = \frac{(1-P)(1-D)(-1)}{(1-G)(1-D)} + \frac{P(1-D-F)(-1)}{(1-G)(1-D)+GF} \tag{5.15}$$

The optimum emergence fraction occurs when the derivative equals zero. After algebraic manipulation of the left-hand side, we obtain

$$G_{opt} = \frac{PF + (1-D)}{F + (1-D)} \tag{5.16}$$

Thus, provided P is less than 1 (and greater than 0, which is obvious since the population would collapse if P were 0), the optimal emergence fraction is less than 1 and selection favors phenotypic variation. Most typically fecundity will not have only two outcomes but many, and the fitness function is written as

$$W_G = \prod_i \left[GF(i) + (1-G)(1-D) \right]^{P(i)} \tag{5.17}$$

where $P(i)$ is the probability of fecundity $F(i)$. Taking logs and differentiating we have

$$\frac{\partial \ln W_G}{\partial G} = \sum_i P(i) \frac{F(i) - (1-D)}{GF(i) + (1-G)(1-D)} \tag{5.18}$$

The above has at most a single maximum: To the left of this maximum the derivative is positive, and to the right it is negative. So when $G = 1$ (100% emergence), the derivative must be negative. Setting $G = 1$ in the above equation gives

$$\frac{\partial \ln W_G}{\partial G} = 1 - \sum_i P(i) \frac{1-D}{F(i)} \tag{5.19}$$

Therefore, an intermediate emergence fraction will be favored whenever

$$1 - D > \frac{1}{\sum_i \dfrac{P(i)}{F(i)}} \tag{5.20}$$

The expression on the right-hand side is the harmonic mean of $F(i)$, which means that an intermediate emergence fraction is favored whenever the proportion surviving the dormancy phase is greater than the harmonic mean of the fecundity of an emergent. Because the harmonic mean is less than the geometric mean, it follows that a sufficient condition for G_{opt} to be less than one is that $1 - D$ is greater than the geometric mean of $F(i)$.

The above model can be expanded to consider, for example the joint evolution of dormancy and flowering time (Ritland 1983), but the basic message remains the same, namely that temporal variation favors "spreading the risk" or "bet-hedging" by the production of variable progeny.

Generalizing the Bet-Hedging Solution

THEORY. To generalize bet-hedging solution we need to define three components (Haccou and Iwasa 1995): (1) $w(X, y)$ = the per-generation fitness if the organism takes trait value X and the environmental state is y; (2) $P(X)$ = the probability density function of X (i.e., the trait value can take on several values, discrete or continuous); (3) $h(y)$ is the probability that the environment is in state y. Fitness, W, is then

$$W = \int_{y_{min}}^{y_{max}} \ln \left[\int_{X_{min}}^{X_{max}} w(X,y)P(X)dX \right] h(y)dy \tag{5.21}$$

If the trait takes only a single value (say X), (there is no bet-hedging) then the

above reduces to $W = \int\limits_{y_{min}}^{y_{max}} \ln[w(X,y)]h(y)dy$. Temporal variation may nevertheless

have an effect on the evolution of the life history because the optimum trait value can be significantly altered relative to that in a constant environment.

To illustrate the application of Equation (5.21) I shall assume both distribution functions are Gaussian

$$w(X,y) = \frac{C}{\sqrt{2\pi\sigma_w^2}} e^{-\frac{1}{2}\left(\frac{X-y}{\sigma_w}\right)^2}$$

$$h(y) = \frac{1}{\sqrt{2\pi\sigma_E^2}} e^{-\frac{1}{2}\left(\frac{y-\mu}{\sigma_E}\right)^2}$$

(5.22)

Without going through further details (see Haccou and Iwasa 1995), bet-hedging will be favored only if $\sigma_w^2 < \sigma_E^2$. In other words, bet-hedging is the optimal strategy when the spread in fitness (σ_w^2) is less than that of the environment (σ_E^2); otherwise, selection will lead to a single strategy. What does this mean in "concrete" terms? Suppose that there is a trade-off between fecundity (X) and survival to reproduce that is dependent on the environmental state (y); survival $= f$ (fecundity, environmental state) $= f(X,y)$, and that the per-generation fitness is the product of fecundity times survival, $w = f(X,y)X$. Now if the above distributions hold (both w and h are Gaussian), then a single fecundity will be optimal when the environmental variance is less than that of fitness.

The above model is strictly phenotypic and is the one that forms the conceptual basis of most of the analyses to follow. Slatkin and Lande (1976) took an alternative approach and asked how selection would affect a locus that could modify the variance in a fitness-related trait. They assumed a symmetric fitness function with "width":

$$\sigma_s^2 = \frac{\int (X-y)^2 w(X,y)dX}{\int w(X,y)dX}$$

(5.23)

(For simplicity I have equated the optimum with the environmental state.) The parameter σ_s^2 is a measure of the strength of selection. Further, Slatkin and Lande assumed that heritability stayed constant over time, and that a parameter, σ_L^2, called the reproductive variance also stayed constant. (**Reproductive variance** is variance in offspring grouped according to the phenotypes of their parents, so if $h^2 = 1$, then σ_L^2 is the within-family variance.) Slatkin and Lande made the further assumption that σ_L^2 did not depend on parental phenotype, an assumption that will be roughly true except for extreme phenotypes. In the absence of fluctuations in selection the phenotypic variance, σ_P^2, is independent of the strength of selection and lies with the region $\sigma_L^2 < \sigma_P^2 < \sigma_L^2/(1 - \frac{1}{2}h^4)$. If a modifier locus is present that changes σ_L^2, the result will be a decline in the variance to zero. If the environment fluctuates such that the optimal trait value fluctuates with amplitude a^2 (which roughly plays the same role as σ_E^2 in the previous model), then variability in trait value is selected provided that $a^2 > \sigma_s^2/(1 + h^2)$. This solution bears a qualitative similarity to the phenotypic model, which required that $\sigma_E^2 > \sigma_w^2$.

AN ILLUSTRATIVE EXAMPLE: NESTING DATE IN THE GREAT TIT. The above principles can be illustrated using the evolution of nesting date in Great tits studied by Nager (Nager and van Noordwijk 1995; Nager et al. 2000). The breeding success of a pair of Great tits appears to be determined in part by their start of nesting date in relation to the peak abundance of caterpillars that form a significant food source for their offspring. Pairs that have seven-day-old young one week before or after the peak caterpillar abundance are not as successful as pairs whose young coincide with the peak (Figure 5.1).

The problem for the Great tit is that there appears to be no reliable cue as to when the peak will occur except that it is roughly normally distributed (Figure 5.1). Further uncertainty is created by variation in caterpillar distribution on relatively small geographic scales (Nager et al. 2000). The question is whether selection will, under these circumstances, favor a single nesting start date or a distribution of dates. Using the framework described above, we can outline the general approach required to answer this question, even if all the necessary parameter values are not available.

For the sake of simplicity I shall "standardize" the nest start date by subtracting the time between the start of nesting and the seventh day of nestling growth. Thus the optimal nest start date would equal the date of the peak caterpillar density. Let the date of peak caterpillar density be normally distributed with mean μ_c and standard deviation σ_c, and the standardized nest start date also be normally distributed with mean μ_n and standard deviation σ_n. The fitness of a given start date declines as the absolute difference between the two dates increases. (This increase is suggested by the decline when the start date is ± 1 week out of synchrony with the peak caterpillar abundance, Figure 5.1.) I shall assume the fitness function for a given combination of dates is Gaussian $w(X,y) = ce^{-\frac{1}{2}\left(\frac{X-y}{\omega}\right)^2}$ where X is the nest start date, y is the date of peak caterpillar density, and c, ω are constants. The fitness function measured over variable time, but not space, is

$$W(\mu_n, \sigma_n) = \int \ln \left[\int ce^{-\frac{1}{2}\left(\frac{X-y}{\omega}\right)^2} \frac{e^{-\frac{1}{2}\left(\frac{X-\mu_n}{\sigma_n}\right)^2}}{\sigma_n\sqrt{2\pi}} dx \right] \frac{e^{-\frac{1}{2}\left(\frac{y-\mu_c}{\sigma_c}\right)^2}}{\sigma_c\sqrt{2\pi}} dy \qquad (5.24)$$

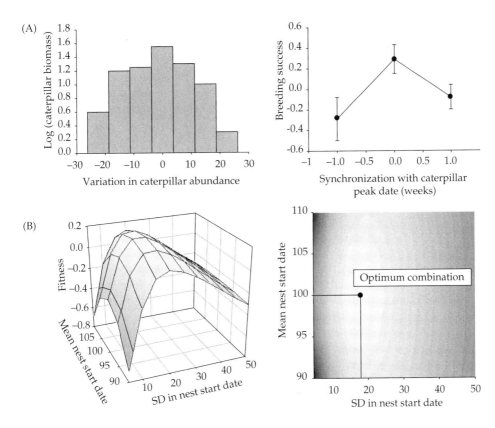

Figure 5.1 (A) The importance for the Great tit of the synchronization of nesting start date with the date of peak caterpillar density. The plot on the right shows the distribution of caterpillar peak dates standardized to a mean of zero for a particular location (data from Nager et al., 2000). On the left is shown the consequences for breeding success (= recruitment rate) of being early, on time, or late in the nest start date relative to the date of peak caterpillar density. (B) The fitness surface for the hypothetical model of nesting start date discussed in the text. Parameter values: $\mu_c = 100$, $\sigma_c = 20$, $\omega = 10$, $c = 4$.

From symmetry, the optimal mean nesting start date will be equal to the mean date of peak caterpillar density (i.e., $\mu_n = \mu_c$, Figure 5.1). If the distributions or fitness function $w(X,y)$ were not symmetrical, then the optimum would depend upon the skewness in the distribution. If a bet-hedging strategy is favored, then the optimal value of σ_n will be greater than zero. This value is shown for particular parameter combinations in Figure 5.1. Without particular values for the Great tit, it is not possible to say if selection is favoring the maintenance of variation, but there is a large parameter space over which bet-hedging is optimal and so the hypothesis is quite tenable.

DIMORPHIC VARIATION AND BET-HEDGING. As discussed in Chapter 2, there are many examples in which two distinct morphs are produced. Suppose, for example that one morph is resistant to attack from a particular predator while another is not, but the unprotected morph has a higher fecundity (e.g., various invertebrates such as *Daphnia*). First consider the relative fitness of two phenotypes, one that produces only the unprotected morph and one that produces only the protected morph, the geometric mean fitnesses being as shown in Table 5.1. Given a temporal frequency of predator-free environments of *f*, the fitness of the unprotected morph will exceed that of the protected morph whenever

$$f > \frac{\ln(w_3) - \ln(w_1)}{\ln(w_3) - \ln(w_1) - \ln(w_2)} \tag{5.25}$$

In the present context the most relevant question is: "Will a mutant that produces both morphs be able to invade the population?" An affirmative answer can be demonstrated by use of a specific case (Roff 1996a): $w_1 = 0$, $w = 1$ (for a general solution see Chapter 6). The unprotected morph now has zero fitness, the protected morph has a fitness of w_2^f, and the dimorphic phenotype has a fitness of $[P + (1 - P)w_2]^f (1 - P)^{1-f}$. Now,

$$\text{As} \quad w_2 \to 0, \quad w(\text{protected}) \to 0$$

But $\tag{5.26}$

$$\text{As} \quad w_2 \to 0, \quad w(\text{dimorphic}) \to P^f (1-P)^f$$

and, therefore, there must be some combination of *P* and *f* for which the dimorphic phenotype is most fit. To examine this in more detail I varied all five parameters within the range 0.1 to 0.9 in increments of 0.1 (Roff 1996a). Of the 13,359 combinations examined, the unprotected morph had the highest fitness in 46.2% of cases, the protected morph in 45.6% of cases, and the dimorphic phenotype in 8.2% of cases. Thus while in this arbitrary parameter space phenotypic variation can be favored, the greatest majority of combinations resulted in a population without phenotypic variation. When the variation is spatial rather than temporal, dimorphic variation is never the most fit life history, unless there is a cue to

---

Table 5.1   Hypothetical fitness values for a dimorphic trait in two environments.

	Environment 1 (no predator)	*Environment 2 (predator present)*	*Geometric mean fitness of morph*
Morph 1 (unprotected)	1	w_1	w_1^{1-f}
Morph 2 (protected)	w_2	w_3	$w_2^f w_3^{1-f}$

the environmental state (Chapter 6). Although this analysis has been developed from an entirely phenotypic perspective, it can equally be developed using the threshold model for dimorphic variation (Chapter 2). By movement of the underlying distribution, any proportion of the two morphs can be achieved and hence the most fit combination can be attained. However, depending on the fitness values associated with each morph and the heritability of the trait, there will be a constant fluctuation in the proportion.

Having developed a general framework, we now turn to specific examples of how temporal variation may determine the evolution of life history traits.

The Evolution of Reproductive Effort
in a Temporally Stochastic Environment

MURPHY'S HYPOTHESIS. Murphy (1968) was the first to consider in any detail the consequences of temporal stochastic variation in life history traits. He proposed that increasing temporal variation will lead to selection for decreased reproductive effort, which is manifested as longer lifespans (p. 392), "Evolutionary pressure for long life, late maturity, and many reproductions may be generated either by an environment in which density-independent factors cause wide variation in the survival of pre-reproductives or by an environment that is biologically inhospitable to pre-reproductives because of intense competition with the reproductives. Conversely, either high or variable adult mortality will tend to generate evolutionary pressure toward early reproduction, high fecundity, and few reproductions, or only one reproduction." The rationale of this hypothesis can be illustrated by comparing a semelparous life history in which m_s female offspring are produced after one year with an iteroparous life history in which the organism lives two years and produces m_i female offspring each year. Assume that the fitness of the semelparous type is greater than that of the iteroparous (i.e., $\lambda_s > \lambda_i$). In a deterministic environment the semelparous life history has the higher fitness. Suppose that the environment fluctuates with "good" and "bad" years, with the survival of offspring in the "bad" years being zero. Even if the probability of a "bad" year occurring is very small (e.g., 1 in 1000), the demise of the semelparous type is certain, while the persistence time of the iteroparous type may be extremely long. (If the probability is 1 in 1000, the probability of two bad years occurring in a row is 1 in 10^6.) Thus in a variable environment iteroparity will be favored.

ITEROPARITY AND SEMELPARITY IN A TEMPORALLY STOCHASTIC ENVIRONMENT. A mathematical basis for the advantage of iteroparity over semelparity in a variable environment can be given as follows (Charlesworth, 1994). Although in an age-structured population there is no formal quantitative genetic model demonstrating that the mean rate of increase (= geometric mean of the finite rate of increase), it seems reasonable to take this as a suitable fitness measure. Consider an iteroparous life history in which adult survival and fecundity are constant (S_I and m_I, respectively), from which we obtain the characteristic equation as

$$1 = l_\alpha m_I \left(e^{-r_I \alpha} + Se^{-r_I(\alpha+1)} + S^2 e^{-r_I(\alpha+2)} + \cdots \right) = \frac{l_\alpha m_I e^{-r_I \alpha}}{1 - Se^{-r_I}} \tag{5.27}$$

where l_α is the probability of surviving to the age of first reproduction, α, and r_I is the rate of increase. A semelparous life history differing only in fecundity, m_S satisfies the equation

$$1 = l_\alpha m_S e^{-r_S \alpha} \tag{5.28}$$

Let the environment be temporally variable, fluctuating between two states for which the probability of surviving to reproduction, which we set at 1 year, is $(1 + d)l_1$ and $(1 - d)l_1$, where d is the mean departure so that the mean survival is l_1. The mean rate of increase for an iteroparous life history using Equation (4.4) is

$$\begin{aligned}\bar{r}_I &= \tfrac{1}{2}\ln\left\{\left[l_1(1+d)m_I + S\right]\left[l_1(1-d)m_I + S\right]\right\} \\ &= \tfrac{1}{2}\ln\left\{l_1^2\left(1-d^2\right)m_I^2 + 2l_1 S + S^2\right\}\end{aligned} \tag{5.29}$$

and the mean rate of increase for a semelparous life history is

$$\bar{r}_S = \tfrac{1}{2}\ln\left[l_1^2\left(1-d^2\right)m_S^2\right] \tag{5.30}$$

The fitnesses of the two life histories will be equal when the ratio of the two rates of increase equals one, from which it can be shown that semelparity will be favored whenever

$$m_S^2 > m_I^2 + \frac{S(2l_1 m_I + S)}{l_1^2(1-d^2)} \tag{5.31}$$

The term $1 - d^2$ must be greater than zero, and hence as the environmental fluctuations increase (i.e., d is increased), the ratio on the right increases. Thus the required semelparous fecundity increases, as expected. Using the same argument, we can also show that increasing fluctuations in adult survival will favor iteroparity.

Bulmer (1985b) extended the analyses of Charnov and Schaffer (1973) using an ESS (nongenetic) analysis to investigate temporal heterogeneity when there is density dependence. He analyzed two density-dependent models, the lottery model and the exponential model. In the lottery model, each plant, annual or perennial, has an equal chance of occupying the available spaces. The exponential model was based on a Ricker recruitment function. Coexistence is not possible in the lottery model but is possible in the exponential model. However, the general message is as found by other workers, namely that variability in juvenile survival rates favors perennials while variability in adult survival rates favors annuals.

REPRODUCTIVE EFFORT IN A TEMPORALLY STOCHASTIC ENVIRONMENT. The argument advanced above can be further extended to the iteroparous condition itself. Increased variability in reproductive success should favor an age schedule of reproductive effort that is extended. The argument can be phrased as the old adage "don't place all of your eggs in one basket." Murphy demonstrated the selective advantage of variation in the degree of iteroparity with two simple simulation models, one using clonal variation and the other a simple Mendelian inheritance of reproductive effort. Hairston et al. (1970) criticized this analysis because the results depended on strong density dependence. They produced a similar genetic model in which the population showed a gradual increase in size over time and no density-dependent effects. In a variable environment, the more iteroparous genotype predominated in the population as in Murphy's simulations. Gurney and Middleton (1996) also using a simulation modeling approach, demonstrated that a graded switch from growth to reproduction can be favored under a high level of environmental variation that is on a time scale approximating an individual's lifetime.

We can use the same general approach as in the case of semelparity versus iteroparity in a fluctuating environment to demonstrate analytically that increased fluctuations select for decreased reproductive effort. This analysis was first presented by Schaffer (1974b) and assumes that reproduction commences one year after birth. In a constant environment the finite rate of increase of the iteroparous life history is $\lambda = m_1 l_1 + S$. Schaffer combined fecundity and survival to first reproduction into a single value which he called "effective fecundity," $B = m_1 l_1$. The environment fluctuates between "good" or "bad" with equal frequency, giving a mean geometric rate of increase $\bar{\lambda}^2 = \lambda_g \lambda_b$, where g and b stand for "good" and "bad" years, respectively. The effective fecundities in good and bad years are represented, respectively, by $B(1 + d)$ and $B(1 - d)$, where d is the departure from the overall average. The geometric mean is

$$\lambda_g = B(1+d) + S, \quad \lambda_b = B(1-d) + S$$
$$\bar{\lambda}^2 = [B(1+d) + S][B(1-d) + S] \tag{5.32}$$
$$= (B+S)^2 - d^2 B^2$$

To find the reproductive effort (E) that maximizes fitness, we differentiate with respect to E, and find the value of E at which the differential is equal to zero,

$$\frac{\partial \bar{\lambda}^2}{\partial E} = 2(B+S)\frac{\partial B}{\partial E} - 2Bd^2\frac{\partial B}{\partial E} + 2(B+S)\frac{\partial S}{\partial E}$$
$$= 0$$

$$\text{when} \qquad \frac{\partial S}{\partial E} = -\left[\frac{\partial B}{\partial E}\left(1 - \frac{Bd^2}{B+S}\right)\right] \tag{5.33}$$

For an intermediate reproductive effort to be optimal, B and S must be concave functions of reproductive effort (Figure 5.2; for the justification of this, see Chapter 4). Increasing values of d, which correspond to increasing variability in repro-

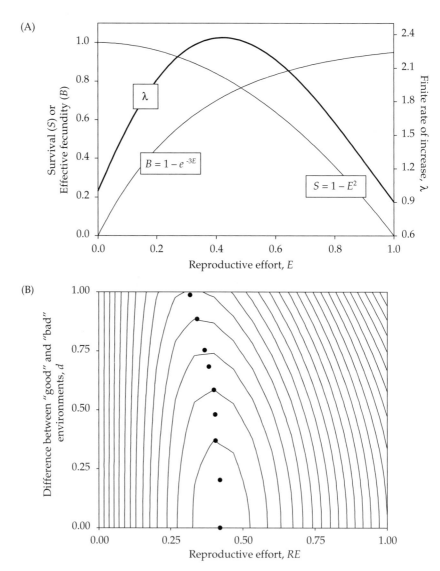

Figure 5.2 An illustration of the effect of variation in adult survival (S) on the optimal reproductive effort. (A) The component functions and the relationship between λ and reproductive effort when there is no variation in survival. (B) A contour plot of λ as a function of reproductive effort and variation in survival. The closed circles indicate the optimal reproductive effort for a given d.

ductive success (juvenile mortality or fecundity), lead to a decrease in the optimal reproductive effort (Figure 5.2B). A similar analysis can be applied to variation in adult survival, S. The optimal reproductive effort occurs when

$$\frac{\partial B}{\partial E} = -\left[\frac{\partial S}{\partial E}\left(1 - \frac{Sd^2}{B+S}\right)\right] \tag{5.34}$$

which leads to the intuitively reasonable result that increasing adult survival favors greater reproductive effort.

The basic selective advantage of increasing iteroparity in increasingly variable environments is that it smooths out the long-term growth rate in the population (Goodman 1984; Bulmer 1985b; Orzack and Tuljapurkar 1989; Tuljapurkar 1989). In this sense, it serves the same function as an increase in phenotypic variance as analyzed by Bull (1987). Indeed, Schultz (1989) has shown that selection for increased phenotypic variation will be reduced as the degree of iteroparity is increased. Whereas there is abundant evidence that the life history components that contribute to reproductive effort and hence degree of iteroparity are heritable, there is no convincing evidence for a genetic basis of random (coin-tossing) phenotypic variation. Thus it seems much more likely that selection in temporally variable environments will cause a change in life history trait contributing to iteroparity rather than random variation. Several simulations have suggested that genetic polymorphisms for iteroparity can be maintained (Hairston et al. 1970; Giesel 1974), but further work needs to be done, particularly with respect to polygenic rather than simple Mendelian models.

IS TEMPORAL VARIATION LIKELY TO HAVE A SIGNIFICANT EFFECT ON REPRODUCTIVE EFFORT? Although theoretical analyses and intuition indicate that temporal heterogeneity will favor decreased reproductive effort, there remains the question of whether such an adjustment is quantitatively significant. Perhaps surprisingly, the answer seems to be no (Sibly et al. 1991; Cooch and Rickelfs 1994). An arbitrary, but biologically reasonable, example of two trade-off functions and the resultant fitness function are shown in Figure 5.2. In the bottom panel is a contour plot of the finite rate of increase as a function of reproductive effort and environmental variation (d). When there is no variation, the optimal reproductive effort is approximately 0.4, and when there is extreme variation, at $d = 1$ the optimal effort is reduced to approximately 0.3, which represents a 25% change in reproductive effort. However, it is unlikely that survival would vary as much as indicated by $d = 1$, since this describes survival beyond the permissible range of 0 to 1. Even $d = 0.5$ represents extreme variation, and at this value there is only a few percent reduction in the optimal reproductive effort (Figure 5.2). Cooch and Ricklefs (1994) examined the consequences of variation in survival or fecundity for life histories mimicking a range of bird species. They never obtained a shift in reproductive effort greater than 4%, and in the majority of cases it was less than 1%. Thus although the theory is correct in predicting a shift in reproductive effort, analyses to date suggest that the effect is likely to be quantitatively small, and operationally very difficult to detect.

Benton and Grant (1999) extended the theoretical analysis of the evolution of reproductive effort in a temporally stochastic environment by determining the optimal effort when there is density dependence in fecundity and/or survival. Their model consisted of two age-classes with a trade-off between fecundity and survival, a density effect, and random variation in one or more vital rates:

$$\begin{pmatrix} (F_1 + \varepsilon_1)Ef(\Delta) & (F_2 + \varepsilon_2)Ef(\Delta) \\ (S_1 + \varepsilon_3)(1 - E^C)f(\Delta) & (S_2 + \varepsilon_4)(1 - E^C)f(\Delta) \end{pmatrix} \begin{pmatrix} n_{1,t} \\ n_{2,t} \end{pmatrix} = \begin{pmatrix} n_{1,t+1} \\ n_{2,t+1} \end{pmatrix} \quad (5.35)$$

where F is maximum average fecundity, ε is a random deviation, E is reproductive effort $(0 < E < 1)$, $f(\Delta) = f(n_1 + n_2)$ is the density-dependent function, S is the maximum average survival, and $1 - E^C$ is a function that determines the trade-off between fecundity and survival. (In all their simulations, $C = 6$, following Cooch and Ricklefs.) The results of simulations using four different density-dependent functions encompassing 15 different life histories are summarized in Table 5.2 and Figure 5.3. When density dependence acted on survival, an increased temporal variance led to a decreased reproductive effort, but the when fecundity was altered by density dependence, the result was not so clear (Table 5.2). Relative to the deterministic effort, the optimal effort expended in a stochastic environment ranged from no change to more than 50%, but there was an overall positive correlation between the two optimal reproductive efforts (Figure 5.3). As the deterministic effort became more semelparous, the percent change decreased and actually became negative (Figure 5.3). A small environmental variance produced a small change in reproductive effort, but a large change could produce anything from no change to a very large change (Figure 5.3). A high change in the optimal reproductive effort was most likely when the deterministic effort was low and the environmental variance high (Figure 5.3C). Cooch and Ricklefs (1994) used a

Table 5.2   **A summary of the effects of temporal variation on reproductive effort in the density-dependent model of Benton and Grant (1999).**

Correlation between ε	Density dependence on	Change in reproductive effort with increased variation in ε
No correlation	Fecundity	Increase or decrease
No correlation	Survival	Decrease
Positive between F and S	Fecundity	Increase
Positive between F and S	Survival	Decrease
Negative between F and S	Fecundity	Increase or decrease
Negative between F and S	Survival	Decrease

F: Fecundity; S: Survival

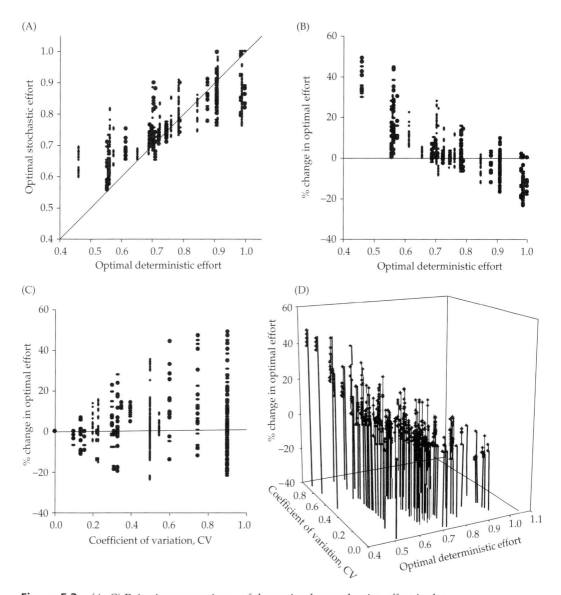

Figure 5.3 (A–C) Pairwise comparisons of the optimal reproductive effort in deterministic and stochastic environments. Values are based on 685 measurements of 15 different life histories, varying in density-dependent functions and correlations between vital rates. (D) Change in RE depends on both CV and optimal deterministic effort. Analysis presented in Benton and Grant (1999).

maximum CV in environmental variation of 20%, which in the present analysis also produced little change in reproductive effort (Figure 5.3). In fact, comparatively few of the situations simulated by Grant and Benton resulted in changes in reproductive effort greater than about 10%. Thus, again, the theoretical results tend to suggest that even though temporal variation can modify reproductive effort, the quantitative change may be slight.

TESTING MURPHY'S HYPOTHESIS. The quantitative results of Cooch and Ricklefs (1994) and Benton and Grant (1999) notwithstanding, the foregoing theory predicts qualitatively an increasing degree of iteroparity with environmental fluctuation. Testing this hypothesis is difficult because it is difficult to adequately quantify environmental variability and to obtain sufficient variation within a single species. Haukioja and Hakala (1978) found a highly significant positive correlation between reproductive lifespan and variation in juvenile survival in the mollusc, *Anodonta piscinalis* (Spearman rank correlation = 0.823, P < 0.001, n = 13). This supports Murphy's hypothesis. The published analysis is biased by a statistical artifact, removal of which does not change the conclusion (Box 5.2). However, the populations differed not only in the variability of the environment but also its mean value, with a significant negative correlation between the average availability of resources and the two components of interest in the present analysis (variation in juvenile survival and reproductive life span). Thus it is possible that the life history variation observed was a result of changes in mean rather than variance.

Murphy (1968) used the ratio of maximum brood strength to minimum brood strength as an index of temporal fluctuation and reproductive lifespan as a measure of the intensity of iteroparity. He estimated the correlation between these two surrogate variables in five species of schooling, plankton-feeding fish of the order Clupeiformes (sardines and herrings). The correlation he obtained was remarkably high ($r = 0.975$), which he advanced as evidence in support of his hypothesis. Armstrong and Shelton (1990) claimed that this correlation was spurious because the use of annual variation ignores differences in reproductive scheduling (i.e., differences in the number of broods per year). Another problem with the data is that, even if we accept annual estimates, the estimate of brood strength variation in the Peruvian anchovy *Engraulis ringens* is inaccurate, because it was made during years when there was not an El Niño event, a climatic phenomenon that can cause drastic fluctuations in the anchovy population. Further data on the anchovy population dynamics indicate that instead of the twofold variation estimated by Murphy, the ratio is at least 11.9 (Roff 1981b). The correlation between reproductive lifespan and brood strength variation is now not significant ($r = 0.776$, $0.05 < P < 0.1$).

Mann and Mills (1979) attempted an analysis similar to that of Murphy (1968), increasing the number of data points to 18 (15 fish species, two species comprising 2 and 3 stocks). They noted that "despite the tendency of data to be biased by the length of the study period, the points generally fit the pattern predicted by Murphy" (Mann and Mills 1979, p. 165–166). However, the correlation is not even close to being significant ($r = 0.03$, $n = 18$, $P > 0.50$)! Roff (1981b) examined the hypothesis using data from a single order, Pleuronectiformes, the flatfishes. I

Box 5.2
Removing the statistical artifact
in the analysis of Haukioja and Hakala

Reproductive lifespan (RLS) was calculated as $1 + t_{50}$, where t_{50} is the time (in years) by which the number of females decreased by half. The rate of decline was estimated from the survivorship curve, that is, the relationship of $\log_e$ (numbers) on age. Thus RLS was estimated as

$$RLS = 1 + \frac{\ln(N) - \ln\left(\dfrac{N}{2}\right)}{M} = 1 + \frac{\ln 2}{M} \tag{5.36}$$

where N is the initial number of females and M is the slope of the survivorship curve (equal to the instantaneous rate of mortality). Note that adult mortality rate could equally well have been used. Variation in juvenile survival was estimated as $1 - r$, where r is the correlation between ln(numbers) and age. The slope of the regression equation of ln(Numbers) and age is equal to $r(\sigma_{Age}/\sigma_N)$, where σ is the standard deviation of age and numbers, respectively. In the present case the slope of this regression is M, and thus the relationship between reproductive lifespan (RLS) and variation in juvenile survival (V) is

$$RLS = 1 + \frac{\ln 2}{r\left(\sigma_{Age}/\sigma_N\right)} = 1 + \frac{\ln 2}{(1-V)\left(\sigma_{Age}/\sigma_N\right)} \tag{5.37}$$

Providing R is not correlated with r, reproductive lifespan will necessarily increase with the variance in juvenile survival as estimated by Haukioja and Hakala. To account for this, we first rearrange the above equation and take logs to obtain

$$\ln(RLS - 1) = \ln(\ln 2) - \ln(\sigma_{Age}/\sigma_N) - \ln(1 - V) \tag{5.38}$$

Thus the predicted slope between $\ln(RLS - 1)$ and $\ln(1 - V)$ due to the algebraic relationship is -1. The observed slope is -3.38 (SE = 0.758), which is significantly different from -1 ($t = 3.145$, $df = 11$, $P < 0.01$). Therefore, providing the index of juvenile survival is appropriate, the hypothesis of a negative correlation between reproductive lifespan and juvenile survival cannot be rejected. These data therefore are consistent with Murphy's hypothesis.

used two measures of brood strength, the ratio of maximum to minimum brood size, and the coefficient of variation in brood strength. Neither measure is significantly correlated with reproductive lifespan ($r = -0.245$, $r = 0.072$, $n = 13$, respectively). Combining the data from all three studies also fails to produce a significant correlation ($r = 0.062$, $n = 29$). Mann and Mills (1979) suggested that variation in the duration of records could account for the reduced correlation. Such a problem is well illustrated by the anchovy data (see also Pimm and Redfern, 1988, for an analysis of this phenomenon with respect to population fluctuations), but the absence of even a trend in the data argues against the hypothesis. Further, given the more recent theoretical analyses indicating that the effect will be quantitatively small, the failure to find a positive correlation when differences among species have not been accounted for is not very surprising. I conclude that brood strength variation is not likely to play an important role in the evolution of the degree of iteroparity when measured across species.

The Evolution of the Optimal Clutch Size

MOUNTFORD'S CLIFF-EDGE EFFECT.   Even if the genotype of the organism dictates a genetic clutch size of X, the phenotypic clutch size will be $C \pm \varepsilon$ simply because of random variation. Mountford (1968) pointed out that this phenomenon will generate an optimum clutch size that is less than the most productive clutch size. To demonstrate this idea, Mountford used data on litter size in guinea pigs (see Roff, 1992a). Here I present a slightly more rigorous example. Let the survival of offspring be a declining function of clutch size, $S(X)$, where X is clutch size. The mean number per clutch, Y, is then $XS(X)$ and the optimum clutch size on a single generation basis, assuming no other effects of clutch size on fitness, is the value of X at which $\partial Y/\partial X = 0$. Now, in reality, clutch size will not be fixed but will show random variation; let the probability density function of clutch size be denoted as $P(X)$. The expected number of offspring, E, is given by $E = \int P(y)yS(y)dy$ and the optimum mean clutch size, μ, (again only from the perspective of a single generation) is obtained by finding the value of μ at which $\partial E/\partial \mu = 0$. According to the hypothesis of Mountford, in general, $\mu < X$. Two plausible survival functions for $S(X)$ are $S(X) = 1 - aX$ and $S(X) = 1 - bX^2$, the latter roughly mimicking the guinea pig data reported by Mountford. I shall suppose that litter size is normally distributed with mean μ and variance σ^2. For the first survival function we have

$$Y = X(1 - aX), \qquad \frac{\partial Y}{\partial X} = 1 - 2aX = 0 \quad \text{when } X = \frac{1}{2a}$$

$$E = \int P(y)y(1 - ay)dy = \mu - a(\mu^2 + \sigma^2) \qquad (5.39)$$

$$\frac{\partial E}{\partial \mu} = 1 - 2a\mu = 0 \quad \text{when } \mu = \frac{1}{2a}$$

In this case the optimum clutch size is also the most productive clutch size, which is contrary to Mountford's suggestion. The principle reason for this result is that the curvature of the survival curve is constant (a straight line) and the distribution of clutch size is symmetric. If the survival curve has an increasingly negative slope, as will almost certainly be the case, then the two optima will not be the same. For example, taking the second survival function,

$$Y = X\left(1 - aX^2\right), \qquad \frac{\partial Y}{\partial X} = 1 - 3aX^2 = 0 \qquad \text{when} \quad X = \sqrt{\frac{1}{3a}}$$

$$E = \int P(y)y\left(1 - ay^2\right)dy = \mu - 3a\mu\sigma^2 - a\mu^3$$

$$\frac{\partial E}{\partial \mu} = 1 - 3a\sigma^2 - 3a\mu^2 = 0 \quad \text{when} \quad \mu = \sqrt{\frac{1}{3a} - \sigma^2}$$

(5.40)

Now the optimum in the presence of variation in clutch size is smaller than when variation is absent. Thus in most populations we should find that the optimum clutch size is less than the Lack clutch size. Because this effect is due to the rapid nonlinear decline in survival with clutch size, Boyce and Perrins (1987) termed this phenomenon the **cliff-edge effect.** Is the optimum derived above the evolutionarily optimum clutch size? If we consider the analysis to apply to a clone in which each individual is subject to variation in clutch size that is independent of other individuals, then if the population size is large, the mean clutch size per generation will remain constant and the solution above is valid. If, however, we follow a single individual each generation, then clutch size will appear to show temporal variation and the optimal clutch size will not be as derived above, because the above solution optimizes the arithmetic average rather than the geometric mean. For the latter, designated E_G, we have $E_G = \int P(y)\ln[yS(y)]dy$. For small populations the type of intrinsic variation described above will reduce the optimal clutch size below that which is optimal in large populations.

THE EFFECT OF VARIATION IN PARAMETER VALUES. A more important source of temporal variation is variation in the parameters determining the trade-off function. To illustrate this point I shall use the first trade-off function described above, $S(X) = 1 - aX$, and assume that a varies from generation to generation with probability density function $P(a)$. The fitness of a particular clutch size is then given by $W(X) = \int P(a)\ln[X(1 - aX)]da$. To proceed I shall simplify the situation further and assume that only two types of years occur, ones occurring with probability P in which $a = a$ ($a > 0$), and ones occurring with probability $1 - P$ in which $a = 0$ (i.e., there is no effect of clutch size on survival). This type of situation has been studied by Tinbergen et al. (1985); they found that adult survival declined linearly with the number of fledglings they attempted to raise when the beech crop was low but showed no relationship with fledgling number when the beech crop was high (see Figure 3.14). Fitness can be defined as

$$W(X) = P\ln\left[X(1 - aX)\right] + (1 - P)\ln X$$

(5.41)

Now

$$\frac{\partial W(X)}{X} = \frac{P(1 - 2aX)}{X(1 - aX)} + \frac{1 - P}{X} = \frac{1 - aX(1 + P)}{X(1 - aX)}$$

(5.42)

and hence the optimal clutch size is

$$X = \frac{1}{a(1+P)} \tag{5.43}$$

If the second trade-off function is used the optimal clutch size is $1/[a(1 + 2P)]$. As the frequency of "bad" years increases, the optimal clutch size decreases. Would it be beneficial in this scenario to produce a range of clutch sizes? Let the probability distribution in clutch size be $h(X)$ with variance σ^2. Fitness is then

$$W(X,\sigma^2) = P\ln\left[\int_{X_{min}}^{X_{max}} X(1-aX)h(X)dX\right] + (1-P)\ln\left[\int_{X_{min}}^{X_{max}} Xh(X)dX\right] \tag{5.44}$$

Suppose that clutch size is normally distributed with mean μ. Then $W(X,\sigma^2) = P\ln[\mu - a\,(\mu^2 + \sigma^2)] + (1 - P)\ln\mu$, and fitness is maximized in this case by minimizing the variance in clutch size, that is, there is no bet-hedging. This analysis is not strictly correct since what we really want to ask is: "Is there a probability density function, $h(X)$ such that optimum from Equation (5.44) is greater than that specified by Equation (5.43)?" That bet-hedging is plausible in some situations was demonstrated by the general example described by Equation (5.22). There are difficulties in adapting this example in the present case. The function $X(1 - aX)$ is dome-shaped, as would be expected for any fitness function. (If the function is not dome-shaped, then selection simply drives the trait value to its physiological extreme.) A Gaussian curve fits this requirement to the extent that it is dome-shaped but it is not easily equated mechanistically with a function of the form $XS(X)$. In particular, the Gaussian curve, when divided by X to give $S(X)$, is not monotonic but still dome-shaped. Whether bet-hedging by producing a range of clutch sizes is ever favored remains to be demonstrated.

Does phenotypic variation in clutch size alter the optimal clutch size in a temporally variable environment? For the above two-state environment with the linear trade-off function, there is no effect of phenotypic variance, but for the quadratic trade-off function, the optimal clutch size is reduced to

$$X = \sqrt{\frac{1 - 3a\sigma_P^2}{a(1+2P)}} \tag{5.45}$$

Figure 5.4 illustrates two trajectories of mean clutch size when there is both genetic and phenotypic variation (model description is given in the figure caption). The long-term equilibrium clutch size should equal that specified by Equation (5.43). For the two cases shown, these values are 9.52 ($P = 0.5$) and 7.94 ($P = 0.8$). The simulations were commenced with a clutch size of 10, and mean clutch size averaged over generations 100 to 200. The means so estimated were 9.25 and

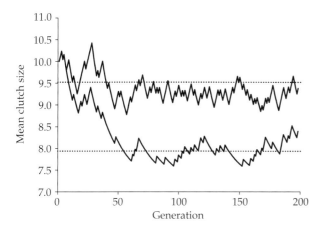

Figure 5.4 A simulation showing the time course of mean clutch size when it is inherited with a heritability of 0.3 and the environment fluctuates between two states that change the parameter of the linear tradeoff function $S(x) = x(1 - ax)$ from 0.07 to 0. The model was constructed as follows: the basic equation is the response equation $X_{t+1} = X_t + h^2(X_{O,t} - X_t)$, where X_t is the mean clutch size in the population before selection, $X_{O,t}$ is the mean clutch size after selection, and h^2 is the heritability. The clutch size after selection is given by $\int XW(X)\varphi(X)dX$, where $W(X)$ is the relative fitness of clutch size X (i.e. $W(X) = XS(X)/\overline{W}$, where $\overline{W} = \int yS(y)\varphi(y)dy$) and $\varphi(X)$ is the probability density function assumed to be normal with mean X_t and variance σ_P^2). Using the linear tradeoff function the foregoing can be solved to give $X_{O,t} = (X_t^2 + \sigma_P^2 - aX_t^3 - 3aX_t\sigma_P^2)/(X_t - aX_t^2 - a\sigma_P^2)$. I assumed values of a of 0.07 (with probability P) and 0, a phenotypic variance of 2 and a heritability of 0.3 (a reasonable value for clutch size and fecundity estimates; see Chapter 2). The upper trace shows that obtained when $P = 0.5$ and the lower trace that obtained with $P = 0.8$. The dotted lines show the predicted optimal clutch size.

7.98, respectively. Even though the long-term means are close to their theoretical values, there is considerable fluctuation about this average (Figure 5.4).

EMPIRICAL EVIDENCE FOR THE IMPORTANCE OF TEMPORAL VARIATION ON THE OPTIMAL CLUTCH SIZE. Boyce and Perrins (1987) invoke temporal heterogeneity in the trade-off between survival and clutch size to explain the mean clutch size of the Great tit, *Parus major,* living in Wytham Wood, Oxfordshire. Three factors suggest that selection is acting on temporal variation in clutch size (Figure 5.5): (1) the observed modal clutch size is smaller than the most productive clutch size, (2) there is a significant trend for the selection differential (as defined in the above simulation) to be positive, (3) additions to clutches suggest that parents could have increased their short-term fitness by increasing their clutches by up to three

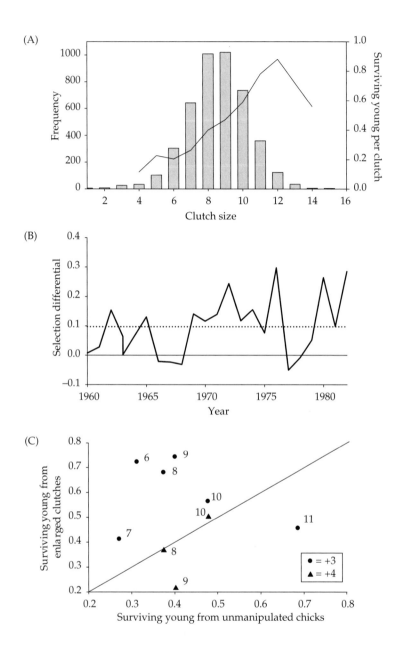

young. An alternative explanation to temporal variation is that adult survival decreases with clutch size. Experimental manipulations of brood size did not show any relationship in the Wytham Wood population, but a significant effect was found by Dijkstra et al. (1990) in a population in the Netherlands. Further, experimental manipulations of clutch size have demonstrated that increased

◀ **Figure 5.5** (A) The distribution of clutch sizes in the Wytham Wood population of the Great tit (histograms) and the number of surviving young from a clutch. Note that the most productive clutch size is greater than the mean or modal clutch size in the population. (B) The selection differential in clutch size, corrected for laying date. The dotted line indicates the overall average. (C) The number of surviving young from enlarged clutches plotted against the number of surviving young from unmanipulated clutches of the same size as the manipulated clutches prior to the addition of young. Numbers beside the symbols indicate the number the original clutch size. Points lying above the 1:1 line indicate that females were able to raise more young than they actually laid. Data from Boyce and Perrins (1987).

clutches decrease the health of the parents as measured by parasite load (Norris et al. 1994; Ots and Horak 1996; Oppliger et al. 1997). Thus it seems likely that adult survival is a function of clutch size, but the relevant question is whether it would be sufficient to account for the discrepancy observed in the Wytham Wood population. Boyce and Perrins address this question by calculating the necessary mortality. These calculations indicate that the mortality of females would have to increase between 10–20%. Such a large effect would probably have been detected in their manipulation experiments though no power analysis was performed to confirm this effect. Boyce and Perrins thus conclude that a trade-off between clutch size and adult survival cannot account for the observed clutch size.

To determine if temporal variation might account for the observed average clutch size in the Wytham Wood Great tits, Boyce and Perrins (1987) developed a model based on geometric mean fitness (Box 5.3). The fitness of different clutch-size phenotypes was computed for the years 1960 to 1982 (Figure 5.6). A quadratic fit to the data gives an optimal clutch size of 9.01, which is not significantly different from the observed clutch size of 8.53 ± 1.79. A clutch size of 9 also corresponds to the observed modal clutch size (Figure 5.5). These results

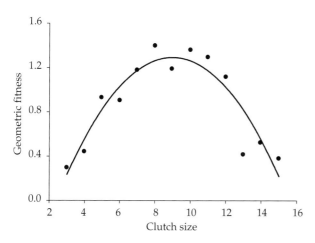

Figure 5.6 Geometric mean relative fitness in Great tits as a function of clutch size over the years 1960–1982. A quadratic curve fit the predicted points very well and indicates an optimum clutch size of 9.01, which is close to the observed mean of 8.53 ± 1.79 (SD). Redrawn from Boyce and Perrins (1987).

Box 5.3
The model of Boyce and Perrins

The model is "clonal" and, although it has age structure, it still assumes that the geo-metric mean is a suitable index of fitness. The model focuses upon a phenotype that produces on average a clutch of size $\overline{X}$, though there is variation from year to year, the clutch size in a particular year being determined from the equation $X = R(X - \overline{X}) + \overline{X}$, where R is the repeatability ($= 0.51$). The number of offspring produced by this phe-notype, $n(\overline{X})$, is estimated as

$$n(\overline{X}, t) = S(\overline{X}, t)\overline{X} + \sum_{i=t+1}^{t+5} S(X, i)l(t, i) \tag{5.46}$$

where $S(\overline{X}, t)$ is the survival of young in year t from clutches of size $\overline{X}$ in year t. (Hence the first term is the number of surviving young.) Also $l(t, i)$ is the probability of the adult surviving from age t to age i. The second term represents the expected produc-tion of offspring from the given phenotype. (Note that, as described above, there is variation in the clutch size of this phenotype.) The assumption is that the maximum lifespan is five years. The survival rates were determined directly from the annual cen-suses, and clutch size was assumed not to influence adult survival. A geometric mean relative fitness was calculated as

$$GM = \left[\prod_{t=1}^{T} \frac{n(\overline{X}, t)}{\overline{n}(t)} \right]^{1/T} \tag{5.47}$$

where $\overline{n}(t)$ is the mean production over all clutch size phenotypes in year t and T is the total number of years observed.

support the hypothesis that in the Great tit population of Wytham Woods, the optimal clutch size is determined by temporal variation in the trade-off between offspring survival and clutch size.

Morris (1992) attempted a similar analysis to that of Boyce and Perrins for a population of white-footed mouse, *Peromyscus leucopus*. The observed distribution of litter sizes has a peak at litter sizes of 4 and 5 although the expected proportion (= number of recruits per litter size x / total number of recruits) has a marked peak at 5 (Figure 5.7A). There was also a significant trend for a positive selection differ-ential, indicating that in most years selection favored a larger litter size (Figure 5.7B). The largest geometric mean fitness was at a litter size of 5, but the arithmetic

Figure 5.7 (A) Observed and expected (= those that would be produced if the number ▶ of litters was proportional to the number of recruits born in each litter size) litter sizes in a population of deer mice. The solid line indicates the observed frequency of recruits per litter size. (B) Selection differentials in litter size. (C) Estimated arithmetic and geometric mean fitnesses. Data source: Morris (1986, 1992).

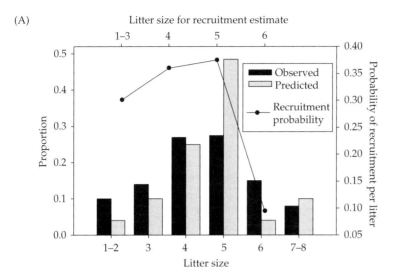

(A)

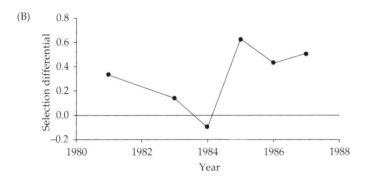

(B)

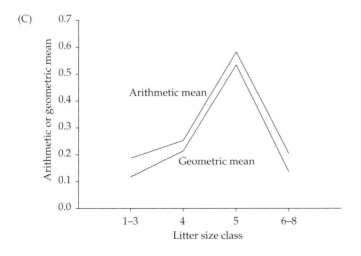

(C)

mean gave essentially the identical result (Figure 5.7C). Data from an earlier study (Morris 1986) showed that the probability of recruitment declined precipitously beyond a litter size of 5 (solid line in Figure 5.7A), leading Morris (1992) to suggest that the most likely explanation for the observed distribution of litter sizes is not temporal variation in the trade-off but Mountford's cliff-edge effect.

The Evolution of Clutch-Size Reduction

HYPOTHESES. In many species of birds the initial brood size is frequently or even always larger than is reared, with mortality of the young occurring early post-hatching (or even pre-hatch in those species in which newly hatched young eject eggs from the nest). Mortality can result from starvation or direct attack by one or more siblings (**siblicide**). Most cases of siblicide involve birds that are predatory, most particularly raptors, pelicans, cranes, gannets, skuas, and herons (Godfray and Harper 1990; Mock et al. 1990). The chick falling victim is usually the smaller individual produced as a result of asynchrony in egg production, a phenomenon that is very common among birds (Stenning 1996). A number of hypotheses have been advanced for the initial overproduction of young (Aparicio 1997):

1. *Bet-hedging hypothesis.* Given uncertainty in the productivity of the forthcoming season, selection favors an "optimistic" clutch size that can be quickly reduced either passively or by active intervention if conditions are not adequate. This hypothesis was first suggested by Lack (1947, 1954).

2. *Ice-box hypothesis.* This is similar to the bet-hedging hypothesis except that it is postulated that the "extra" offspring serve as a food storage site if conditions are poor (Alexander 1974). Although cannibalism does occur (Bortolotti et al. 1991), it does not appear to be general, and this hypothesis does not provide a general explanation.

3. *Progeny-choice or selective abortion hypothesis.* Kozlowski and Stearns (1989) considered this as a possible component in the general overproduction of zygotes, but no evidence has been put forward that the individuals that die are genetically inferior to the survivors. The "victims" are almost always the youngest (Mock and Parker 1986), smallest individuals but this reflects the environmental effect of being last-born.

4. *Insurance-egg hypothesis.* This hypothesis is really just a variant of the bet-hedging hypothesis focusing only upon the possibility that one or more eggs will fail to hatch (Doward 1962; Stinson 1979; Lundberg 1985). Thus in the case of many raptors that lay two eggs but raise only one young, the second egg is insurance against hatching failure of the first egg. If the first egg hatches, the chick invariably kills its younger sibling.

IS EGG PRODUCTION COSTLY? The assumption underlying these hypotheses is that the cost of producing the extra egg or supporting the newly hatched young at least for a short period is less costly than the advantages that accrue if the offspring survives. For passerines and many nonpasserines, a single egg represents

---

Box 5.4
Egg production costs in terms of biomass

What are the costs of producing an extra egg or offspring that is doomed to die? Rahn et al. (1985) estimated the relationship between egg mass and body mass in passerines as

$$Egg\ mass\ (g) = 0.258\ (body\ weight)^{0.73}$$

($r = 0.96$, $n = 1,244$), and for nonpasserines

$$Egg\ mass\ (g) = 0.399\ (body\ weight)^{0.72}$$

($r = 0.94$, $n = 557$). Rearranging these equations (in this instance the correlation coefficients are so high that this is statistically justified) gives,

$$Relative\ egg\ mass = 0.258\ (body\ weight)^{-0.27}$$

for passerines and

$$Relative\ egg\ mass = 0.399\ (body\ weight)^{-0.28}$$

for nonpasserines. Passerine body weights recorded by Rahn et al. (1985) range from 4 g to 1200 g, giving a variation in relative egg mass from 18% to 4%, with the highest density of body sizes between 20 g to 60 g, which translates into a typical relative egg mass of 11% to 9%. For nonpasserines, body weights range from a 3-g hummingbird to the 92,000-g ostrich, from which the range in relative egg mass is 29% to 2%. The majority of nonpasserines fall in the weight range from 100 g to 1,200 g, for which the range in relative egg mass is 11% to 5%.

---

a fairly substantial investment of biomass (Box 5.4). The basal metabolic rate of passerines increases allometrically with body weight, with virtually the same exponent (0.71) as the egg mass/body weight relationship (0.73). From this, Rahn et al. (1985) estimated that the cost of producing an average passerine egg is equal to 41% of the daily metabolic rate of the adult. The figure is typically much larger for nonpasserines: galliformes (126%), ducks (180%), shorebirds (149%), gulls and terns (170%), hawks and owls (39%; data from Ricklefs 1974). The cost in time of producing an egg can be quite considerable, for seabirds ranging from 10 to 40 days, and can significantly influence the potential clutch size (Goodman 1974; Grau 1984; but see Anderson 1990 for a contrary example).

The principal energetic cost of reproduction in birds is in supplying food to the offspring prior to weaning/fledging. Various predictive equations have been suggested for the daily energy expenditure of birds during the period of nestling care, a simple one that appears to be satisfactory being that of Drent and Daan (1980: see Table 5 in Tinbergen and Dietz 1994 for a review of the equations with respect to their ability to predict energy expenditure in the Great tit). This equation states that the average daily energy expenditure will be approximately 4 times basal metabolic rate. The available data show that birds increase their rate of activity from approximately 2.6 times their basal metabolic rate (BMR) when nonreproductive to between 3.1–3.9 times BMR when raising young (Peterson et al. 1990; Daan et al. 1990; Bryant and Tatner 1991; see Table 5.7 in Roff 1992a). This rate is approximately equivalent to that of heavy human labor (Drent and Daan 1980). Drent and Daan (1980) speculated that above 4.0 times basal metabolic rate, a bird would experience severe physiological problems. This level is achieved by a 53% increase, a value exceeded by the house martin and the long-eared owl (Table 5.7 in Roff 1992a), by tree swallows, *Tachycineta bicolor* (Williams 1988) and by some seabirds (Roby and Ricklefs 1986). In addition to these costs, there are incubation costs which may not be negligible (Smith 1989; Heaney and Monoghan 1996), and nutritional constraints on egg production that lead to smaller, less viable offspring (Bolton et al. 1992; Monaghan et al. 1995; Ramsey and Houston 1997; Vinuela, 1997). These data clearly indicate that the production and rearing of nestlings is energetically costly. Hence the laying of an additional egg is probably not an insignificant cost to the female parent.

MODELS OF BROOD REDUCTION. All of the hypotheses advanced for the evolution of brood reduction involve an element of uncertainty and therefore the most appropriate model will involve some element of geometric mean fitness. In fact, most models developed have ignored this component (O'Connor 1978; Dickins and Clark 1987; Parker and Mock 1987; Kozlowski and Stearns 1989; Parker et al. 1989; Godfray and Harper 1990; Lamey et al. 1996). Models that have incorporated the geometric mean concept have been produced by Temme and Charnov (1987), Forbes (1991) Ford and Ydenberg (1992) and Konarzewski (1993). All have shown that selection will favor the overproduction of offspring provided that the costs in bad years are more than balanced by the gains in good years. The model by Temme and Charnov illustrates the general approach. The model focuses upon "parental fitness" and maximizes its geometric mean. It is not clear if this is an appropriate index of fitness, but the index is valid if we assume that each bird breeds only once, since in this case it represents the geometric mean rate of increase for the given life history. The model components are as follows:

1. All eggs hatch in all years. This assumption precludes the egg-insurance hypothesis from the analysis.

2. Offspring survival, S, decreases with clutch size, X. Citing Ricklefs (1977), they used the function $S = ae^{-bX^c}$, which has a wide range of shapes depending on the parameter values a, b, c. Following the usual technique, the optimal clutch size in a constant environment is

$$X = \left(\frac{1}{bc}\right)^{\frac{1}{c}} \tag{5.48}$$

3. As is typical of many analyses, years are considered to be either "good," with probability p, in which case the clutch size is X or "bad," with probability $1 - p$, in which case the clutch size is adjusted after hatching either actively or passively but invariably by a factor R (i.e., clutch size is RX).

4. There are no reliable cues at the time of egg laying as to the state of the present year and hence parents cannot make an "informed decision" prior to egg hatch. Either the environment (which includes siblings) eliminates the fraction R, or the "wise" parents do so. The latter condition would make this life history trait a norm of reaction. (These ideas are discussed in detail in the next chapter.) However, for the sake of continuity I shall consider both options here for the phenomenon of brood reduction.

5. The survival of offspring for a given clutch size is less in a bad year than a good year, so $\ln(S_B) = \ln(a_B) - b_B X^c$, and $\ln(S_G) = \ln(a_G) - b_G X^c$, respectively.

6. There is a cost associated with producing the extra offspring even if they die shortly after hatching. In the present case this cost is incorporated in the survival of the offspring. Temme and Charnov use the explicit cost function, R^q, where q is a shape parameter. So in a good year $\ln(\lambda)$ is $\ln(\lambda_G) = \ln(XS_G) = \ln(Xa_G) - b_G X^c$, and in a bad year, $\ln(\lambda_B) = \ln(R^{q+1}Xa_B) - b_B RX^c$.

   The log of the geometric mean rate of increase is, as given previously, $\ln(\overline{\lambda}_g) = p\ln(\lambda_G) + (1-p)ln(\lambda_B)$ and hence in the present case we have

$$\ln\left(\overline{\lambda}_g\right) = p\left(\ln X + \ln a_G - b_G X^c\right) + (1-p)\left(\ln X + [q+1]\ln R + \ln a_B - Rb_B X^c\right) \quad (5.49)$$

The optimal clutch size is

$$X_{opt} = \left\{ \frac{1}{c\left[pb_G + (1-p)b_B R\right]} \right\}^{\frac{1}{c}} \quad (5.50)$$

Thus the optimal clutch size is affected by the frequency of good and bad years and by the relative decline in clutch size in bad years as specified by R but not by the extra cost specified by q. It is possible not only that clutch size can evolve but also that R can evolve, since it depends upon the actions of siblings or parents. This implies that there is some cue that initiates the action. However, this need only be something as simple as sibling aggression being a function of the amount of incoming food. As noted above, the evolution of R is really the evolution of a reaction norm, but because the response may also be largely passive it shall be discussed here. To find the optimal R we proceed as usual by differentiating Equation (5.49) with respect to R and setting the result to zero, which gives

$$R = \frac{1+q}{b_B X^c} \quad (5.51)$$

Note that the optimal reduction factor depends upon the cost parameter q. No overproduction of young will be favored when the optimal value of R is 1. Overproduction will be selected if

$$p < \left(\frac{X_{opt,G}^c}{X_{opt,G}^c - X_{opt,B}^c} \right) \left(\frac{q}{1+q} \right) \tag{5.52}$$

where $X_{opt,J}$ is the optimum clutch size in the Jth environment: $X_{opt} = (1/ac)^{1/c}$. Although this model and the others noted above demonstrate that in theory there are conditions under which brood reduction will evolve, no researchers have tested the model predictions nor demonstrated that the models make such predictions for reasonable parameter values.

TESTS OF HYPOTHESES. Apacaricio (1997) performed a manipulation experiment on the lesser kestrel (*Falco naumanni*) in which he removed a chick egg at random from clutches in which all eggs hatched and added a chick to those clutches in which one egg failed to hatch. "Control" clutches, in which there was no loss of chicks or loss of a single chick, were also measured. These are not strictly controls since such clutches may not have lost eggs nonrandomly. The four hypotheses make different predictions according to the environmental conditions (Table 5.3). In 1993 nestling survival in the control clutches was twice as great as in 1992, which Apacaricio argues is evidence that 1993 was a "good" year and 1992 was a "bad" year. Although this is correct, it does not demonstrate that the difference between years was great enough to cause a shift in the optimal clutch size. (For a general discussion of this issue, see Forbes 1994.) In both years the experimentally reduced clutches produced more fledglings than those nests with complete clutches (Figure 5.8). The same pattern was observed in the "controls" but the difference was not great in 1993 (the good year). Based on the experimentally altered clutches, the data are most consistent with the insurance-egg hypothesis (Table 5.3), although the bet-hedging hypothesis cannot be ruled out, particularly given the results from the "control" nests. Apacaricio rejected the ice-box hypothesis because in no case were dead nestlings eaten.

Table 5.3   **Predictions by the four hypotheses on the consequences of brood reduction in "good" versus "bad" years.**

Hypothesis	Good year	Bad year
Bet-hedging	Redundant chick (full) clutch	Randomly reduced clutches
Ice-box	Equal fitness of clutches	Redundant chick (full) clutch
Progeny-choice	Redundant chick (full) clutch	Redundant chick (full) clutch
Insurance-egg	Randomly reduced clutches	Randomly reduced clutches

There are two types of clutches, those with a "redundant" chick and those in which the clutch has been reduced by random selection of a chick. The table shows which clutch has the higher fitness under the two environmental conditions.

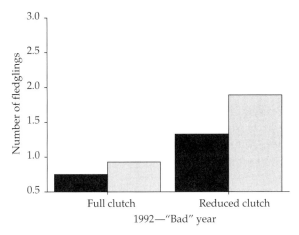

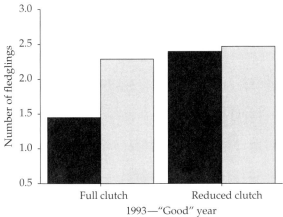

Figure 5.8 Results of clutch manipulation in the lesser kestrel. Black histograms show the results from experimental manipulation, gray histograms show the results for "control" nests in which the same effect was achieved without intervention. Data from Aparicio (1997).

Anderson (1990) argued for the importance of the egg-insurance hypothesis for those species in which obligate siblicide has been observed. He specifically examined reproductive variation among species of the genus *Sula* (boobies and gannets), testing the hypothesis that in those species with obligate siblicide, hatching success would be relatively low. In four species that lay a single egg, hatching success ranged from 85%–98%, whereas in the two species laying 1–2 eggs and showing obligate siblicide, the hatching success was only 53–60% (masked booby) and 51–61% (brown booby). Hatching success in other bird species ranged from 70% to 90%, although Anderson did not distinguish between birds with and without siblicide (see Table 9 in Anderson, 1990). Mock and Parker (1986) divided the reproductive value of the additional chick into two components: (1) **extra reproductive value** (RV_{extra}), which is the additional value obtained if the extra chick survives in addition to the others, and (2) **insur-**

ance reproductive value ($RV_{insurance}$), which is the value when the extra chick replaces another chick. These two measures are estimated from

$$RV_{extra} = P_{All}P_{Y1}$$
$$RV_{insurance} = (1 - P_{All})P_{Y2} \tag{5.53}$$

where P_{All} is the proportion of broods in which the youngest was not predeceased by an elder sib, P_{Y1} is the proportion of youngest chicks that survive in this group, and P_{Y2} is the proportion of youngest chicks that survive in the group in which an elder chick dies first. Data on a number of species in which siblicide is known to occur show that the extra reproductive value can be high (e.g., osprey, Figure 5.9) whereas in other species it is the insurance component that is most important (e.g., black eagle, masked booby). These data are the most compelling evidence that the extra egg serves both a bet-hedging and egg-insurance role, the relative importance of which varies among species and clutch sizes.

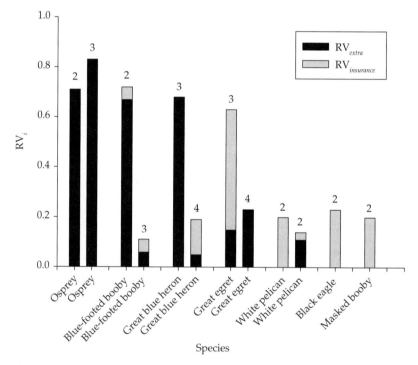

Figure 5.9 Estimates of the extra and insurance reproductive values as a function of clutch size for a range of species in which siblicide is known to occur. The number above the bar indicates the clutch size. Data from Mock et al. (1990).

The Evolution of Propagule Size in a Temporally Stochastic Environment

Under the Smith-Fretwell framework we may imagine that the relationship, $f(X)$, between expected progeny fecundity and propagule size is an asymptotic function that varies according to environmental quality as illustrated, for example, in Figure 5.10. Experimental evidence suggests that as environmental quality increases, so does propagule and clutch size (see Chapter 4) although this relationship is not mathematically necessary (Figure 5.10). Following the recipe given above, the general equation for the geometric mean rate of increase of a semelparous organism in a temporally variable environment is

$$\ln\left(\overline{\lambda}_g\right) = \int_{y_{min}}^{y_{max}} \ln\left[\int_{X_{min}}^{X_{max}} \frac{I}{X} f(X)Q(X)dX\right]P(y)dy \tag{5.54}$$

where y is the environmental state (e.g., the value of the parameter in $f(X)$ that denotes environmental variation), I is the total allocation to offspring, $Q(X)$ is the frequency distribution of offspring size, and $P(y)$ is the frequency distribution of environmental states. Note that the formulation used here assumes that the generation interval is fixed (i.e., that varying propagule size does not change the time to maturity). Kaplan and Cooper (1984) examined the hypothesis that selection in a temporally fluctuating environment would favor variation in propagule size among generations, assuming that there was no variation within generations (i.e., there is a constant for the enclosed integral in the above equation). Their simulation analysis showed that variation would be selected. However, as argued by McGinley et al. (1987), a more realistic scenario is for there to be variation within generations, that is, for females to produce a range in offspring size. Using a simulation model they also showed that variation would be favored under certain parameter values. Without providing a mathematical proof, Yoshimura and Clark (1991) suggested that mixed strategies will be superior whenever two environmental states differ greatly. I provide here a simple analytical demonstration.

As in all the above analyses, I shall assume that there are two environmental states, good and bad, and that the $f(X)$ function takes the extreme form of a step function (Figure 5.11). It is immediately obvious that in a good environment the optimal propagule size is X_G and in a bad environment it is X_B. If there is temporal variation, then either the optimal propagule size is X_B or it is a mixture of the two sizes. Let the probability of a good environment be p and the proportion of resources devoted to propagules of size X_G be q: the fitness function is then

$$\ln\left(\overline{\lambda}_g\right) = p\ln\left[\frac{I}{X_G}Gq + (1-q)\frac{I}{X_B}G\right] + (1-p)\ln\left[(1-q)\frac{I}{X_B}B\right] \tag{5.55}$$

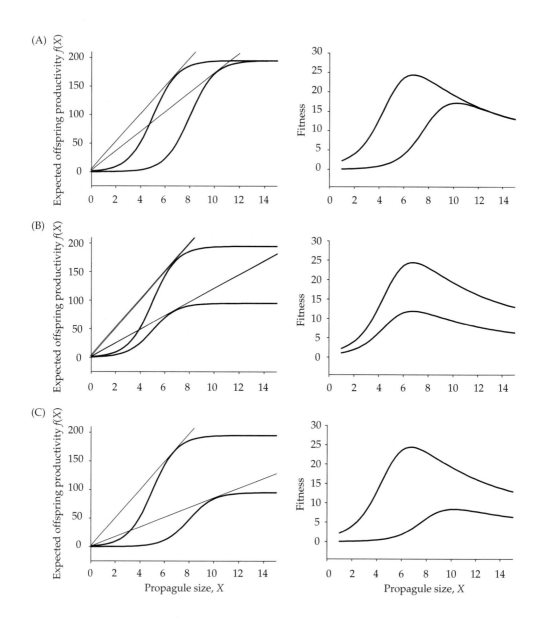

What is the optimal value of q? From differentiating with respect to q and setting the result to zero, we find that

$$q_{opt} = \frac{p - c}{1 - c} \qquad (5.56)$$

◄ **Figure 5.10** Hypothetical Smith-Fretwell curves for two environments, one "good" and the other "bad." The expected offspring fecundity, $f(X)$ was set using the sigmoidal function

$$f(X) = \left[\frac{a}{1 + e^{-\left(\frac{X - x_0}{b}\right)}} \right] - \left[\frac{a}{1 + e^{\left(\frac{x_0}{b}\right)}} \right]$$

the last term to ensure that $f(X) = 0$ at $X = 0$. In all cases the parameters in the good environment are $a = 195$, $b = 1$, $x_0 = 5$. In (A), the bad environment is characterized by a change in the rate of change ($x_0 = 8$) but not the maximum value. The optimal propagule size in the bad environment is increased. In (B), the bad environment is characterized only by a change in maximum value ($a = 95$), which leads to no change in the optimal propagule size. (C) A bad environment in which both slope and maximum value are changed. The optimal propagule size is increased and is the same as the first case.

where $c = \dfrac{X_B}{X_G}$. Thus, as c diminishes, the difference between the two environments increases because the difference between the two optimal propagule sizes (X_G, X_B) increases. The optimal proportion decreases as c increases (Figure 5.12). Over the entire parameter space 50% of the combinations favor only the conservative life history of producing a single propagule size (X_B). I know of no experimental or observational analysis that has examined this question. In practice, such an analysis would also have to take into account spatial variation.

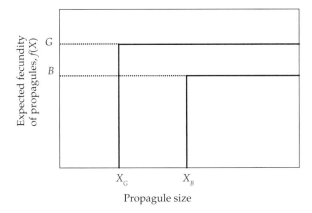

Figure 5.11 Hypothetical curves for the Smith-Fretwell model. In the "good" (*G*) environment the optimum sized egg is smaller than in a "bad" (*B*) environment.

Figure 5.12 (A) Isoclines of optimal proportion of propagules of size X_G for variation in the proportion of "good" years and the ratio, $c = X_B/X_G$. (B) The 3D plot.

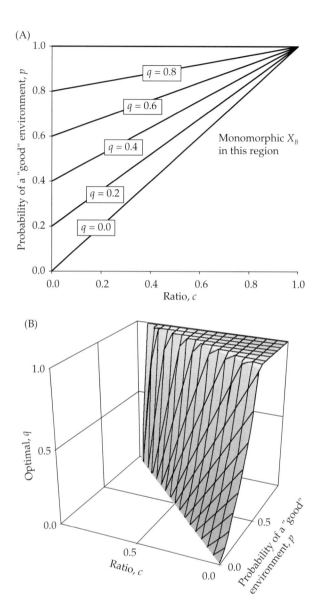

(A)

Monomorphic X_B in this region

$q = 0.8$
$q = 0.6$
$q = 0.4$
$q = 0.2$
$q = 0.0$

Probability of a "good" environment, p

Ratio, c

(B)

Optimal, q

Ratio, c

Probability of a "good" environment, p

Evolution in Spatially Stochastic Environments

In this section I consider environments that are spatially variable but not temporally variable. **Migration** in the present context is restricted in meaning to permanent movement from one site to another (sometimes called **dispersal**).

Evolution of migration in spatially variable environments

AVOIDING KIN COMPETITION. In the absence of any cues as to the state of an environment, selection would, in most cases, simply select for the absence of migration and hence evolution to the optimum trait combination for each separate habitat. An exception to this would be if all available resources were taken up and density dependence were operating. In this case migration may be selected for because it is evolutionarily better to compete against nonrelatives than relatives. This obvious fact has been the subject of detailed mathematical analysis by Hamilton and May (1977), Comins et al. (1980), Comins (1982), Motro (1982a,b), Frank (1986a), and Taylor P. D. (1988). Because of their relative simplicity and generality, I shall present the results of Motro (1982a) and Frank (1986a).

The general set of assumptions is that :

1. Population size is infinite.

2. A fraction *m* of the progeny migrate while the remainder stay close to the parent.

3. Generations are discrete and nonoverlapping.

4. Each site is occupied by a single individual and all sites are occupied.

5. The successor at a site is chosen at random from those individuals in the vicinity.

6. Because of the costs of migration in terms of resources and/or survival, individuals that have migrated to a site have a reduced probability of acquiring the site.

Motro (1982a) considered a haploid organism in which migration was determined by a single locus with two alleles. This is equivalent to considering a clonal model with two clones. If we let the two types be designated by the subscripts 1 and 2, the analysis proceeds by the following steps: The frequency of the *i*th type in the population in the next generation is

$$p' = pP_{ii} + qP_{ij} \tag{5.57}$$

where p is the original frequency of the *i*th type, $q = 1 - p$, P_{ii} is the probability that the site will be occupied by the same type, and P_{ij} is the probability that the site will change types (note that $P_{ij} \neq P_{ji}$). The expected number of immigrants per site, I, is

$$I = nS(m_1 p + m_2 q) \tag{5.58}$$

where n is the number of offspring per adult, and S $(0 < S \leq 1)$ is the probability of a migrant surviving. From the above we find

$$p' = p\frac{n(1-m_1) + nSm_1 p}{n(1-m_1) + I} + q\frac{nSm_1 p}{n(1-m_2) + I} \tag{5.59}$$

and hence

$$\Delta p = p' - p = \frac{n^2 S^2 pq(m_1 - m_2)^2(p - \bar{p})}{\left[n(1-m_1) + I\right]\left[n(1-m_2) + I\right]} \tag{5.60}$$

where $\bar{p} = (1 - m_1 - (1 - S)m_2)/[S(m_2 - m_1)]$. If the migration fraction is $m_{opt} = 1/(2 - S)$, then Δp is greater than zero for all migration fractions, and the type with migration fraction m_{opt} becomes fixed in the population.

Frank (1986a) took a more quantitative genetic approach (Box 5.5). He assumed that mating occurred prior to migration. Migration was controlled by the maternal genotype, from which he derived the expression for the optimal migration fraction as

$$m_{opt} = \frac{\rho - (1 - S)}{\rho - (1 - S)^2} = \frac{\rho - C}{\rho - C^2} \tag{5.61}$$

where ρ is the relatedness coefficient, defined as

$$\rho = \frac{V(p_s)}{V(p_s) + \sum f_s V_s(p_{si})} \tag{5.62}$$

ρ is the expected among-subpopulation variance divided by the total population variance, S is, as previously defined, the probability of the migrant surviving (in which case $1 - S = C$ can be considered the "cost" of migration). The relatedness coefficient is a function of the inbreeding and coancestry coefficients (Perrin and Goudet 2001). The optimal proportion of migrants increases as the cost of migration, C, decreases (= $1 - C$, the increasing relative worth of a migrant) and/or the

---

Box 5.5
Frank's model

Frank made use of the relationship

$$\Delta p = \frac{Cov(w_s, p_s) + \sum_s f_s R_s(w_{si}, p_{si}) V_s(p_{si})}{\bar{w}} \tag{5.63}$$

For a full discussion of the derivation of the above, see Price (1972), Wade (1985), and Frank (1986b). The model posits a number of subpopulations indexed by the letter s, with individuals indexed by the letter i. The relative contribution of an s-type group to the next generation is w_s, and the frequency of a particular genotype within an s-type subpopulation is p_s. The covariance between these two components is $Cov(w_s, p_s)$. The second term in the above numerator is the summed product of the frequency of the groups (f_s = frequency of the s-type group), the slope of the within-group fitness (= number of progeny) on the additive genotypic value for individuals in an s-type group (= $R_s(w_{si}, p_{si})$) and the within-group variance in additive genotypic values (= $V_s(p_{si})$).

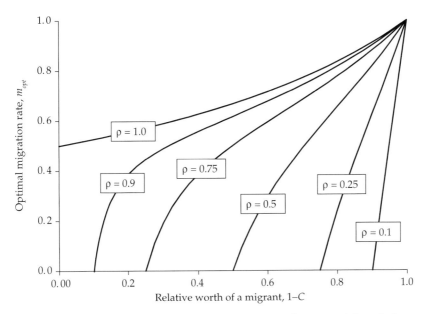

Figure 5.13   The optimal migration proportion as a function of the relative worth of a migrant, $(1 − C)$, and the regression coefficient of relatedness of controlling genotypes onto competing individuals, ρ.

relatedness (ρ) increases (Figure 5.13). When the number of individuals per site is 1, then $\rho = 1$, and the above reduces to the same result as Motro. If a mother mates only once, her offspring are full sibs and $\rho = 0.5$, whereas with multiple mating, her offspring are half sibs and $\rho = 0.25$. Thus as the relatedness of potentially competing individuals increases the advantage of migration increases, even if the environment is saturated.

Equation (5.61) can be rearranged to give (Taylor, P. D. 1988; Perrin and Mazalov 1999)

$$C = 1 - S = \rho\left(\frac{1 - m_{opt}}{1 - m_{opt} + Sm_{opt}}\right) = \rho\left(\frac{1 - m_{opt}}{1 - Cm_{opt}}\right) = \rho P_f \qquad (5.64)$$

The term in parentheses ($= P_f$) is the proportion of breeding females in the patch that are local (that did not disperse). Thus, at equilibrium, migration costs balance kin-competition costs. Further, the optimal migration rate is the same for males and females (Perrin and Goudet 2001).

AVOIDING INBREEDING. In addition to the above advantage of not reducing inclusive fitness by competing with one's kin, migration may also be selected because

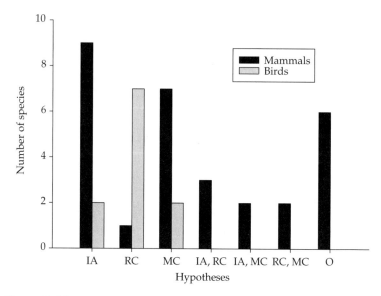

Figure 5.14 A survey of proposed hypotheses for the evolution of dispersal in mammals and birds. Each point represents a separate species and may comprise several studies. Hypotheses are IA, inbreeding avoidance; RC, resource competition; MC, mate competition; O, other. Data from Table 2 of Johnson and Gaines (1990).

it diminishes the reduction in fitness brought about by inbreeding. The avoidance of inbreeding by itself or in conjunction with other factors has been hypothesized to be an ultimate cause for the evolution of migration in a wide range of mammals and birds (Figure 5.14). Although there is abundant evidence that inbreeding is generally detrimental under natural conditions (Crnokrak and Roff 1999, Chapter 2), and while it is obvious that migration will generally lead to a reduction in inbreeding, there is no good evidence that migration has evolved as a consequence of this effect. However, theoretical analysis does point to trends that are found birds and mammals. There are two primary situations to be considered: first, the evolution of migration as a result of inbreeding avoidance, and second, the evolution of migration as a result of both the effects of inbreeding and kin competition. These cases have been analyzed by Perrin and Mazalov (1999, 2000. For a summary of these papers, see Perrin and Goudet 2001); I present only the final ESS equations.

With inbreeding only, a stable equilibrium is attained when

$$C_f = \delta P_m + \rho(1 - P_f)P_{m/f}$$
$$C_m = \delta P_f + \rho(1 - P_m)P_{f/m}$$

(5.65)

where the subscripts *f* and *m* refer to female and male, respectively, *C* is the cost of migration, δ is the inbreeding depression coefficient ($= 1 - W_I/W_O$, where W_I is the trait value of the inbred individual and W_O is the trait value of the outbred individual, see Chapter 2), P_i is the proportion of the *i*th sex that is local—for example, $P_m = (1 - m_m)/(1 - m_m C)$ and $P_{i/j}$ is the relative fraction nondispersing members of the *i*th sex to the *j*th sex—for example, $P_{m/f} = (1 - m_m)/(1 - m_f C)$. The first pair of terms in the above equations is the cost of an inbred mating weighted by its probability of occurrence, and the second triplet is a measure of the increase in inclusive fitness of the nonmigrant due to the decrease in risk of inbreeding by the emigration of the migrant. The only stable equilibrium in this case is for one sex alone to disperse, though because of symmetry, which sex is a matter of chance. The important message of this analysis is that migration by both sexes is not favored simply to avoid inbreeding. Since migration occurs typically in both sexes, although it may be frequently sex-biased, the avoidance of inbreeding is not a sufficient explanation for the evolution of migration. Thus inbreeding avoidance is not a viable hypothesis to explain migration in the nine mammalian and two bird species shown in Figure 5.14, where inbreeding alone was posited as the causative evolutionary agent.

AVOIDING THE JOINT EFFECTS OF KIN COMPETITION AND INBREEDING DEPRESSION. With inbreeding plus local kin competition the equilibrium condition is

$$
\begin{aligned}
C_f &= \delta P_m \left[1 + \rho \left(1 - P_m \right) \right] + \rho P_f (1 - \delta P_m) \\
C_m &= \delta P_f \left[1 + \rho \left(1 - P_f \right) \right] + \rho P_m (1 - \delta P_f)
\end{aligned}
\tag{5.66}
$$

The first product on the right-hand side measures the effect due to inbreeding and the second product measures the effect due to kin competition, but in both cases there is an interaction between effects. Depending upon parameter values migration in both sexes can be favored.

Various elaborations on the above model can be made to address hypotheses previously put forward (Perrin and Goudet 2001).

The major alternative or additional factors to inbreeding avoidance that have been suggested are (1) resource competition, and (2) mate competition (Figure 5.14). Resource competition has been most frequently suggested for birds, whereas mate competition is favored as an explanation for mammalian migration (Figure 5.14). The resource competition hypothesis was put forward by Greenwood (1980, 1983) to explain the observation that the philopatric sex in mammals is typically the female and the male in birds (see Table 7.1 in Greenwood 1983). Greenwood hypothesized that selection would favor the sex that holds a territory (female in mammals, male in birds), because it is this sex that benefits most from "prior ownership." This hypothesis can be incorporated into the mathematical model by the addition of a coefficient that reduces the value of an immigrant: so, whereas before we had

$$P_i = (1 - m_i)/(1 - Cm_i) = (1 - m_i)/(1 - m_i + m_i S) \qquad (5.67)$$

we now write

$$P_i = (1 - m_i)/(1 - m_i + a_i m_i S) \qquad (5.68)$$

Asymmetry between a_f and a_m now produces sex-biased migration in conformity with Greenwood's hypothesis. However, it is important to note that this model shows that Greenwood's contention that inbreeding avoidance was the source of selection is incorrect; rather it is kin competition that is the driving factor (Perrin and Goudet 2001). Perrin and Mazalov (2000) examined the mate competition hypothesis by assuming that males compete for mates but females do not. They further assumed that the females did not experience local resource competition. Under these conditions, kin competition will favor the evolution of male-biased migration.

In summary, inbreeding avoidance can favor the evolution of migration but does not produce a sex-bias in rates. Kin competition acting on the sexes differentially by itself or in conjunction with inbreeding can favor the evolution of the types of migration patterns found in mammals and birds.

The Evolution of Other Life History Traits in a Spatially Stochastic Environment

Given that there is some level of migration, what effect will this have on the optimal life history? The appropriate mathematical framework was enunciated by Houston and McNamara (1992) and Kawecki and Stearns (1993). As described in Chapter 2, it is necessary to compute the rate of increase of the population (= clones) as a whole, which assuming a stable age distribution means solving the equation

$$\int p(h) \int l(x,h)m(x,h)e^{-rx}dxdh = 1 \qquad (5.69)$$

where $p(h)$ is the probability of habitat h occurring, $l(x\,h)$ is the probability of survival to age x in habitat h, and $m(x,h)$ is the number of female births at age x in habitat h. To illustrate how this analysis would proceed, I shall examine the consequences of a two-patch environment on the semelparous life history with the $l(x,h)m(x,h)$ function $l(x,i)m(x,i) = e^{-M\alpha}c\alpha^{b_i}$, where neither the mortality rate nor the parameter c varies between patches, but parameter b does so. If we let the proportion of habitat patches in which $b = b_1$ be p, Equation (5.69) can be written as

$$\left[pe^{-M\alpha}c\alpha^{b_1} + (1-p)e^{-M\alpha}c\alpha^{b_2} \right]e^{-r\alpha} = 1 \qquad (5.70)$$

To find the optimal age at maturity we rearrange the equation to give

$$r = \alpha^{-1} \ln\left[pe^{-M\alpha} c\alpha^{b_1} + (1-p)e^{-M\alpha} c\alpha^{b_2} \right]$$
$$= \alpha^{-1}\left\{ \ln\left[pc\alpha^{b_1} + (1-p)c\alpha^{b_2} \right] - M\alpha \right\} \qquad (5.71)$$

The above is then differentiated with respect to α and the turning point located by setting $\partial r / \partial \alpha = 1$. The result is too messy to display here. In a constant environment with a single value of b,

$$r = \alpha^{-1} \ln\left(e^{-M\alpha} c\alpha^{b} \right) = \alpha^{-1}(\ln c + b \ln \alpha) - M \qquad (5.72)$$

and, as demonstrated in Chapter 4, the optimal age at maturity is $\alpha = \exp(1 - 1\text{n}\ c/b)$. If the environment varies temporally rather than spatially, the appropriate equation would be

$$\ln \overline{\lambda}_G = \overline{r} = \alpha^{-1}\left\{ p\left[\ln\left(e^{-M\alpha} c\alpha^{b_1} \right) \right] + (1-p)\left[\ln\left(e^{-M\alpha} c\alpha^{b_2} \right) \right] \right\}$$
$$= \alpha^{-1}\left\{ p\left[\ln\left(c\alpha^{b_1} \right) - M\alpha \right] + (1-p)\left[\ln\left(c\alpha^{b_2} \right) - M\alpha \right] \right\} \qquad (5.73)$$

The relationship between r and the age at maturity for the spatial and temporal cases are shown in Figure 5.15. At the two extremes of $p = 0$ and $p = 1$ the two sets of curves are obviously the same, but the trajectories between these two point differ dramatically. When $p = 0$, the optimal age at maturity for the parameter values chosen ($M = 0.2$, $a = e^1 = 2.17$, $b_1 = 1$, $b_2 = 5$) is $e^{(1-1/5)} = 2.23$ and when $p = 1$, the optimum value is $\exp(1 - 1/1) = 1$. Rather remarkably, in the case of spatial variation, as the proportion of the poor habitats increases, the optimal age at maturity increases. At approximately $p = 0.98$, two equilibria appear, one tending to infinity and the other tending to 1 as p approaches 1. The optimal age at maturity is entirely unpredictable from the optimal values in either homogeneous habitat! In contrast, in the temporally variable environment the optimal age at maturity declines monotonically with p. The optimal α lies above the simple arithmetic mean of the two separate optimal αs (Figure 5.15).

Kawecki and Stearns (1993) analyzed a similar but slightly more complicated model to that described above. They assumed the same life history except that their hypothetical organism was iteroparous, with fecundity $m(x) = c\alpha^b$, and the environment (= parameter value) varied uniformly. The most important finding in the present context was that whereas in a constant environment, the optimal age at maturity was independent of the mortality rate (M), in a spatially heterogeneous environment the optimal age at maturity declined with an increase in the average mortality rate. Kawecki (1993) extended the analysis under the assumption of exploitation competition and found that selection favors a delayed maturity and larger size.

The foregoing analyses assume that there is a single optimal trait value. As in the temporal case, are there situations in which it would be advantageous to pro-

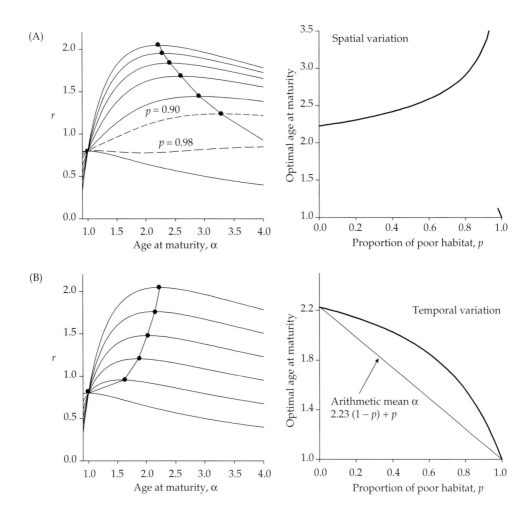

duce several types of offspring? Single locus genetic polymorphisms can be maintained by frequency-dependent selection, and several quantitative genetic models have been proposed that also lead to the maintenance of variation (see beginning of this chapter). A detailed analysis of the maintenance of phenotypic variation in a spatially variable environment is lacking.

Evolution in Temporally and Spatially Stochastic Enviroments

It is highly unlikely that an environment that is spatially variable is also not temporally variable. Considering the two components separately allows the separation of the two effects, which as we have seen are actually quite different. When habitat patches vary in both space and time it is intuitively reasonable to sup-

◀ **Figure 5.15** The optimal age at maturity for spatial (A) and temporal (B) variation in a two-patch universe. Parameter values: $M = 0.2$, $a = e^1 = 2.17$, $b_1 = 1$, $b_2 = 5$. The leftmost graphs show the relation between r and the age at maturity for values of $p = 0$ to 1 in steps of 0.2 (the dashed line shows p as indicated). The filled circles indicate the optimal age at maturity. For the case of spatial variation, two equilibria appear at approximately $p = 0.98$.

pose that, for many states of variation, selection will favor the evolution of migration, because habitats are likely to become totally uninhabitable at some time or another. The importance of this effect on the evolution of migration has been long hypothesized as the prime factor (Southwood 1962; den Boer 1968; Johnson 1969; Dingle 1980, 1984), although the avoidance of inbreeding has also been suggested (Johnson and Gaines 1990). The interest in the evolutionary and ecological importance of migration has spawned a considerable number of books devoted to the subject (e.g., Johnson 1969; Rabb and Kennedy 1979; Gauthreaux 1980; Lidicker and Caldwell 1982; Swingland and Greenwood 1983; Dingle 1996; Clobert et al. 2001). Here I shall discuss only the highlights of this extensive research. As in the previous section, I shall restrict the word "**migration**" to mean the permanent movement from one habitat patch to another (i.e., the effective movement of genes between subpopulations).

The Evolution of Migration as a Response to Habitat Variability: Theory

MIGRATION AND THE GEOMETRIC MEAN RATE OF INCREASE. Southwood (1962; p. 172) put forward the hypothesis that for terrestrial arthropods, "The prime evolutionary advantage of migratory movement lies in its enabling a species to keep pace with the changes in the locations of its habitats." The hypothesis can equally well be extended to cover all organisms. The theoretical basis for the hypothesis has received extensive mathematical treatment (e.g., Reddingius and den Boer 1970; Van Valen 1971; Roff 1974a,b, 1975, 1994b,c; Jarvinen 1976; Jarvinen and Vepsalainen 1976; Kuno 1981; Levin et al. 1984; Klinkhamer et al. 1987; Venable and Brown 1988; Cohen et al. 1991; Ludwig and Levin 1991; Leimar and Norberg 1997; Olivieri et al. 1997; Levin and Muller-Landau 2000). All of these analyses have confirmed the hypothesis that selection favors migration in heterogeneous environments. The essential reason for this is that migration increases the geometric mean rate of increase of clones showing such behavior. Consider two scenarios, one in which a population is divided among k patches with migration among patches, and one in which there is no migration and a single patch. Assuming no spatial or temporal correlation in the finite rates of increase per patch, the two populations obey the following recursive equations:

$$N_t = N_0 \prod_{j=1}^{t} \left(\frac{1}{k} \sum_{i=1}^{k} \lambda_{ij} \right) \qquad \text{with migration}$$

$$N_t = N_0 \prod_{j=1}^{t} \lambda_{ij} \qquad \text{without migration}$$

$$(5.74)$$

The expected generation growth rate $\left((N_t/N_0)^{\frac{1}{t}} \right)$ for the population with migration is approximately the arithmetic mean of the finite rate of increase, whereas without migration the expected generation growth rate is the geometric mean of the finite rate of increase. Now in general $\left(\sum x_i \right)/n \geq (x_i)^{\frac{1}{n}}$ and hence the population with migration will have a higher long-term growth rate. If we assume that $r(= \ln\lambda)$ is normally distributed with mean μ and variance σ^2, then the mean and variance for the population as a whole (μ_T and σ_T^2) is (Kuno 1981)

$$\mu_T \approx \mu + \frac{1}{2}\left(1 - \frac{1}{k}\right)\sigma^2, \quad \sigma_T^2 \approx \frac{1}{k}\sigma^2 \quad \text{with migration}$$

$$\mu_T = \mu, \qquad\qquad\qquad \sigma_T^2 = \sigma^2 \quad \text{without migration}$$

(5.75)

Thus the effect of migration is to increase the expected population rate of increase and decrease its variance. Lande (1993) showed that the expected time to extinction, T_ω for a population obeying the rules $N_t = N_{t-1}\lambda_{t-1}$ if $N_t < K$, else $N_t = K$, is

$$\ln T_\omega \approx \left(\frac{2\mu}{\sigma^2}\right)\ln K - \ln\left(\frac{2\mu^2}{\sigma^2}\right)$$

(5.76)

Thus the substantial reduction in variance produced by migration will produce a dramatic increase in the expected time to extinction, as shown in numerous simulations (e.g., Roff 1974a,b).

Although simulation models have shown that the above effect of migration exists for a wide range of migration models, it has proven very difficult to arrive at any analytical solutions. I shall consider two such models, one using an ESS approach (Levin et al. 1984) and the other a genetic modeling approach (Roff 1975, 1994b).

AN ESS APPROACH. The analysis by Levin et al. (1984) actually examined the combined effect of dormancy and migration. Here I consider only the latter phenomenon. The basic model consists of a set of n uncorrelated habitat patches in which the population at generation t is given by

$$N_{i,t+1} = N_{i,t}\lambda_{i,t}(1-m) + \frac{sm}{n}\sum_{j=1}^{n}N_{j,t}\lambda_{j,t}$$

(5.77)

where m is the probability of an individual migrating, s is the probability of a migrant surviving, and λ is the per-generation rate of increase, which depends upon population density. If there are x types of migrant strategies (i.e., x different values of m) in the population, the population size in the ith patch is simply $N_{i,t} = \sum_{l=1}^{x} N_{l,i,t}$, from which a general density-dependent function is

$$\lambda_{i,t} = K_{i,t} f\left(\sum_{i=1}^{x} N_{l,i,t}\right) = K_{i,t} f(N_{i,t}) \tag{5.78}$$

where $K_{i,t}$ is a random variable and f denotes the density-dependent function. As a starting point in their analysis, Levin et al. assumed that each patch was saturated each generation by letting $f(z) = 1/z$, in which case the population reaches its carrying capacity (K) each generation. The population equation thus becomes

$$N_{i,t+1} = K_{i,t}(1-m) + sd\bar{k} \tag{5.79}$$

where $\bar{k} = \dfrac{1}{n}\sum_{i=1}^{n} K_i$ as $n \to \infty$. The **ESS migration rate**, m_{ESS}, is that rate at which the expected yield from migrants is the same as the expected yield from nonmigrants. The ith habitat patch will have, prior to density-dependent effects, $K_i(1 - m_{ESS}) + sm_{ESS}\bar{k}$ individuals competing and hence the expected yield will be $\bar{k}/[K_i(1 - m_{ESS}) + sm_{ESS}\bar{k}]$. The expected yield of a migrant is the average

$$E_m = s\sum \frac{\bar{k}P(k)}{k(1-m_{ESS}) + sm_{ESS}\bar{k}} \tag{5.80}$$

where $P(k)$ is the probability density function for k. The average yield must be 1, and since the average of migrants must equal that of nonmigrants, the ESS is the value of m_{ESS} at which $E_m = 1$. The optimal migration fraction tends to 1 as s tends to 1. That is, in the absence of any cost to migration, all individuals should migrate. To obtain the ESS value, we must supply the probability distribution for k. Suppose k takes only two values, 0 and K with probability p and $q = 1 - p$, respectively, then

$$m_{ESS} = \frac{p}{1-qs} \tag{5.81}$$

If k takes values k_1 with probability p and k_2 with probability q (= $1 - p$) then

$$m_{ESS} = \frac{p}{1-s\bar{k}/k_2} + \frac{q}{1-s\bar{k}/k_1} \tag{5.82}$$

To examine the effect of variance in k, we can write their two values as $k_1 = \bar{k} - q\varepsilon$ and $k_2 = \bar{k} + p\varepsilon$, where $k_2 - k_1 > \varepsilon$. The variance in k is $\sigma_k^2 = pq\varepsilon^2$ and Equation (5.82) can be written as

$$m_{ESS} = \frac{1 + s(1-s)\bar{k}^2}{\bar{k}^2(1-s)^2 + (2p-1)\varepsilon\bar{k}(1-s) - \sigma_k^2} \tag{5.83}$$

If the variance in k alone is increased, the optimal migration rate declines. However, a change in variance will mean a change in either p or ε. If p is kept constant, then there are some parameter values for which the optimal migration increases with the variance and some for which it decreases (Figure 5.16). This

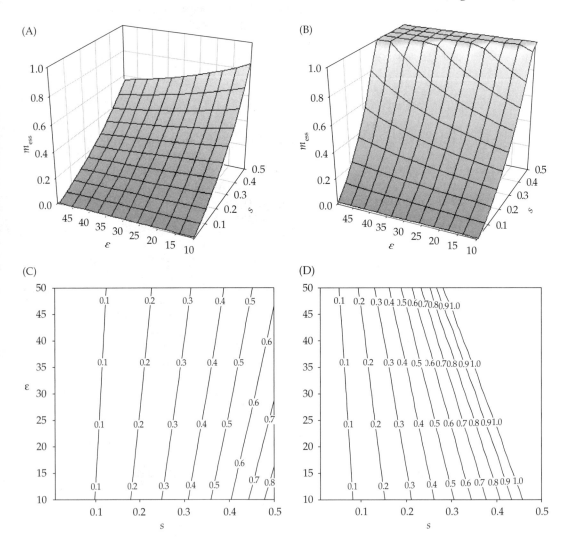

Figure 5.16 The effect of variation in ε, s, and p on the optimal migration rate for the saturated version of the Levin model. The plots on the left (A,C) show the relationship when the "good" habitat (k_1) occurs with a probability of $p = 0.9$, and the plots on the right (B,D) show the case for $p = 0.1$. In both cases, $\overline{k} = 100$. Because the variance in environmental value is $p(1-p)\varepsilon^2$ and p is fixed for each plot, increasing values of ε mean that the variance in carrying capacity increases, while the value of k_1 actually decreases. When $p = 0.9$, the optimal migration rate decreases with increasing variance (and concomitant decrease in k_1), whereas the opposite occurs when $p = 0.1$.

result stems from the fact the optimal migration fraction depends upon the size of the environmental variance relative to the mean carrying capacity and the cost of migration. Further, as will be discussed in more detail below, in the Levin model the effects of environmental carrying capacity and rate of increase are confounded, but they have separate effects on the optimal migration rate.

Levin et al. (1984) were not able to determine analytical solutions for the nonsaturating case, and they turned to simulation. They found that, provided s was not too small, the results from the saturated model fitted well with those of the nonsaturated model.

A GENETIC MODELING APPROACH. Roff (1975) analyzed a model that differed from Levin et al. (1984) in the following aspects:

1. The Roff environment consisted of a grid of habitats with migration only to the neighboring patch (this change is minor),

2. Unlike the Levin model, the Roff model separated the rate of increase, λ, from carrying capacity, K, so that there were three stages:

$$
\begin{aligned}
\text{Reproduction} \quad & N_{i,t+1}^{*} = N_{i,t}\lambda_{i,t} \\
\text{Migration} \quad & N_{i,t+1}^{**} = N_{i,t+1}^{*} - \text{Emigrants} + \text{Immigrants} \\
\text{Regulation} \quad & N_{i,t+1} = N_{i,t}^{**} \quad \text{if} \quad N_{i,t}^{**} < K_{i,t+1} \\
& N_{i,t+1} = K_{i,t+1} \quad \text{if} \quad N_{i,t}^{**} > K_{i,t+1}
\end{aligned} \tag{5.84}
$$

3. Migration rate was determined by either a simple single-locus model in which the three genotypes M_1M_1, M_1M_2, M_2M_2 migrated with probabilities, $m_1, m_2, 0$, respectively, or by a quantitative genetic model in which the heritability was set at 1.

Changing the probability of migration changed the equilibrium genotype frequencies but had little effect on the equilibrium proportion of migrants. Similarly, the type of genetic determination had little effect on the equilibrium proportion of migrants. One of the advantages of the use of a genetic model is that there is no need to directly define a fitness measure that will be maximized. Attainment of equilibrium is also generally very quick, making the location of the equilibrium migration rate computationally very efficient. In my analysis I varied means and variances independently and found that the equilibrium migration rate decreased with (1) an increase in the mean carrying capacity, (2) a decrease in the variance of the carrying capacity, (3) a decrease in the variance of the rate of increase, and (4) an *increase* in the mean rate of increase. Thus the equilibrium migration rate could increase with an increase in variance if the mean rate of increase also increased.

HABITAT PERSISTENCE AND THE EVOLUTION OF MIGRATION RATE. The foregoing analyses show that the evolution of migration rate depends upon the relative value of means and variances. Unfortunately, the environmental parameter values have not, to my knowledge, been measured in a real situation. Thus these analyses and others of the same ilk are important in coming to grips with the potential complexity of the evolutionary process, they are not at present subject to testing. To produce such a model, we need an environmental parameter that has been measured and relates to environmental stability, variation, or persistence. A simple such parameter would be the **persistence time of a habitat patch**. The hypothesis advanced by Southwood and others would then be that the proportion of migrants in a population would be inversely related to the persistence time of habitat patches. Such a measure was estimated by Denno et al. (1991) in their analysis of the evolution of wing dimorphism in planthoppers. This analysis and other empirical investigations, described in the next section, supported the persistence time/migration rate hypothesis. Therefore, I constructed a model in which the main environmental component was habitat persistence time to address two questions: (1) Would such a model provide theoretical justification for the persistence time/migration rate hypothesis? (2) Could such a model correctly predict the quantitative relationship reported by Denno et al. (1991)?

An operational difficulty in many species is that it is not possible to morphologically distinguish migrants from nonmigrants. However, there are a group of taxa, both animal and plant, in which two morphs occur, one that is incapable of migration and another that is capable of migration, but may not choose to do so. For example, in paedomorphosis in salamanders or wing dimorphism in insects, there is a morph that clearly cannot migrate to the same extent as the other morph. (Thus the neotenic form of the salamander cannot leave its natal pond, and the flightless insect morph is generally restricted to migration by walking, which is clearly more limiting than flight.) In these cases we can call one morph the "nonmigrant" and the other morph a "potential migrant." The data analyzed by Denno et al. (1991) consisted of the incidence of wing dimorphism in planthopper species, and thus the model I developed was predicated on this nonmigrant/potential migrant dichotomy. It is, however, a simple matter to generalize the results to migration in general.

To illustrate how a dimorphism for wing morph (or some other morphological trait that enhances/prevents migration) can evolve and be maintained in an environment that is spatially and temporally variable, consider an environment consisting of discrete patches, with each patch persisting for some finite time period and the intervening habitat being such that the "nonmigrant" form cannot move from one patch to another. Patches are coming into and out of existence, and hence a population consisting solely of nonmigrants will not persist. Initially, therefore, we suppose that the population consists of only winged individuals, which is the ancestral state of all wing-dimorphic insects. Now a mutation arises that produces a flightless morph. Because of the initially low frequency of the mutant allele, it will occur overwhelmingly in the heterozygous state. If a mutant allele is recessive, it will not be expressed in the heterozygous condition, and hence such mutations will increase initially only by genetic drift. However, if the allele is dominant, the flightless (nonmigrant) morphology will be immediately expressed and any advantage it confers will be realized. In this example, therefore, I assumed that the mutant allele is dominant, which is consistent with the observation that where wing dimorphism is determined by a single locus, two-allele mechanism, almost without exception the flightless allele is dominant (Roff 1986, 1992b).

As the flightless morph is not diverting energy into the maintenance of flight muscles and associated structures, the flightless form will be assumed to have a higher fecundity than the winged morph, evidence for which is plentiful (Roff and Fairbairn 1991). Therefore, the flightless allele will increase in frequency in the patch in which it arises. This increase will be enhanced if, as is likely, the emigration of the winged morph exceeds immigration. Because the flightless morph cannot move between patches and the flightless allele is dominant, the flightless allele can be passed to another patch only if a flightless male mates with a winged female prior to migration. If mating always occurs after migration, the flightless allele will never move to another patch and will be eliminated from the population when the patch in which it arose ceases to exist. Assuming random mating within a patch, the frequency of mating between a volant (flight-capable = winged) female that subsequently migrates and a flightless male increases with the frequency of the flightless allele (which itself must increase with the persistence time of the patch). Because of its fecundity advantage and relatively greater survival rate within a patch (in other words, no emigration), the flightless allele will spread though the population. However, it cannot replace the winged allele, since a population homozygous for the flightless allele will never colonize new patches and hence will become extinct as each patch eventually ceases to exist. Thus an equilibrium frequency will be established that is dependent upon the persistence time of the patch, the cost of being winged, the relative survival rate of the winged morph, the probability that a winged female migrates, and possibly the genetic determination of the trait (single locus versus polygenic).

The above verbal model provides a logical foundation for the persistence time/migration hypothesis but gives no quantitative basis for it. To achieve a quantitative approach, I examined the following model. The environment is a set of patches that last a fixed interval of T generations and there is a stable-age distribution of patches that can be seen as units of patches, each with T patches aged 1, 2, 3, . . ., T. In the next generation all patches are advanced one unit, the patch aged T units at the previous generation being eliminated, along with all its inhabitants, and a new patch aged 1 unit coming into existence. This model environment is computationally simple to analyze, and the results are essentially identical to an environment in which the probability of persistence for a patch is set such that the expected persistence time is T units (Roff 1994b).

As discussed above, the flightless morph is restricted to the patch in which it is born, but the winged morph migrates with some probability m. This is not the same as the migration fraction in the Levitan model as all individuals in this model are potential migrants. The comparable migration rate is the product of the proportion of winged individuals in the population and *m*. As in the Levin model a relative proportion *s* of the migrants locate new patches. Unlike the Levin model, I assumed a cost to being winged even if the female did not migrate. Thus a winged female had a potential rate of increase of *c*λ (0 < *c* < 1) while a flightless female had a rate of increase of λ. I examined the consequences of a single locus determination of wing morph with flightlessness as the dominant allele and a polygenic mode of inheritance using the threshold concept of quantitative genetics (see Chapter 2). The equations for the single-locus case are straightforward, but those for the polygenic case require some explanation (Box 5.6).

The model contains five parameters: *m*, *c*, *s*, λ, and *K*, of which the first three have significant influences on the equilibrium value. The quantitative genetic model requires the heritability of wing dimorphism. Estimates from five species range from 0.30 to 0.98 (Roff 1994b). The heritability determines the rate at which the flightless morph increases within a patch. Within the range $0.25 < h^2 < 0.75$, the equilibrium proportion winged macropterous was independent of heritability. It can be shown both analytically and through simulation that neither of the last two parameters has much influence on the equilibrium proportion of migrants. Very large changes in λ increase the equilibrium proportion of migrants (e.g., a fivefold change in λ increases the proportion of migrants by about 30%). The lack of an effect of *K* is contrary to results from other models (Roff 1975; Jarvinen 1976; McPeek and Holt 1992; see above). The reason for this is that *K* in these previous models was a randomly distributed variable, whereas in the present case *K* takes the value of 0 with probability $1/T$ or *K* with probability $1 - 1/T$. The analysis of Levin et al. (1984, above) showed that under this circumstance, and if each patch is saturated each generation, then the ESS migration rate is independent of *K*, and from Equation (5.81) is equal to

$$m_{ESS} = 1 / \left\{ T \left[1 - s \left(1 - \frac{1}{T} \right) \right] \right\} = 1 / \left\{ T(1-s) + s \right\} \tag{5.85}$$

If in the present model *K* is made a random variable, the same results are obtained as for the other models. If patch persistence is finite and *K* is also a random variable, the former effect generally dominates (Roff 1994b).

To examine the time course of invasion by a dominant flightless allele, I ran the single-locus model with only the winged allele, thereby producing a monomorphically winged population. After the population had attained equilibrium, I introduced a single flightless allele into one patch. The simulation was run with a variety of parameter values, all of which gave the same qualitative result (Figure 5.17). The flightless allele spreads through the population very rapidly and reaches an equilibrium frequency in about 50 generations. Population size settles into a new equilibrium that is below the equilibrium value of the monomorphic population (Figure 5.17A). This depression of population size following the invasion of the flightless allele was observed in all simulations and is consistent with the earlier simulations described above (Roff 1975). However, though population size is

Box 5.6
Description of the quantitative genetic model

According to the threshold model the dimorphic trait is actually the phenotypic expression of a threshold effect acting on a normally distributed trait termed the liability. Individuals that lie above the threshold develop into one morph, while individuals below the threshold develop into the alternate morph (e.g., Figure 2.27). Without loss of generality, I shall assume that individuals that lie above the threshold develop into the flightless morph and that the threshold is set at 0 and the variance of the liability is 1. The proportion above the threshold, P, is then

$$P = \int_0^\infty \phi(x - \mu)dx \qquad (5.86)$$

where $\phi(x) = \exp(-\frac{1}{2}x^2)/\sqrt{2\pi}$, and μ is the mean value of the liability. The mean phenotypic value of the flightless form, μ_f, is $\mu_f = \mu + \phi(\mu)/P$ and that of the winged morph, μ_w, is $\mu_w = \mu - \phi(\mu)/(1 - P)$. The mean offspring value, assuming random mating prior to migration, can be calculated from the standard offspring on mid-parent value, $Y = (1 - h^2)\mu + h^2X$, where Y is the mean offspring value, h^2 is the heritability of the liability, μ is the population mean, and X is the mid-parent value. Using this equation the mean phenotypic values of offspring from flightless, Y_f, and winged, Y_f, females are

$$Y_f = P\left[\mu(1 - h^2) + h^2\mu\right] + (1 - P)\left[\mu(1 - h^2) + h^2\frac{\mu_f + \mu_w}{2}\right]$$

$$= \mu + \frac{\phi(\mu)h^2}{2P}$$

$$Y_w = \mu - \frac{\phi(\mu)h^2}{2(1 - P)} \qquad (5.87)$$

The mean phenotypic value of offspring in the ith habitat patch at time $t + 1$ is

$$Y_{i,t+1} = \frac{\lambda N\left[Y_f P + Y_w(1 - P)(1 - m)c\right] + Y^*/T}{\lambda N\left[P + (1 - P)(1 - m)c\right] + N_M/T} \qquad (5.88)$$

For convenience of display, the subscripts have been omitted from the right-hand side of Equation (5.88). The first term in the numerator is the weighted contribution due to the flightless females and the winged females that do not migrate. The second term is the contribution from migrant females, assuming that an equal number colonize each patch:

$$Y^* = \lambda csm \sum_{j=1}^{T} N_j(1 - P_j)Y_{w,j} \qquad (5.89)$$

The denominator is the number in the patch resulting from nonmigrants plus the number of immigrants, where $N_M = \lambda csm \sum_{j=1}^{T} N_j(1 - P_j)$.

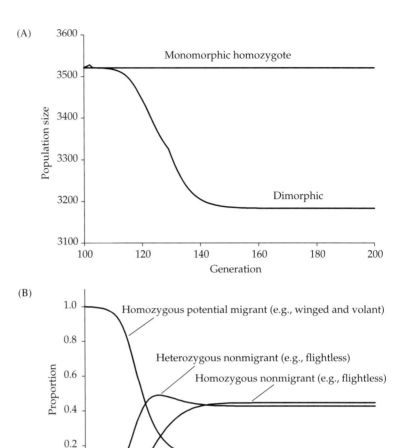

Figure 5.17   (A) Population trajectories of a population comprising only homozygous volant (= able to migrate) individuals (top line) and one in which a single dominant nonmigrant allele is introduced into a single patch in generation 100 (lower line). (B) Time course of the genotypic frequencies. $T = 20$, $c = 0.6$, $s = 0.4$, $m = 0.4$.

depressed, the range of environments in which the population can persist is increased. For example, for the parameter values shown in Figure 5.18, a monomorphically winged population cannot persist when the probability of migration exceeds 0.7, whereas a dimorphic population, though smaller in size, can persist for $m > 0.6$ (Figure 5.18B; note that actual migration rate = mP_3). Similarly, if the survival rate of migrants, s, is below 0.3 a monomorphic population becomes extinct, but a dimorphic population persists (Figure 5.18A).

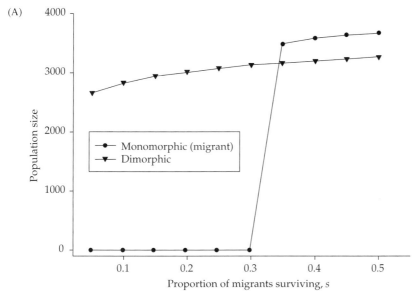

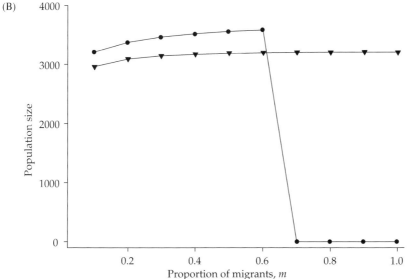

Figure 5.18 Population size versus s (A) and m (B) for a monomorphic (winged) population ($\bullet$) and a dimorphic population ($\blacktriangle$) at equilibrium. The monomorphic population cannot persist when $s < 0.3$ or $m > 0.6$. (A, parameters kept constant are $c = 0.6$, $m = 0.6$; B, parameters kept constant are $c = 0.6$, $s = 0.4$.)

The collapse of the winged population with particular combinations of c, s, or m can be explained as follows: Total population size at time $t + 1$, N_{t+1}, is given approximately by

$$N_{t+1} = \lambda N_t \left[(1-m)c + mcs\right] = \lambda N_t c(1-m+ms) \tag{5.90}$$

This is only an approximation because it does not take into account the loss of one subpopulation at each generation. A necessary condition for the population to be stable is $\lambda c(1 - m +ms) = 1$. For the case of the populations shown in Figure 5.18, persistence requires $m < 0.74$ (B), and $s > 0.26$ (A), as roughly observed. The instability arises because the number leaving and finding another patch exceeds replacement. Population persistence is enhanced by the invasion of the flightless allele because it reduces the proportion migrating. Using the same approach as above, population size is given very approximately by

$$N_{t+1} = \lambda N_t \left\{1 - P_w\left[1 - c(1 - m + ms)\right]\right\} \tag{5.91}$$

where P_w is the mean proportion of winged genotypes in the population. The condition for stability is now

$$\lambda\left\{1 - P_w\left[1 - c(1 - m + ms)\right]\right\} \geq 1 \tag{5.92}$$

In the case of $\lambda = 3$, $c = 0.6$, $s = 0.4$, $m = 0.7$, where the monomorphic winged population is unstable (Figure 5.18B), the dimorphic population is stable, with the overall frequency of the winged genotype being 0.086. This frequency satisfies the foregoing inequality, the left-hand side being 2.83. The average proportion actually migrating is considerably reduced from 0.7 to 0.06 ($P_w m$), leading to a population size ($N = 3,204$) very similar to that produced by a winged population with $m = 0.06$ ($N = 3,059$). The general depression of population size, observed in all simulations, is a direct consequence of the reduction in average migration rate.

In summary, the invasion of the population by a mutant flightless (nonmigrant) allele causes population size to decline but permits the organism to spread into habitats previously unavailable. For a discussion of the general stabilizing effect of migration on population dynamics see Doebelli (1995), Doebelli and Ruxton (1997), and Hanski (1999).

For a given combination of c, s, and m, the equilibrium proportion of winged individuals predicted by the single-locus model are very similar for that predicted by the polygenic model, the regression between the two predictions being $y = 0.016 + 0.815x$ ($r = 0.95$, $n = 120$, $P < 0.0001$), where y is the frequency predicted by the single locus model and x is that predicted by the polygenic model. Thus the evolutionary equilibrium does not appear to depend critically upon the mode of inheritance. This finding indicates that the results are robust to the assumption about the genetic mechanism.

Equation (5.91) gives the requirement for existence; rearrangement of this equation, assuming an equilibrium population gives $P_W = 1/[1 - c(1 - m + ms)]$. This equation does not take into account the effect of persistence time and should only be taken as implying that the equilibrium proportion of winged individuals will be approximately inversely proportional to $1 - c(1 - m + ms)$. Using the rationale developed for the model of Levin et al. (1984) described above suggests that the proportion winged should be inversely proportional to the persistence time of the patch, $P_W = 1/T$. Putting these two approaches together produces the hypothesis that $P_W \propto 1/\{T[1 - c(1 - m + ms)]\}$, or $\log(P_W) \propto -\log(X)$, where X is the denominator. This hypothesis is supported by the data, provided the predicted proportion winged is less than about 0.95. The regression equation is $\log(P_W = 0.154 - 0.998\log(X)$, $r = 0.996$, $n = 1,291$, $P < 0.0001$ (Roff 1994b).

The Evolution of Migration as a Response to Habitat Variability: Empirical Tests

The theoretical analyses presented above support the logic of the hypothesis that migration evolves in response to the persistence time of the habitat. (A second hypothesis, that it is a mechanism in some cases to avoid inbreeding, was discussed above.) Southwood (1962) provided qualitative support for, but no statistical analysis of, the hypothesis by an examination of the incidence of migratory movements in arthropods in relation to habitat persistence, which he classified as either "temporary" or "persistent." Analysis of wing dimorphism and habitat characteristics in gerrids (Brinkhurst 1959, 1963; Vepsalainen 1978; Dingle 1979, 1985), planthoppers (Denno 1978, 1979; Denno and Grissell, 1979; Denno et al., 1980, 1991, 1996; McCoy and Rey 1981) and carabid beetles (den Boer 1970, 1971; den Boer al. 1980) have all given further support for the hypothesis. Roff (1990c, 1994c) strengthened the general analysis by greatly increasing the database and presenting a quantitative analysis that showed that the proportion of winged, flightless, or dimorphic populations or species varied according to the persistence properties of the habitat. An example from this analysis is shown in Figure 5.19, in which is plotted the distribution of winged, dimorphic, and flightless species of North American Orthoptera among different habitat types. As predicted by the persistence-time hypothesis, winged species predominate in habitats that are at the "temporary" end of the persistence scale (e.g., open and cultivated habitats). The apparent anomaly of a high frequency of winged species in arboreal habitats is explicable both because flight is selected for organisms moving in three-dimensional space (see Roff 1990c for tests) and because in many instances the arboreal species inhabit vines and shrubs in early successional habitats, whereas flightless species predominate in the "permanent" woodland habitats.

Probably the most convincing empirical data supporting the persistence time hypothesis is that of Denno et al. (1991) on variation among planthoppers. Planthoppers are typically either long-winged (macropterous) and capable of flight or short-winged (brachypterous) and flightless. Persistence time was estimated in terms of planthopper generations and matched against the proportion of winged individuals in the particular population or species. In total there were 35

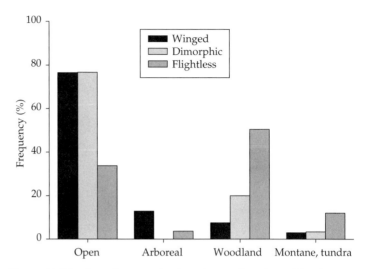

Figure 5.19 Distribution of volant (winged and capable of flight), dimorphic, and flightless morphs of North American orthoptera. To minimize phylogenetic effects, the data have been scored using genus rather than species. The original analysis in Roff (1994c) was done at several taxonomic levels and used seven rather than the four categories shown. The conclusion from this analysis was that winged and dimorphic taxa occurred more frequently than flightless taxa in relatively temporary sites of open areas, including cultivated areas. (The present grouping lumps three previous habitats—open, dry; open, wet; cultivated.) than in the more permanent and contiguous woodland habitats. (The present grouping combines surface and subterranean habitats.)

species and 41 populations. In both males and females there is a highly significant negative relationship between habitat persistence and proportion winged (Figure 5.20). The theoretical model presented in the previous section will, given appropriate parameter values, predict the equilibrium proportion winged. Can it predict the planthopper/habitat persistence relationship?

The persistence times of habitats reported by Denno et al. (1991; Figure 5.20) fall into two groups: 1–110 years, and 2,000–12,000 years (corresponding to T values, as measured by habitat duration × generations per year, of 2–330, and 2,000–30,000). Habitats that are reported to persist for at least 2,000 years are salt marshes, freshwater marshes, bogs, and mangrove swamps. It seems highly likely that during this period these habitats have suffered several catastrophic episodes, such as incursions of seawater, that would have eliminated planthopper populations from them. While habitats recorded with persistence times in excess of 2,000 years can be considered to be very long lasting, it is unlikely that an adequate quantitative measure can be placed upon the exact duration. This

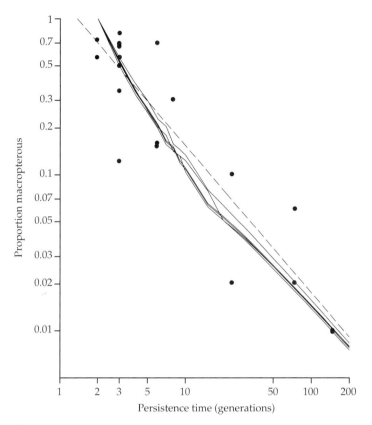

Figure 5.20 Relationship between predicted and observed proportion macroptery (= long winged and capable of flight) in various species of planthoppers in relation to habitat persistence. Dots show observed values; dashed line shows fitted regression line, $\log Y = 2.129 - 0.942 \log X$, $r = 0.92$, $P < 0.0001$; solid lines indicate predictions using five combinations of parameter values determined by fitting the model to the observed mean proportion at a persistence time of three generations. (See Table 5.4 for parameter values and Roff 1994b for further details.)

does not detract from the analysis presented by Denno et al. (1991), but does limit the utility of the data for the present analysis. For this reason I restricted my analysis to the range $T = 2$–200. The largest number of data points are at $T = 3$. For a persistence time of three generations, the proportion of macropterous females in planthoppers ranges from 0.12 to 0.80, with a mean and median of 0.53 and 0.50, respectively ($n = 10$). To test the ability of the single-locus and polygenic models to predict the relationship between T and proportion macropterous, I chose five disparate combinations of c, s, and m that give an

Table 5.4   Combinations of *c, s,* and *m* that give an equilibrium proportion
macropterous (*P*) of about 0.5 when *T* = 3.

c	*m*	*s*	*P* Single-locus model	*P* Quantitative genetic model
0.056	0.856	0.195	0.47	0.50
0.219	0.847	0.052	0.51	0.52
0.720	0.965	0.062	0.58	0.56
0.092	0.068	0.519	0.55	0.54
0.107	0.557	0.161	0.53	0.53

equilibrium proportion of about 0.5 when *T* = 3 (Table 5.4). For each combination
I obtained the relationship between proportion macropterous and *T* by running
the simulation for 24 values of *T* ranging from 2 to 200 generations.

The predictions of the polygenic model are less sensitive to the actual combinations of *c, s,* and *m* than the single-locus model. However, for both the single-
locus and polygenic models there is excellent agreement between the predictions
of the models and the observed data. (Figure 5.20 shows the results for the quantitative genetic model.)

The Evolution of Migration in Island Populations

One of the predictions of the theoretical analysis is that the proportion of winged
individuals should decrease over time within a local habitat patch. Observations
on various weevil species and an hemipteran support this prediction (Stein, 1977
and Wallace, 1953 discussed in Roff 1990c). Darwin (1876) put forward the observation that insects on oceanic islands are more likely to be flightless than on the
mainland and suggested that this resulted from the same basic factors as
advanced above: "[D]uring many successive generations each individual beetle
which flew least, either from its wings having been ever so little less perfectly
developed or from indolent habit, will have had the best chance of surviving
from not being blown out to sea; and, on the other hand, those beetles which
most readily took to flight would oftenest have been blown to sea, and thus
destroyed" (Darwin 1876, p. 109). The association between flightlessness in
insects and oceanic islands gained widespread acceptance (Brues 1902; Huxley
1942; Schmalhausen 1949; Hemmingsen 1956; Carlquist 1965; Mathews and
Mathews 1970; Powell 1976; Davis 1986; Liebherr 1988), although several
authors did question both the presumed association and the underlying hypothesis (Bezzi 1916; Jackson 1928; Scott 1935; Darlington 1943; Byers 1969) but none
tested either.

What is the probability of an insect leaving an island? Let us make the conservative assumption that any individual that moves across the land/sea boundary

is lost. For simplicity I shall also assume that the island is circular with radius R. The probability of leaving the island is equal to the probability of being at some point that is a distance r from the center of the island times the probability that the distance and direction of travel are such as to take the animal to the edge of the island. The former probability is $(2\pi r)/(\pi R^2)$ and the latter is

$$\int_{R-r}^{R+r} \frac{e^{-dx}}{2r} dx \tag{5.93}$$

where e^{-d} is the probability of moving a unit distance. Multiplying these two together and solving the integral gives

$$\left(\frac{2\pi r}{\pi R^2}\right)\left(\frac{e^{-dR}}{2dr}\right)\left(e^{dr} - e^{-dr}\right) \tag{5.94}$$

To obtain the probability of leaving the island we integrate this expression from $r = 0$ to $r = R$ to give

$$P_{Loss} = \frac{1 + e^{-2dR} - 2e^{-dR}}{m^2 R^2} \tag{5.95}$$

Darwin developed his hypothesis originally upon reading Wollaston's monograph on the beetles of the islands of Madeira, in which Wollaston commented upon the large number of flightless species (Wollaston 1854). Darwin drew particular attention to the fact that the proportion of flightless beetles is less on the Madeiran island of Desertas than on the main island of Madeira. Desertas has an area of approximately 8 square kilometers and 54 species of which 52% are flightless. In contrast, Madeira has an area of approximately 829 square kilometers and 242 species of which 40% are flightless. Now suppose that the probability of moving a unit distance is the same on both islands. If the proportion leaving Madeira per generation/time period were 5%, then the proportion leaving Desertas would be 65%. On the other hand, if the proportion leaving Desertas per generation/time period were 5%, then the proportion leaving Madeira would be 0.05%. The difference in selective pressure in the two scenarios is surely so great that the difference in proportion flightless should be much greater than observed. The problem with Darwin's hypothesis is that it does not take adequately into account the question of scale: Madeira is so large, that for all intents and purposes it likely represents a "mainland" area to most beetles.

The weak theoretical basis for Darwin's argument raises the question of whether his "observation" that flightlessness is typical of islands is indeed correct. Analysis of a wide range of taxa indicates that contrary to Darwin's comments, insects are no more likely to be flightless on islands than on the mainland

Figure 5.21   Relationship between the incidence of flightlessness on oceanic islands and mainland areas. Data from Figure 4 (shown as % not transformed values) of Roff (1990c).

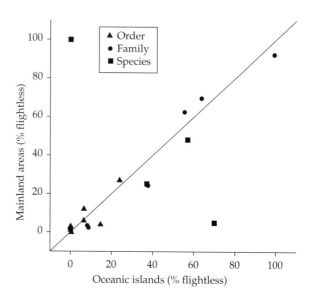

(Figure 5.21). There are islands that have high proportions of flightless species but, as exemplified by the ground beetle family, the Carabidae, this increase is associated with latitude and elevation, the proportion of flightless species increasing with latitude and elevation (Figure 5.22). Further the high proportion of flightless species on the island of Madeira can be accounted for by the high relief of the island (Table 5.5). The effect of altitude and latitude can be understood under the habitat persistence hypothesis (see Roff 1990c, p. 405–406 for a full discussion).

   If islands are small and migration relatively undirected, as in wind-dispersed seeds, then Darwin's hypothesis is plausible. Such a situation is found near-shore islands off the coast of British Columbia. Cody and Overton (1996) examined the migration capabilities of wind-pollinated plants on near-shore islands off the coast of British Columbia that ranged in area from a few square metres to 1 km³. They found evidence for the rapid evolution of reduced migration on these islands (Cody and Overton 1996).

   Wallace (1880), citing Darwin, hypothesized that island existence would lead to flightlessness in birds. In this case, the hypothesis seems more tenable as birds almost certainly move at larger scales than individual insects and hence for them an island might indeed be a "small" patch. The problem in testing the hypothesis for birds is that, whereas flightlessness has evolved in at least 10 orders, it is very patchily distributed, and the question of phylogenetic independence rears its ugly head (Roff 1994c). For example, all 18 species of penguins have probably evolved from a single common flightless ancestor. The one group that has a reasonably large number of flightless species that can be considered independent evolutionary transitions is the rails. Flightless rails are found only on islands, which are typically geographically widely separate. For example, "in the case of

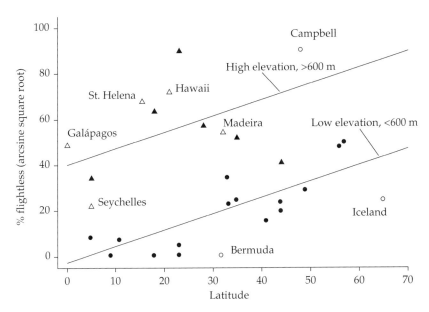

Figure 5.22   Percentage of flightless carabid species as a function of latitude, altitude (circle < 600 m, triangle > 600 m), and location type (open symbols = oceanic islands; closed symbols = continents). There was no significant effect attributable to location type but significant effects of latitude and altitude (lines show the two regression lines); for details, see Roff (1990c).

Atlantisia . . . it is quite within reason to hypothesize that a single species could have given rise to three entirely new species, all flightless and all on islands separated from each other and the nearest mainland by vast expanses of ocean" (Olson 1973, p. 36). Of the 122 species listed in Olson (1973) and Ripley (1977), 87 species are capable of flight, only 18 of which occur on islands. There are no

Table 5.5   **The altitudinal distribution of winged and wingless coleopteran species of Madeira.**

Altitude (m)	Number of Species		% Wingless
	Winged	Wingless	
0–300	73	22	23.2
0–900	20	4	16.7
300–900	14	13	48.2
300–1800	22	24	52.2
900–1800	17	33	66.0

Data from Wollaston (1854).

flightless rails on continental landmasses and 17 flightless species, all on separate islands. The statistical association between flightlessness and insularity in rails is overwhelmingly significant ($\chi^2 = 49.1$, $df = 1$, $P < 0.0001$). For a discussion of other avian groups see Roff 1994c).

The Coevolution of Traits with the Evolution of Migration

The model presented above for the evolution of migration is based on the assumption that there is a suite of trait values associated with migrants and a suite of traits associated with nonmigrants. For very general analyses, such as those presented in the previous sections, the dichotomy assumption is reasonable for describing large-scale patterns. (Note that the axes in Figure 5.20 are log scales.) This assumption is likely to be unrealistic when considering smaller scale variation, as occurs between populations of the same species subject to differing degrees of temporal and spatial heterogeneity. As first pointed out by Fairbairn (1986, 1994; Fairbairn and Desranleau 1987; Fairbairn and Butler 1990) this assumption is incorrect, because suites of traits will coevolve as a result of both correlational selection and genetic correlations between traits. This observation applies not only to migratory dimorphism but dimorphic variation in general, where the typical approach has also been to assume only two suites of traits (e.g., Lively 1986b; Moran 1992; Pfennig 1992; Rowell and Cade 1993).

The genetic correlation between a dimorphic trait and another trait is equivalent to the genetic correlation between the liability for the dimorphic trait and the other trait (Figure 5.23). Selection and breeding experiments in the sand cricket have shown (Figure 5.23) that wing morph is (1) genetically correlated with juvenile hormone levels (as assessed by the titre of JH esterase, which breaks down JH: Fairbairn and Yadlowski 1997; Roff et al. 1997), (2) the propensity to initiate flight is positively correlated with wing morph (Fairbairn and Roff 1990), (3) the rate of wing muscle histolysis is negatively related to wing morph (Roff 1994d), (4) fecundity is positively correlated with muscle histolysis (Roff 1994b), negatively correlated with wing morph (Roff 1994d; Roff et al. 1997, 1999; Stirling et al. 2001) and negatively correlated with the titre of JH esterase in the last nymphal instar (Roff et al. 1997). Similar findings have been made with respect to calling in male sand crickets (Crnokrak and Roff 1998, 2000). The importance of these results is that they demonstrate that the traits associated with any particular macropterous (= long-winged) individual vary not only as a function of the environmental component but also according to its genetic constitution. A macropterous females from a population of primarily micropterous (= short-winged) females will show an age schedule of fecundity that is more like a micropterous female than a macropterous female from a population of primarily macropterous females. So selection that changes the incidence of the two-wing morphs in the population will also change the life history trait values within each morph. Such correlated changes in fecundity and flight propensity have been observed in selection experiments (Fairbairn and Roff 1990; Roff 1994d; Roff et al. 1999) and in a comparison of different geographic populations differing in proportion macropterous (Roff et al. 2001). Without a quantitative genetic

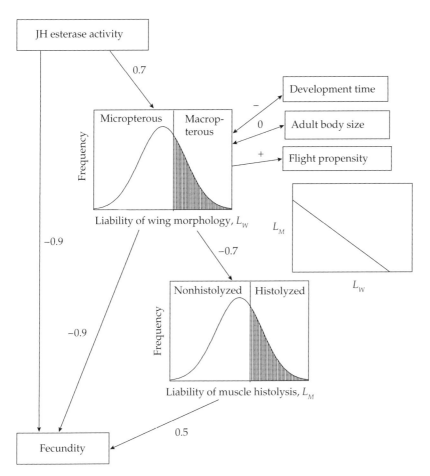

Figure 5.23 Genetic correlations between migratory and related traits in the sand crick-et, *Gryllus firmus*. Arrows indicate the hypothesized direction of influence (e.g., wing morph is determined in part by the level of JH esterase). Double arrows indicate that the direction of influence is uncertain. The inset three graphs show schematically that an increased wing morph liability (= increased proportion long-winged females) is negatively genetically correlated with a decreased rate of wing muscle histolysis (i.e., macropterous females retain their flight muscles longer). Figure modified from Roff and Fairbairn (2001).

understanding of the interrelationships between traits, these correlated respons-es would have been inexplicable.

An alternative to migration for some habitats is dormancy. The incorporation of this as an alternative to migration would be expected to decrease the optimal migration fraction. Although this is generally the case, there are mathematic cir-cumstances under which the optimal migration fraction increases with dorman-cy (Levin et al. 1984; Cohen and Levin 1991). Whether such conditions are bio-logically realistic or common was not addressed.

Ronce and Olivier (1997) explored the question of the optimal reproductive effort in an environment that is spatially and temporally variable. Migration was allowed but was kept at a fixed level. A reduction in the expected patch persistence time led to an increased reproductive effort. However, a reduction in patch persistence time should also select for an increased migration rate, which would have an impact on the optimal reproductive effort. The simulations of Ronce and Olivier indicated that increased migration rates favored a decreased reproductive effort. As a consequence, the combined effects of an evolving migration and reproductive effort are difficult to predict.

Summary

This chapter examines the evolution of life history traits in environments that are variable but predictable in the sense that a probability distribution can be attached. Temporal variation tends to erode additive genetic variance, whereas spatial variance coupled with migration is more likely to maintain, or at least greatly retard, the rate of loss of genetic variation.

In a temporally varying environment, selection—in the absence of density or frequency dependence—maximizes the geometric value of r or its equivalent in an age-structured population. This process is well illustrated by the evolution of dormancy, where "bet-hedging" in the form of a variable proportion of the population entering dormancy is readily obtained. Temporal variation has been hypothesized to favor the evolution of iteroparity and the reduction of reproductive effort. There is little empirical evidence for this hypothesis and, further, theoretical analyses using realistic levels of temporal variability suggest that the effect is likely to be weak. There is better evidence that clutch size evolves in response to temporal variation and in particular, that siblicide, common in some bird taxa, is an evolutionary response to laying an "insurance" egg.

In spatially variable environments, migration will evolve if there is sib competition within a patch or mating within a patch results in inbreeding depression. The importance of these mechanisms has yet to be determined.

Most environments are both spatially and temporally variable. In such environments one of the most important life history responses is migration. Considerable theoretical study has shown that migration can enormously increase fitness. Specifically the level of migration is expected to be an inverse function of habitat persistence. This prediction has been confirmed in a number of studies of variation in wing dimorphism among populations, species, and higher taxa of insects. The long-standing claim by Darwin that insects on islands are more often flightless than on mainland areas is wrong both theoretically and empirically. However, rails do show such a pattern. Finally, the evolution of migration produces a host of correlated responses in other life history traits such as fecundity. To understand how life history traits evolve in heterogeneous environments we need to understand the relationships (trade-offs) between migratory capability and critical life history characters.

CHAPTER 6

Evolution In Predictable Environments

I n the last chapter I examined how evolution is predicted to proceed in a world that is variable and unpredictable. But in many situations there is a degree of predictability in the environment, beyond simply a general probability statement. For example, seasonality is a common feature in both the tropics and temperate regions. Predictability might be absolute in the sense that in a seasonal environment the year can be divided into a fixed period suitable for growth, reproduction, and development and a fixed period during which the organism cannot reproduce and must adopt an appropriate response, such as hibernation or migration, to survive this inclement period. More typically, whereas there will always be a "favorable" period each year, its length will be variable, with cues that can be used to predict the probability of its length. Given the existence of predictability in the environment, we might reasonably expect organisms to vary their life histories according to environmental cues. For example, the presence of low food early in the breeding season might signal a general lack of food and hence signal that the clutch size should be reduced. Another example is the protective dimorphism known as **cyclomorphosis** in *Daphnia* and other zooplankters. The helmeted/protected morph has reduced susceptibility to predation but reduced fecundity (Chapter 3 and later in this chapter) and hence in the absence of predators the optimal morphology would be the unprotected morph. The presence of predators can be detected by chemical cues and, as expected, such cues induce the production of "helmets" and other protective morphologies. See Roff (1996a) and Tollrian and Harvell (1999) for reviews, and later in this chapter.

Selection favors interactions with the environment that increase fitness. Suppose there are two types of habitats, designated E_1 and E_2. The most fit phenotype/genotype is that which is able to perceive the environmental value and react in such a manner that its trait values in habitats 1 and 2 are the optimal values (X_1^* and X_2^*, respectively). That is, selection will favor the evolution of some response $f(E)$ such that $X_i = f(E_i) = X_i^*$. Thus, for example, in the two-patch spatial model discussed in the previous chapter (Figure 5.16), the optimal ages at maturity if the organism were able to perceive what type of patch it was in would be 1 and 2.23, not the longer duration that is most fit in the absence of any cues (Figure 5.16). The production of different phenotypes according to the environmental conditions is known as **phenotypic plasticity** and the function, $f(.)$, is termed the **norm of reaction**. An implicit assumption in the preceding description is that there is a single genotype. (In other words, the different phenotypes are not a consequence of selective mortality of particular genotypes.) A general definition of phenotypic plasticity is "a change in the average phenotype expressed by a genotype in different macro-environments" (Via 1987, p. 47). In a similar manner the norm of reaction can be defined as the following: "A reaction norm as coded for by a genotype is the systematic change in mean expression of a phenotypic character that occurs in response to a systematic change in an environmental variable" (de Jong 1990a, p. 448). This definition does not exclude discrete environments since they can be subsumed under the definition by the statistical approach of dummy variables. In some instances the same phenotype can be produced by several different genotypes or in the face of environmental variation. This phenomenon is termed **canalization** (Waddington 1942).

Phenotypic plasticity has been divided into two categories:

1. Graded responses to the environment, termed **dependent development** by Schmalhausen (1949), and **phenotypic modulation** by Smith-Gill (1983). Examples include changes in photosynthetic rate with temperature and light level, changes in flowering time and flowering height, and variation in life history traits with density or morphology with temperature.

2. Discrete variation produced in different environments, termed **autoregulatory morphogenesis** by Schmalhausen (1949) and **developmental conversion** by Smith-Gill (1983). Examples include cyclomorphosis, paedomorphosis, wing dimorphism, and diapause.

This twofold classification of phenotypic plasticity is not particularly meaningful as the second type can be subsumed under the first using a threshold model (Chapter 2). The underlying continuously distributed trait varies in a graded manner with the environment, but the phenotypic shift between morphs occurs only when the trait value exceed the threshold (Figure 2.27). At the population level there is a graded response in the proportion of the morph with the environment (Hazel et al. 1990). I shall refer to both types of variation simply as reaction norms.

Not all reaction norms will be adaptive (Scharloo 1984; Schlichting 1986; Sultan 1987; Stearns 1989), but their prevalence and the obvious advantages of such responses argues very strongly that selection has molded many, if not most, of them. The interest in reaction norms goes back to the beginning of this century with the work of Woltereck (1909) on cyclomorphosis in *Daphnia*. It was emphasized as an important factor in evolution by Schmalhausen (1949), but it has been only comparatively recently that a concerted effort has been made to understand how reaction norms evolve. An excellent review of the history of the concept is given by Schlichting and Pigliucci (1998).

Comparing the Two Perspectives of Phenotypic Plasticity

Theoretical and empirical developments have proceeded typically by adopting either a character-state approach to phenotypic plasticity or a norm-of-reaction approach, the latter being the one most frequently used in life history analysis. An understanding of the two perspectives is particularly important with respect to the analysis of the genetic basis of phenotypic plasticity.

The Character-State-Approach

Falconer (1952) noted that conceptually the same trait measured in two environments can be considered as two traits that are genetically correlated (Figure 6.1). In Figure 6.1 if the lines joining the phenotypic values in the two environments (E_1 and E_2) intersect at a common point between E_1 and E_2 the genetic correlation between the two traits is −1 (Figure 6.1D). If the lines intersect outside the range E_1 to E_2, the genetic correlation is +1 (Figure 6.1A; note that parallel lines also produce a genetic correlation of +1, since mathematically their point of intersection is at the point of infinity point beyond E_1 or E_2). The genetic correlation will differ from ±1 if there is no common point of intersection (Figure 6.1B,C). The above description is based on clones; for a sexually reproducing organism, genetic variation can be roughly visualized by using family means.

The phenotypic plasticity of a genotype can be defined as the difference in the phenotypic value between the two environments. Note that genetic variation in phenotypic plasticity (i.e., differences among genotypes in plasticity) does not necessarily mean that selection is not constrained. Genetic variation for plasticity exists in all cases shown in Fig 6.1 but is unconstrained in only two of the cases.

The two-state approach can be extended readily to a finite number of cases, but in many cases, such as variation in, say, temperature or photoperiod, the number of character-states is infinite. To solve this problem Kirkpatrick and Heckman (1989) introduced their **infinite-dimensional model**. This approach is appropriate for any continuous reaction norm, such as growth trajectories (Kirkpatrick et al. 1990), age-related traits (Kirkpatrick et al. 1994), or variation across environments (Gomulkiewicz and Kirkpatrick 1992). An alternative method is to view phenotypic plasticity as a norm of reaction.

Norm-of-Reaction Approach

Whereas the character-state approach views the phenotype as two points in state space, the norm of reaction approach sees a line (Figure 6.2; de Jong 1990b; de Jong and Stearns 1991). Thus, for two environments the phenotype can be equated to the environment by the function $X(E) = a + bE + \varepsilon$, where the two parameters a and b are considered to be traits, and ε is an error term, which is normally distributed with mean 0 and standard deviation σ_ε. Evolutionary change in trait

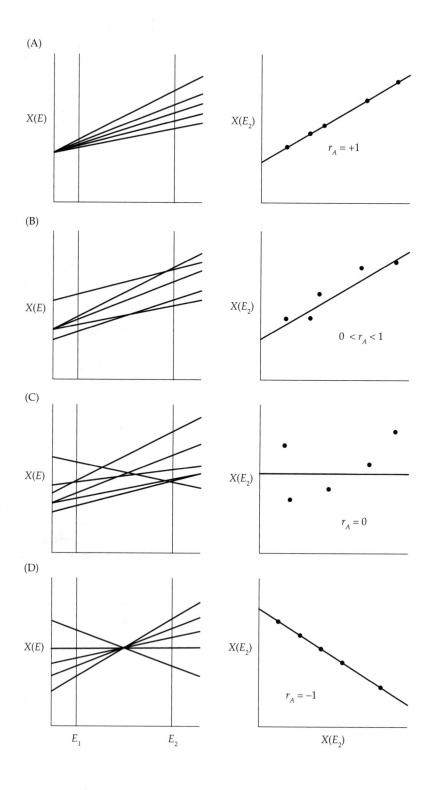

◀ **Figure 6.1** Schematic illustration of the character state approach to phenotypic plasticity. The panels on the left show the character states in the two environments E_1 and E_2,
each line joining the trait values of a single genotype (the reaction norms of the genotype). The panels on the right show the regression of the trait value from the second
environment, $X(E_2)$, on the trait value from the first environment, $X(E_1)$. (A) The reaction
norms meet at a single point beyond the range of the two environments. The correlation
between trait values is +1. Note that parallel reaction norms also give a genetic correlation of +1 (mathematically this is because the lines meet at infinity). (B,C): The reaction
norms cross at several points within the range of the two environments. Depending on
the distribution of intersections the genetic correlation will be positive but less than +1
(B), zero (C), or negative but greater than –1 (not shown). (D) The norms of reaction
intersect at a single point between the two environments. In this case the genetic correlation is –1. Because in cases A and D all the points lie on a single line ($r = \pm1$) only combinations that lie on this line can be achieved. In all other cases, because in principle the
distribution about the regression line is normal (i.e., no value is excluded), selection can
move the population to the joint optimum. Figure modified from Via (1987).

X is a function of the heritabilities of a and b and the genetic correlation between
them. If the genetic correlation is ±1, then the line is genetically fixed and corresponds to the cases A and D in Figure 6.1. The reaction norm is an alternate formulation of the character-state approach (Van Tienderen and Koelwijn 1994; De
Jong 1995). Woltereck (1909), who introduced the concept of reaction norms,
emphasized that it is actually the reaction norm that is inherited. The conceptual
advantage of this approach is that it extends quite naturally to continuously distributed environmental variables such as temperature. Though, in general, a lin-

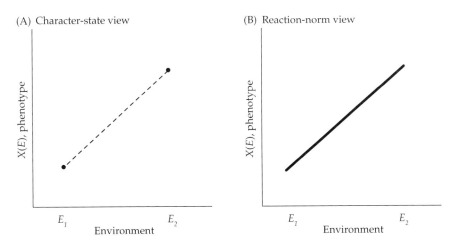

Figure 6.2 A pictorial representation of the two viewpoints of phenotypic plasticity.
(A) The character-state approach focuses attention on the trait values in the two environments, whereas the reaction-norm approach focuses on the line joining the trait
values(B). Redrawn from de Jong (1990a).

ear reaction norm is the easiest to work with, there is no reason why a more complex function should not be used. Below I discuss one based on the logistic function. For other complex functions, see Windig et al. (1994), Delpuech et al. (1995), and David et al. (1997). The more terms in the reaction norm, the more heritabilities and genetic correlations are required to predict evolutionary trajectories. For example, the quadratic function $X(E) = a + bE + cE^2 + \varepsilon$ requires three heritabilities and three genetic correlations.

For the linear reaction norm, an index of phenotypic plasticity is the slope b, but there is no obvious index for nonlinear reaction norms. de Jong (1995) suggested the first derivative of the function, but this describes only part of the reaction norm shape and hence may be misleading. For any given environment, the use of the first derivative may be useful, but it must be remembered that plasticity will vary with the environment. Since the potential for considerable change in shape is possible with nonlinear reaction norms, I suggest dispensing with any formal quantitative definition of phenotypic plasticity, except for linear reaction norms.

Scheiner and Goodnight (1984) proposed the alternative statistical definition of plasticity, $\sigma_{PL}^2 = \sigma_E^2 + \sigma_{G \times E}^2$, where σ_{PL}^2 is phenotypic plasticity, σ_E^2 is the contribution from the environment, and $\sigma_{G \times E}^2$ is the genotype by environment variance. The remaining genotypic variance σ_G^2 was assumed by Scheiner and Goodnight (1984) to be independent of the environment. For a linear reaction norm this is equivalent to the separation of a and b and is a valid partition, but if the reaction norm is nonlinear, there are additional coefficients and σ_G^2 is no longer independent of the environment (de Jong 1990b; Muir et al. 1992; Scheiner 1993).

Genetic Analysis of Phenotypic Plasticity

I present an overview of methods in Roff (1997); here I give what I think are the best methods (if the design permits), illustrating them with an analysis of diapause induction in the cricket *Allonemobius socius* (Roff and Bradford 2000; Roff 2001).

Allonemobius socius is a small common ground cricket found in wet grasslands in the southeastern United States from Florida to New Jersey (Howard and Furth 1986). In the northern part of its range *A. socius* is univoltine becoming bivoltine in Virginia and possibly multivoltine in Florida (Howard and Furth 1986). The transition from a univoltine to bivoltine phenology occurs between latitudes 34°–37°N, in which region voltinism, as shown by common garden experiments, is due primarily to phenotypic plasticity and not a simple genetic polymorphism (Bradford and Roff 1995). In the transition area, overwintering eggs hatch in May and first-generation adults appear in July and early August. Females of the first generation produce mixtures of nondiapausing and diapausing eggs (Mousseau 1991). Eggs laid in August or later typically diapause. Thus second-generation females produce only diapausing eggs. Individuals used in the experiment described here were from a transitional site (Danville, Virginia, 36°40′ N).

A full-sib design was employed to estimate genetic parameters. Although estimates from such a design are potentially biased by nonadditive genetic effects, space limitations and difficulties with mating prevented us from using the preferred half-sib design. Full details of the experimental protocol are presented in Roff and Bradford (2000). I present only a very brief overview here. Mousseau and Roff (1989) estimated the heritability of diapause propensity, assuming that it was a trait of the offspring. Further research suggested that it might more properly be considered a maternal trait (Tanaka 1986; Mousseau 1991; Bradford and Roff 1995), and thus in the present experiment we used the proportion of eggs diapausing as a trait of the female, not the egg itself. We used two environments corresponding to early and late summer but I present here only the analysis of the early environment. For a given environment, 6–8 females were chosen haphazardly from each family and mated and allowed to reproduce for the estimation of diapause proportion as a function of age since eclosion into the adult state. Four batches of eggs were collected at four-day intervals from each female, beginning on the ninth day after the final moult (i.e., covering days 9–12, 13–16, 17–20, 21–24). Eggs were incubated in the same environment as the mothers for 14–18 days, at which point diapause, direct-developing, and infertile eggs were scored. Infertile eggs were not used in the calculation of diapause proportion, and batches of fewer than four eggs were excluded from the analysis. As the likelihood of an offspring being able to complete a generation must decline with the age of the adult female we would expect that the proportion of diapause eggs would increase with age. Except for an initial dip (which I suspect is an experimental artifact) this prediction is upheld (Figure 6.3).

Estimating the Genetic Parameters of the Reaction Norm

Because it is conceptually easier, I begin with the reaction norm method. We can define the reaction norm as some general function, $f(\theta_0, \theta_1, \ldots \theta_k, X)$, where X is the trait in question and the θs are the coefficients in the function (e.g., for a linear function $\theta_0 = a$, $\theta_1 = b$). The task is to estimate the heritabilities of the coefficients and the genetic correlations between them. In the simplest case, such as a growth trajectory, the function can be estimated for an individual and the parameters can be estimated in the same manner as any trait (Chapter 2). One potential difficulty is that the parameter estimates are not statistically independent of each other, and in this way they differ from, say, two morphological traits such as head width and tarsal length. It is possible to estimate all parameters simultaneously, though whether the computational headache is worthwhile is not clear.

In principal, the proportion of diapause eggs produced by a female *A. socius* should start at some value and asymptote at 1. The estimation of such a function is hindered by the fact that for each female there are only four data points. For the present illustrative purposes I fitted for each female a two parameter logistic function

$$P(t) = e^{a+bt} / \left(1 + e^{a+bt} \right) \tag{6.1}$$

where $P(t)$ is the proportion of diapause eggs at age t (subscripts denoting family and individual have been omitted). The coefficient "a" changes the position of the curve along the time axis (right or left shift), while the coefficient "b" changes

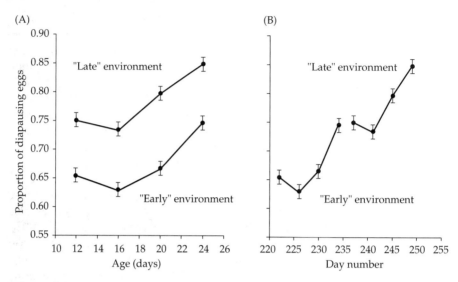

Figure 6.3 Population diapause reaction norm in *Allonemobius socius*: mean proportion of diapausing eggs (±1 SE) as a function of age and environment (A) or approximate time of year mimicked by the environment (B), which shows that there is approximately a single norm of reaction.

the rate at which the curve approaches 1. Given estimates of a and b for each female, it is a simple matter to estimate their heritabilities and the genetic correlation between them. The heritability of a was significant ($h^2 = 0.32 \pm 0.12$) but that of b was not ($h^2 = -0.02 \pm 0.19$). The two coefficients were phenotypically correlated ($r_p = 0.33 \pm 0.09$), but the negative estimate of additive genetic variance for b precluded the estimation of the genetic correlation, which almost certainly was close to zero.

If the trait cannot be measured on the same individual, then there is no simple solution, unless the organism is clonal, in which case at least broad-sense heritabilities and genetic correlations can be estimated. For an example, see Van Tienderen and Koelewijn 1994.) For the restrictive case of a linear reaction norm and two environments, Gavrilets and Scheiner (1993a) developed a procedure based on a half-sib mating scheme (Box 6.1).

Estimating the Genetic Parameters using the Character-state Perspective

The basic elements of this perspective are the heritabilities of the trait in each environment and the genetic correlation between environments. At present the best approach appears to be that of Fry (1992; see also Shaw and Fry 2001), which consists of a mixed model ANOVA to estimate the genetic covariance plus, in general, two separate ANOVAs to estimate the genetic variances. (Estimation of these parameters can also be done using restricted maximum likelihood using a VARCOMP procedure.) The method is described in detail in Chapter 3.

Box 6.1
Estimating the genetic parameters of a "two-environment" reaction norm

Each family is divided between the two environments and there are an equal number of offspring per full-sib family per environment (n). Let $\bar{x}_{ij} = \sum x_{ij}/n$ be the mean value in full-sib family i of trait x measured in environment j ($j = 1$ or 2). Without loss of generality, the environments are scaled such that $E_2 = -E_1 = \frac{1}{2}$. The intercept a for the ith family is estimated by the overall mean for family i, $a_i = (\bar{x}_{i1} + \bar{x}_{i2})/2$, and the slope b bythe difference between the two means, $b_i = \bar{x}_{i1} - \bar{x}_{i2}$. If we use the mean values, the design can now be viewed as a half-sib design in which each sire is mated to several dams and each dam produces *one* offspring with two traits a_i and b_i. The analysis of variance is then exactly the same as for the full-sib with multiple offspring per family, replacing the term "among families" with "among sires" and "among progeny within families" with "among progeny within sires." There are two such analyses, one for each "trait" (a, b). The additive genetic variance for each is estimated as $\sigma_A^2 = 4(MS_{AF} - MS_{AP})/k$, where MS_{AF} is the mean square among families, MS_{AP} is the mean square among progeny within families, and k is the number of dams per sire. Note that the multiplier is 4, not 2, since these are half-sib families. The heritability estimates are biased due to the use of the full-sib means, and hence a correction factor is required for the phenotypic variances (Gavrilets and Scheiner 1993a):

$$\sigma_P^2(a) = \sigma_P^2(a^*) + \left[(\sigma_{e1}^2 + \sigma_{e2}^2)/4\right][(n-1)/n] \tag{6.2}$$

and

$$\sigma_P^2(b) = \sigma_P^2(b^*) + \left(\left[\sigma_{e1}^2 + \sigma_{e2}^2\right]/4\right)([n-1]/n) \tag{6.3}$$

where the values superscripted with * are the unadjusted phenotypic variances and σ_{ej}^2 is the error variance from an ANOVA within environment j. If the number per family differs between the environments the term $1/n$ is replaced by $2/(n_1 + n_2)$, where n_i is the number of offspring per full-sib family per environment. The covariance between a and b is obtained in the same manner as above by using ANOVA (see Table 3.3 in Roff 1997). Standard errors can be estimated using the jackknife.

The heritabilities, phenotypic, and genetic correlations of diapause proportions in the cricket *A. socius* were estimated using the above approach (Roff and Bradford 2000). At all ages there was a significant heritability for diapause proportion and little difference among time periods (Table 6.1). Likewise, the phenotypic correlations are similar to each, as are the genetic correlations, though the latter are considerably larger than the former (Table 6.1). Notice that the char-

Table 6.1   Heritabilities (diagonal), phenotypic (above diagonal) and genetic correlations (below diagonal) between age-specific diapause proportions,
P(t), in the cricket *Allonemobius socius*.

	P(1)	P(2)	P(3)	P(4)
P(1)	0.40 (0.10)	0.67 (0.04)	0.61 (0.04)	0.53 (0.05)
P(2)	1.00 (0.06)	0.43 (0.11)	0.75 (0.03)	0.64 (0.04)
P(3)	0.99 (0.07)	1.01 (0.04)	0.49 (0.11)	0.74 (0.03)
P(4)	0.95 (0.17)	0.90 (0.06)	0.95 (0.06)	0.48 (0.13)

Standard errors are shown in parentheses. Data from Roff and Bradford (2000).

acter-state approach leaves us with many more parameter values than the reaction-norm approach. The number of parameters could be reduced for the character-state approach by using the infinite-dimensional model of Kirkpatrick and
Heckman (1989), but its implementation is not trivial. The disadvantage of the
reaction-norm approach in the present case is that there are really too few data
points per female to adequately describe the individual reaction norms. Given
this potential inadequacy, the character-state approach is to be preferred. From
this analysis we can discern that diapause proportion can evolve ($h^2 \approx 0.45$), but
the genetic correlation is close to or equal to 1, which could greatly retard evolution in some directions. In particular, a genetic correlation of 1 between time
periods suggests that the shape of the reaction norm cannot be easily changed.
This suggestion is consistent with the observations from the reaction norm
analysis that the slope parameter has no genetic variation and hence variation in
slope (= shape) is primarily phenotypic. The two approaches should be seen as
complementary (Via et al. 1995).

The Evolution of Phenotypic Plasticity

General Analysis

The evolution of phenotypic plasticity implies environmental variation in time
or space. The simplest case to consider is that in which there are two habitat
patches that require different trait means, but these do not vary over time. Thus
the environment is predictable but variable in space and time. In Chapter 2 I
presented the equation describing the change of a suite of traits, $\Delta \overline{X} = \mathbf{GP}^{-1}\mathbf{S}$.
The same equation applies to phenotypic plasticity when viewed from the char-
acter-state approach, with an additional component that accounts for migration
between patches each generation

$$\begin{pmatrix} \Delta X_1 \\ \Delta X_2 \end{pmatrix} = \begin{pmatrix} \sigma^2_{A11} & \sigma_{A12} \\ \sigma_{A21} & \sigma^2_{A22} \end{pmatrix} \begin{pmatrix} w_1 \sigma^{-2}_{P1} S_1 \\ w_2 \sigma^{-2}_{P2} S_2 \end{pmatrix} \tag{6.4}$$

where ΔX_i is the change in trait X in the ith environment, σ^2_{Aii} is the additive genetic variance in the ith environment, σ^2_{Aij} is the additive genetic covariance between the two environments, and w_i is a weighting function that depends upon whether selection is hard or soft (see Chapter 2 for definitions). With soft selection, w_i is the proportion entering the ith patch; with hard selection, the formula is a little more complex, and since the overall result is the same, it is omitted here (see Via and Lande 1985). The situation is exactly the same as the case of two traits in a single environment, namely, that if the genetic correlation is not equal to ±1, the joint equilibrium will be achieved (Figure 6.4). This result can be easily derived using the reaction norm approach and optimality analysis (Box 6.2).

Limits to Evolution and Multiple Fitness Peaks

In the simple spatial model described in Box 6.2 there is no limit to plasticity under any but the most unusual circumstances. For n traits, the optimal combination will not be achieved if the average correlation is $-1/(n-1)$ (Dickerson, 1955). Thus, as the number of patches increases, and hence also the number of genetic correlations, the rate of progress to the optima may become extremely slow, and even low correlations may impede progress.

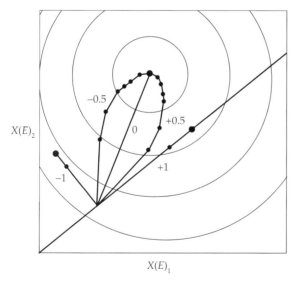

Figure 6.4 The evolution of two traits subject to the influence of the constraint due to their norm of reaction. Contours indicate combinations of mean fitness as a function of the average phenotypic value in each environment. The heavy lines show the trajectories given the genetic correlations indicated beside the lines, each dot representing the change after about 50 generations. The three large dots show the equilibria reached by the population. When $r_A = 0$ the population moves along the line of steepest ascent to the equilibrium. Deviations of r_A from zero cause the trajectory to be "bent," and at the extremes of ±1, the path is fixed and the fitness peak will not typically be achieved. Modified from Arnold (1992, after Via and Lande, 1985).

Box 6.2
In two environments a joint optimum can be achieved provided $r \neq \pm 1$

Without loss of generality, code the environments as 0 and 1, in which case the trait values specified by the reaction norm for the environments 1 and 2 are $X_1 = a$, $X_1 = a + b$, respectively. Via and Lande (1985) assumed a gaussian fitness function

$$W_i = \exp\left\{-\tfrac{1}{2}\left[(X_i - \mu_i)/\omega_i\right]^2\right\} \tag{6.5}$$

where μ_i is the optimal trait value in the ith environment and ω_i is an environment-specific constant that measures the strength of stabilizing selection. Under soft selection, the overall fitness is given by the geometric mean over the per patch fitnesses

$$W(a,b) = \sum_{i=1}^{2} w_i \ln\left[e^{-\tfrac{1}{2}\left(\frac{X_i - \mu_i}{\omega_i}\right)^2} \right] = \sum_{i=1}^{2} -\tfrac{1}{2} w_i \left(\frac{X_i - \mu_i}{\omega_i}\right)^2 \tag{6.6}$$

To find the values of a and b that maximize fitness, we differentiate the above with respect to a and b and set the results equal to zero:

$$\frac{\partial W(a,b)}{\partial a} = \sum_{i=1}^{2} -w_i \left[\frac{a - b(E_i) - \mu_i}{\omega_i}\right]$$
$$\frac{\partial W(a,b)}{\partial b} = \sum_{i=1}^{2} w_i E_i \left[\frac{a - b(E_i) - \mu_i}{\omega_i}\right] \tag{6.7}$$

Both of the above equal zero when

$$w_1(a - \mu_1) + (1 - w_1)(a - b - \mu_2) = 0 \tag{6.8}$$

and this is satisfied when $a = \mu_1$ and $b = \mu_1 - \mu_2$. Substituting these values in the reaction norm we get that the trait values in the two environments are μ_1 and μ_2, respectively. Now if the genetic correlation between a and b is ± 1, all trait values lie along a single line and the two parameters can vary only according to the relationship $a + bE_i = C$, where C is a constant. Because of the scaling of the environments to 0 and 1, the constraint is $a + b = C$. Substituting for b in Equation (6.8) and rearranging gives the optimal value of a (which is also the optimal value in environment 1) as

$$a = \left[(1 - w_1)(C + \mu_2) + w_1 \mu_1\right]/(2 - w_1) \tag{6.9}$$

The optimal trait values in the two environments are no longer the two means of the stabilizing fitness curve because the genetic constraint does not permit both to be satisfied. The optima now depend upon the proportion of each habitat and the constraint implied by C.

The character-state approach allows considerable evolutionary flexibility, but this flexibility can be overlooked by a simplistic use of the reaction norm model. If, for example, one assumes a linear reaction norm, $X(E) + bE$, then, as shown above, in a two-patch universe the constraint on reaching the optimum combination is the same as for the character-state model. (For a quantitative genetic analysis demonstrating this result, see Gavrilets and Scheiner, 1993a,b.) However, if the reaction norm is linear, then, in general, an optimal reaction norm can evolve only in a two-patch environment, because the reaction norm cannot be made to pass through all points in a multi-patch universe. For a habitat consisting of three patch types, a quadratic relationship is required. As the number of patch types increases, so also will the general complexity of the reaction norm. This result is implicit in the character-state approach and can be captured in the reaction-norm approach by assuming a Taylor series approximation (de Jong 1995).

The model described above assumes that there is no penalty to plasticity. The incorporation of a cost can produce multiple equilibria. Like Via and Lande (1985), Van Tienderen (1991) also assumed stabilizing selection acting in a two-patch universe but added a second component, $W_c(E)$, which incorporated a cost to being a generalist in the sense of being adapted to both habitats

$$W_c(E) = \exp\left(-[X_2 - X_1 - \mu_{2-1}]^2 / g(E)^2\right) \tag{6.10}$$

where X_i is the mean trait value in ith environment, μ_{2-1} is the cost-free reaction (a function of the difference between habitats in average response), and $g(E)$ is a parameter that is inversely related to the strength of selection against deviating from the cost-free reaction in habitat E. When there is no cost to adaptation, then as observed by others, there is a single peak. But, the cost function generates a ridge in the fitness surface so that now there is no single optimum. Under soft selection, the two selective forces acting together produce a single peak that is shifted away from the peak when there is no cost to adaptation. With hard selection, the combined action can generate several peaks, and, as a consequence, the combination to which selection will drive a population depends upon the initial conditions.

The Consequences of Uncoupling Development and Selection on the Reaction Norm

An important feature of the models thus far described is that migration is accomplished by the zygotes, meaning that development takes place in the same habitat in which selection occurs. But suppose that selection occurs after migration and hence not necessarily in the habitat of development. In this situation the reaction norm that will evolve will not be that in which the trait value equals the mean for the habitat of selection. A genetic analysis of this scenario is very complex (Scheiner 1998; de Jong 1999) but is very easily analysed using an optimality approach assuming that the reaction norms are not constrained (Box 6.3): As the genetic analyses assumed non-zero genetic variances and covariances, this assumption is justified. Not only will the reaction norm that evolves not be that

Box 6.3
Constraints on evolution when development and selection are uncoupled

Consider the case in which organisms distribute themselves among habitats after development but before selection. Focus upon an organism that develops in habitat h but is selected in a habitat with value y. Let the optimal trait value for a habitat of value y be specified by the linear function $f(y) = A + By$. The organism has a reaction norm function $X(h) = a + bh$. Following de Jong (1999), I shall use a quadratic fitness function

$$w(X) = 1 - \omega\left[f(y) - X(h)\right]^2 \tag{6.11}$$

where ω is a constant. As generations are discrete and nonoverlapping, fitness is maximized by the combination of a,b that maximizes

$$W(a,b) = \int_{y,h} p(y,h)\left(1 - \omega\left[f(y) - X(h)\right]^2\right)dydh$$
$$= 1 - \omega\int_{y,h} p(y,h)(A - By - a + bh)^2 dydh \tag{6.12}$$

where $p(y,h)$ is the probability of developing in habitat y and migrating to habitat h. Differentiating, we obtain

$$\frac{\partial W(a,b)}{\partial a} = \omega\left(-A + B\mu_y + a - b\mu_h\right)$$
$$\frac{\partial W(a,b)}{\partial b} = \omega\left[A\mu_h - B\left(\sigma_{yh} + \mu_y\mu_h\right) - a\mu_h + b\left(\sigma_h^2 + \mu_h^2\right)\right] \tag{6.13}$$

where μ denotes the mean value, σ^2 the variance and σ_{yh} the covariance. Setting the above equal to zero (and substituting for a in the second equation), we find the optimal combination of a,b is

$$a_{opt} = A - B\mu_y + b_{opt}\mu_h$$
$$b_{opt} = \frac{B\sigma_{yh}}{\sigma_h^2} \tag{6.14}$$

as found by de Jong (1999). Because of the unpredictability of the habitat of selection relative to the habitat of development, the optimal reaction norm, *given that it is constrained to be linear*, does not have coefficients that would be optimal if development and selection occurred in the same habitat. We can reduce the constraint on the shape of the reaction norm to a degree by adding a quadratic term, giving the fitness function

$$W(a,b,c) = 1 - \omega\int_{y,h} p(y,h)\left(A - By - a + bh + ch^2\right)^2 dydh \tag{6.15}$$

If we follow the same procedure as above, the optimal combination is given by

$$a_{opt} = A - B\mu_y + b_{opt}\mu_h + c\left(\sigma_h^2 + \mu_h\right)$$

$$b_{opt} = \frac{B\sigma_{yh} + c\mu_h\left(\sigma_h^2 + \mu_h^2\right) - c_{opt}E\left(h^3\right)}{\sigma_h^2}$$

$$c_{opt} = \frac{B\sigma_{yh} + b_{opt}\sigma_h^2}{E(h^3) - \mu_h\left(\sigma_h^2 + \mu_h^2\right)}$$

$$(6.16)$$

where $E(h^3)$ is the expected value of h^3. If both distributions are symmetric and $\mu_y = \mu_h$ $= 0$ then, $c_{opt} = 0$ otherwise, selection will favor a curved reaction norm despite the fact that the locally optimal reaction norm is linear.

which produces the optimal mean trait value in a given environment, but it will have a shape that is different from that predicted by any single environment. Thus spatial heterogeneity and the timing of selection relative to development can have profound effects on the evolution of the reaction norm.

Evolution in a Seasonal Environment

Many environments are seasonal with a period during which no reproduction or growth can occur. This may have a profound influence on the phenology of an organism, because there may be only a single life stage that can pass through the inclement season. For example, all but the eggs and seeds of many invertebrates and plants are killed by cold weather. We begin by examining the simple case of an annual organism in a seasonal environment that is deterministic.

The Optimal Timing of Maturation and Reproductive Effort in Annual Species

FITNESS DEFINITION. By definition an annual species must have nonoverlapping generations, and, hence, if density dependence can be ignored, selection will maximize the expected production of offspring (e.g., seeds, eggs) that are able to pass through the upcoming period of unfavorable conditions

$$W(X) = \int_\alpha^\omega l(t, X)m(t, X)s(X)dt \qquad (6.17)$$

where X is the trait under selection, α is the start of reproduction, ω is the end of reproduction (equal to or less than the end of the growing season), and $s(X)$ is the probability of individuals with trait X surviving the unfavorable period. In the first instance I shall examine evolution in a seasonal environment that has a deterministic length. Given the certainty of the timing of the end of the season, selection will not favor the evolution of phenotypic plasticity, except possibly when there are density-dependent processes.

THE OPTIMAL ALLOCATION TO GROWTH AND REPRODUCTION. For a given life history, when should an individual mature and how much should it allocate to reproduction versus growth? Cohen (1971) appears to have been the first to give serious attention to this question. His model is based on an annual plant, but it can equally well be applied to any univoltine organism. Body size at age $x + 1$, $B(x)$, is equal to $B(x) + A(x)A(1 - c_1)$, where $A(x)$ is the amount of energy assimilated from $x - 1$ to x and c_1 is the proportion allocated to reproduction. Assimilation is proportional to body size: $A(x) = c_2c_3B(x)$, with the proportionality being the product of c_2, the proportion of the body allocated to energy gathering (such as through leaves), and c_3, the net assimilation rate per unit of energy gathering organ. Body size can thus be written as $B(x) = B(0)[1 + (1 - c_1)c_2c_3]^x$. Offspring (female) production at age x, $m(x)$, is also proportional to body size $m(x) = B(x)c_1c_3$ with some fraction, c_4, of the body mass being recovered at the end of the season and converted into offspring. Notice that this model does not include mortality. Given a fixed season length, the optimal life history is to allocate either all energy to growth or all energy to reproduction, sometimes referred to as a "bang-bang" life history (Cohen 1971, 1976; Vincent and Pulliam 1980; Ziolko and Kozlowski 1983; Kozlowski and Weigert 1986). Introducing a constant mortality rate into the model changes the optimal age at maturity, but not the sudden switch from growth to reproduction. However, if mortality changes at maturity then for particular parameter combinations, selection favors a period of both growth and reproduction following maturation (Engen and Saether 1994). As might be expected from the analysis of age and size at maturity presented in Chapter 4, a delayed age at maturity and increased size at maturity are selected for by (1) a high growth rate, (2) a high percentage increase in reproductive rate with increases in body size (If $m(x) = aB(x)^b$, then increasing b selects for a delay in maturation), (3) a decreased mortality rate, or (4) a mortality rate that decreases with size (Cohen 1976; Ziolko and Kozlowski 1983; Kozlowski and Weigert 1986, 1987). An increased season length, T, will also favor an increased age at maturity. These effects are illustrated with the simple model described in Box 6.4.

TESTING THE PREDICTION THAT AN INCREASED SEASON LENGTH FAVORS AN INCREASED SIZE. Because of declining temperatures, the season length available for univoltine ectotherms declines with altitude and along a latitudinal gradient (Masaki 1967; Roff 1983c; Nylin and Svard 1991). This phenomenon permits an inter-populational test of the prediction that an increasing season length will be accompanied by an increased size and age at maturity, provided that we make the assumption that life history parameters stay more or less the same. In princi-

Box 6.4

A simple model for the evolution of an annual species

Assumptions are: (1) $m(x)$ is rectangular, (that is, there is a constant egg production per time interval), and $m(x)$ is proportional to a power function of the age at reproduction, $m(x, x + 1) = c\alpha^b$, where α is the age at maturity, and c, b are constants, (2) survival rate is a constant, so $l(x) = e^{-Mx}$. Given a season length, T, the expected offspring production by the end of the year is

$$R_0 = \int_{x=\alpha}^{T} c\alpha^b e^{-Mx} dx = \frac{c\alpha^b}{M}\left(e^{-M\alpha} - e^{-MT}\right)$$

(6.18)

Differentiating with respect to α and setting the result to zero gives the optimal age at maturity to be that which satisfies $e^{-M(T-\alpha)} = 1 - (\alpha M)/b$. Increasing T decreases the right-hand side (RHS) and hence the optimal α must also increase. Increasing the parameter b, which increases the marginal value of delaying maturation, decreases the left-hand side (LHS) and hence a delayed maturation is selected. Increasing the mortality rate decreases the LHS, favoring an earlier age of maturation.

ple we require the following relationships: (1) a negative relationship between development time and latitude, (2) a positive relationship between body size and development time, (3) a positive relationship between body size and fecundity, and in addition, (4) genetic variation for these components. Because season length declines with latitude, we can predict that in natural populations a cline in body size/development time (measured in degree-days) should be evident. Rearing individuals from different latitudes in a common garden experiment is required to show that such variation is in part genetically determined. Latitudinal clines in body size in field-caught univoltine insects are common (Alexander and Bigelow 1960; Masaki 1978a,b, 1996; Brennan and Fairbairn 1995; Mousseau 2000). With respect to laboratory-reared individuals, variation in body size consistent with that observed in field populations has been demonstrated in a variety of insects, including *Teleogryllus emma* (Masaki 1967, Figures 6.5 and 6.6), *Pteronemobius fascipes* (Masaki 1973), *Allonemobius socius* (Mousseau and Roff 1989), *Lasiommata petropolitana* (Nylin et al. 1996; Gotthard 1998, Figure 6.7), *Aquarius remigis* (Blanckenhorn and Fairbairn 1995) and numerous European butterflies (Nylin and Svard 1991).

EFFECTS OF COMPETITION ON THE DISTRIBUTION OF EMERGENCE AND DEVELOP-MENT. If there is competition for resources the optimal time of hatching and maturity may not be the same for all individuals in a population. Iwasa (1991) examined this question assuming that resources were not evenly distributed

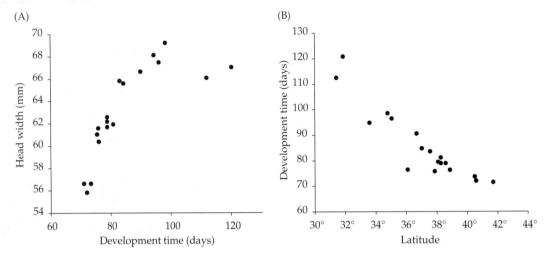

Figure 6.5 Development time in relation to body size (A) and latitude (B) in the Japanese field cricket *Teleogryllus emma*. Data from Masaki (1967).

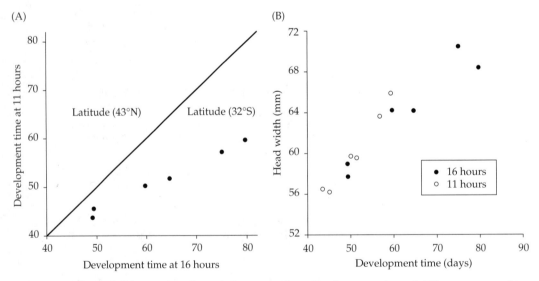

Figure 6.6 (A) The effect of photoperiod on development time of different geographic populations of the Japanese field cricket *Teleogryllus emma*. Each point represents a different site, with the northernmost having the shortest development time. (B) The changing development time results in a changed body size. Data from Masaki (1967).

(A)

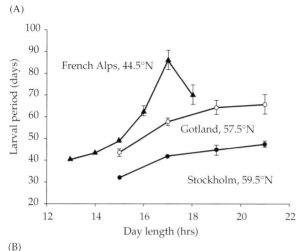

(B)

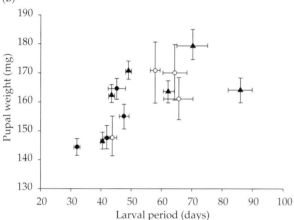

Figure 6.7   Larval period in relation to day length, latitude (A) and pupal weight (B) in the butterfly *Lassiomata petropolitana*. Data from Nylin et al. (1996) and Gotthard (1998).

through the favorable season but showed a peaked shape, an assumption that is entirely reasonable. Let this resource distribution be described by the function, $R(t)$, where t is time. Iwasa assumed a density-dependent growth rate and a density-independent mortality rate, giving the change in body size, X, as

$$\frac{dX}{dt} = \frac{R(t)X}{1 + c\sum_{i=1}^{N(t)} X_i(t)} - MX = \left[\frac{R(t)}{1 + cS(t)} - M \right]X \tag{6.19}$$

If the quantity in brackets is equal to 1 then there is no fitness gain even if a range of body sizes exist. Such a range can result from variation in hatching date and/or in maturation/pupation date. Suppose that a population consists only of one type that hatches on day t_1^* and pupates on day t_2^*. If there is no competition between individuals then $c = 0$ and the optimal combination of hatch and maturation (= pupation in Iwasa's paper) is the same for all individuals (t_1^*, t_2^*). Now add density dependence to the population. With a single type in the population, $S(t)$ drops to zero on day t_2^*, but resource availability is a continuous function, and hence a mutant that continues to grow after day t_2^* will be able to invade the population but not to fixation. The same argument can be applied to hatching date. So when resources availability changes continuously and there is density dependence there will be parameter combinations under which the optimal life history will be to produce offspring that hatch and pupate/mature asynchronously. An important point to note is that this is not a case of bet-hedging.

EFFECTS OF VARIATION IN EMERGENCE TIME ON DEVELOPMENTAL EVENTS. The above model postulates variation in hatching time as an adaptive response to density dependence. In many cases, however, variation in hatching time may simply be a consequence of stochastic factors. Given this variation, each individual potentially faces a season length that is not of a fixed length. An individual that emerges relatively late has three options: (1) increase the growth trajectory so that it achieves the appropriate life stage by the end of the season, (2) keep the same growth trajectory and enter the appropriate life stage at a smaller size, (3) increase the growth trajectory less than required to achieve the "typical" size. Option 1 must logically carry a cost, because otherwise selection would increase the growth rate of individuals emerging earlier. A likely cost to increased growth rate is an increased mortality by virtue of the individual foraging more and hence being subject to increased predation (Weissburg 1986; Godin and Smith 1988; Lima and Dill, 1990; Gotthard 1999). Increased growth rate may be energetically costly and decrease starvation resistance (Gotthard et al. 1994) or detract from the production of defenses against other sources of mortality such as viruses (Boots and Begon 1993) or pesticides (Carriere et al. 1994). Parasitoids preferentially selected the most rapidly growing individuals of the leaf-galling sawfly (*Pontania*); Clancy and Price (1987) speculated that this was due to these being superior resources. It has been hypothesized that slow growth rate leads to an increased development time and hence an increased window of opportunity for predators, thus making slow growth subject to high mortality. This was demonstrated in the cabbage butterfly (*Pieris rapae*) under attack from the parasitoid *Cotesia glomerata* (Benrey and Denno 1997). The optimal growth rate is clearly a compromise among a suite of factors; the important point here is that as the time available for growth decreases, then overall selection is likely to favor some increase in growth rate.

Rowe and Ludwig (1991) and later Abrams et al. (1996) examined the evolution of development patterns and concluded that insects emerging later should decrease their development time and emerge at a smaller size. The general nature of these analyses can be quite readily obtained with a simple semelparous life history model in which (1) fecundity is proportional to body size, X, which is itself proportional to development time, t, giving $M(x) = ct$, where c is a constant and (2) survival is a function, $M(c)$, of the growth rate c, $l(t) = e^{-M(c)t}$. The net reproductive rate is $R_0 = c\alpha e^{-M(c)\alpha}$ and the optimal development time is $\alpha = 1/M(c)$. Now if the season length, T, is less than $1/M(c)$, the optimal life history is to develop until the very end of the season, that is, the optimal development time is T. We can now ask what is the optimal growth trajectory in relation to T. We differentiate R_0 with respect to the growth parameter c,

$$\frac{\partial R_0}{\partial c} = cTe^{-M(c)T}\left[1 - T\frac{\partial M(c)}{\partial c}\right] = 0 \quad \text{when} \quad \frac{\partial M(c)}{\partial c} = \frac{1}{T} \qquad (6.20)$$

The only thing that we need to know about the $M(c)$ function is its shape (Figure 6.8). If the curve is decelerating (concave down), the derivative is decreasing with growth rate and, therefore, from the above equation the optimal growth rate decreases as the season length decreases. Likewise if the curve is accelerating (convex, concave up) the optimal growth rate increases as the season length decreases. Abrams et al. (1996) suggest that an accelerating curve is the most likely; I concur with this suggestion but do not think that the other alternatives can be discounted. Regardless of the growth rate, the final size will decrease with season length or remain constant.

ADAPTIVE REACTION NORMS. To adapt to temporal variation in season length, the organism must have the ability to perceive some cue that indicates at least the time of year. The most obvious cue is photoperiod. If the development of the organism takes place entirely before or after the summer solstice, then the organism merely has to be able to detect the length of the photoperiod, which is a trait found in a vast number of species (Harper 1977; Beck 1980; Danks 1987,

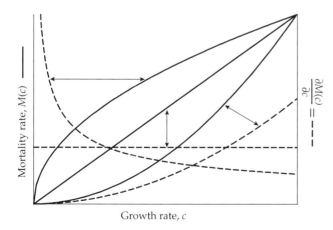

Figure 6.8 The three types of curves connecting mortality rate with growth rate. The arrows indicate the appropriate set of paired lines.

1994; Dingle 1996). If the developmental period spans the summer solstice, the same photoperiod could be indicative of either early or late summer and a second cue is needed, such as the rate of change in photoperiod (increasing = early summer, decreasing = late summer) or possibly temperature if there it has sufficient asymmetry . Developmental responses to photoperiod have been observed in crickets and butterflies and support the hypothesis of an adaptive response, though in no case have the appropriate measurements taken to show that the response is optimal.

In the northern part of Japan the band-legged ground cricket *Pteronemobius fascipes* is univoltine and develops post summer solstice (Figure 6.9). Therefore, we predict that the development time should show a positive correlation with photoperiod, which indeed it does (Figure 6.9). The reduced development time results in a reduced body size (Masaki 1973), though there are no data on the effect of photoperiod on the growth trajectory. The univoltine butterfly *Lasiommata petropolitana* also develops after the summer solstice and, as predicted, development time shows positive covariation with photoperiod (Figure 6.7). Body size increases with development time and growth trajectories appear to accelerate (i.e., weight per unit time increases) at the shorter photoperiods (Nylin et al. 1996; Gotthard 1998). These data also illustrate that the photoperiodic reaction varies with latitude as would be expected, given that the relationship between day length and remaining time will vary with latitude. Variation in photoperiodic reaction is also to be expected along an altitudinal gradient, as has been observed in the cricket *Allonemobius fasciatus* (Tanaka and Brookes 1986). Another species of *Lasiomatta*, *L. maera* is univoltine in Sweden, diapauses as a larva, with adults appearing and laying eggs from July to August (Gotthard et al. 1999). As a consequence of this phenology, the larvae upon hatching develop during a period of declining photoperiod, but after diapause in the spring they complete development when the photoperiod is increasing. Thus during spring they should respond to photoperiod such that development time decreases with increasing photoperiod. On the other hand, during the autumn the opposite pattern should hold, though since the larvae are not completing development, merely entering diapause at this time, the response may be muted, unless survival in diapause is correlated with body size, in which case there could be selection on increased development rate. As predicted, there is a strong negative relationship between photoperiod and development time in the post-diapause (= spring) larvae but a positive relationship in the pre-diapause (= autumn) larvae (Nylin et al. 1996; Gotthard et al. 1999). Development time and adult size are positively correlated, and in the pre-diapause larvae there is an increase in the growth trajectory with decreased photoperiod, suggesting that as the time remaining becomes shorter selection, favors an increased rate of growth. Similar photoperiodic responses in other univoltine insects have been observed (Masaki 1978b, 1996; Nylin 1992; Gotthard et al. 1999).

These data are consistent with the hypothesis of adaptive variation but are somewhat indirect in that they did not measure growth rate in the field. Carriere et al. (1996) attempted a direct test by removing diapausing eggs of the cricket

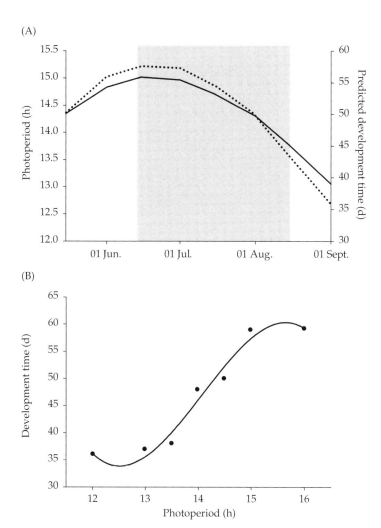

Figure 6.9   (A) The change in daylength (photoperiod) during the summer months at 40°N (data from Beck, 1980), with the phenology of the band-legged ground cricket superimposed (hatched region). The dotted line shows the predicted development time using the cubic curve fitted to the data shown in (B), which consists of the median development times at different, constant photoperiods (data from Masaki, 1973).

Gryllus pennsylvanicus from cold temperature conditions maintaining diapause at different times. In the field, eggs hatch in the middle of June. To simulate early and late hatch, eggs were removed and placed in diapause-terminating conditions such that nymphs hatched at the end of May (26th), the middle of June (16th) and the beginning of July (7th). These nymphs were raised outside, in cages, at the same location from which their parents were collected and so experienced ambient temperature and photoperiod. Additionally, a control group was raised in the laboratory under constant temperature and photoperiod. There was no relationship between survival during the first two weeks post hatch and time of hatching under the control conditions, but survival increased from the early to the late cohort under field conditions. However, the cohort that was made to emerge on July 7th did not have sufficient time to complete development, and only a single female eclosed into the adult form before a frost killed the nymphs and adults. There was no significant difference in development time between the first two cohorts, whether expressed in physiological time (day degrees, Figure 6.10; See Box 6.5 for an explanation of physiological time) or real time (cohort 1 took 103.1 ± 1.9 days whereas cohort 2 took 98.2 ± 2.9 days). However, there was a highly significant decrease in body size of the second cohort (Figure 6.10). There was a weak relationship between the time of hatch and adult size in the laboratory stock, suggesting that time in diapause might influence adult size, but the effect was insufficient to account for the relationship observed in the field. Using the observed survival rates and the body size/fecundity relationship, Carriere et al. (1996) estimated the fecundity of the three cohorts to be approximately 386, 332, and ≈ 0. Thus these data suggest that the crickets would improve their fitness by hatching earlier, which, though it increases mortality, leads to a larger body size. There are clearly ways that the original hypothesis can be salvaged (e.g., higher mortality in the free-ranging crickets). The important point is that the demonstration of the adaptive significance of the photoperiodic or other mechanisms for adjusting development time is not a simple task, but its clear predictions make it a very attractive subject for study.

REACTION NORMS MAY COMPRISE MULTIPLE CUES. Photoperiod is a reliable indicator of the time of the year and although it may indicate the time remaining for the average season, it is not sufficient to predict the duration of a particular season. How can an organism such as an insect use multiple environmental cues to adjust its development time so that it makes optimal use of the variability in season length? A simple rule is; "Add an instar if the number of degree-days exceeds Y on Julian date X." This rule is not beyond the physical capabilities of an insect since degree days can be measured indirectly by size achieved and Julian date by photoperiod. The rule then becomes: "Add an instar if size exceeds Z when the photoperiod is W." The red-legged grasshopper *Melanoplus femurrubrum* appears to follow this type of rule. A useful feature of this species is that the number of instars can be estimated from the number of antennal segments in the adult (Bellinger and Pienkowski 1987). The number of instars increases with rearing temperature, and in the field the number varies both temporally and geographically. This variation is correlated with environmental conditions, the number of instars being correlated with the number of degree-days (see Box 6.5) achieved in May. Thus, in warmer years or at warmer locales, the grasshopper increases the number of instars and presumably its final size (the latter has not been verified).

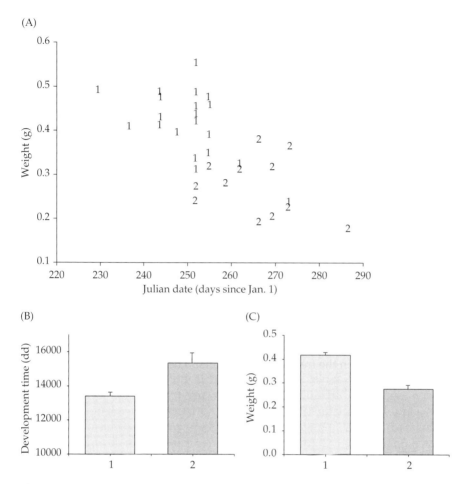

Figure 6.10   (A) Relationship between adult weight and date of eclosion for cohorts hatching earlier than usual (cohort 1, 26th May) and at the typical time (cohort 2, 16th June) in the cricket *Gryllus pennsylvanicus*. (B,C) Development time (dd = degree days) and adult weight of the two cohorts are compared. Data from Carriere et al. (1996).

It remains to be demonstrated that this norm of reaction is optimal and that there is genetic variation for the trait.

The Optimization of Developmental Events in Organisms with Generations Greater than One Year

ALLOCATING ENERGY INTO REPRODUCTION, GROWTH, AND STORAGE. Many plants and animals mature after one year and may live for a number of years. There are potentially a large number of decision points in such a life history, only a few of which have been investigated. For a perennial organism, an important

Box 6.5
The degree-day concept

The degree-day (or day-degree) concept is based on the observation that, except at extreme temperatures, development time, x, measured on an absolute scale (e.g., days) is related to the temperature, y, according to the equation

$$y = T_0 + c\frac{1}{x} \tag{6.21}$$

where T_0 is the threshold temperature for development (i.e., that temperature at which development ceases and $1/x = 0$) and c is a constant. Changing the temperature changes the development time on an absolute scale but not on the degree-day (dd) scale. So, for example, suppose that the threshold temperature for development is 10°C and development time is 100 degree days. At 20°C each day equals 10 degree days and development takes 10 days, whereas at 30°C each day equals 20 degree days and development takes 5 days.

decision is the allocation of energy into reproduction, growth, and storage for overwinter survival. Iwasa and Cohen (1989) examined the question from the perspective of a perennial plant. The analysis is detailed and sets up a framework for further modeling but does not generate any major testable generalities. Kozlowski and Uchmanski (1987) developed a model based on the assumption that within each favorable season the organism grew until a critical switch point after which it channeled all further energy into reproduction. Using a dynamic programming approach and assuming that (1) selection maximizes the expected lifetime reproductive success, (2) mortality rate is constant, (3) fecundity is size-dependent and (4) growth rate is size-dependent, they developed a formula for determining the optimal time of switching in any given year. A general prediction of this model is that the amount of time devoted to growth each season decreases with age, which is consistent with the findings for a constant environment that reproductive effort should, in general, increase with age (see Chapter 4). Unfortunately, the data available are insufficient to test this prediction, though Kozlowski and Uchmanski are able to show that their model is at least consistent with the growth patterns in a gastropod, a scallop, and the arctic charr.

For organisms that live more than a single year, a potentially important selective factor is survival during the unfavorable period. In plants this may depend upon the amount of storage material (Iwasa and Cohen 1986), whereas in animals size (which may reflect the amount of stored reserves) often appears to be critical (Henderson et al. 1988; Post and Evans 1989; Gunnarsson 1988; Ydenberg 1989; Conover 1990; Preziosi and Fairbairn 1997; Schultz and Conover 1997, 1999; Schultz et al. 1998). In the case of an annual species, we saw that a decreasing season length is likely to favor a decreased adult size but an increase in the growth trajectory (i.e., change in size per unit time). Now consider a perennial organism that does not mature in its first year of life and whose survival over the winter months increases with the size at the end of the growing season. Under this scenario, what should we expect? Overwinter survival can be equated at least roughly with fecundity in a semelparous annual species, and hence we would expect the same general result, namely that a decline in season length would favor an increased growth rate but a typically smaller size at the end of the season.

DISTINGUISHING EFFECTS OF TEMPERATURE PER SE FROM SEASON LENGTH. It is important to distinguish between effects of temperature per se and effects of season length. A decrease in temperature shortens the available period of growth, but for ectotherms that period should be measured in degree days (Box 6.5) not absolute time.

Comparisons of life history patterns that do not use the appropriate scale can easily draw the wrong conclusion. In their examination of developmental patterns in the frog *Rana clamitans,* Berven et al. (1979, p. 619) concluded: "All the observed variation in green frog life history characteristics, including duration of the breeding season, length of the metamorphic period, size of developing larvae at all stages, and rate of development, was interpretable as the direct effect of temperature alone. The relatively cold environment of the mountain tops shortens the available season, directly slows down developmental rates of tadpoles, and prolongs the metamorphic period one or two years relative to the lowlands. . . . The various observed states in morphology and life history observed in the green frog across this altitudinal gradient cannot be specific adaptations to local environments."

This conclusion is misleading because it gives the impression that the appropriate time scale is days rather than physiological time. From a life history analytical perspective we should proceed by asking if there are effects of temperature once time is scaled in physiological units. If the effect of temperature is simply to change the scale, then from the perspective of the organism nothing may have changed. Of course, increases in latitude or altitude decrease temperature and also decrease the available time for growth. This is predicted to have an effect on life history characteristics, particularly growth trajectories. However, one cannot use growth trajectories plotted in real time taken from the field because these will differ by virtue of the change in measurement scale. In this sense I agree with the conclusion of Berven et al. (1979). But we must inquire whether on a degree-day scale we would expect changes. To do this requires a life history model and appropriate analysis, which was not attempted by Berven and his colleagues. They did, however, calculate their parameters on both an absolute and degree-day scale. The lowland and montane populations had the same threshold temperature for larval development (15.58°C and 15.54°C for the lowland and montane population, respectively). In the lowlands the total season length in degree-days was 1624 dd and in the highlands it was 726 dd. The lowland population metamorphosed in one season, taking 1100 (±31) dd. Based on the previous analysis, we would predict that the larval period would be shortened in the montane population, which indeed it was, the average larval period being 818 (±29) dd. This shortening of the larval period was not sufficient to permit metamorphosis until the second summer. In the field, montane frogs metamorphosed at a larger size than lowland frogs, but the pattern in the lab was more complex. Grown in the lab under constant temperature, the lowland population metamorphosed at a larger size than the montane population at 18°C and 23°C but at a smaller size at 28°C. Size at metamorphosis was highly correlated with development time measured in days but not when measured in degree-days (analysis of data in Table 3 of Berven et al. 1979). These results imply an interaction between temperature and growth rate. Without more information on such factors as trade-offs it is impossible to determine if the observed differences are adaptive or not. A similar pattern of variation was found in another frog *Rana sylvatica* (Berven 1982a; Berven and Gill 1983) and the presence of additive genetic variation for development and metamorphic size (Berven 1982b, 1987) indicates that these traits could respond to selection.

SIZE-RELATED WINTER MORTALITY AND THE OPTIMAL TIMING OF LIFE HISTORY EVENTS. Although in many fish species the size achieved during the first year of growth decreases with latitude and presumably season length, there are a number of interesting exceptions in which body size shows no latitudinal cline (reviewed in Conover 1990). Laboratory experiments with the Atlantic silverside, a species showing the lack of a latitudinal cline, showed that at the same temperature northern fish grew more rapidly than southern fish (Conover 1990). Conover hypothesized that size-selective mortality (starvation and/or predation) during the winter months was the primary factor selecting for increased growth rate. Supporting of this hypothesis was a significant latitudinal cline in size-related mortality as indexed by the difference in mean size between fall and autumn silversides. Overwinter mortality has also been hypothesized to be an important determinant of the size and timing of fledging in seabirds, for which a specific model was developed by Ydenberg (1989).

I present the model of Ydenberg to give the general flavor of the analyses. The model is predicated on the assumption that overwinter survival increases with the body size achieved at the end of the summer. Parameter values were selected to match the life history of the common murre. Mortality rate in the nest is less than that at sea, but the rate of growth is potentially higher at sea. Therefore, there will be an optimal time to depart the nest that maximizes the first year body weight. Growth in the nest can be described empirically by a logistic curve in which the daily change in weight, ΔX, is given by $\Delta X = 0.2X(1 - X/220)$, which means that the chick cannot do better than weight 220 g if it stays in the nest. No data were available for the growth rate at sea and Ydenberg "guessed" that it would also be asymptotic and used the function

$$\Delta X = 30[1 - (X/1000)^2] \tag{6.22}$$

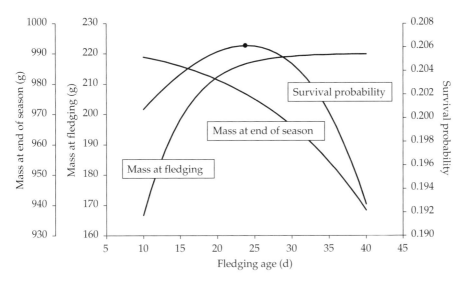

Figure 6.11 Relationship between mass at fledging and at the end of the season as a function of fledging age. Survival shows a curvilinear relationship with a maximum at the large dot. Data from Byrd et al. (1991).

which means that at sea the fledging can attain a maximum weight of 1000 g. (Note that in the paper there is a typographical error in the formula.) Daily mortality in the nest was set at 0.005 and in the sea 0.01 and the weight-dependent overwinter survival as $0.00204(X - 700)$. Ydenberg found the optimal fledging date by dynamic programming, but a simpler method is direct optimization (Roff 1992a). The mass at the end of the season declines with fledging age but because of the differences in survival probability between the nest and ocean the overwinter survival probability has a maximum at 25 d, 217 g (the estimates given in Ydenberg are in error, Figure 6.11), which is at the upper end of the observed fledging ages of 18–25 d (Table 1 in Ydenberg 1989). If parents are able to provision their young at a higher rate (indicated by an increase in the parameter previously set at 0.2), the model predicts an earlier fledging date at a higher fledging mass. Such a negative correlation in association with increased provisioning has been observed in the thick-billed murre, providing qualitative support for the model (Ydenberg, 1989).

NICHE SHIFTS. The idea that life history decisions involving switches between two habitats that differ in the opportunities for growth and the probabilities of survival has been a central feature of hypotheses for the evolution of complex life cycles (Wilbur 1980; Werner 1988). Amphibian metamorphosis has been a particular focus of study. Wilbur and Collins (1973) presented what is essentially a verbal argument for the optimal timing of amphibian metamorphosis based on relative growth and mortality rates but did not consider the potential importance of a time constraint. Without time constraints and fixed rates in the two habitats, the rate of increase is given by (Werner 1986)

$$r = \frac{F_x}{V_x} + g_x \frac{\partial M_x}{\partial g_x} - M_x \tag{6.23}$$

where x is age, F is fecundity rate (eggs per unit time), V is reproductive value, g is growth rate, and M is the mortality rate. Maximizing r and hence fitness requires that the right-hand side be maximized for all ages. For the pre-reproductive ages $F = 0$ and in a stable population $r = 0$, giving the quantity to be maximized, W, as

$$W = \frac{g_x}{M_x} \tag{6.24}$$

An important assumption underlying this result is that time is not limiting. When there is a fixed time interval at the end of which the organism must initiate some other activity such as reproduction, then selection can favor a switch to a habitat in which growth is faster but survival less than would be predicted by Equation (6.24) (Ludwig and Rowe 1990). An analysis of the optimal timing of life history switches such as amphibian metamorphosis was developed by Rowe and Ludwig (1991). They assumed that growth and survival varied between two habitats but were fixed within these, that is, there was no permitted trade-off between growth rate and survival within a habitat. This restriction did not allow an individual to change its growth pattern depending on the time of hatching. Thus the only phenotypic plasticity with respect to metamorphosis was variation along a fixed size/time trajectory. They showed, as found for the analyses of annual organisms, that size at metamorphosis is expected to vary through the season. Depending on the relative values of mortality and growth in the two habitats, tadpoles emerging later in the season might metamorphose at a reduced or enlarged size.

THE OPTIMAL TIMING OF REPRODUCTION. The foregoing models focussed on the "decision" of an individual offspring (chick, tadpole etc.) to leave one environment for another. For the parent an important decision is when in a season to commence reproduction. The problem faced by the parent is that a start date that is too early may force the parent to commence with few resources or bring the offspring into the world at a time when resources are relatively low, but a delay may give the offspring insufficient time to accumulate sufficient resources to survive to the next season. In their analysis of the causes and consequences of seasonal decline in reproductive success of the Great tit, Verhulst et al. (1995) estimated that 87% of the decline could be attributed to the effect of timing per se, whereas 13% was a result of differences in quality between early and late breeders. Thus there would appear to be strong selection on the date of breeding in the Great tit. A similar selection was observed by manipulation of start date in the common guillemot (*Uria aalge*), but in this instance it appeared that it was the start date in relation to other members of the colony rather than calender date, as in the Great tit (Hatchwell 1991). A seasonal decline in reproductive success is a common feature of bird populations (Perrins 1970; Daan et al. 1989; Brinkhof et al. 1993). If there is such strong selection on start date, why has it not responded? One possibility is that start date is not heritable (Price et al. 1988). Another is that those birds commencing later are being optimal given other factors, such as condition. In the great tits, for example, parents of low quality bred later, which may have given them the largest fitness under the circumstances.If they had attempted to breed earlier, their reproductive success would have been even lower. Finally, if there is variation in the optimal start date, then the date that is observed to be optimal in one year may be suboptimal in another.

Loman (1982) and Rowe et al. (1994) produced theoretical analyses of the optimal start time. The latter analysis is very general, whereas the former is more specific and is subjected to a preliminary test. I shall discuss the Loman model. Loman (1982) assumed that, because the female must accumulate resources, clutch size, C, is proportional to the start date, $C(t) = at$, where time is measured from the beginning of the season. Incubation does not commence until the last egg is laid (one egg per day) and offspring survival, S, is a linear function of the time when incubation starts, $S(t) = d - b(t + at)$. Here d is a constant that is limited in its maximum value such that at the earliest possible start date survival cannot exceed 1. (Note that this definition differs from the slightly more complex reasoning of Loman); it is related to the maximum survival. Assuming no cost to the parent, the fitness of start date t, $W(t)$, is $W(t) = C(t)S(t) = adt - ab(1 + a)t^2$, which is parabola, for which the turning point and hence optimal starting date is $t_{opt} = (1/2d)/(b + ab)$. Substituting this in the clutch-size equation gives the optimal clutch size as $C_{opt} = (1/2ad)/(b + ab)$. Selection favors a delay in the start of breeding if maximal survival (d) is increased, the rate of accumulation of resources (a) decreased or the mortality rate (b) decreased. The model does not take explicitly into account the amount of time remaining in the season relative to the start, though this could be accommodated by making d a function of time remaining. (For example, d decreases as the difference between the season length and the start date diminishes.) If there is variation among individuals in resource gathering ability, expressed as variation in the parameter a, then the optimal reaction norm is one in which clutch size and start time negatively covary. If parental quality can increase offspring survival by decreasing b or increasing d, then the optimal reaction norm will be one in which there is a positive covariance between start date and the optimal clutch size. Thus different measures of parental quality can lead to opposite predictions concerning the optimal reaction norm. There is no fundamental contradiction here, but an important message that qualitative labels such as "parental quality" can be misleading and should be precisely defined.

Loman attempted to test his model using a clutch manipulation experiment in the hooded crow (*Corvus cornix*). Previous observations (Loman 1977) showed that clutch size declined over the season (4.5, 4.4, 4.0, with the season divided into three parts) as did the survival of young (0.28, 0.12, 0.08, respectively). Loman either removed or added an egg from nests, arguing on the basis of his model that the crows should compensate for the addition or deletion of eggs by 0.5 eggs. The crows, apparently unfamiliar with the model, showed no compensation (Loman 1982). While compensation could have been taken as support for the model, it is more difficult to decide what failure to compensate actually means. It is possible that the egg-laying regime is "decided" in advance and cannot be easily manipulated by the crows. Alternatively, though Loman argued that one in two females should have compensated by a single egg (0.5 eggs per female), it could be that the manipulation really wasn't sufficient to induce a change. Further study and testing is warranted.

The Optimization of Development and Reproduction in Organisms with More than One Generation per Year

OPTIMAL SIZE AND DEVELOPMENT TIME. For a multivoline phenology, analysis of life history traits such as the optimal age and size at reproduction must take into account the optimal number of generations per season. Many invertebrates, some plants, and many small mammals have several generations per season, the unfavorable season being passed in a state of diapause/dormancy/hibernation (or in the case of some tropical and semi-tropical habitats, aestivation). Therefore, in addition to the interactions between size, development time, fecundity, and survival that are important in determining the optimal age at maturity in a continuous environment (Chapter 4), there is the additional constraint that at the end of the season the descendants of a female alive at the beginning of the favorable season must be in the appropriate stage for overwintering (or oversummering).

For simplicity I shall first assume a season length of fixed duration with all individuals hatching/emerging from their overwintering sites at the same time. To illustrate the general approach, I shall compare the fitness of a univoltine strategy against a bivoltine. I assume the following life history characteristics:

1. Fecundity increases with development time (based on the assumption that fecundity and body size covary).

2. The organism is semelparous.

3. The overwintering stage is the egg.

4. Egg-to-hatching time is included in development time.

5. There is a constant mortality rate of M, except for the overwintering portion, which is irrelevant in the present analysis. (To permit the use of R_0 as the fitness measure I shall standardize overwintering motality to 0.)

For a univoltine phenology the fitness, $R_0(D)$, is $R_0(D) = F(D)e^{-MT}$, where T is the length of the season, D is generation time, and $F(D)$ is the number of female eggs laid, and for a bivoltine phenology, $R_0(D_1, D_2) = F(D_1)F(D_2)e^{-MT}$, where D_i is the length of the ith generation. The first question is whether the optimal phenology uses the entire season. Because the mortality is constant and fecundity increases with development time, the answer is yes. (A simple geometric argument can be used by noting that the product $F(D_1)F(D_2)$ defines the area of a rectangle with sides D_1, D_2, and this is greatest when the rectangle is a square.) A univoltine phenology will be favored over a bivoltine phenology whenever $R_0(T) > R_0(1/2T)$. Suppose fecundity increases in proportion to development time. Then

$$R_0(T) = aTe^{-MT} \tag{6.25}$$

and

$$R_0(\tfrac{1}{2}T) = (\tfrac{1}{2}T)^2 e^{-MT} \tag{6.26}$$

and so the univoltine is most fit whenever $T < 4/a$. As T increases, the optimal development time increases with the season length until a season length at which the foregoing inequality is no longer satisfied, when the bivoltine phenology is favored and there is a sudden shift as the optimal development time is halved (Figure 6.12), producing a saw-tooth change in development time and body size (Roff 1980b, 1983c).

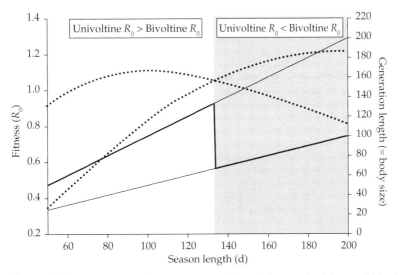

Figure 6.12   Fitness in relation to season length for a univoltine and bivoltine semelparous life history. The dotted lines show the fitness functions for the univoltine $R_0 = 0.03Te^{-.01T}$ and for the bivoltine $R_0 = (0.03T/2)^2 e^{-0.01T}$. The thin solid lines indicate the development times for the two phenologies and the bold portions delineate the most-fit phenology. At approximately 134 units there is a switch from the univoltine to the bivoltine phenology.

While the saw-tooth jump is a general prediction when there is a shift in the phenology, the above predictions that (1) the whole season will be utilized and (2) that the generations will be of equal length are not specific to all models. Suppose, as might be likely, that the egg is a "safe harbor" and suffers a much lower mortality than the nymph or adult; in this case the best univoltine semelparous phenology might be to lay eggs prior to the end of the season. The prediction of equal development times can be shown not to be general by a simple counterexample (Box 6.6).

Box 6.6
Optimal development times may differ between generations

Let the organism be iteroparous with a fecundity that is constant with the daily production a function of development time. For the univoltine phenology, R_0 is

$$R_0(D) = \int_{x=D}^{T} F(D)e^{-Mx}dx = \frac{F(D)}{M}\left(e^{-MD} - e^{-MT}\right)$$

(6.27)

To obtain the equation for the bivoltine life history, first consider a female from the first generation of some adult age t, which corresponds to a time $D_1 + t$ since hatching in the spring. Her probability of survival to this age is e^{-t-D_1} and the number of eggs produced on this day is $F(D_1)$. After D_2 days there are on average $F(D_1)e^{-M(D_1+t+D_2)}$ offspring from the female that hatched in the spring. These offspring lay each day $F(D_2)$ eggs. Integrating the equation over the time from the second adult eclosion to the end of the season gives the fitness component for the second-generation individual laid on day t, $W_2(D_1, D_2, t)$:

$$
\begin{aligned}
W_2(D_1, D_2, t) &= \int_{x=D_1+t+D_2}^{T} F(D_1)e^{-M(D_1+t+D_2)}F(D_2)e^{-M[x-(D_1+t+D_2)]}dx \\
&= \int_{x=D_1+t+D_2}^{T} F(D_1)F(D_2)e^{-Mx}dx \\
&= \frac{F(D_1)F(D_2)}{M}\left[e^{-M(D_1+D_2+t)} - e^{-MT}\right]
\end{aligned}
$$

(6.28)

To obtain the fitness for the entire bivoltine life history we have to integrate from the day the first-generation female ecloses ($= D_1$) to when any eggs she lays no longer have sufficient time to complete development ($= T-D_2$)

$$R_0(D_1, D_2) = \frac{F(D_1)F(D_2)}{M}\left\{\frac{e^{-M(D_1+D_2)}}{M}\left[e^{-MD_1} - e^{-M(T-D_2)}\right] - e^{-MT}(T - D_1 - D_2)\right\}$$

(6.29)

To find the optimal values of D_1 and D_2, we differentiate the above with respect to these variables and find the combination at which both derivatives are equal to zero. An example fitness surface is shown in Figure 6.13. Although the season length is 100 units in length, the sum of the two optimal development times is only about 35 units long, with the second generation being more than twice as long as the first. Iwasa et al. (1992) examined a model in which resources varied seasonally such that the maximum growth rate occurred in the middle of the season; depending on the details of the model, he found also that the optimal development times were either equal or the second-generation longer than the first.

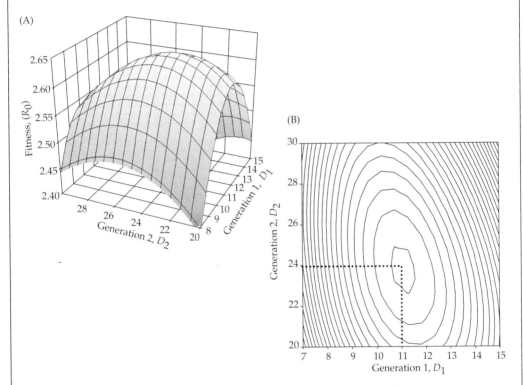

(A)

(B)

Figure 6.13   Fitness surface for an iteroparous organism with a bivoltine phenology. (A) The fecundity curve is uniform, with a height that is proportional to development time, giving a fitness function

$$R_0(D_1, D_2) = \frac{aD_1 aD_2}{M} \left\{ \frac{e^{-M(D_1+D_2)}}{M} \left(e^{-MD_1} - e^{-M(T-D_2)} \right) - e^{-MT} \left(T - D_1 - D_2 \right) \right\} \qquad (6.30)$$

with parameter values $a = 0.04$, $M = 0.01$, $T = 100$. (B) The dotted lines in the contour plot intersect at the optimal combination of approximately $D_1 = 11$, $D_2 = 24$.

The saw-tooth pattern of changing body size predicted above for changing season length should be evident in latitudinal clines where there is a shift in voltinism. Such patterns have been observed in crickets (Figure 6.14; Masaki 1978a,b, 1996; Mousseau and Roff 1989) and butterflies (Nylin and Svärd 1991). These patterns of necessity are those observed in field-collected specimens. In *Allonemobius socius* there is a high correlation between the population mean size of crickets caught in the field and the mean size grown in the lab under constant conditions, indicating a genetic basis to the variation (Mousseau and Roff 1989). Nevertheless, in the field the organism must have a cue that informs it of the time of year and hence whether it is in the first or second generation. As with the univoltine case, photoperiod may be a sufficient cue. The adaptive response to photoperiod is exemplified by the cricket *Pteronemobius fasciipes*. At 40°N the species is univoltine and the reaction norm of development time on photoperiod is monotonic, causing a decrease in development time as the photoperiod indicates less time remaining (Figure 6.15A,C). At 35°N the species is bivoltine and the reaction norm takes on a very different shape, being peaked (Figure 6.15B,D). The first generation develops during a period of lengthening photoperiod and

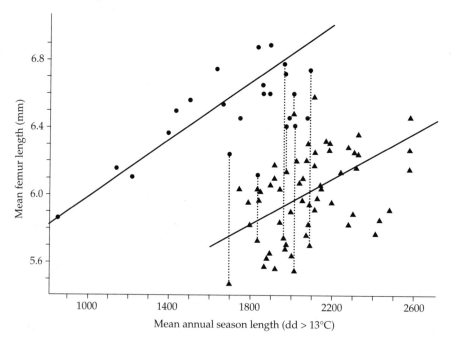

Figure 6.14   An example of a "saw-tooth" pattern in body size as a function of increased season length. The data shown are macropterous males of the cricket *Allonemobius socius/fasciatus*. • = univoltine populations sampled in September. ▲ = bivoltine populations sampled in either July or September. Symbols connected by vertical dotted line show bivoltine populations sampled in July (•) or September (▲). Solid lines fitted by eye to indicate trends in body size. From Mousseau and Roff (1989).

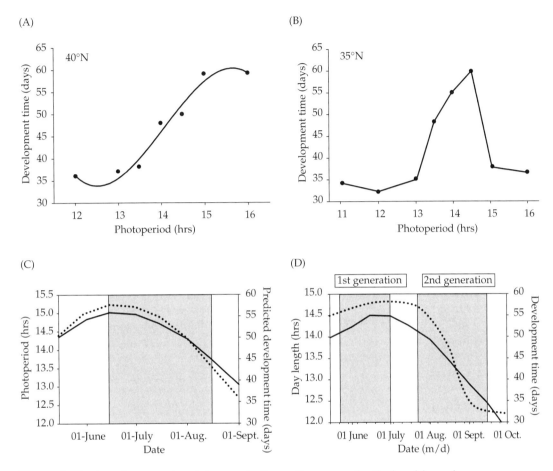

Figure 6.15 Development time/photoperiodic reaction norms for the band-legged ground cricket, *Pteronemobius fascipes* from 40°N (A: line shows fitted cubic) and 35°N (B). (C,D) Day length (solid line) as a function of date (data from Beck, 1980) and the development time (dotted) of *P. fascipes* predicted using the observed reaction norms. Hatched regions show phenology. Data from Masaki (1973).

the development time/photoperiod reaction norm dictates a more or less constant development time. The second generation encounters decreasing day lengths that do not overlap with those experienced by the first generation, and the reaction norm creates a shortening development time as the season progresses. The predicted development time of the first generation is longer than the second as is also that for *A. socius* (Figure 6.14). This is opposite to that predicted by the simple models described above and indicates that there is an important missing component.

The length of the growing season at any particular location will vary from year to year and, as a result, depending on the particular form of the $l(x)m(x)$

function, selection may favor variable generation numbers and/or variable development time (Roff 1983c). If the generation time is very short relative to the length of the growing season, fitness will be maximized by the addition of an extra generation in years with longer than average growing seasons. Contrariwise, variation in development time will be favored in those species in which the generation length is long relative to the season length, that is, in univoltine and bivoltine species.

THE OPTIMAL TIMING OF BREEDING. The onset of breeding typically coincides with the onset of favorable weather and food conditions. However, there is a continuous interplay between the advantages of early reproduction and high rate of increase versus the potential catastrophic mortality that could occur due to the uncertain onset of favorable weather. This principle is amply illustrated by Fairbairn's analysis of reproduction in the deermouse *Peromyscus maniculatus*. Only 25% of females that attempted to breed in early spring survived a six-week period, whereas 69% of nonreproductive females survived (Fairbairn 1977). Females breeding early in the season did not represent merely the tail of the breeding distribution, for the early breeding episode was separated from the main breeding period by at least four weeks, during which no females were lactating (Britton 1966; Fairbairn 1977). In contrast to the pattern displayed by females, males continued to be in breeding condition throughout this time, though the proportion breeding during the early phase was small.

Two advantages to breeding early are, first, that the female can breed twice in the season and, second, that the young of the first brood may also be able to breed in their first year rather than waiting until the next year. A difficulty in the analysis of this problem is that there is an overlap of generations making a simple "clonal" model inappropriate. Fairbairn (1977) dealt with this by estimating fitness as the number of genes at the end of the main breeding season that are descended from either an early-breeding female or a late-breeding female. This approach takes into account the fact that offspring of early-breeding females that themselves breed, leave offspring (grandchildren of the early breeding female) that carry only one-half of the genes of the early-breeding female. A full description of the model and parameter values is given in Table 6.2. Briefly, the reproductive success of late-breeding females is the sum of the number of offspring produced by a late-breeding female herself, plus the number of offspring produced by the offspring of this female, divided by one-half to account for the decrease in the number of genes descended from the original female. The reproductive success of an early-breeding female is the sum of three components: the number of offspring per litter that recruit into the population, the expected number of grandchildren produced by these offspring, and the relative success of an early-breeding female in the late-breeding peak.

Substituting estimated values gives the estimated ranges for the reproductive success of early- and late-breeding females as $1.64 < R_{late} < 1.87$ and $1.72 < R_{early} < 2.10$. There is substantial overlap in the two estimates, and hence the hypothesis

Table 6.2   **Description and parameter values of the model by Fairbairn (1977) of reproductive phenology in the deermouse.**

Symbol	Description	Value
S_1	Probability that a female that does not breed early will survive to the main breeding period	0.51
n_1	Number of juveniles per litter born in the late peak of breeding that recruit into the population	2.81
P_i	Probability of a female having at least i litters during the late peak of breeding. The maximum number of litters is 5 and for $I = 0$ to 5 are 0.35, 0.65, 0.24, 0.09, 0.04, 0.02.	over
B_1	Probability that a juvenile from the late peak of breeding will breed during the summer of its birth	0.07–0.18
R_{late}	Reproductive success of a late-breeding female $$\left(S_1 n_1 \sum_{i=1}^{5} P_i\right) + \frac{1}{2}B_1 n_1\left(S_1 n_1 \sum_{i=1}^{5} P_i\right) = \left(S_1 n_1 \sum_{i=1}^{5} P_i\right)\left(1 + \frac{1}{2}B_1 n_1\right)$$	
S_2	Four-week survival rate of females lactating during the early peak of breeding	0.18
n_2	Number of juveniles per litter born in the early breeding period that recruit into the population	4.33[a]
B_2	Probability that a juvenile from the early peak of breeding will breed during the summer of its birth	0.33–0.60
R_{early}	Reproductive success of an early-breeding female $$S_2 n_2 + \frac{1}{2}B_2 n_1 S_2 n_2 + \frac{S_2}{S_1}R_1$$	

[a] The higher number that can recruit into the population from the early-breeding females compared to offspring born in the late peak is due to the low frequency of breeding males in the population at this time. Breeding males are very aggressive to juveniles and probably account for the higher mortalities of juveniles born in the late peak (Fairbairn 1977).

that the fitnesses of the two patterns of reproduction are equal cannot be rejected. The potential advantages of more children and grandchildren combined with the good survival of offspring born in the early breeding period compensate for the high mortality of early-breeding females.

Why does there exist a period between the early and late breeding periods when no female attempts to breed? Fairbairn (1977, p. 868) suggested the following reason: "A female that breeds early has the potential advantage of having more children and grandchildren within that summer than females that do not breed early. This is a time-dependent advantage which decreases as the later peak approaches. Further, as the late peak approaches and the proportion of adult males breeding increases, litter survival declines. Thus, there will be a period when the advantages of early breeding have diminished, and litter survival is poor. During this period, females that become pregnant will have little success. They will do better to breed later, when their offspring will have a better chance of survival. Thus, we can expect a bimodal distribution of pregnancies."

The bimodality should disappear with an increase in the survival of young. Two studies from the Gulf Islands (British Columbia, Canada) fulfill this prediction (Fairbairn 1977). On one island, Samuel Island, males are significantly less aggressive than on the mainland and juveniles are recruited into the population throughout the breeding season (Sullivan 1976). As predicted, there is no bimodality in breeding. On Santurna Island the survival rates are intermediate between those on the mainland and Samuel Island. There is no bimodality, but many females begin breeding before a high proportion of males have entered breeding condition.

The Evolution of Dormancy

THEORY. For many organisms the only way to survive the period of inclement conditions is to enter a state of **dormancy** (= diapause = hibernation). In the previous sections I have examined how developmental processes should be cued to photoperiod and other environmental variables to ensure that the organism is at the appropriate stage for diapause at the end of the season. In this section I shall examine the theory and its application of the optimal photoperiodic reaction norm for the induction of diapause. This is a particularly good trait to examine because there is in many cases a very clear "end of season" point determined by the onset of a killing frost. Hence, historical temperature data can be used to actually construct the optimal reaction norm. Further, there is abundant evidence for rapid evolutionary change in introduced species and in laboratory stocks that demonstrate that the trait is a useful model for a study of adaptation in real time (Danilevskii 1965; Tauber et al. 1986).

We begin by considering a season of fixed duration and a life history defined by (1) a constant mortality rate, (2) a uniform fecundity function, (3) a nymphal diapause stage, and (4) a sensitive period during which the nymph is sensitive to an environmental cue and either enters diapause or completes development. There is a time in the season beyond which an individual should enter diapause rather than completing development and attempting to produce another generation. This point is the time at which the expected future contribution of the female is less than 1, since at this point she can enter diapause, complete development in the next season, and have an expected contribution greater than 1. Letting this critical switch point be t_{crit} the previous argument gives $t_{crit} = T - (\alpha + t_r + \delta)$, (Taylor 1980), where T is the season length, α is the age at first reproduction, t_r is termed the replacement time and is the time from the beginning of reproduction required by the female to produce enough offspring such that on average at least one survives to age α, and δ is the difference in age between the diapause age and the end of the sensitive period. If the diapause occurs in a stage other than the nymphal period or the sensitive period is indeterminate, the formula for t_{crit} must be modified somewhat (Taylor and Spalding 1988). For a given life history, we can construct a fitness function for the switch time. Depending on the length of the season, such a function can have a plateau (multiple switch dates with the same fitness) or a sharp peak (Taylor 1986b). The shape of the fecundity function can also affect the shape of the fitness function (Taylor 1986b), but realistic juvenile survival rates seem to have little impact (Taylor 1986c).

GEOGRAPHIC (CLINAL) VARIATION. Along a south-to-north latitudinal gradient, the length of the season should decline, and hence the optimal switch date will vary. However, there is also a change in the photoperiodic curve such that the end of the season is cued by increasing daylength after the summer solstice (June 21) and decreasing day length before. For most temperate insects, the critical day length will occur after the solstice, and hence there should be positive covariation between latitude and critical day length, a prediction verified by Taylor and Spalding (1986) for a wide range of insect species (see also Masaki 1999). A particularly fine example is found in the pitcher-plant mosquito *Wyeomyia smithii*. This mosquito develops only in the water-filled leaves of pitcher plants in North America. In the south there may be five generations a year, whereas in the north the number is reduced to two or only one (Bradshaw and Holzapfel 1990). The cold winter months are passed over by diapause in the third or fourth larval instar. Changes in latitude and/or altitude reduce the season length, and hence there should be a corresponding shift in the critical photoperiod inducing diapause. An extensive investigation of diapause induction over a wide range of latitudes and altitudes has confirmed this expectation (Bradshaw 1976; Bradshaw and Lounibos 1977; Figure 6.16). The northward expansion of the mosquito has required adaptation in diapause induction, for which experiments have demonstrated abundant additive genetic variation—$h^2 \approx 0.5$ (Hard et al. 1993). In contrast to the strong covariation between critical photoperiod and latitude of origin, there is no relationship between development time and latitude of origin (Bradshaw and Holzapfel 1990; Hard et al. 1993). This implies that the optimal development time is governed not simply by season length but some other factors. One such factor is density dependence, which is strong in the southern populations and weak or nonexistent in the northern populations (Bradshaw and Holzapfel 1983, 1990). Whether this factor is sufficient to eliminate the decline in development time favored by a change in season length under density-independent conditions has yet to be determined. Development time does have a significant but low heritability of approximately 0.05 (Hard et al. 1993; Bradshaw and Holzapfel 1996; Bradshaw et al. 1997). Istock et al. (1976) estimated a heritability of 0.22, but this was under less natural conditions than used by Bradshaw and his colleagues. Despite the apparent independent evolution of critical photoperiod and development time, the two traits are positively genetically correlated (Hard et al. 1993). Hard et al. (1993, p. 719) argue: "Because, for a given photoperiodic switchpoint, fast-developing genotypes are better able to replace themselves and enter diapause before the end of season arrives, natural selection should

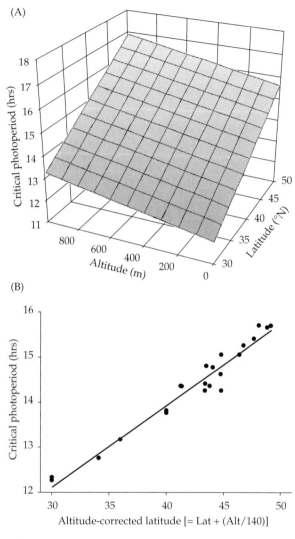

Figure 6.16 (A) Critical photoperiod in the pitcher-plant mosquito, *Wyeomyia smithii* as a function of latitude and altitude: critical photoperiod = 6.52 + 0.0013Alt + 0.185Lat. (B) The altitude-corrected relationship. Data and analysis from Bradshaw and Lounibos (1977).

favor such a correlation in seasonal insects. . . . Thus, a positive genetic correlation between these traits is indicative of genetic coordination of demography and phenology."

WHEN OPTIMALITY AND QUANTITATIVE GENETICS DISAGREE. In some cases a genetic correlation between diapause induction and a life history trait may be maladaptive, at least temporally. An example of this is the evolution of diapause and insecticide resistance in the lepidopteran, the oblique-banded leafroller (*Choristoneura rosaceana*). This moth is a pest of apple orchards and has been subjected to extensive insecticidal sprayings. In the area of Montreal, Canada, the species has two generations a year, with a larval diapause. The phenology is facultative, with some larvae from the first generation entering diapause and others completing development to produce a second generation, the offspring of which all enter diapause. All individuals are potentially subject to insecticidal sprays applied early in the summer, but diapausing individuals can escape the sprayings applied in late summer. Therefore, we can make the following predictions: (1) insecticidal spraying should select for physiological resistance to the insecticide and a propensity to enter diapause in the first generation, (2) as physiological resistance evolves selection, will favor a reversal of the diapause proportion. A simple model following the spirit of Cohen's model (Cohen 1966) illustrates this process (Carriere et al. 1995). Assume that there are two types of summer, "good" and "bad," which occur with probabilities P and $1-P$, respectively. In a bad year there is insufficient time for the second generation. The fitness for a female emerging in the spring and producing a diapause proportion D is $W(D) \propto [(1 - D)S^n F + Ds]^P (Ds)^{1-P}$, where S is the probability of surviving a single late summer spraying, n is the number of sprayings in late summer, F is the expected number of offspring produced by a second-generation female, and s is the survival of larvae that diapause in the first generation. Differentiating $W(D)$ with respect to d and setting the result to zero gives the optimal proportion as

$$D_{opt} = 1 - \left(\frac{PS^n F - s}{S^n F - s} \right) = \frac{S^n F(1 - P)}{S^n F - s} \tag{6.31}$$

The optimal diapause proportion is a monotonically decreasing function of the survival to the insecticidal spraying (S^n). In the absence of spraying or complete resistance, the optimal diapause proportion is

$$D_{opt} = \frac{F(1 - P)}{F - s} \tag{6.32}$$

If F is much larger than s, then the optimal proportion is approximately equal to the proportion of bad years. In the presence of spraying, insecticide resistance increases, and as the survival rate increases, the optimal proportion declines back to the optimal proportion in the original population (Figure 6.17). While this model gives a qualitative picture, only a quantitative genetic model can give a quantitative measure of the trajectory (Box 6.7).

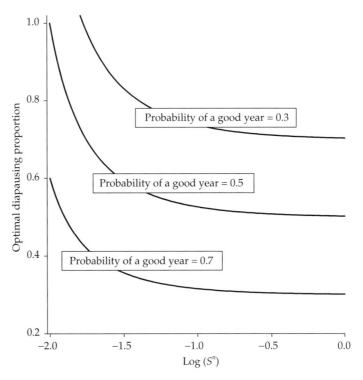

Figure 6.17   Predicted optimal proportion of diapausing larvae derived using the optimality model described in the text, $D_{opt} = \dfrac{S^n F(1-P)}{S^n F - s}$, where in the plot $F = 200$, $s = 1$.

Box 6.7

A quantitative genetic model for the joint evolution of diapause and insecticide resistance

We use the threshold model as our starting point for both diapause propensity and insecticidal resistance, both of which are all-or-none phenomena (in this model). The liabilities of diapause propensity and resistance are, as assumed by quantitative genetics, normally distributed. The proportion entering diapause, d is then

$$D = \int_0^\infty \frac{1}{\sqrt{2\pi}} e^{-\frac{1}{2}(x-\mu)^2} dx = \int_0^\infty \phi(x-\mu)dx \qquad (6.33)$$

where μ is the mean diapause liability (= propensity on the underlying scale). For convenience, and without loss of generality, the threshold has been set to zero and the variance to 1. (This simply means a rescaling of the mean liability.) Similarly, the proportion of resistant larvae is $I = \int_0^\infty \phi(x-v)dx$. As an example, suppose that in the initial population the proportion diapausing is 0.5 and resistance is essentially nonexistent. Then $\mu = 0$, $v = 0$. We now need the mean liabilities of the diapausing, μ_D, and nondiapausing larvae, μ_{ND}, which, using standard probability theory, are

$$\mu_d = \mu + \frac{\phi(\mu)}{D}, \quad \mu_{ND} = \mu - \frac{\phi(\mu)}{1-D} \tag{6.34}$$

For the mean liabilities of resistance we have $v_r = v + \phi(v)/I$ and $v_{nr} = v - \phi(v)/(1-I)$ for resistant and nonresistant larvae, respectively. To keep the model reasonably simple I shall assume that spraying occurs only in the second generation and that diapausing larvae are completely safe from the insecticide while nonresistant larvae are killed. Further, I shall assume that diapause propensity and resistance are genetically uncorrelated. The mean value of survivors in the presence of spraying is then

$$\text{Good year} \quad \mu^* = \frac{\mu_D Ds + \mu_{nd}(1-D)FI}{Ds + (1-D)FI}$$
$$\text{Bad year} \quad \mu^* = \mu_D \tag{6.35}$$

The mean liability of resistance is given by

$$\text{Good year} \quad v^* = \frac{vDs + v_I(1-D)FI}{Ds + (1-D)FI}$$
$$\text{Bad year} \quad v^* = v \tag{6.36}$$

Note that the mean value of the liability of resistance in a good year depends upon the proportion entering diapause. The mean offspring liabilities of diapause and resistance are, respectively,

$$\mu_{t+1} = \mu_t(1-h_D^2) + \mu_t^* h_D^2$$
$$v_{t+1} = v_t(1-h_I^2) + v_t^* h_I^2 \tag{6.37}$$

The results of applying the quantitative genetic model for a particular set of values are shown in Figure 6.18. As indicated by the optimality analysis, the proportion of diapausing larvae initially increases but returns to the pre-spraying level as resistance spreads through the population. For the particular parameter values chosen, the spread of resistance is relatively slow, both because of a rela-

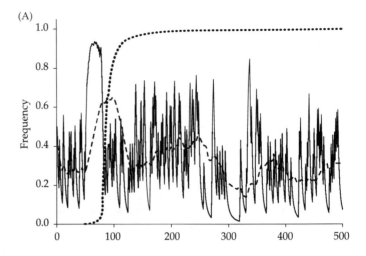

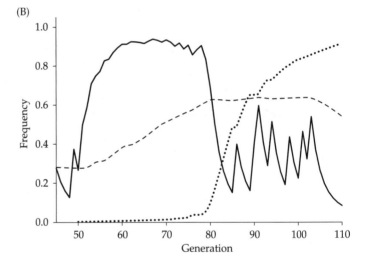

Figure 6.18 The evolution of diapause proportion and insecticide resistance using the quantitative genetic model described in the text. (A) A 500-year time trace and an expanded view of the period 45–110th generation (B) For the first 50 generations no spray was applied. The solid line shows the proportion of diapausing larvae each year, the dashed line the 50-year running average, and the dotted line the proportion of resistant larvae. Note that as resistance spreads in the population, the diapause proportion returns to its prespraying value. Parameter values: Probability of a good year = 0.7, $h_D^2 = 0.5$, $h_I^2 = 0.2$, $s = 1$, $F = 200$.

tively low heritability of resistance and because the high diapausing proportion. If, as in the oblique-banded leafroller, both generations are subject to spraying, the rate of spread of resistance is very much higher.

The prediction that orchards sprayed with insectide would have a higher diapause fraction and greater insectide resistance was supported very clearly (Figure 6.19). To test the second prediction that there will be a negative correlation between diapause propensity and insecticide resistance, Carriere et al. (1995) estimated the probability of survival to spraying for five orchards for which the spraying history was known. Contrary to prediction, there was a highly signifi-

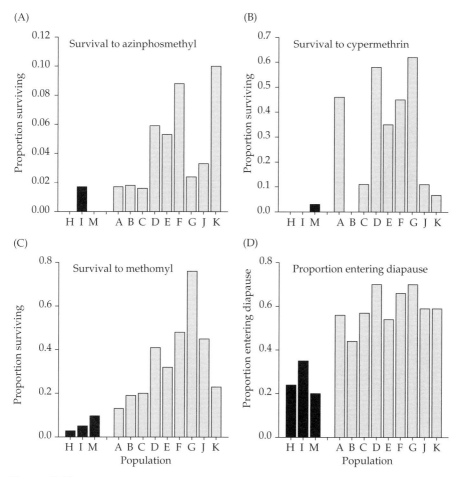

Figure 6.19   Resistance to three insecticides (A–C) and diapausing propensity (D) of *C. rosaceana* in insecticide-treated (grey bars) and insecticide-free (black bars) orchards.

cant positive correlation between diapause fraction and survival to spraying ($r =$ 0.92, $P = 0.026$, $n = 5$). One possibility is that the spraying had not been done long enough (in the initial stages there is a positive correlation as both evolve in concert), but this is unlikely since resistance had progressed quite far (Figure 6.19). An alternative reason is that one of the model assumptions is incorrect. An important assumption is that the two traits are uncorrelated. However, the breeding data showed that, in fact, the two traits are positively genetically correlated (Carriere and Roff 1995b). The positive genetic correlation would result in the two traits evolving together for a longer period than if uncorrelated, although an equilibrium would be eventually achieved.

PREDICTING THE OPTIMAL SWITCHING CURVE: CASE STUDY #1. As noted above, one of the advantages of studying diapause is that historical data can be used to predict the optimal switching dates. Bradford and Roff (1997) used 37 years of seasonal temperature data to estimate this date for the cricket *Allonemobius socius* from the area of Danville, Virginia, where the cricket is facultatively bivoltine. To estimate the optimal switching date in each year we need to estimate the day in the year at which the fitness of a direct-developing egg equals that of a diapausing egg. Fitness is defined as the number of diapausing eggs descendant from a female emerging at the beginning of the year. To estimate this, we made the following model: A diapausing egg laid by a first-generation female has a constant probability, S_d, of surviving to the winter. Therefore the fitness of a diapausing egg produced by a female in the first generation is

$$W_d = S_d \tag{6.38}$$

The fitness of a nondiapausing egg laid on day t in the first generation (all second generation eggs diapause) is the product of the egg to adult survival and the expected fecundity:

$$W_{nd}(t) = S_{nd}F(t) \tag{6.39}$$

The fecundity curve could be well fitted by a parabolic function (coefficients a, b, c). Assuming an adult mortality rate of M, the expected fecundity is equal to

$$F(t) = \int_0^{T-(t+\alpha)} m(y)l(y)dy \tag{6.40}$$

where $m(y) = a + by - cy^2$ and $l(y) = e^{-my}$

The integration covers the time available (in degree-days), given a season length of T and an egg to adult development time of α. If the available time was such that the fecundity function declined below zero, then $m(y)$ was set to zero. Development time varies and the probability density function was modeled with an inverse normal function:

$$P(\alpha) = \frac{1}{\sqrt{2\pi\sigma_\alpha^2}} e^{-\frac{1}{2}\left(\frac{\alpha^{-1}-\mu_\alpha}{\sigma_\alpha}\right)^2} \tag{6.41}$$

where $P(\alpha)$ is the probability of a development time of α, μ_α is the mean, and σ_α is the standard deviation. The fitness of a nondiapause egg in the first generation is then

$$W_{nd}(t) = S_{nd} \int\limits_{-\infty}^{\infty} P(\alpha) \int\limits_{0}^{T-(\alpha+t)} m(y)l(y)dyd\alpha \qquad (6.42)$$

Parameter values for the fecundity function and development times were estimated using laboratory data and survival estimates from field data for other orthopterans. The end of the season was defined arbitrarily as the first day on which temperatures dropped to –8°C. Because the threshold temperature for development is 13°C, the actual value of the end-of-season temperature is not critical. For each of the 37 years we estimated the day on which the fitness of diapause and nondiapause eggs was equal (Figure 6.20). The date fluctuated between day 203 (22 July) and day 224 (12 August), with an average of day 215 (3 August) and a standard deviation of 5.7 d (Figure 6.20). If females had complete knowledge, they would shift without transition between nondiapause and diapause eggs. In practice the remaining season length is uncertain, and hence there will be a period during which it will be advantageous to produce both diapause and nondiapause eggs. The spread in the switching curve will depend upon the variability in season length. Using the historical data we determined that the mean time remaining declined linearly with the day of the year ($\mu(t)$ = 3225 – 12.41t, n = 15, r = 0.99) as also did the standard deviation ($\sigma(t)$ = 160 – 0.41t, n =15, r = 0.98). Thus for any time of the year we could assign a probability distribution for the remaining time, $T–t$:

$$f(T - t, t) = \frac{1}{\sqrt{2\pi}\sigma(t)} e^{-\frac{1}{2}\left[\frac{T-t-\mu(t)}{\sigma(t)}\right]^2} \qquad (6.43)$$

Negative season lengths were eliminated but these were very rare as the mean season length was 10 standard deviations from zero. Fitness is now estimated as the geometric mean

$$\ln\left[W(P_d, t)\right] = \int\limits_{-\infty}^{\infty} f(x, t) \ln\left[P_d W_d + (1 - P_d)W_{nd}(t)\right]dx \qquad (6.44)$$

where $W(P_d, t)$ is the fitness of a female that produces a proportion P_d of diapausing eggs on day t. For each day t, the optimal proportion of diapausing eggs was found numerically (Figure 6.20). The median switching day was day 214, one day earlier than from the analysis assuming complete knowledge. Further, the switch takes place over a very short space of time, a complete shift occurring in about 10 days.

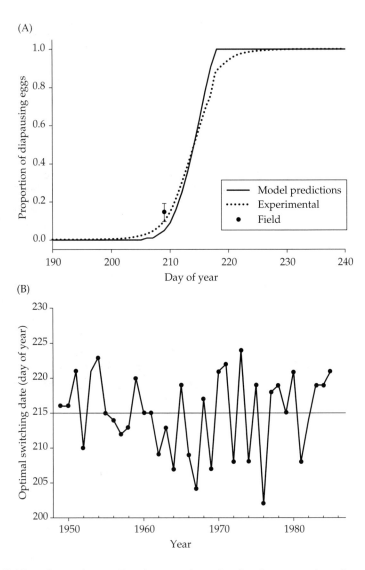

Figure 6.20 (A) Solid line shows the predicted norm of reaction for the proportion of diapause eggs produced by first-generation females of *A. socius* when the season length varies. The dotted line shows the observed mean female response curve centered over the median diapause proportion to permit comparison of the shapes of the two curves. The model predicted a median date of day 214 and the laboratory data gave a median date of day 215. Filled circle (± SE) shows the mean proportion of diapausing eggs produced from field collected females. Samples collected between days 255 and 265 laid only diapausing eggs. Data from Bradford and Roff (1997). (B) Time series of optimal switching dates for the cricket *Allonemobius socius*, estimated from historical temperatures at Danville, Virginia. Horizontal line drawn at the mean switching date (day 215).

Females collected from the field on days 208–210 produced few diapausing eggs, whereas females collected on day 255–260 (September) produced only diapausing eggs. Both of these observations were in accord with the model predictions (Figure 6.20). A second test was made by estimating the switching curves for females raised in the laboratory under conditions mimicking the average seasonal temperature/photoperiod cycle of Danville. The average median date of switching was day 215, close to the model prediction of day 214, and the shape of the curve closely matched that predicted (Figure 6.20). Given the uncertainty in the parameter estimates, these results are encouraging. The most uncertain estimates were the survival rates; large variation in these changed the median switch date by up to 3 days but did not affect the shape. Although there is good overall agreement there was considerable variation among females in the switch curves. Further study is needed to address the question of the maintenance of this variation.

PREDICTING THE OPTIMAL SWITCHING CURVE: CASE STUDY #2. For the species discussed thus far, the end of the season is determined by a drop in temperature. In copepods the end of the season may be determined by the drying up of a pond or the onset of heavy seasonal predation by fish (Hairston and Olds 1984, 1987). The switch from laying nondiapausing to diapausing eggs is, however, still mediated by a temperature/photoperiod reaction norm (Hairston and Olds 1986; Hairston and Kearns 1995). The selective effects of fish predation on diapause in copepods was studied in detail by Hairston and his colleagues. The springtime switch to diapause in one pond studied (Bullhead pond) coincided very closely to the increased activity of fish predators (Hairston and Munns 1984). A natural experiment in which all sunfish were killed in one pond (Little Bullhead pond) provided very strong evidence for the effect of fish predation. In both Little Bullhead pond and Bullhead pond, the dates of switching were identical prior to the loss of fish in Little Bullhead, but afterwards the switch date in Little Bullhead was very much later in the year, as would be predicted if fish predation were selecting switch date (Figure 6.21). The response was extremely rapid, indicating that fish predation was acting as a very strong selective factor. Hairston and Dillon (1990) estimated the change in the mean timing of the switch as a function of fish density (measured as a deviation from the long-term average) and showed, first, that there is a strong negative relationship and, second, that the response in one year can be as high as 0.3 standard deviation units. As might be expected from such a marked response, the heritability of the photoperiodic response is large, being approximately 0.6 (Hairston and Dillon 1990).

To assess their hypothesis that fish predation acts to determine the optimal switch date, Hairston and Munns (1984) produced a simulation model that mimicked a clonal version of copepods. An important feature of the copepod life history is that diapause eggs form a "seed bank" with some eggs potentially not hatching for decades, depending on the depth they reside in the sediment (Hair-

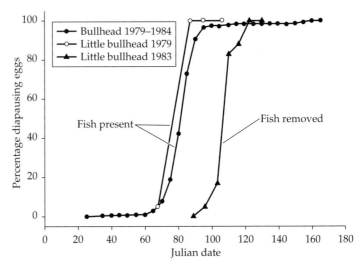

Figure 6.21   Cumulative frequency distributions of the proportion of *D. sanguineus* switching from making nondiapausing to diapausing eggs in two ponds (Little Bullhead and Bullhead). In 1979 fish were present in Little Bullhead pond but were extirpated by a severe freezing in 1981. The data for Bullhead pond is averaged for the period 1979–1984. Data kindly supplied by Dr. N. Hairston.

ston et al. 1995). This characteristic establishes an age structure to the population. Thus the number of eggs at time t, $n_{0,t}$ is given by

$$n_{0,t} = F \sum_{i=\alpha}^{\omega} n_{i,t-1} \tag{6.45}$$

where F is the fecundity of an adult, α is the first reproductive age, and ω is the maximum age. Mortality was assumed to be constant and hence

$$n_{i,t} = S_i n_{i-1,t-1} \tag{6.46}$$

In the absence of competition, S was equal to 1 for immatures and 0.9 for adults. Based on observation, mature females were allowed to lay four clutches per time period, each clutch consisting of a maximum of 20 subitaneous (nondiapausing) eggs or 10 diapausing eggs. Density-dependent survival and fecundity were introduced using the functions

$$S_{i,t} = \frac{S_i}{1 + aN_t}$$

$$F_t = \frac{F}{1 + bN_t} \tag{6.47}$$

where N_t is the total number of individuals and a, b are constants set such that the equilibrium population size approximated that in Bullhead pond. To determine the optimal date to switch from nondiapause to diapause eggs, Hairston and Munns used pairwise contests starting "contestants" off at different numbers to establish both if a particular type could invade and could resist invasion. Depending on the environmental values, they found four situations: (1) types that outcompeted all others, (2) those that coexisted with others, (3) those that occasionally invaded, (4) those that never invaded.

Hairston and Munns estimated that the day on which 50% of females in Bullhead pond switched to producing diapause eggs was day 83.8 with 99% confidence limits of 80.9 to 86.7. Depending upon whether the simulation was run with a single hatching date or a distribution of hatching dates, the optimal switching dates are 84–88 (with 84 winning most often) or 78–88 (with 83 winning most often). These predictions are in excellent agreement with the observations and give further evidence (in addition to the "natural" experiment) that the onset of significant fish predation is the factor selecting the timing of the switch to diapause eggs. As in the case of *A. socius,* there is a large variance among females in switch date that is not directly predicted by the foregoing model. The maintenance of this variation can be explained by the overlap of generations in the copepods (Ellner 1996; Ellner et al. 1999). However, while this hypothesis is consistent with the data, it cannot explain the large variance in *A. socius* in which there is no overlap of generations. A large heritability for diapause propensity is a characteristic of insects and other invertebrates (Roff 1996a) and it seems likely that we need to invoke mechanisms other than overlapping generations, at least in most cases.

The Evolution of Reaction Norms in Variable and Predictable, but Not Necessarily Seasonal Environments

The Optimal Adjustment of Parental Care

THEORY. Parental care, as understood in common parlance, is frequent in vertebrates and occurs in some invertebrates. The term **parental care** can be extended to cover reproductive effort in general, since "care" could refer to the amount of material placed within an egg or seed. This may seem to be stretching the term, but the same conceptual framework applies equally to the case of a female bear abandoning its cubs as to the abortion of seeds by a plant. The general question to be asked here is: "How should an organism alter its reproductive effort (desertion, abortion, defense) in the face of variation in offspring number (such as loss of offspring before the entire clutch is independent of the parent) or environmental quality (such as changed food supply)?"

It clearly will profit a parent to care for its offspring if this increases the survival rate of its offspring without incurring excessive costs such as increased mortality of the parents, or missed breeding opportunities. In the case of ectotherms, Clutton-Brock (1991, p. 101) suggested that parental care would "be expected where environmental conditions are harsh, predation is heavy, or competition for resources is intense-or where the costs to the parent of providing care are reduced." Though this hypothesis is supported by anecdotal evidence, an adequate statistical analysis is lacking. Parental care to some degree is found in all endotherms and is undoubtedly necessary, given the relatively undeveloped or vulnerable condition of neonatal birds and mammals.

With respect to desertion and sexual selection, Trivers (1972) concluded (p. 146): "At any point in time the individual whose cumulative investment is exceeded by his partner's is theoretically tempted to desert, especially if the disparity is large." This proposition could also be applied to a single parent deserting its offspring. The perspective of looking at past investment is incorrect. The critical factor is the evaluation of present and future fitness, not past investment (Dawkins and Carlisle 1976; Boucher 1977). In general, there is a point in time when the fitness gain from extending more effort in the offspring is less than can be obtained by abandoning those offspring. The hypothesis that selection acts on past rather than future investment has been dubbed the **Concorde fallacy**, after the rationalization given by certain politicians for continued investment in the development of the supersonic aircraft, the Concorde (Dawkins 1976). Because, mathematically, the two measures are likely to be highly correlated (Sargent and Gross 1985), an analysis based on past investment will generally yield the correct answer. However, the possibility of reaching the wrong conclusion is sufficient grounds for not using this approach.

Maynard Smith (1977, 1982) developed some simple game-theoretic models but empirical investigations testing these models appear to be largely lacking. Winkler (1987) developed a very general model for the analysis of parental care but provided no tests. The predictions concerning optimal parental care can either be shown using some very simple models or are sufficiently obvious that no mathematical model is required to make qualitative preditictions. A relatively easy subject to analyze is abandonment when only a single parent is involved. This situation also has very general applicability since it applies equally to the situation of optimal investment into offspring.

From the present viewpoint of evolution in predictable but variable environments, I wish to focus not on the optimal initial investment (e.g., clutch or propagule size) but rather on the response to a change in the value of the young, as might result from the changing age of the young, a loss of young, or a change in environmental conditions that changes the value of the young. The fitness returns of parental investment, which I shall refer to as the **brood fitness function**, can be expected to increase with time spent with the brood or, more generally, amount allocated to the brood, with the rate of returns eventually diminishing (Figure 6.22). The appropriate time (or allocation amount) to desert is dependent upon the time between broods and is found by finding the tangent to the brood fitness function as shown in Figure 6.22. As the time between broods increases, then so does the amount of effort it is worthwhile to expend on the

(A)   (B)

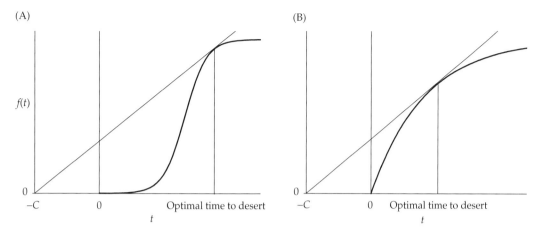

Figure 6.22 Two hypothetical functions relating the brood fitness function, $f(t)$, to the time spent in parental care. The time required to remate is C and fitness is defined as the number of offspring per unit time, $W(t) = \dfrac{f(t)}{C+t}$. The maximum fitness is obtained where a line drawn from $-C$ is tangent to $f(t)$. (A) The optimal parental care requires the parent to remain with the eggs to the point were the fitness function shows very little change, whereas in (B), the parent deserts the offspring when the rate of fitness gain still appears large.

present brood (Grafen and Sibly 1978). The graphical method presented in Figure 6.22 illustrates the conceptual solution very well but is a rather cumbersome method for determining the value of the optimal desertion time. The graphical model can be converted into an analytical model as follows. First we note that the decision to desert is made when the production of young per unit time is maximized. Let the brood fitness function be designated $f(t)$, where t is time spent in parental care, and C the time between broods. Then fitness, $W(t)$, is

$$W(t) = \frac{f(t)}{C+t} \tag{6.48}$$

Differentiating $W(t)$ with respect to t and setting the result to zero gives the optimal time for desertion,

$$\frac{\partial W(t)}{\partial t} = \frac{f'(t)}{C+t} - \frac{f(t)}{(C+t)^2} = \frac{1}{C+t}\left[f'(t) - \frac{f(t)}{C+t} \right]$$

$$= 0 \quad \text{when} \quad f'(t) = \frac{f(t)}{C+t} \tag{6.49}$$

where $f'(t) = \partial f(t)/\partial t$. A critical feature of this model is that the brood fitness function is concave down. Lack of this feature may make desertion unprofitable, as illustrated by Townsend's analysis of brood care in the neotropical frog *Eleuthero-dactylus coqui*. Male *E. coqui* provide parental care in elevated, terrestrial nests, protecting their eggs from predators and desiccation. Eggs of this species undergo direct development, the young completing metamorphosis and hatching as tiny, mobile frogs. The male parent remains with the clutch until hatching and for several days thereafter (Townsend et al. 1984). Townsend defined the brood fitness function as the probability of hatching following abandonment at any given age and the cost of caregiving being the time required to remate. The best fitting brood fitness function was a parabola $f(t) = 0.42t^2 - 0.26t$, which is a convex (concave up) function (Townsend 1986), and hence there is no intermediate point at which desertion is favored. Fitness is maximized by remaining with the young at least until hatching. (Consequences of desertion beyond this age were not empirically investigated.)

Although a mathematical model such as the one developed above is necessary to make quantitative predictions, it is possible to make qualitative predictions without having to explicitly define the brood fitness function. Suppose a female loses part of her brood: She now has two choices, continue to care for the brood or abandon the brood, remate, and produce another brood. The appropriate reaction norm will depend upon the worth of the present brood and the worth of another brood. Obviously, if brood size is reduced early in development the parent is more likely to abandon the brood and start again. This is not because there has been little investment in the present brood but because its present and future worth is low. Tait (1980) examined the selective advantage of desertion in grizzly bears, *Ursus arctos*, by means of a state-dependent simulation model

$$F(t,i) = S(t,i)\left\{ \sum_{j=1}^{7} P_{ij}\left[F(t+1,j) + V_{ij} \right] \right\} \qquad (6.50)$$

where $F(t,i)$ is the expected number of offspring for a female aged t in state i, $S(t,i)$ is the probability of survival of a female of age t in state i, P_{ij} is the probability of a surviving female in state i going to state j, and V_{ij} is the average number of recruits or value of a transition from state i to state j. There are seven states, state 1 being "no cubs", states 2–4 being 1–3 cubs, respectively and states 5–7 being 1–3 yearlings, respectively. A female could nurse a lone cub for the two years required to reach independence or abandon the cub, mate again the same year, and produce a new brood of perhaps two or three young. In this framework, it is obvious that there will exist conditions under which desertion will be the optimal behavior. The important question is whether there are conditions that are realistic with respect to the grizzly bear life history. Tait was able to show that, for realistic parameter values, abandonment could be optimal. The model may also be applicable to black bears and polar bears (Bunnell and Tait 1981).

EXPERIMENTAL TESTS. Mock and Parker (1986) advanced the same hypothesis for high desertion rates of singleton broods by egrets and great blue herons. An experimental test of the desertion hypothesis can be made by artifically reducing clutch size. At some level of reduction the parents should abandon their young and renest. If there is no time to renest, then abandonment would not be advantageous. Despite the large number of clutch manipulation studies, there appear to be few examples where the parents desert and renest. Armstrong and Robertson (1988) found that the probability of desertion of their clutch by blue-winged teal, *Anas discors*, is a direct function of reduction in clutch size. Artificial reduction of clutches caused earlier desertion by one of the parents in the snail kite, *Rostrhamus sociabilis* (Beissinger 1990). Ectoparasites such as the hen flea, *Ceratophyllus gallinae*, decrease the health of young birds and in the great tit, *Parus major*, reduce egg hatch (Oppliger et al. 1994). Therefore, the value of clutches in infested nests is reduced and desertion rates should be higher than in noninfested nests, which has indeed been found (Oppliger et al. 1994). The success of nests is frequently affected strongly by the start date, and hence the cost of desertion may be too high unless there are very large changes in clutch size. The importance of the size of clutch reduction was demonstrated by the experimental reductions carried out on tree swallows, *Tachycineta bicolor*. A reduction of 50% led to only a 21% desertion whereas an 80% reduction (one egg remaining) led to 100% desertion, with most pairs renesting (Winkler 1991).

An alternative experimental manipulation is to alter the relative worth of the offspring by handicapping one or both parents. Handicapping the parent by removal of feathers or the addition of weight reduces the feeding rate to chicks and leads to a lowered growth rate and fledging weight (Table 6.3). The effect on the parent appears to be a function of the average adult survival rate. Species with low adult survival rates (small passerines) show an adverse response (loss of weight), whereas the two larger, longer-lived sea birds adjust their parental care such that they show no loss in condition (Mauck and Grubb 1995; Table 6.3).

Table 6.3 **Handicapping studies in birds in which the effects on both parent and offspring were determined.**

Species	Annual adult survival	Handicap	Chicks adversely affected	Parents adversely affected
Pied flycatcher	0.30–0.70	Removal of feathers	Yes	Yes
Blue tit, Coal tit	0.35–0.50	Removal of feathers	Yes	Yes
Starling	0.33–0.66	Weight added	Yes	Yes
Leach's storm petrel	>0.78	Feathers shortened	Yes	No
Antarctic petrel	>0.80	Weight added	Yes	No

Modified from Mauck and Grubb (1995).

This difference in response can be explained by the relative worth of the chicks, which will be much greater for the shorter-lived species.

The relative worth of an offspring increases as it ages. Therefore, we might expect a parent to expend more effort in defense of offspring as they age. Similarly, the worth of a clutch increases as the number increases and hence the vigor with which parents defend their clutch should go up with clutch size. Both of these predictions assume that there is a chance of future reproduction. Andersson et al. (1980) and Carlisle (1982) developed theoretical models for parental defense, but these have not been explicitly tested, and so I shall consider just the qualitative and intuitively obvious predictions made above. Male threespine stickleback, *Gasterosteus aculeatus*, guard eggs and young, keeping the eggs aerated and defending them against predators such as other sticklebacks, other species of fish, and invertebrate predators (Giles 1984). The fitness value of a brood increases with the number and age of eggs, the latter not because of the amount invested but because of the reduction in the amount still required to be invested. Therefore, males should defend broods more vigorously as age or number of eggs increase. Pressley (1981) tested this prediction using a dummy of a potential predator of both eggs and parent, the prickly sculpin, *Cottus asper*. He applied the test to two separate populations in British Columbia (Canada). In accord with theory, males that remained within the nest area and attacked the sculpin had more and older eggs than those that deserted (Figure 6.23). Lachance and FitzGerald (1992) attempted to repeat this experiment using a population in Quebec (Canada), but instead of using a dummy predator, either waved a hand over the nest or used a model of a conspecific male. They found no relationship between parental defense and either the number or age of eggs. One possible explanation for the absence of results is that the stimuli were inappropriate. Another possibility, suggested by the authors, is that, because the environment is highly unstable (the pools dry up) and males probably do not get a second chance, the optimal behavior is to defend the nest with full vigor from the very beginning.

An increase in defensive behavior with brood size and/or egg age has been observed in a wide range of fish and bird species, in cases in which variation was natural and manipulated (Table 6.4). A notable exception is the willow ptarmigan, *Lagopus lagopus*, which did not increase its defensive behavior with artificially increased or decreased brood sizes. Willow ptarmigan chicks are precocial and disperse upon hatching, and it is hypothesized that a predator cannot threaten the entire brood and hence the efficacy of defense may not increase with brood size (Sandercock 1994). This study illustrates the importance of the biological details in molding the life history. Nevertheless, overall there is substantial evidence that the qualitative prediction is supported—of an increased defensive behavior with the relative worth of the brood (increased number or age).

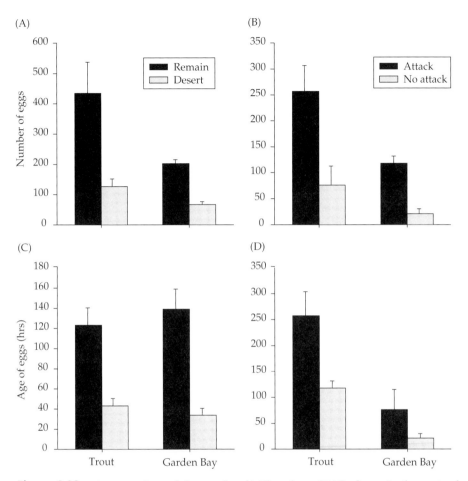

Figure 6.23   A comparison of the number (A,B) and age (C,D) of eggs in the nests of sticklebacks in relation to defensive behaviors. As predicted, males that deserted or did not attack had fewer and younger eggs than males that remained at the nest and attacked the dummy predator. Data from Pressley (1981).

The Optimal Adjustment of Clutch Size

TEMPORAL VARIATION IN RESOURCES. One of the most obvious factors that affect the optimal clutch size is the amount of resources accumulated by the parent(s). Given an increase in resources an organism has three options: (1) increase fecundity, (2) increase propagule size, (3) maintain fecundity and propagule size and use the extra resources to increase its own survival probability. The options (or a specific mix of them) that produce the greatest increase in fitness will depend

Table 6.4 Studies examining parental defense in relation to brood size and offspring age.

Species	Type of study[a]	Brood size	Offspring age	Reference
Fish				
Trichogaster trichopterus	Expt	Yes[b]	Not studied	Kramer (1973)
Gasterosteus aculeatus	Nat	Yes	Yes	Pressley (1981)
Gasterosteus aculeatus	Nat	No	No	Lachance and FitzGerald (1992)
Pomatoschistus microps	Nat	Not studied	Yes	Magnhagen and Vestergaard (1991)
Lepomis macrochirus	Expt	Yes	Not studied	Coleman et al. (1985)
Aequidens coeruleopunctatus	Expt	Yes	Not studied	Carlisle (1985)
Cichlasoma nigrofasciatum	Expt	Yes	Yes	Lavery et al. (1990)
Birds				
Agelaius phoenicieus	Expt	Yes	Not studied	Robertson and Biermann (1979)[c]
16 passerine, 2 non-passerine species	Both	Not studied	Yes[d]	Mongomerie and Weatherhead (1988, p 178)
Carduelis tristis,	Expt	Not studied	Yes	Knight and Temple (1986)
Tachycineta bicolor	Expt	No effect[d]	No effect[e]	Winkler (1991)
Falco columbarius,	Expt	Yes	Not studied[f]	Wiklund (1990)
Lagopus lagopus	Expt	No effect	Not studied	Sandercock (1994)

[a]Nat: Experiment used naturally ocurring variation in brood size; Expt: Brood sizes were manipulated.

[b]Parental behavior was assessed, but not overt defensive behavior.

[c]For a discussion of this experiment, see Nur (1983) and Bierman and Robertson (1983).

[d]Not corrected for possible seasonal effects. As the probability of renesting diminishes, the worth of present offspring increases.

[e]Although nest defense did not change, there was an increase in desertion rate.

[f]Young in replacement clutches were defended less vigorously than young in original clutches. This result is predicted on the assumption that replacement young have a lower expected survival than the original young (Wiklund 1990).

upon the particular biological circumstance. Numerous experiments providing supplemental rations to birds have found that reproduction commences earlier in the year and clutch size is increased (Nager et al. 1997). Two analyses by Nager et al. suggested that the effect of supplemental feeding is less when the year is more favorable than average. The strongest data set consists of experiments that have been conducted in several different years. Nager et al. (1997) took the conservative approach and pooled data such that each species was represented by a single data point and found a significant effect for both laying date and change in clutch size (Figure 6.24C,D). In their second analysis they used data from 32 studies and transformed the data as follows: independent variable = environmental quality index = $(C - A)/A$, where C is the control clutch size and A is the long-term clutch size; dependent variable = relative change in experimental clutch size = $(E - C)/C$, where E is the clutch size in the food-supplemented birds. The problem here is that the dependent variable equals $E/C - 1$ and the independent variable equals $C/A - 1$. Hence there is the potential for a statistical artifact due to the regression of x vs $1/x$. To remedy this problem, I computed the residuals from the regression of control clutch size on the long term average ($y = -0.292 + 0.996x$) and the residuals from the regression of experimental clutch size on control-clutch size ($y = 0.620 + 0.940x$), the first representing environmental quality and the second the amount of change resulting from the treatment. There is a highly significant interaction between the type of year (good vs bad) and relative environmental quality ($F_{1,28} = 8.96$, $P = 0.006$), indicating that the relative clutch size changes depending on both the year status and the relative environment. Coding the type of year as 0 (= bad) and 1 (= good) and running a multiple regression gives $y = 0.27 - 0.54x$ and $y = -0.44 + 0.83x$, for bad and good years, respectively. The second equation would appear to suggest that relative clutch sizes increased as the environmental quality increased. However, simple regressions of relative clutch size on environmental quality for the two types of years separately gives a highly significant negative relationship for bad years ($t = 2.72$, $df = 11$, $P = 0.02$) but a nonsignificant result for good years ($t = 1.98$, $df = 17$, $P = 0.064$). Thus when food is relatively scarce there is a decline in the treatment effect as environmental conditions improve. But if conditions are already good then further amelioration of the environment has little or no effect on clutch size. These results can be used to derive the qualitative relationship between ration and the optimal clutch size. The optimal reaction norm is a concave curve that shows a strong decline in the rate of change of clutch size with ration (Figure 6.24). This reaction norm makes intuitive sense, for we can suppose that the marginal benefit of food quality or quantity must be a saturating curve.

SPATIAL VARIATION IN RESOURCES. In a spatially variable environment in which there was migration among habitat patches, selection might be expected to favor the evolution of a reaction norm that produced the optimal clutch size in each patch. Dhondt et al. (1990) investigated this hypothesis for two tit species (*Parus major* and *P. caeruleus*) inhabiting habitats of differing quality. For *P. major,* the most productive clutch size was 8 in the "poor" habitat and greater than 12 in the "good" habitat, but the most frequent clutch size in both habitats was 9–10. For *P. caeruleus* the most productive clutch size was 9 in the "poor" habitat and greater than 14 in the good habitat, but the most frequent clutch size was 11–12 in both habitats. From these results Dhondt et al. (1990, p. 723) concluded that "the results support the 'nonadaptive' hypothesis that because of gene flow between

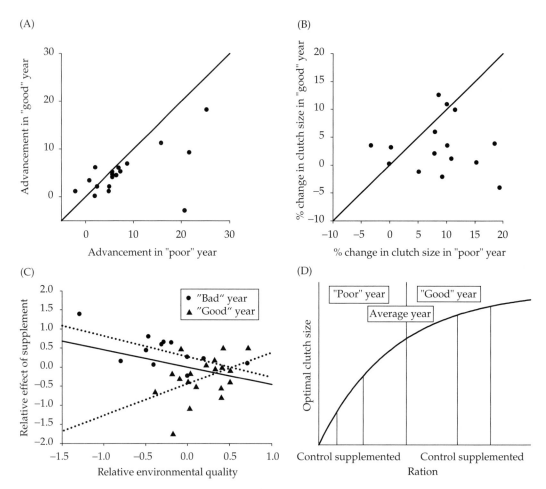

Figure 6.24 (A,B) For 14 species of bird (multiple populations in some species), the relationship between laying dates and clutch sizes relative to controls when years are either more or less favorable than average. (C) The relationship between relative clutch size (= residual from experimental clutch on control clutch) and relative environmental quality (= residual from control clutch on long-term average). Data from Nager et al. (1997). The solid line indicates the overall regression and the dotted lines the separate regressions for the two types of years. (D) The reaction norm for clutch size that is indicated by the experimental results of supplemental feeding on clutch size.

habitats of different quality as a result of birds moving from one habitat to another, birds can be unable to produce a clutch size adapted to the habitat in which they breed." While I agree that these data suggest that no reaction norm evolved, I think that the term "nonadaptive" is too strong. In the absence of appropriate cues, we would expect evolution to favor a compromise clutch size, which could be estimated using the approach of Kawecki and Stearns (1993; Chapter 5). It remains to be demonstrated whether the observed clutch sizes are optimal or suboptimal under the constraint that no reaction norm is possible.

In Chapter 4 I discussed the question of optimal clutch size in organisms that show no parental care beyond selecting the oviposition site and provisioning their offspring by modifying propagule size. Assuming that propagule size is kept constant, how should a female adjust her clutch size in the face of patches of varying quality? This question was addressed in Chapter 4 and a theory developed based on the assumption that fitness is maximized by maximizing productivity per unit time, where productivity is defined as the product of clutch size, offspring survival, and offspring fecundity—the last because clutch size typically affects adult offspring size (in addition to survival) which then affects the offspring fecundity. In many, but not all models, the optimal clutch size increases with site quality. These models are predicated on a single type of patch, but it has been observed in several studies (Chapter 4) that females vary their clutch size in accord with site quality. Here I shall discuss two different models that address this question, but from rather different perspectives. The first is the one developed in Chapter 4 and described above, namely that selection maximizes productivity per unit time, whereas the second ignores time and looks at the optimal placement of clutches by examining the consequences for each egg.

To illustrate the general approach of the first type of model, I shall use the study by Vet et al. (1993, 1994) on the allocation of clutch size in the hymenopteran, endoparasitoid *Aphaereta minuta*. Females of this species lay their eggs on the larvae of various species of flies, most particularly the first and second instars. Clutch size ranges from 1 to 14 eggs on the larvae of *Drosophila hydei* and *Delia antiqua,* with the number of eggs per clutch increasing with host size and differing between the two host species (Figure 6.25). Using lab data, Vet et al. (1994) estimated the productivity functions for *A. minuta* laying on first and second instars of *D. hydei*. The productivity function was constructed by multiplying together a series of regression equations. This presents two problems: First it was necessary to extrapolate in some cases beyond the range of the data; second, the error accumulation can be extreme since the final parameter values are the product of several estimates, each with its own variability. For the present purpose I shall ignore these difficulties. The productivity function, $W(X)$, was estimated as

$$W(X) = XS(X)S_r(X)F(g(X)) \qquad (6.51)$$

where X is clutch size, $S(X)$ is survival as a function of clutch size, $S_r(X)$ is the proportion of females, which varies with clutch size, and $F(g(X))$ is the fecundity of the offspring which depends on body size, which is itself a function of the number of emerging adults, $g(X) = f(XS(X))$. Only females are considered in this

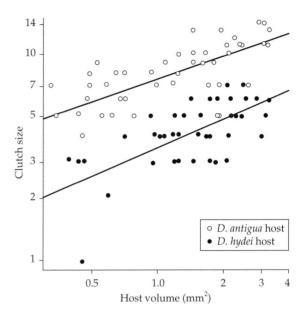

Figure 6.25   Clutch size versus host volume (*Drosophila hydei* or *D. antiqua*) in the parasitoid *Aphaereta minuta*. Data from Vet et al. (1993).

analysis, though it is clear that the variation in sex ratio with clutch size adds another dimension to the problem. If the survival or sex ratio exceeded 1 according to the linear regression model, Vet et al. (1994) set the value equal to 1. The relevant equations are given in the caption of Figure 6.26. Note that survival of first instar larvae increased with clutch size rather than decreasing. The multiplication of the regressions in principle produces a quartic equation, but because of the limitation of some regressions to 1, I calculated the productivity function using the regression model and then fitted a cubic equation to the points. As can be seen, the fitted cubic is an excellent representation of the curve (Figure 6.26). The optimal productivities (the Lack clutch size) are 12.88 and 17.27 for first and second instars, respectively, (Productivities were estimated using the full regression model not the fitted cubic.) The observed mean clutch sizes were 5.25 ± 1.99 and 8.50 ± 2.04. The Lack clutch size is greater on the second instar (= larger) larvae, but this does not indicate that the optimal clutch size will also necessarily be greater. The optimal clutch size is given by (see Equation 4.91)$X = W(X)/W'(X) - t$, where t is search time measured in egg units. We can rearrange this equation to obtain the search time under the assumption that the clutch size is optimal. For first instar larvae, the search time that corresponds to the observed clutch size is 0.092 and, for the second instar larvae the search time is 5.11. In Figure 6.26B, I calculate the optimal clutch size for varying search times. The optimal clutch size is always greater on the second instar larvae, but the optima converge as search time increases. If hosts consisted only of these two types, then females should ignore first instar larvae and search only for second instar larvae, because their

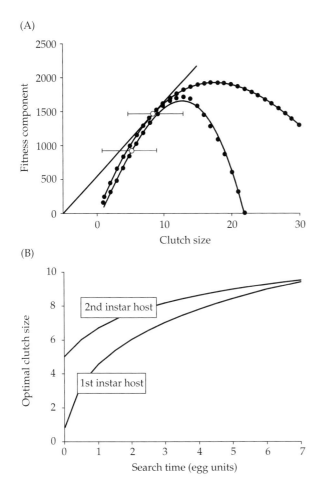

Figure 6.26   Estimates of the productivity function for female *Aphaereta minuta* laying eggs on first and second instar larvae of *Delia antiqua*. The component regression equations are:

Function	First instar	Second instar
Survival[a], $S(X) =$	$0.794 + 0.018X$	$0.951 - 0.008X$
Sex ration[a], $S_r(X) =$	$0.557 + 0.032X$	$0.738 + 0.007X$
Size of daughters, $g(X) = f(XS(X)) =$	$4.92 - 0.134XS(X)$	$5.04 - 0.117cS(c)$
Fecundity, $m(g) =$	$-220.6 + 111.9g$	

[a]Set to 1 if predicted value exceeds 1

(A) The dots show the predicted productivity using the product of the above functions, while the solid lines show the empirically fitted cubic equations $W(X) = -114 + 211X - 0.35X^2 - 0.42X^3$ for first instar hosts and $W(X) = -6 + 245X - 9.03X^2 + 0.08X^3$ for second instar hosts. The open circles show the observed clutch sizes (±2SE). For the second instar host, the tangent to the curve at the point of the observed clutch size is drawn, illustrating graphically the method of obtaining the search time required such that the observed clutch size is the optimal clutch size under the particular set of assumptions described in the text. (B) The predicted optimal clutch size for varying search time.

productivity per unit time will be greatest. Females do not ignore hosts of different sizes, though whether females do show selectivity and pass over some potential hosts is not indicated in the published papers. Obviously, this model is too simplistic, although it is probably a reasonable kernel from which to develop further models. The general problem is much the same as that of optimal foraging, which has a very well developed theory (Stephens and Krebs 1986; Kramer 2001).

TESTING OPTIMAL CLUTCH SIZE MODELS WITH *CALLOSOBRUCHUS MACULATUS*. If a female faces only a single type of environment, then the models discussed in chapter 4 are appropriate for describing the optimal norm of reaction. Table 6.5 gives the formulas for the optimal clutch size under various scenarios, each of

Table 6.5   **Optimal clutch size for various models of oviposition behavior under different constraints.**

Constraint	Optimal clutch size		Notes
Eggs	1		Female suffers no mortality
Sites	$X_{opt} = X_S = \dfrac{-P(X)}{P'(X)}$		Female suffers no mortality
Adult mortality	$X_{opt} = X_T = \left(\dfrac{-P(X)}{P'(X)}\right)\left(\dfrac{T}{T+Xt}\right)$		T = Search time t = Time to lay an egg
Trade-off between fecundity and adult mortality	$\left(\dfrac{-P(X)}{P'(X)}\right)\left(\dfrac{BT}{BT+X}\right)$		Fecundity = $A + Bx$, x = Adult lifespan
Eggs and adult mortality	if $\dfrac{L}{F} > T+t$ $X_{opt} = 1$ if $T+t \ge \dfrac{L}{F} \ge T + \dfrac{t}{X_T}$ $X_{opt} = \dfrac{T}{L/F - t}$ if $\dfrac{L}{F} \ge T + \dfrac{t}{X_T}$ $X_{opt} = X_T$		No trade-off, L = Lifespan F = Maximum fecundity
Eggs and sites	if $\dfrac{F}{N} < 1$ $X_{opt} = 1$ if $1 \le \dfrac{F}{N} \le X_T$ $X_{opt} = \dfrac{F}{N}$ if $\dfrac{F}{N} > X_S$ $X_{opt} = X_S$		Uncorrelated, N = Number of sites

The basic model is $W(X) = XP(X)$, where $W(X)$ is fitness for clutch size c and $P(X)$ is the product of offspring survival and their expected fecundity from a clutch of size X. Modified from Wilson and Lessells (1994).

which makes a unique set of predictions on the optimal clutch size when parameter values are increased singly. The predictions (Table 6.6) apply, in general, to the case of only one female ovipositing or the last female when a number of females oviposit, but in the latter case the number of eggs is obviously modified by those already present. To distinguish among these different models, Wilson, K. (1994) analyzed the oviposition behavior in the seed beetle *Callosobruchus maculatus*. The eggs of the bean weevil are laid on the outside of beans, the larva burrowing into the bean upon hatching and developing entirely within the bean, emerging as an adult. From a series of experiments Wilson obtained the following component equations. Adult size, Y, varied linearly with clutch size (= number of eggs on the seed): $Y = -31.18 + 17.49Y$. Lifetime fecundity increased linearly with body size, $F = 4.86 - 0.04X$. Combining these two equations gives lifetime fecundity as a function of clutch size: $F(X) = 53.82 - 0.70X$. This equation was used to estimate the fecundity of offspring emerging in those experiments in which offspring fecundity was not directly measured. Over the range in larval densities measured, the number of emerging larvae versus clutch size could be described by a linear or exponential function: Wilson used both models in generating the set of predicted values. A phenotypic trade-off between lifespan and egg production was highly significant and indicated that each egg decreased lifespan by 3.9 hours. Variation in travel time was obtained simply by moving females between eggs at preset intervals of 1, 10, 120, or 1440 minutes. The time to lay each egg was independent of clutch size and estimated to lie between 3.6 and 10 minutes. Using these data, Wilson predicted the optimal clutch size for different travel times, the two extreme values of the estimated parameter compo-

Table 6.6   Qualitative predictions of the clutch-size models given in Table 6.5.

| | | | | Effect on optimal clutch size, c_{opt}, of increasing | | | | | |
| | | | | | | | X[b] | | n[c] | |
Model	X_{opt}	Site quality	T	t	F	L	Lin	Exp	Lin	Exp
1	1	+[d]	0	0	0	0	−[d]	0	−[d]	0
2	X_{Lack}[a]	+	0	0	0	0	−	0	−	0
3	$\leq X_{Lack}$	+	+	−	0	0	−	0	−	0
4	$\leq X_{Lack}$	+	+	0	0	0	−	0	−	0
5	$\leq X_{Lack}$	+	+	−	+	−	−	0	−	0
6	$\leq X_{Lack}$	+	0	0	+	0	−	0	−	0

Modified from Wilson and Lessels (1994).

[a]The Lack clutch size

[b]Current egg load of site. "Lin" refers to a linear larval fitness function, whereas "Exp" refers to an exponential function.

[c]Number of ovipositing females

[d]Probability of laying on site

nents, and single versus multiple ovipositions. For the case of single ovipositions, the three models based on limited number of sites (#2), adult mortality (#3), or both (#6) all give absurdly high predictions compared to the observed when travel time is extreme. The entirely unconstrained model (#1) predicts one egg for all travel times, but there is a clear increase in clutch size as travel time increases (Figure 6.27). Depending on the value of the ratio L/F, the model based on the assumption of limited fecundity and adult mortality (#5) gives either a clutch size that is approximately one half (2 vs 5.5) or two times (10 vs 5.5) the observed clutch size at the most extreme travel time. Depending on parameter values, the model predicated on a trade-off between lifespan and fecundity (#4) can give a reasonable prediction for the clutch size at the two longest travel times but tends to underestimate clutch size at the shorter times (Figure 6.27). To obtain an overall measure of fit, I calculated the statistic (observed–predicted)^2/observed for each separate travel time and summed the individual values. Whether we consider the separate travel times or the combined statistic, the result is the same: Model #4, the trade-off model, with a particular set of parameter values gives the best fit to the observed clutch sizes (Figure 6.27). A better fit can be obtained if we assume that the females have been selected in an environment in which there are 5–7 other females ovipositing (Wilson, K. 1994). This analysis shows that the models can be distinguished, but we shall probably have to increase the complexity of the models and obtain more information on oviposition behavior in natural populations.

DYNAMIC MODELS OF CLUTCH-SIZE ADJUSTMENT. Myers (1976) and Ives (1989) examined the consequences of several females potentially laying on the same site. Obviously, the optimal clutch size will be a function of the number of eggs already present. One of the most important findings of the analysis by Myers (1976) is that we should pay considerable attention not only to the mean clutch size, the focus of the models discussed by Wilson and Lessels (1994), but also to the distribution of clutch sizes. Females do not typically have knowledge of the distribution of egg batches in the habitat. Therefore, the decision on the number of eggs to be laid at a particular location is most likely made using some "rule of thumb." This is the second perspective from which the optimal clutch-size reaction norm can be approached. The study by Mitchell (1975) of the egg oviposition rules used by the bean weevil *Callosobruchus maculatus* and that of Jones (1977, 1987) on the rules used by the cabbage butterfly *Pieris rapae* are excellent examples of this approach.

To obtain differences in site/host density, Mitchell (1975) exposed 10 g of beans to 200 female *C. maculatus* for three different time periods. With increasing exposure time the mean number of eggs per bean increased (0.2, 1.1, 1.8, respectively). Note that the density of eggs in this experiment is far below that used by Wilson, K. (1994). Females did not lay their eggs at random (i.e., there is a significant departure from a Poisson distribution), but laid more eggs on larger beans, resulting in a more uniform distribution of eggs per bean. Among beans containing only one egg, there was a significant and clear positive regression between survival rate and bean weight. For beans containing 2 or 3 eggs, there was con-

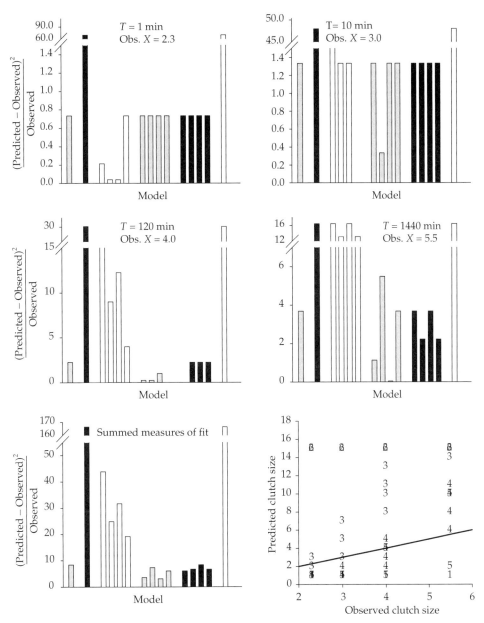

Figure 6.27   A comparison of the observed and predicted clutch sizes in *Callosobruchus maculatus* given different travel times and varying model assumptions. In all models it is assumed that only a single female oviposits. The panel in the lower right corner shows the model predictions on the observed clutch sizes at four travel times (1, 10, 120, 1440 minutes), with the symbols on the graph indicating the model. The other panels show an index of fit for each model (the sequence of differently shaded bars indicates the different models, some models being run with several parameter values) and the overall fit (lower left).

siderably more scatter and no significant correlation was found, but the overall survival rate was lower. Mitchell assumed that the decrease in survival was paid by the second and third eggs and estimated these as follows. The mean survival with one egg is 0.62, and the mean survival with two eggs is 0.464. Hence the survival of the second egg must be 0.308 ([0.62 + 0.308]/2 = 0.464). By the same logic he obtained the survival of the third egg to be also 0.308.

Ignoring search costs, a female *C. maculatus* should lay no more than one egg on each bean, since survivorship is highest at this density (regardless of how individual larval survival rates are estimated), and should select *beans in decreasing sequence of size*. Thus, in the experiment yielding 0.2 eggs per bean, the optimal oviposition behavior, ignoring all but survival rates of larvae, would be to lay the 46 eggs laid in this experiment in the 46 largest beans. If the number of eggs exceeds the number of beans, the second egg should be laid on beans *in ascending order of bean weight*. The reason for this is that the survival rate of two larvae is independent of bean size (at least over the range in bean size examined). The survival rate of a single larva on the smallest bean is approximately 50% compared to 46% for two larvae while for a single larva on the largest bean, it is approximately 72%. Therefore, the expected yield from laying two larvae on the smallest bean is hardly changed, while there is a great decrease in yield if the second egg is laid on the largest bean. The rule for laying the second egg presupposes that the first egg belongs to the present female. If it does not, the female should lay her egg on the largest bean since that gives the highest larval survival.

Note that the above analysis does not address the question of whether the second egg should be laid at all, but begins with the proposition that a specific number of eggs will be laid. Based on this assumption, the above gives the maximum yield possible in terms of surviving offspring. The observed distribution of eggs gave a yield averaging 0.94 of the maximum possible, whereas a random distribution of eggs gave a yield averaging only 0.67 of the maximum. Thus the females were using a rule that gave them a considerable fitness advantage over random dispersion of eggs. Although we cannot say that the behavior adopted is the best possible, it certainly comes very close. An important assumption is that the number of surviving offspring is an appropriate measure of fitness. The data presented by Wilson, K. (1994) on *C. maculatus* and the general findings on density-related changes in body size indicate that, in general, body size is decreased with increasing density. Therefore, the fitness measure used by Mitchell is only an approximation.

A female weevil obviously cannot inspect all the beans in her "universe" and hence she must use some rule of thumb to determine whether she lays an egg on the bean presently under inspection. Mitchell proposed that females use a simple decision rule based on the size of the previous bean encountered and the number of eggs laid upon it, relative to the size and egg complement of the current bean (Figure 6.28). Specifically the rule was: "Oviposit *if* the present number of eggs is less than on the previous bean *or* the present number equals the previous and the present bean is not larger than the previous." Mitchell compared the pre-

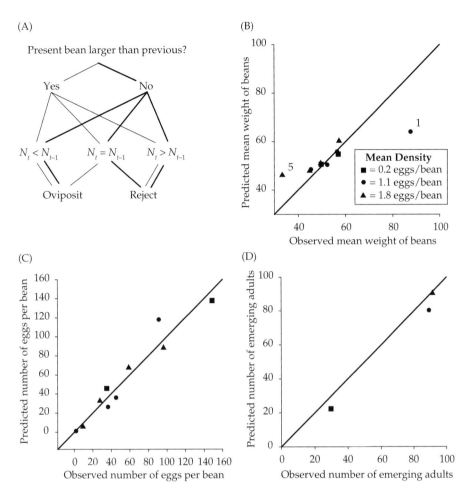

Figure 6.28 (A) The oviposition decision tree of *Callosobruchus maculatus* used to generate the predicted egg distribution. (B–D) Comparisons of the observed and predicted values. Sample sizes of less than 10 are indicated by the actual sample size next to the data point (5 and 1). Data from Mitchell (1975).

dicted and observed results with a χ^2 test and a *t*-test. Even though these can be used as measures of goodness of fit, they should not be used as statistical tests, because if the simulation is made large enough, one is almost guaranteed to find a difference between the two data sets. Here I use a plot of predicted on observed without a statistical comparison of the data (Figure 6.28). There appears to be a discrepancy between the mean weight of beans selected to have a given number of eggs, with the model overestimating and underestimating the weight at extreme values. However, the two deviant points are based on only five and one observation(s), which makes them suspect. Ignoring these two points, the predicted and observed points are reasonably close. The model also does a good job of predicting the frequency distribution of eggs per bean. It slightly underestimates the yield of weevils, though there are only three data points (Figure 6.28). Contrary to the claim by Mitchell, the simulated distribution is actually more uniform than the observed distribution. The reduced selectivity of the weevils in the simulation leads to the mean size of beans in each egg class (0–3) being less variable than in the observed population. Despite this shortcoming, the model illustrates how the oviposition behavior of the weevil can be accomplished by a very simple decision rule.

Quantitative genetic analyses have shown that oviposition behavior has a nonzero heritability (Messina 1989; Fox 1993b) and hence can evolve to adapt to changing distributions of seeds. Demonstration of adaptive evolution of oviposition behavior is strongly suggested by Jones's analysis of oviposition behavior in the cabbage butterfly *Pieris rapae*. Jones (1977) set out pots of plants in grids with varying spacing and then monitored the movement of female cabbage white butterflies that naturally entered the experimental area. Using a range of behavioral parameters, Jones then constructed a simulation model to mimic the observed behavior. The model consisted of the following set of "rules":

1. STOP is the probability (0–1) that a butterfly newly arrived at a grid point will land. This probability increased linearly with the age of the plant and varied among plants.

2. ZERO is the probability that a female will move to another leaf on the same plant. This probablity also varies with the type of plant.

3. LAY is the probability that a female lays an egg. Like the previous two probabilities, it depends on plant type and age.

4. MOVE was a vector of length 8 that defined the directionality of travel in the sense of whether females tended to fly straight or make more frequent turns. MOVE(1) is the probability of moving in the preferred direction.

5. DIR was a vector of length 4 defining a preferred direction of travel. DIR(J) is the probability that the preferred direction is to position J.

6. MISS is the probability of being attracted to a plant. It was found to vary with the butterfly's current fecundity.

7. CONT is the probability that in the absence of a plant the female will continue in the same direction as previously.

The values of these parameters differed between Canadian and Australian females (Table 6.7), suggesting an evolutionary change in oviposition behavior, though such differences are not by themselves evidence of adaptive evolution.

The model so defined by the foregoing parameters did mimic the movement and oviposition patterns of the females quite satisfactorily. Of course, if such fits are based on the same data set as used to estimate the parameters, the fit only

Table 6.7   **A comparison of movement parameters in ovipositing females of the cabbage white butterfly.**

Parameter	Description	Function of	Australian	Canadian
MOVE	Directionality		HIGH MOVE(1) = 0.6	LOW MOVE(1) = 0.3
DIR	Preferred direction		VARIABLE DIR(1) < 0.45	LESS VARIABLE DIR(1) > 0.7
STOP	Probability of landing	Plant species, size, age: current fecundity, F	STOP(G) = 0.015^a	STOP(G) = $0.012 + 0.003F^b$
MISS	Responsiveness to host plants	F	LESS MISS = 1–0.03F	MORE MISS = 1–0.07F
CONT	Probability of maintaining flight direction in absence of plant		0.6–0.8	0.6–0.8
ZERO	Probability of not moving	Plant species, size, age	LOW 0.2–0.3	HIGH 0.5
LAY	Probability of laying an egg	Plant species, age	0.7–0.9	0.7-0.9

Modified from Jones (1977).

^{a}STOP(G) are all grid points that do not have a plant. The probabilities shown are the probability of stopping at a nonplant site after contact with a host plant.

^{b}Current number of mature eggs

demonstrates that the model is a good description of the data, not that it has any predictive power. The latter was demonstrated when Jones used the model to predict egg distributions in field plots not used to construct the model (Figure 6.29). In experiment 1, Jones varied the quality of the host plant ("Young" kale, "Old" kale, "Old-stunted" kale) and found, as predicted, that the "Old-stunted" kale was least preferred (predicted = 0.14, observed = 0.15). The model predicted that "Old" kale would be preferred over the "Young" kale but the reverse was observed, though the differences were not great (predicted vs observed: "Young", 0.40 vs 0.48; "Old", 0.46 vs 0.37). In the second experiment, Jones varied host quality by using three different brassicae, cabbage, kale, and radish. The general pattern of observed and predicted was similar (cabbage, kale, radish: 0.63, 0.28, 0.09 for predicted and 0.49, 0.40, 0.11 for observed). The third experiment consisted of making a grid that was composed of plants at 2 m and 1 m spacing using either kale or radish. For this experiment the fit between prediction and observation was excellent (predicted vs observed at 2 m spacing: 0.61 vs 0.68 for kale and 0.61 vs 0.57 for radish. The results for the 1 m spacing are simply 1 minus the previous results). Taken together, the overall fit of the model to

Figure 6.29   A comparison of predicted and observed proportions of eggs laid by female *Pieris rapae* under different environmental conditions. Symbols indicate the experiment: 1 = kale of three qualities, 2 = cabbage, kale, and radish, 3 = varying grid spacing using kale, 4 = varying grid spacing using radish. Data from Table 7 of Jones (1977).

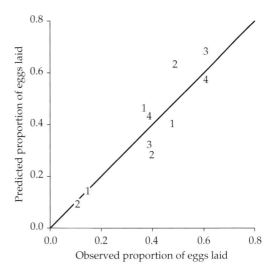

the data is good (Figure 6.29) and provides sufficient grounds for accepting that the model has captured the essential elements of the oviposition behavior of the cabbage white butterfly. Using the model, Jones asked how the two populations differed in the resulting egg distribution. Australian females disperse their eggs widely and in a less contagious distribution than the Canadian females. The adaptive reasons for this are not known but Jones and Ives (1979) speculated that it was a consequence of differences in temperature and local population densities. In the Australian (Canberra) population, caterpillar densities frequently reached levels where density-dependence effects were observed (see also Jones et al. 1987) and the weather was generally suitable for flight, thus permitting females to distribute their eggs more widely to reduce the density effects. In contrast, in the Canadian (Vancouver) population, densities were generally so low that density dependence was an unlikely occurrence, and the weather was such that females had a relatively short flight period. Therefore, in the Vancouver population, selection would favor short flights and a greater clumping of eggs. Similar conditions probably occur in the United Kingdom and in a later series of experiments Jones (1987) found that U.K. females showed behavioral characteristics and egg distribution patterns similar to those of the Vancouver population and significantly different from the Canberra population.

The Optimal Adjustment of Propagule Size

Although many animals can actively provide parental care by defense of their young, there are yet many more, such as plants and most invertebrates, for which such an option is not readily available. However, the future survival and success of offspring does depend upon the initial start given the young, which can be typically measured as the initial size of the propagule (see Chapter 4).

Models for the optimal egg size under different environmental conditions such as variation in ration and site quality were discussed in Chapter 4. These models apply equally well to plastic responses as well as to fixed responses in a constant environment.

EGG-SIZE PLASTICITY IN STATOR LIMBATUS: USE OF A DIRECT CUE. One of the best studied cases of phenotypic plasticity in response to site quality is the analysis of egg-size plasticity in the seed beetle *Stator limbatus* (Fox et al. 1997, 1999). *Stator limbatus* is a generalist seed parasite with a wide geographic distribution from northern South America to the southwestern United States. The larvae have been raised on over 70 species of plants, although at any given location only a few species may be available. Females oviposit directly onto exposed seeds, and the larvae burrow into the seed upon hatching. Larval development and pupation takes place entirely within the seed. Hence female choice of seed and maternal provisioning of the egg are critical components for survival and adequate growth. Larval survivorship is low (< 40–50%) on seeds from most populations of *Cercidium floridum,* and there is a positive correlation between survival and egg size (Fox and Mousseau 1996). In contrast, larvae survive very well (95–99%) on seeds of *Acacia greggii* and there is no correlation between egg size and survival. Given the high survival rate this finding is unsurprising (Fox and Mousseau, 1996). Thus females laying on *C. floridum* should lay large eggs, whereas females laying on *A. greggi* should maximize fecundity by laying small eggs. As both plants occur sympatrically, selection will favor the evolution, if possible, of maternally adjusted egg size. In the field, eggs collected from *C. floridum* are larger than those from *A. greggi* (Fox et al. 1995, 1997; Fox and Mousseau 1996), but this could reflect population division or female choice rather than adaptive phenotypic plasticity. In an ingenious set of experiments Fox et al. (1997) demonstrated that (1) females adjusted their egg size according to the host seed, (2) that larval survival varied according to host and egg size, and (3) that there was a trade-off between egg size and egg number.

Phenotypic plasticity was demonstrated first by placing females with one or both seed types following eclosion into the adult stage. When placed with only *A. greggii* seeds, females laid small eggs, whereas when placed only with *C. floridum* seeds, females laid large eggs (Figure 6.30). If the seeds of both species are presented simultaneously, females lay relatively large eggs that approximate the size of eggs laid if only *C. floridum* are presented, though eggs laid on *A. greggi* seeds are smaller than those laid on *C. floridum* seeds (Figure 6.30). A possible reason for this response is that, as shown by the following experiment, there is a time delay in switching egg size. A time lag in the response was shown by a transfer experiment in which females were presented with a single type of seed until they had laid at least one egg, at which time they were switched to the seeds of the alternate plant species. When transferred from *A. greggii* to *C. floridum,* females increased the size of egg laid over a three-day period (Figure 6.30). Females switched in the reverse direction showed a decline in egg size but

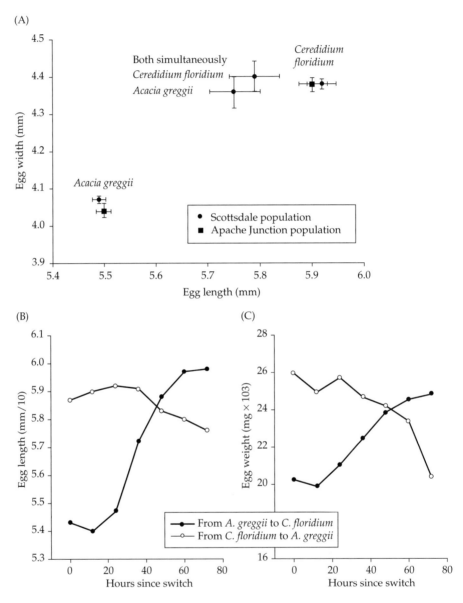

Figure 6.30   Phenotypic plasticity in egg size of *Stator limbatus*. (A) The egg size for females from two populations when presented either with seeds from a single species or from both species simultaneously. (B,C) The temporal change in egg size produced by females when shifted from one host to another after having laid at least one egg on the first host. Data from Fox et al. (1997) and unpublished data from C. Fox.

not to the egg size produced if females first encounter *A. greggii* (Figure 6.30B). However, a subsequent experiment did show a complete reversal in both cases (Figure 6.30C). When presented with both species, a female *S. limbatus* cannot simultaneously produce eggs of the appropriate sizes and, therefore, she must select a size that is optimal for both. Let survival as a function of egg size, X, on *A. greggii* be designated $S_A(X)$ and on *C. floridium* let it be designated $S_C(X)$. The fitness given p proportion of *A. greggii* seeds is

$$W(X) = \left(\frac{B}{X}\right)\left[pS_A(X) + (1-p)S_C(X)\right] \tag{6.52}$$

where B is a constant and B/X is the fecundity for an egg of size X. Both previous experiments and the present indicated that survival varied with host and egg size (Fox and Mousseau 1996; Fox et al. 1997). Survival was found to be linearly related to egg size over the range of egg sizes studied. A linear survival function cannot be appropriate over all egg sizes since this would lead to the optimal egg size being as small as possible. Suppose, for example, we let $S(X) = a + bX$, where a, b are constants. Then for a single host, $W(X) = (Ba/X) + Bb$, and fitness is maximized when egg size is zero! For the present models I selected functions that were biologically realistic and for which the optimal egg size on the two hosts was as observed (i.e., 5.5 on *A. greggii* and 5.9 on *C. floridium*). To find the optimal egg size on a single host we differentiate with respect to X and find the value of x at which the derivative is zero:

$$\frac{\partial W(X)}{\partial X} = \left(\frac{B}{X}\right)\frac{\partial S(X)}{\partial X} - \left(\frac{B}{X^2}\right)S(X) \tag{6.53}$$

where, for notational simplicity I have used the prime notation for the derivative: $\partial S(X)/\partial X \equiv S'(X)$. For both hosts, the same procedure is used,

$$\frac{\partial W(X)}{X} = \left(\frac{B}{X}\right)\left(pS_A'(X) + (1-p)S_C'(X)\right) - \left(\frac{B}{X^2}\right)\left(pS_A(X) + (1-p)S_C(X)\right)$$
$$= 0 \quad \text{when} \quad p\left[S_A(X) - XS_A'(X)\right] + (1-p)\left[S_C(X) - XS_C'(X)\right] = 0 \tag{6.54}$$

The three functions and their derivatives used here are given in the caption of Figure 6.31. For the first pair of survival functions, the optimal egg size declines rapidly as the proportion of *A. greggii* seeds increases. When the proportion is approximately 0.3, the optimal egg size is essentially the same as that if only *A. greggii* is present (Figure 6.31). In the second pair of survival functions (which differ with respect to the survival function for *C. floridium*) the optimal egg size remains more or less at that appropriate for *C. floridium* until the proportion of *A. greggii* seeds exceed about 0.8. The second case matches what was found when *S.*

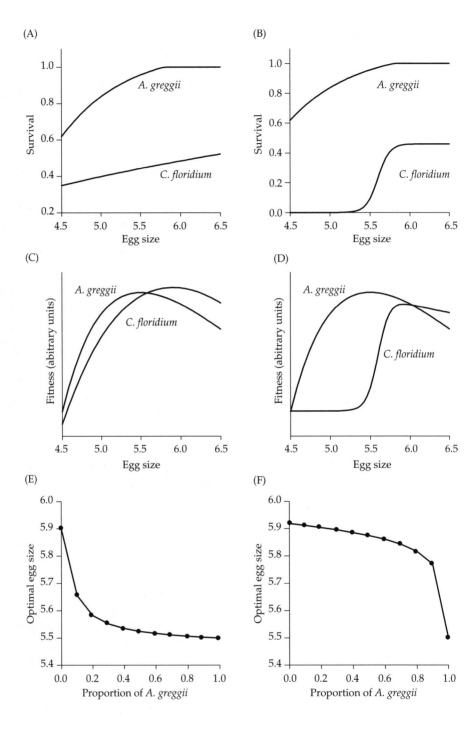

◀ **Figure 6.31**   (A,B) Two models demonstrating the optimal egg size when both host species of the seed beetle *Stator limbatus* are present. (C,D) The fitness curves for each individual species using the fitness function $W(X) \propto S(X)/X$, where the survival functions are as shown below. In both cases the optimum on *Acacia greggii* is 5.5 and on *Cercidium floridium* it is 5.9. (E,F) The optimal egg size when both seed types are present and the beetle cannot change rapidly enough to change egg size according to the seed type encountered.

Species	S(X)	S'(X)	Parameter values
A. greggii	$a(1-be^{-cx})$ 1 if $S(x) \geq 1$	bce^{-cx}	$a = 1.1$ $b = 100$ $c = 1.207$
C. floridium (left panels)	As above	As above	$a = 1.0$ $b = 1.3$ $c = 0.154$
C. floridium (right panels)	$\dfrac{a}{1+e^{-\frac{X-b}{c}}}$	$\dfrac{ae^{-\frac{X-b}{c}}}{c\left(1+e^{-\frac{X-b}{c}}\right)^2}$	$a = 0.46$ $b = 5.6$ $c = 0.0723$

limbatus was presented with both types of seed simultaneously. An interesting feature of this model is that the parameter *a* plays no role in the optimal egg size when there is only a single host but does contribute when there is a mixture of seed types.

The above experiments demonstrated that female *S. limbatus* are capable of varying their egg size in relation to their host species. These experiments do not discriminate between a single phenotypically plastic genotype and a distribution of genotypes within a population. For phenotypic plasticity to evolve, the heritability of egg size in the two environments (= hosts) must be greater than zero and there must be a genetic correlation of less than 1 between the environments. Fox et al. (1999) determined these parameters in two populations ("Apache" and "Del Rio") using a half-sib breeding design. The heritabilities on *A. greggii* estimated from the sire component were 0.91 and 0.44 for the Apache and Del Rio populations, respectively. A parent-offspring regression gave estimates of 0.58 ± 0.12 and 0.44 ± 0.10, respectively. On *C. floridium* the sire estimates were 0.66 and 0.22, respectively. All estimates were significantly greater than zero and indicate that egg size can readily change in response to selection. The genetic correlations between the two hosts were 0.71 ± 0.18 and 0.61 ± 0.29 for the Apache and Del Rio populations, respectively. The former estimate is significantly less than one but the latter is not. The two estimates are very similar, and the lack of difference in the second estimate is probably simply a consequence of the larger standard error. Therefore, the data indicate that there is abundant genetic variation for egg size per se and for phenotypic plasticity in egg size. An important question that remains unanswered is: "What maintains the genetic variation for egg size and plasticity?"

Another invertebrate that also appears to adjust its egg size in accord with the perceived conditions to be experienced by its offspring is *Daphnia*. Females of the two species *D. pulicaria* and *D. hyaline* produce large eggs when reared under abundant food but eggs of reduced size when given limited ration (Gliwicz and Guisande 1992). Clutch size declines with food ration, presumably as a consequence of the decreased availability of food and the increased allocation per egg. The survival time of offspring is positively correlated with their hatching size. Hence a plausible explanation for the phenotypic plasticity is that females are assessing the future conditions for their offspring and adjusting their egg size and clutch size accordingly.

EGG SIZE PLASTICITY IN *PARNARA GUTTATA*: USE OF AN INDIRECT CUE. *Stibor limbatus* probably uses chemical or physical cues from the seed to modulate the size of the egg it produces. Similarly, the two *Daphnia* species can assess food conditions directly. In other cases a female may use an indirect cue such as photoperiod, an example of this being egg size in the butterfly *Parnara guttata*. This species changes its habitat over the course of a year, the adults of the first two generations laying their eggs on grasses in wet lowland areas and the adults of the third, overwintering generation laying their eggs in dry upland areas. Host plants of the first two generations are *Oryza sativa* (rice), *Echinochloa Crus-galli* (barnyard grass), *Phragmites karka* (reed), and *Carex olivacea* (sedge), with by far the largest proportion of larvae being found on rice (91%; Nakasuji 1982). The third generation larvae feed primarily on *Imperata cylindrical* (cogon grass, 35%), *Miscanthus sinensis* (Eulalia, 37%), and *Festuca ovina* (Sheep's fescue, 31%; Nakasuji 1982). Cogon grass and Eulalia are found in abundance in both habitats but rice is absent from the dry upland areas and hence is unavailable to the overwintering (third-generation) offspring. The survival of larvae is strongly influenced by egg size and plant species (Figure 6.32). On rice, survival is high for both small and large eggs, but on cogon grass, larvae hatching from small eggs do not survive. Let us take as a measure of fitness the ratio of larval survival/egg size measured as a half ellipsoid = $(0.25)(.5)(4/3)(\pi)$(egg length)(egg diameter^2). Fitness is maximized by laying small eggs on rice (Figure 6.32). Thus selection should favor the laying of small eggs on rice but large eggs on cogon grass. The latter host should be selected only if rice is not available. These predictions match what *P. guttata* females actually do (Nakasuji and Kimura 1984). The presence and absence of rice is precisely determined by the generation and hence, if there exists a cue that "informs" the female in which generation she has emerged, then she can use this cue during either her larval development or during the development of her eggs to set egg size. One such cue is photoperiod, and Nakasuji and Kimura (1984) determined that egg size is a function of the photoperiod under which the

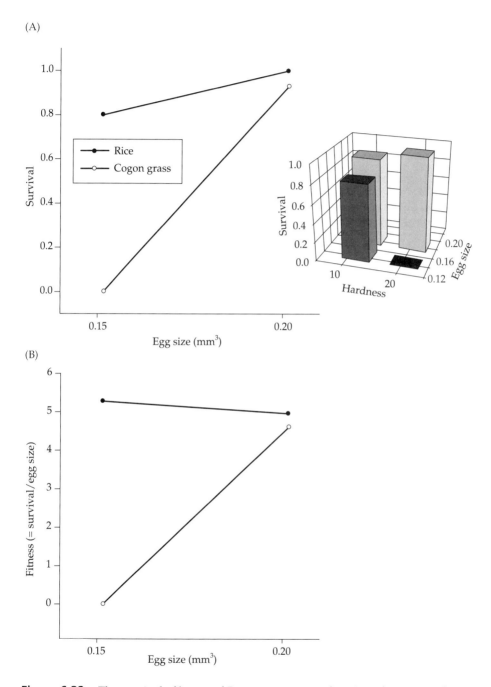

Figure 6.32 The survival of larvae of *Parnara guttata* as a function of egg size volume and host plant. (A) The inset shows a 3D bar graph of survival as a function of egg size and leaf toughness. (B) Fitness as a function of egg size and host plant. Data from Naka-suji and Kimura (1984).

larvae were raised, though the effect was no great enough to produce the range in naturally occurring eggs.

The reason for the low survival of *P. guttata* larvae on cogon grass is likely to be a consequence of the different toughness of the leaf cuticle. Cogon grass is twice as tough as rice; toughness is measured by the weight required to push an insect pin through the leaf. For rice this weight is 11.4 ± 0.65 g and for cogon grass it is 20.2 ± 0.71 (Nakasuji 1987). If this hypothesis is correct, and given the major effect on survival of toughness, we would expect, all other things being equal, that among closely related species there should be a correlation between host plant toughness and egg size. There is a strong correlation within the Hesperidae of Japan between egg size and leaf toughness (Figure 6.33). A predicted consequence of increased egg size is a decreased fecundity, which will favor an increased size at maturity (see Chapter 4). As predicted, adult female size among species is positively correlated with egg size (Figure 6.33).

Like *P. guttata*, many organisms show seasonal variation in propagule size, though the trend may be either upward or downward. Propagule size typically declines in plants, fish, and birds but increases in invertebrates and some reptiles (Figure 6.34). This response might represent a reaction norm that is programmed to environmental factors, indicating season length as in diapause, or be essentially independent of season length and be a reaction to some environmental factor that is correlated with seasonal events. Given the diversity of responses, it is unlikely that a single model will be sufficient to account for all cases.

MORE EXAMPLES OF THE USE OF INDIRECT CUES: SEASONAL VARIATION IN EGG SIZE.
In many species, propagule size increases with ration (Figure 6.34). Thus one pos-

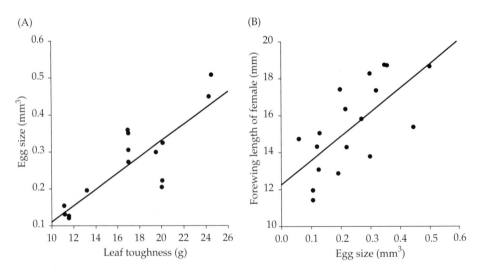

Figure 6.33   The relationship between egg size and leaf toughness (A), and between adult size and egg size (B) among species of Hesperidae. Redrawn from Nakasuji (1987).

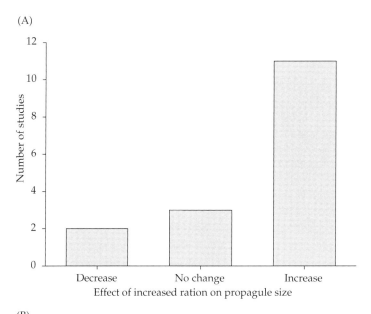

(A)

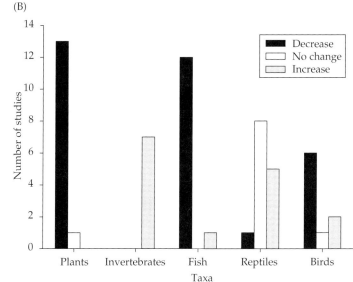

(B)

Figure 6.34   A summary of studies examining variation in propagule size in relation to ration size (A) and time of year (B). Data taken from Roff (1992a, Tables 10.8, 10.10).

sible explanation for seasonal variation is that resources show seasonal variation; of course, this would imply that in some cases resources increase with the season and in other cases they decrease. In fact, one would generally expect resources to show some sort of peak in the middle of the season rather than at either end.

Based on the observation that female isopods (*Armadillidium vulgare*) produce larger eggs when fed a reduced ration, Brody and Lawler (1984) suggested that the observed increase in offspring size between spring and summer is adaptive because it increases the survival chances of the offspring in times of food scarcity. A different explanation probably applies to the increase in size of the fall eggs of the moth *Orgyia thyellina* relative to those laid by the spring generation (Kimura and Masaki 1977). Spring eggs develop directly, whereas the fall eggs pass through a long diapause. The increased size probably reflects increased provisioning. Nussbaum (1981) advanced the hypothesis that seasonal variation is a bet-hedging strategy, conditions at the end of the season being more variable than at the beginning. Ware (1975b) advanced the hypothesis that evolutionary changes in the size of fish eggs are a consequence of variation in incubation time. This hypothesis postulates that, as temperatures fall, incubation time increases but mortality rates do not, and hence size-selective mortality rates increase on eggs. Selection therefore favors a larger egg size but a lower number per female. Such a mechanism predicts a decrease in egg size as the season progresses and temperatures increase, as observed in many fish species (Bagenal 1971; Figure 6.34).

An alternate hypothesis to explain seasonal variation in size of fish eggs is that the size spectrum of food available for the newly hatched larvae shifts over the season, and hence selection favors females that alter their egg size such that their larvae can take advantage of the change in the mean size of their food. The mean egg diameter of the Atlantic mackerel *Scomber scombrus* declines during the summer and is correlated with both the mean sea surface temperature and the mean particle size in the plankton (Ware 1977). Although this is circumstantial evidence for the food-size hypothesis, it remains to be shown that the size of the larvae is optimal in relation to the size of prey. Since adult mackerel are piscivorous, the proximal cue used by the female mackerel to adjust her egg size is unlikely to be the size of the plankton, but more likely the water temperature, which correlates with particle size (Ware 1977). The incubation-time and food-size hypotheses are not mutually exclusive but could operate in conjunction, increasing the selective pressure on egg size.

The hypothesis that seasonal changes in propagule size take advantage of the changing spectrum of prey size can be inverted to postulate that changing propagule size of prey species is a response to the changing size spectrum of predator species. Kerfoot (1974) used this argument to account for shifts in egg size of the cladoceran *Bosmina longirostris*. In this species, females produce small eggs in the summer, then switch to large eggs in late fall. Winter generations produce large offspring that mature at larger sizes, then shift to producing small eggs in the spring. Although during the summer *Bosmina* suffer heavy predation from fish, these same predators reduce competition from larger invertebrate planktivores. In the fall, after the warm-water fishes cease feeding, *Bosmina* is subjected to predation from two invertebrates, *Cyclops* and *Chaoborus*. Predation by these two invertebrates is known to favor large size in *Bosmina* (reviewed in Kerfoot 1974). The changing pattern of heavy visual predation by fishes in the summer to predation by grasping invertebrates in the winter favors a changing pattern of size in *Bosmina*, from small to large phenotypes. This hypothesis has received theoretical support from an analysis by Lynch (1980b), but there are insufficient empirical data on the benefits and costs of different sized eggs to adequately test it. In the present context the changing pattern reflects changes in response to selection rather than phenotypic plasticity, though the hypothesis could be recast in this framework.

Responses to Increased Mortality Regimes

THE EVOLUTION OF INDUCIBLE DEFENSES: INTRODUCTION. Selection will obviously favor the evolution of traits than increase the survival, $l(x)$, function. Countering such evolution is the cost of such traits, manifested in other fitness components such as fecundity, $m(x)$, or delayed maturation, α. If there is a cost to a defensive trait, we might expect it to be phenotypically plastic, appearing when some environmental stimulus indicates a need. Many traits do indeed show such plasticity, this plasticity consisting of either a primarily dimorphic, irreversible condition or a reversible, largely quantitative response. In the first group are many animal examples, such as helmets and neckteeth (cladocera), spines (rotifers, protozoans), and cryptic morphs (lepidoptera). Plants seem to fall mainly in the second group, though there are also animal examples (Table 6.8). The induction of the defensive structures may require only a chemical cue released by the potential predator or be a response to a physical assault. For example consumption of leaves by insect herbivores induces the production of defensive compounds in many plant species. Examples are presented in Table 6.8 for cases in which both the mode of induction and costs of the defenses have been measured. The overall finding is clearly in support of the hypothesis that inducible defenses are costly and reduce fitness in the absence of the predator. In general, defended morphs show a reduced rate of growth, a reduced fecundity, and in some cases an elevated mortality rate in the absence of the predator.

Riessen (1992) constructed a demographic model (Table 6.9) of the two morphs of *Daphnia pulex* to examine the effects of varying food ration and predator density on the relative fitness of the two morphs. In the presence of chemicals released by the predator *Chaoborus*, the first three instars of *Daphnia pulex* produce spines or "neckteeth" on their "necks," which serve to hinder the handling ability of *Chaoborus* and hence increase the probability of the *Daphnia* surviving and escaping the attack (Tollrian 1995). The principal cost of this defensive structure is a delay in maturity (Riessen and Sprules 1990). Riessen examined the effect of changing the amount of food by varying clutch size. With decreasing ration the predator density at which the protected morph gained an advantage decreased. The *Chaoborus* density at which the protected morph gained a fitness advantage over the unprotected morph increased with food ration and reduction in the overlap between predator and prey. These results show that if *D. pulex* can detect the density of *Chaoborus*, then selection will favor a reaction norm for the development of neckteeth, as observed.

Table 6.8 **Examples of inducible defenses in various species (mainly animals) and their costs.**

Taxon	Defensive morph induced	Inducing factor produced by	Costs[a] α , $l(x)$, $m(x)$	Ref
Primarily Dimorphic Variation (Unprotected and Protected Morph)				
Daphnia spp, Cladoceran	Helmeted	Invert. and vert. Predators[b]	+, −, −	1
Daphnia pulex, Cladoceran	Necktoothed	Invert. predators	+, −, nd	1
Brachionous calyciflorus,	Spined	Invert. predators	nd, −, −	1
Keratella spp, Rotifers	a) Cruciform	a) α-tocopherol released by algal cells	nd, nd, −	1
Asplanchna spp, Rotifer	b) Giant	b) Dietary α-tocopherol + large prey		
Protozoa, *Onychodromus quadricornutus, Euplotes* spp,	Spined	Giant morph, predatory ciliates	Reduced growth	1, 2
Membranipora membranacea, Bryozoan	Spined	Grazing by nudibranch	Reduced colony growth	1
Chthmalus anisopoma, Gastropod	Bent	Predatious gastropod	+, nd, −	1
Thais lamellosa, Gastropod	Larger apertural teeth	Predatory crab	+, nd, nd	3
Littorina obtusata, Gastropod	Thicker shell	Predatory crab	Possibly reduced growth	4
Corals and sea anemones	Catch tentacles	Proximity of competitors	Reduced number of feeding tentacles	1
Papilionidae, swallowtail butterflies	Pupal color	Photoperiod, substrate color, foodplant odor	Increased mortality on wrong background	1
Acyrthosiphon pisum, aphid	Macroptery	Predatory beetle	Reduced fecundity, increased development time	5
Nemoria arizonaria, caterpillars	Twig mimic	Tannin concentration	+, −, −[c]	6
Hyla chrysoscelis, tadpoles	Inactive, larger more brightly colored tailfin	Dragonfly larvae[d]	Lowered survival in absence of predator	7

Table 6.8 (continued)

Induced Defenses that are Primarily Quantitative and Reversable

Species	Defense	Inducing agent	Effect[a]	Ref
Carassius carassius, Carp	Deep bodied	Predatory fish (Pike)	+, nd, nd[e]	8
Harmonia axyridis, ladybird beetle	Reflex bleeding	Predators	+, nd, nd	9
Gossypium thurberi, plant	Resistance	Natural damage by leaf miners early in season	None (survival, growth, reproduction)	10
Hordeum vulgare, plant	Resistance(?)	Avirulent strain of powdery mildew[f]	Grain yield, kernel weight, grain protein	10
Lycopersicon esculentum, plant	Resistance	Chitin injection[g] (simulating caterpillar attack)	None (survival, growth, reproduction)	10
Nicotiana sylvestris, plant	Resistance	Artificial damage (simulating caterpillar attack)	Total plant mass, number of fruits	10
Brassica rapa, plant	Cyanide compounds	Moth larvae and fungal pathogen	Decreased seed production	11[f]
Pastinaca sativa, plant	Furano-coumarins	Attack by herbivores or pathogens grazing by	Decreased fruit production[g]	12
Phytoplankton spp	Colony form, spines	small zooplankters	Higher sinking rate	13

[a]Columns indicate: Development time (in some cases growth was slowed but time to maturity not presented), Mortality in absence of predator, Fecundity. +; increased. –; decreased. 0; no effect. nd; no data.

[b]Induction is initiated by the presence of chemicals released by the predator.

[c]The dimorphism is seasonal with catkin mimics being produced early in the season and twig mimics later in the season, when catkins are no longer present.

[d]Dragonfly larvae kept in cage inside tadpole container.

[e]Effect found under food stress.

[d]Review data from Tables 4.1, 5.5.

[e]Induction in laboratory.

[f]Costs assessed from lines genetically altered to increase production of defensive compounds.

[g]Negative genetic correlation between production of compound and fruit set.

References 1) Roff (1996a); 2) Kusch and Kuhlman (1994), Kusch (1995), Wiackowski and Szkarlat (1996), Wiklow (1997); 3) Appleton and Palmer (1988); 4) Trussell (1996); 5) Weisser et al. (1999); 6) Greene (1989); 7) McCollum and van Buskirk (1996); 8) Bronmark and Miner (1992), Bronmark and Pettersson (1994), Nilsson et al. (1995), Pettersson and Bronmark (1997); 9) Grill and Moore (1998); 10) Karban and Baldwin (1997); 11) Siemens and Mitchell-Olds, 1998 ; 12) Berenbaum and Zangerl (1999); 13) Van Donk et al. (1999).

Table 6.9 Description of the demographic model analyzed by Riessen (1992) comparing the population growth rates of normal and spined *Daphnia pulex* in the presence of a predator, *Chaoborus*.

Trait	Symbol/function	Values for morph	
		Unprotected	Spined (protected)
Clutch size	—	Same for both morphs	—
Age at maturity (days)	—	4.75	5.35
Encounter probability	$P(x) = E(x)e^{-E(x)}$	—	—
Encounter rate for body size x and *Chaoborus* density d	$E(x) = (0.55x - 0.35)d/3600$	For both morphs	—
Probability of predator striking	—	0.5	0.5
Probability of capture given a strike	$S(x) = -0.154x + 0.375)$	Same for all instars	Instars 1–3[a] 0.25–0.5$S(x)$, thereafter as for unprotected
Survival	$= 1 - 0.5P(x)S(x)D = 1 - 0.5P(x)(-0.154 + 0.375)D$ where $D = 1, 0.25 - 0.50$		
Body size for each instar	Instars 1–4 $x = 0.455 + 0.189y$, where $y =$ instar number Instars 5–19 (adult) $x = 0.899 + 0.096y - 0.00169y^2$		

[a]Neckteeth are produced only by the first three instars.

A number of theoretical analyses of the evolution of **inducible defenses** have been undertaken, varying from models that focus on dimorphic variation in spatially heterogeneous environments (Lloyd 1984; Lively 1986b), dimorphic variation in which there is a time lag between stimulus and morph realization (Padilla and Adolph 1996), to models in which there is induction of a defensive compound or structure that varies in a continuous fashion (Clark and Harvell 1992; Adler and Karban 1994). To my knowledge there has been no analysis that has attempted to match a theoretical model of induction of defense with an empirical study. The general findings of all of these models can be demonstrated with either very simple models or from logical considerations alone.

The basic model for an inducible defense is based on the assumption that survival probability can be assessed by some environmental stimulus. This survival probability includes time lags due to development of the morph following perception of the stimulus; time lags have been explicitly examined by Padilla and Adolph (1996) but do not add anything conceptually to the following analyses. Now, suppose that there are two morphs—an unprotected morph and a protected morph—and that the strength of the stimulus indicates the probability of

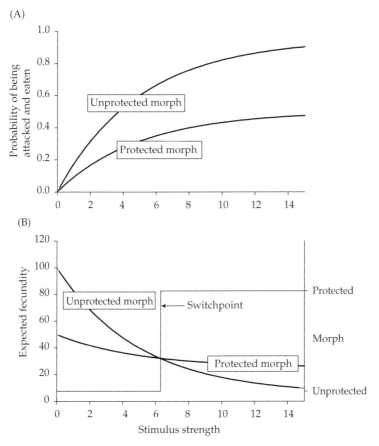

Figure 6.35   A hypothetical example of an inducible trait. Some cue (= Stimulus strength) is present in the environment that indicates the probability of being attacked and eaten (A). This probability is greater for the unprotected morph, but its higher fecundity gives it a higher expected fitness (= survival × fecundity) for low levels of the stimulus (B). However, at some value, the expected fecundity of the protected morph exceeds that of the unprotected morph and the organism should switch to the protected morph.

being attacked and eaten. As shown in Figure 6.35, the probability for the protected morph will lie below that of the unprotected morph. As shown by many empirical studies, the protected morph suffers a cost in another fitness component, say, fecundity. Taking lifetime reproductive success as our fitness measure (for simplicity only), then at low-stimulus strength the fitness of the unprotected morph will be higher than that of the protected morph, but eventually the curves cross and the fitness ranks are reversed (Figure 6.35). The immediate prediction from this model is that the optimal reaction norm is a step function (Figure 6.35; Frank [1993] also found a sharp transition between induced and noninduced states in a host parasite model). If there is genetic variation for the threshold of induction, we have the standard threshold model of quantitative genetics and a population reaction norm that is a cumulative normal (Myers and Hutchings 1986; Hazel et al. 1990).

THE EVOLUTION OF INDUCIBLE DEFENSES: SPATIAL VARIATION. Consider now the situation in which there is spatial variation consisting of two patches in which an animal lives throughout its life but random migration between generations. Let the frequency of patches, the fitness in each patch, and the probability of developing into the protected morph be as listed in Table 6.10. To avoid trivial solutions I shall assume that, in the patch with few predators, the most fit type is the unprotected morph and, in the patch with many predators, the most fit type is the protected morph. There are three possible life histories: monomorphically unprotected, monomorphically protected, and dimorphic according to the given probabilities. The fitness associated with each type are

Unprotected: $\qquad W_U = f + (1-f)w_1$

Protected: $\qquad W_P = fw_2 + (1-f)w_3$ $\hspace{3cm}$ (6.55)

Dimorphic: $\qquad W_D = f\left(1 - P_F + P_F w_2\right) + (1-f)\left([1 - P_M]w_1 + P_M w_3\right)$

The first question we can ask is: "Will a general bet-hedging response—that is, a response that is independent of any environmental cue that is a function of the prospective fitness—have a higher fitness than both monomorphic types?" Mathematically this implies $P_F = P_M = P$. To find the value of P that maximizes the dimorphic fitness, we set $\partial W_D / \partial P = 0$. However, $\partial W_D / \partial P = -1 + w_2 + (1-f)(-w_1 + w_3)$, which is a constant, and, therefore, fitness is maximized either at $P = 0$ or $P = 1$, depending on parameter values. Thus spatial heterogeneity will by itself never select for a bet-hedging response (Lloyd 1984; Lively 1986b; Moran 1992).

Table 6.10 **Parameter definitions for the evolution of an inducible defense in a two-patch universe.**

	Patch type	
	Few predators	*Many predators*
Frequency of patches	f	$1-f$
Fitness of the unprotected morph	1	w_1
Fitness of the protected morph	w_2	w_3
Probability of developing into the protected morph	P_F	P_M

The maximum fitness is 1, achieved by the unprotected morph in the environment with few predators.

Given that there is an environmental cue that predicts fitness, will a reaction norm evolve? We can demonstrate that this is possible under some conditions as follows: First it is obvious that if the organism has "complete knowledge," then in the few-predators patch it should develop into the unprotected morph, meaning $P_F = 0$ and in the many-predators patch it should develop into the protected morph, meaning $P_M = 1$. If the relationship between the cue and prospective fitness is very slightly less than perfect, the dimorphic morph will still, in general, have a higher fitness than either monomorphic type. At some point the relationship between the cue and fitness will be so poor that the fitness of the dimorphic morph will drop below that of one of the monomorphic types. To find the conditions under which a dimorphic reaction norm is favored we simply have to compare the pairwise fitnesses. After a little algebra

$$W_U < W_P \quad \text{if} \quad f < \frac{\left(w_3 - w_1\right)}{\left(1 - w_2\right) + \left(w_3 - w_1\right)}$$

$$W_U < W_D \quad \text{if} \quad f < \frac{P_M\left(w_3 - w_1\right)}{P_F\left(1 - w_2\right) + P_M\left(w_3 - w_1\right)} \qquad (6.56)$$

$$W_P < W_D \quad \text{if} \quad f < \frac{\left(P_M - 1\right)\left(w_3 - w_1\right)}{\left(P_F - 1\right)\left(1 - w_2\right) + \left(P_M - 1\right)\left(w_3 - w_1\right)}$$

The superiority of the reaction norm response depends on the frequency of the two patches relative to the difference in fitness between the two morphs within patches and the probability of developing into the protected morph. An important point to note is that a reaction norm is not guaranteed to have a higher fitness than a monomorphic type; it all depends on the accuracy with which the cue predicts the prospective fitness. Another important point is that the most-fit morph is not determined by the patch with the highest frequency Thus it can happen that the most-fit morph is actually determined by a highly infrequent patch that has a high fitness cost for making the wrong decision. Therefore, it could appear to the casual observer who fails to record these infrequent patches that the organism is maladapted to its environment. To examine how frequent the three life history responses (monomorphic protected, monomorphic unprotected, reaction norm) are the most fit, I ran the following simulation. Within-patch fitness values were generated at random subject to the constraint (Table 6.10) $0 < w_1 < w_3 < w_2 < 1$. The probability of adopting the protected morph in the patch with the most predators, P_M, (which is a measure of the reliability of the signal) was also selected at random from the interval 0–1 and P_F selected at random from the interval $0–P_M$. For each combination of parameters and given patch frequency, f, the most-fit life history response was determined. One thousand replicates were generated for each value of f (Figure 3.36). When most patches contain many predators (approximately $0 < f < 0.1$), a reaction norm is favored most often, but when the frequency of patches with few predators

(A) Spatial variation

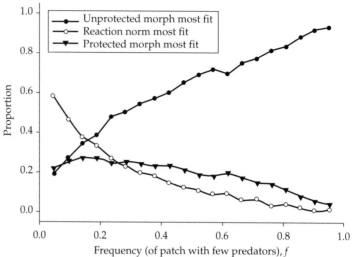

(B) Temporal variation

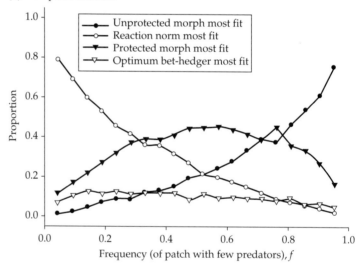

Figure 6.36 The proportion of random parameter combinations for which particular life history responses are the most fit in a two-patch, spatially variable (A) or temporally variable (B) environment. Fitnesses are as defined in Table 6.10. See text for details of the calculation of the fitnesses of the three (spatial) or four (temporal) types of responses.

exceeds approximately 0.2, the unprotected morph most frequently has the highest fitness.

THE EVOLUTION OF INDUCIBLE DEFENSES: TEMPORAL VARIATION. It has already been demonstrated for a particular case that in a temporally variable environment a bet-hedging response can be favored (see Chapter 5). Here I show that this is a general result and present the result for the reaction norm model. The three fitnesses are

$$
\begin{aligned}
\text{Unprotected:} & \quad \ln W_U = (1-f)w_1 \\
\text{Protected:} & \quad \ln W_P = f \ln w_2 + (1-f)\ln w_3 \\
\text{Dimorphic:} & \quad \ln W_D = f \ln(1-P_F + P_F w_2) + (1-f)\ln\big([1-P_M]w_1 + P_M w_3\big)
\end{aligned}
\tag{6.57}
$$

Fitness is now a function of the logarithms rather than simple sums. As before, we can address the question of the fitness of a general bet-hedging response by setting $P_F = P_M = P$ and taking the derivative,

$$
\begin{aligned}
\frac{\partial \ln W_D}{\partial P} &= \frac{f(w_2 - 1)}{(1-P)+Pw_2} + \frac{(1-f)(w_3 - w_1)}{(1-P)w_1 + Pw_3} \\
&= 0 \quad \text{when } P = \frac{1-f}{1-w_2} - \frac{w_1 f}{w_3 - w_1}
\end{aligned}
\tag{6.58}
$$

Now there is a maximum and hence an optimal value of P, though this does not guarantee that the associated fitness exceeds the fitness of one or both of the monomorphic types. The important point is that it is possible for temporal variation to favor bet-hedging (i.e., a dimorphism). To examine how often each of the four possible life histories is favored, I generated parameter combinations as in the case of spatial variation. For the bet-hedging life history, I calculated the optimum P for each combination. As previously, when the frequency of patches with few predators was low (approx. $0 < f < 0.2$), the reaction norm life history was most frequently favored (Figure 6.36). At intermediate frequencies (approx. $0.3 < f < 0.7$), the protected morph had the highest fitness most often, and at high frequencies ($f > 0.8$), the unprotected morph dominated most often. Optimal bet-hedging had the highest fitness in approximately 10% of the combinations, regardless of patch frequency, but never had the highest overall probability of being the most fit. These results demonstrate (1) that for particular parameter values, any of the four life histories might be the most fit (when all possible life histories were observed) and (2) that a reaction norm is much more likely that an optimal bet-hedging response.

INDUCED CHANGES IN LIFE HISTORY TRAITS. In the previous section I examined the evolution of defensive reaction norms in response to an actual or perceived mortality threat. Here I consider changes in reproductive effort, age at maturity, and so forth. An actual increase in mortality rate could come about because of increased predation, increased parasitism, increased disease, or other factors. The organism might respond only following predacious attacks, parasitism or infection. On the other hand, cues that indicate that environmental conditions have changed in a manner that indicate that predation rates, parasitism, or infection rates are likely to increase may be sufficient to initiate life history responses. For example, exposure to water in which predators have resided causes *Daphnia hyaline* to mature earlier, at a smaller size, and to increase its reproductive effort (Stibor 1992; Stibor and Luning 1994). Other examples are given in Table 6.11. These examples fall into two broad classes: first, changes in life history traits, such as age and size at maturity, that result from actual or perceived exposure to a predator and, second, behavioral responses to the perceived presence that result in changes in, for example, growth rate, which then affect other life history traits.

Werner and Anholt (1993) examined models predicting the optimal foraging activity in the presence of predators, but there have been no quantitative tests of these models. Whether particular responses to the presence of predators or parasites are optimal depends upon the details of the model (Forbes 1993; Figure 6.37). If the trade-off between current and future reproduction is concave down, there will be a particular combination that will maximize fitness, defined graphically as that combination at which a fitness isocline is tangent to the trade-off curve (Figure 6.37B; see also Chapter 3). Changes in the mortality regime will shift the trade-off curve leftwards. If the trade-off curve remains concave, the optimum allocation to present reproduction may stay the same (6.37B), increase (6.37C) or decrease (6.37D). If the trade-off function changes its shape to being convex (concave up), then the optimum will be to either allocate all to present reproduction (6.37E) or cease present reproduction (6.37F). Thus it is possible to "explain" any response to increased mortality as "adaptive." Such explanations are without merit unless the trade-off functions can be specified.

To illustrate the construction of the optimal norm of reaction, I shall use the life history response of the snail *Biomphalaria glabrata* to actual and perceived infection by the parasite *Schistosoma mansoni*. Infection by the parasite severely reduces fecundity and survival of the snail (Minchella and Loverde 1981; Gerard and Theron 1997; Figure 6.38). Snails that were exposed to the parasite but did not become infected showed a surge in early reproduction, though this surge reduced lifespan and expected lifetime fecundity (Figure 6.38). A plausible hypothesis for this response is that a female that responds to the parasite cue has a higher lifetime fecundity *if parasitized* than a female that does not respond. The lower lifetime fecundity of responding females when they do not become infected is then a cost to the response that is offset by the benefits when they are infect-

Table 6.11   Some examples of presumed adaptive reaction norms involving
changes in life history traits in response to actual or perceived
increased mortality.

Species	Source of mortality cue	Reaction norm[a]	Reference
Daphnia hyalina	Predator-treated water	Decreased age and size at maturity Increased RE	Stibor (1992); Stibor and Luning, 1994
Biomphalaria glabrata (snail)	Exposure to parasite	Increased early $m(x)$ Decreased late $m(x)$	Minchella and Loverde (1981); Gerard and Theron, 1997
Cerithidea californica (snail)	Exposure to parasite	Decreased age at maturity	Lafferty (1993)[a]
Ambystoma spp (salamanders)	Predator-treated water	Delayed hatching, larger size at hatching	Sih and Moore (1993)
Clethrionomys glareolus (vole)	Predator odor	Breeding suppressed	Ylonen and Ronkainen (1994); Mappes and Ylonen (1997)
Littorina sitkana (snail)	Predator presence and feeding but not direct attack	Habitat shift, reduced growth	Yamada et al. (1998)
Lepomis macrochirus (fish)	Presence of predator	Habitat shift, reduced growth	Werner et al. (1983)
Gerris remigis (water strider)	Presence of predator	Reduced mating	Sih et al. (1990)
Notonecta hoffmanni (backswimmer)	Presence of adults (= predators)	Habitat shift, reduced growth	Sih, (1980 1982)
Poecilia reticulata (guppy)	Presence of predator	Change in mating behavior	Endler (1987)
Drosophila nigrospiracula	Parasitized[b]	Increased mating speed	Polak and Starmer (1998)

[a]As discussed by Lafferty (1993), it is not clear if this is a case of phenotypic plasticity. Field collected data showed a negative correlation between maturation size and parasitism after correction for environmental variables. A reciprocal transplant experiment between areas with high and low parasitism demonstrated differences between the populations that could be due to early experience of the snails or genetic differences.

[b]Males were infected, not simply exposed to a cue

ed. To locate the parameter space in which a responder has a higher fitness, we must consider the four combinations resulting from the expected lifetime fecundities following exposure to the parasite. Using the data of Minchella and Loverde as a guide, I set the expected lifetime fecundities for the four combina-

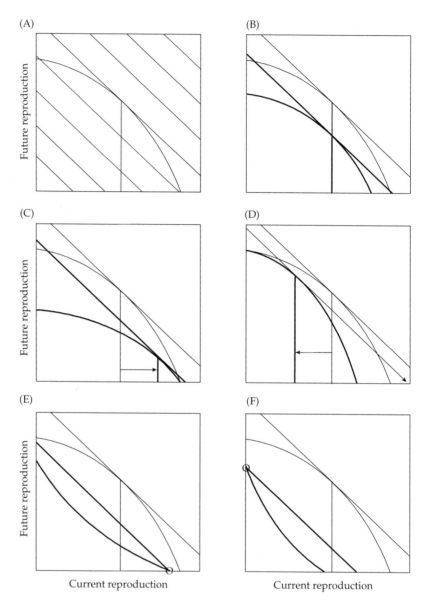

Figure 6.37   Optimal responses to changes in the trade-off between current and future reproduction. (A) The initial conditions, a concave down trade-off curve. Onto this surface is drawn a series of fitness isoclines, values increasing with increasing values of current or future reproduction. The optimal combination of reproductive effort is that point at which a fitness isocline is tangent to the trade-off curve. (B–F) Various curves that could result from an increased source of mortality. Where the optimum combination changes, the new optimum is indicated by an arrow or open circle. Modified from Forbes (1993).

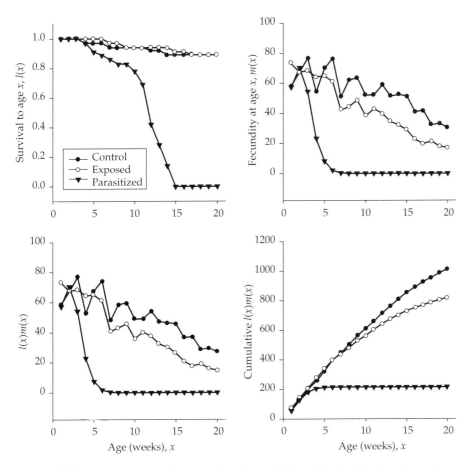

Figure 6.38   The $l(x)$ and $m(x)$ curves for the snail *Biomphalaria glabrata* when not exposed to the parasite *Schistosoma mansoni* ("Control"), when exposed but not infected ("Exposed"), and when infected ("Parasitized"). Note that exposed snails show an increased initial fecundity but a reduced lifetime fecundity. Data from Minchella and Loverde (1981).

tions as shown in Table 6.12 If the probability of becoming infected when exposed is P, the expected lifetime fecundities of the two types is

$$
\begin{aligned}
\text{Nonresponders} \quad & W_N = P(214)c + (1-P)1014 \\
\text{Responders} \quad & W_R = P(214) + (1-P)819
\end{aligned}
\tag{6.59}
$$

Table 6.12 **Expected lifetime fecundities for a hypothetical organism that has the capability of responding to a parasite cue by adjusting its *l(x)m(x)* function.**

Life history type	R_0 if infected	R_0 if exposed but not infected
Nonresponder	$214c^a$	1014
Responder	214	819

[a] *c* is a constant between 0 and 1.

The values are based on the data of Minchella and Loverde (1981) for the response of the snail, *Biomphalaria glabrata* to parasitism by the trematode parasite *Schistosoma mansoni* (see Figure 6.38).

The difference between the two fitnesses is

$$W_N - W_R = 214P(c-1) + 195(1-P) = 214Pc - 195P - 19 \qquad (6.60)$$

When Equation (6.60) is positive, the nonresponder has the higher fitness, whereas if it is negative, the responder is more fit. For the particular set of parameter values (Table 6.12) the proportion of parameter space over which the fitness of a responder exceeds that of a nonresponder is greatest when the probability of infection is high ($P > 0.5$) and the ratio of expected lifetime fecundities is low ($c < 0.4$; Figure 6.39). The surface shown in Figure 6.39 defines the optimal norm of reaction surface for environments characterized by P and c. The optimal reaction is a threshold response, either showing a response or not. In the present model, a bet-hedging response is never favored, although it might appear so since some individuals respond and suffer a "cost" because they are not infected.

The above model assumed just two options, respond or not respond, but, as with inducible defenses, in some cases the optimal response may be continuous rather than discrete. So long as the changed mortality regime applies on the same scale as the generation length, then we can compute the optimal suite of life history traits separately for each environment and from this generate the optimal norm of reaction (Stearns and Koella 1986). If there is spatial variation, then the optimal reaction norm will depend upon the amount of migration among patches (Houston and McNamara 1992; Kawecki and Stearns 1993, see Chapter 2, "Defining fitness when selection is hard and the environment spatially stochastic") and it will depend on the timing of migration relative to the cue that can be used to predict the future environment (de Jong 1999, see the section "The evolution of phenotypic plasticity"). The empirical examination of reaction norms in life history traits in conjunction with quantitative tests of their adaptive value is sadly lacking. This is an area in which there remains great scope for further research.

(A)

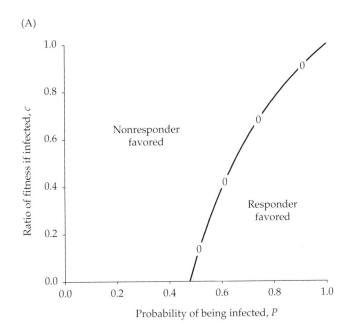

(B)

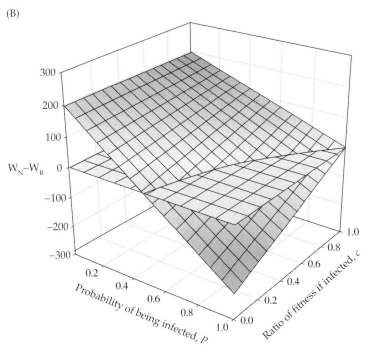

Figure 6.39 (A) A comparison of the fitnesses of responders and nonresponders for variation in the probability of infection, and the ratio of lifetime fecundities if infected (using values derived for the snail *Biomphalaria glabrata*). Parameter values are given in Table 6.11 and the relevant equations in the text. (B) The horizontal surface denotes the zero isocline, above which nonresponders are favored and below which responders are favored.

Summary

Phenotypic plasticity can be viewed and analyzed from two viewpoints: the character state approach, which makes use of the concept of a genetic correlation between environments, and the reaction-norm approach, which considers the functional relationship between the trait and the environment. In general, both approaches are mathematically equivalent, but their slightly different perspectives make both frameworks useful for analysis, when possible. The estimation of genetic parameters for the reaction norm method is a straightforward extension of the analysis of several traits, as discussed in Chapter 2, where in this case the "traits" are the coefficients of the reaction norm. The parameters for the character state approach can be estimated using a mixed model ANOVA.

Seasonal environments can have significant effects on the evolution of phenology, modifying, for example, the age at maturity and the timing of dormancy. The timing of these events is frequently cued by photoperiod, and there is evidence that the relationship between life history traits, such as development time, and photoperiod is adaptive. However, to my knowledge there has been no demonstration that a particular reaction norm is optimal.

Parental care can be provided in terms of the initial investment into the egg or seed or also at a later stage in which the parents provision and defend the growing young. The optimal allocation depends upon the environmental state and hence is expected to evolve via a reaction norm. Variation in desertion rates in bears, several taxa of birds, and fish supports this hypothesis, but there have been no experimental evaluations demonstrating that particular patterns of desertion are the most fit. Similar conclusions can be reached for phenotypic plasticity in propagule and clutch size. There is good theoretical reason to expect the evolution of reaction norms and experimental evidence that they occur, but there is inadequate testing of the hypothesis that the observed reaction norms are optimal.

Mortality regimes are variable and can lead to the evolution of a variety of responses, from the induction of defensive structures to changes in the $m(x)$ function. Spatial heterogeneity can favor the evolution of a reaction norm (i.e., inducible defense) but not to bet-hedging. In contrast, temporal variation can favor either the evolution of an inducible defense or to bet-hedging, depending upon the reliability of the cue.

CHAPTER 7

Topics for Future Study

In the first chapter I presented an overview of life history analysis as described in the later chapters. Therefore, in this chapter I shall look ahead rather than providing yet another summary of the book. I hope that two general messages have emerged from this book: first, that life history theory consisting of both phenotypic and genetic elements has been extraordinarily successful in providing a framework within which to understand evolutionary change, and second, that evolution can only be understood within the context of the environment. In a constant environment, phenotypic plasticity is unlikely to be favored, whereas in a variable but predictable environment, reaction norms are to be expected.

Growth, reproduction, and death are characteristics of all living organisms. These process can be subsumed into two functions, the age schedule of reproduction, $m(x)$, and the age schedule of survival, $l(x)$, which can themselves be brought together in the characteristic equation, $\int l(x)m(x)e^{-rx}dx = 1$. Components such as size that influence these processes are themselves components of these equations For example, the $m(x)$ function can be expanded to $m(g(x))$, where $g(x)$ specifies growth (and hence size) with age. Similarly, all of the processes discussed in this book (e.g., genetic variability, size, age-structure, environmental variability, reaction norms) can be analyzed by the appropriate modification of the characteristic equation. By keeping this equation central to an analysis, we can be reasonably assured that our assumptions, which must limit the scope of our predictions, are plainly stated.

In the remainder of this chapter I outline some questions and topics for future research that I find particularly interesting and in need of attention. I have no doubt that this list is far from exhaustive, which is what makes the study of life history evolution such an exciting "growth" area.

The Prevalence and Importance of Nonadditive Genetic Variance

Because of its relation to the immediate response to selection, quantitative genetic analyses of life history traits have most frequently focused upon the additive component. There is a growing interest in the important role that nonadditive variance plays in evolutionary change. With regard to life history evolution, nonadditive variance is important because of (1) its potential role in maintaining genetic variability via antagonistic pleiotropy (Chapter 3), (2) the potential for the creation of additive variance when populations pass through bottlenecks (Chapter 2), (3) the effect that non-additive genetic variance can have on trade-off functions (Chapter 3) (4) the deleterious nature of inbreeding depression (Chapter 2), the avoidance of which might result in assortative mating or changes in migratory (dispersal) patterns (Chapter 5). The estimation of nonadditive variance components is challenging, but given its importance, such estimates are sorely needed. We certainly have an abundance of estimates of additive genetic variance of life history traits and there is no doubt that, in general, there is enough for rapid evolutionary response (Chapter 4). But there is a dearth of estimates of dominance variance much less epistatic.

Response to Selection on Multiple Traits

Many artificial selection experiments on single traits have been conducted and there is general agreement between the predicted and observed responses. However, there are virtually no experiments in nondomestic species in which two or more traits were simultaneously selected. Those experiments that have been conducted have produced results that are rather ambiguous (Chapter 2). It is far from clear whether if the failure to fit theoretical predictions is a result of poor initial estimates of genetic parameters or a real discrepancy. Some well-conducted experiments are needed. In particular, does the ability of the simple quantitative genetic models to predict the multivariate response depend upon the quadrat into which selection drives the traits (see Figure 2.21)? If this is the case, as suggested by present data, what refinements to the basic model are required?

Under the multivariate genetic model we might expect that evolution will tend to move in a direction biased by the size of the eigenvalues, that is, in the direction of the major axes (Chapter 2). There are insufficient case studies to yet determine if this is generally true, although the initial studies suggest that this could well not be a general finding. More empirical and theoretical studies are required on the trajectories in multivariate space that suites of traits are likely to follow.

Over What Phylogenetic Range Do Genetic Variance-Covariance Matrices Remain Constant?

The assumption of a constant **G** (and **P**) matrix is clearly false at some taxonomic level (Chapter 2). In fact, it may not even be true between populations. There is a need to develop and understand the statistical methods of analysis that can test the hypothesis of equality and proportionality. Perhaps even more important are methods of adequately describing how matrices differ in a way that is of evolutionary significance. Most analyses have been conducted in a hypothesis-testing framework, whereas it is perhaps more appropriate to be conducting such studies within a parameter-estimation framework designed to ask how differences accumulate at the different levels from population to species, to genera, to family, and so on.

Trade-offs Are Generally Not Bivariate

Most studies examining trade-offs generally consider the bivariate correlation between traits and conclude that if the genetic correlation is not equal to ±1, the traits are not constrained. Such a conclusion is not warranted, because if there are more than two traits involved in the trade-off the genetic correlation between any pair will be typically not equal to ±1 (Chapter 2 and 3) even though the set of possible values is constrained.

The Shape of a Trade-off Function is Critical

An important prediction of life history theory is that the evolution of traits is not only due to trade-offs but that the functional form of the trade-off function can be critical (Figure 4.21; Chapter 4; Figure 6.37). Despite the importance of knowing the shape of the trade-off function, we have very few, if any, studies that have estimated with any accuracy, its form. We also need studies that examine how suites of traits interact to form a multivariate trade-off function. A particularly interesting and important case is that of growth and reproduction. How is growth rate constrained and how is it related to development time, adult size, and reproductive success (fecundity or mating advantage)? Such relationships are at the core of life history variation but are very poorly documented (Chapter 4).

Phenotypic Correlations Are Important

The conventional wisdom that two traits must be genetically correlated to be evolutionarily important is correct only if selection acts on one trait alone. When selection acts on both traits, their joint evolution will be modulated by their phenotypic correlation even if the genetic correlation is zero (Chapter 3). Thus, given that life history traits are probably all under both direct selection to a greater or lesser degree, the measurement of phenotypic correlations is important. It is also clearly important by virtue of the fact that selection acts directly on the phenotype not the genotype, and hence the first stage in evolutionary change is phenotypic selection. It is this phenotypic selection that determines the fitness surface and hence the optimal suite of trait values (Chapter 3).

Establishing the Genetic Basis of Trade-offs Is Important, But Establishing Mechanisms May Be More So

The omission of an important variable can lead to a misinterpretation of the presence or absence of a trade-off and hence the construction of an incorrect fitness surface (Figure 1.3; Chapter 3). The probability of making such errors can be reduced by determining the underlying mechanism that is generating the trade-off (Chapter 3). Further, the dissection of the trade-off functions into mechanistic components (physiological, behavioral, and so on) is likely to give insight into how constrained evolutionary change may be.

Field Studies Integrating Reproductive Effort with Other Life History Components, Such as Survival, to Produce and Test Predictive Models Are Lacking

Reproductive effort is a primary life history variable, and there have been a relatively large number of theoretical studies examining its evolution in relation to age, resources, season length, and so on (Chapter 4). But though there are some field studies that have measured these variables, we are in need of field studies that are designed to explicitly test the theoretical models or are conducted in such a manner that sufficient information is gathered to construct a quantitative, predictive model.

The Importance of Density-Dependent Interactions Requires More Study

Virtually all of the analyses presented in this book have ignored density-dependent interactions. How frequent are such interactions in natural populations? Further, how important are they in determining the combination of mean trait values? Also, how do such interactions affect how and are they affected by genetic variance in the population?

The Smith-Fretwell Model Is Still in Need of Testing

The Smith-Fretwell model is probably one of the simplest in life history theory (Chapter 4) but, despite its frequent use in theoretical models, there are very few tests of its adequacy (Chapter 4).

What Is an Invariant?

Attempts to produce very general models of life history evolution are laudable, and the suggestion that there may be quantities (e.g., αk) that remain invariant is very intriguing (Chapter 4) but there needs to be a more rigorous criterion for what is an invariant quantity. How much variation is permitted? Must all this variation be statistical error or do we expect some variation in an invariant?

Persistence or Natural Selection?

Most life history models are predicated on the assumption that fitness is maximized. However, an alternate hypothesis is that patterns of variation are simply a consequence of the requirement that r must be equal to or greater than zero (Chapter 4). This alternate hypothesis should form the null hypothesis against which that of natural selection is compared (Chapter 4).

The Importance of Environmental Heterogeneity

Most analyses have centered upon evolution in constant environments. This has been fruitful because of the relative simplicity of such models. But it is obvious that the world is spatially and temporally complex. The incorporation of this complexity is a necessary step in generalizing the models. At the same time, we should remain cognizant of the fact that this additional complexity may be unnecessary, at least for some questions. For example, bet-hedging may be favored but the amount might be so small that it can be ignored.

How Important Is Environmental Heterogeneity in Preserving Genetic Variation?

Theoretical analyses indicate that temporal variation is unlikely to preserve genetic variation when generations are nonoverlapping (Chapter 5) but could be an important factor in age-structured populations (Chapter 5). Spatial variation should be a potent factor in maintaining genetic variation. The only test of the ability of environmental heterogeneity to preserve or retard the loss of genetic variation was inconclusive. Given the possible role of environmental variation further experimental evaluations are warranted.

In What Circumstances Is Bet-hedging Important?

There is now a rich theory for the evolution of bet-hedging (Chapter5) but its prevalence in the natural world is still based largely on anecdotal evidence or "just-so" stories. The history of Murphy's hypothesis on the evolution of reproductive effort in a temporally stochastic environment (Chapter 5) is a sobering example. Although it was relatively easy to show that Murphy's hypothesis is logically consistent later analyses showed that, given realistic parameter values, the effect of temporal stochasticity on the optimal reproductive effort will be slight. It is, therefore, not surprising to find that empirical tests of Murphy's hypothesis have been negative. The potential importance of bet-hedging needs to be reconsidered by theoretical examination using appropriate parameter values of possible cases in which it can occur. Only when such an analysis establishes that bet-hedging is predicted should the more time- and labor-consuming task of testing the hypothesis in the real world be attempted.

Does Temporal Variation Influence the Optimal Trait Value?

There is considerable room for more theoretical and empirical treatment of the influence of temporal variation on the optimal trait value. Some work has been done on clutch and propagule size (Chapter 5) and there is empirical evidence that temporal variation determines in part the optimal clutch size in the great tit. Numerous models of brood size reduction have been developed (Chapter 5) but experimental tests are generally lacking.

Does Spatial Variation Influence the Optimal Trait Value?

Spatial variation alone is an unlikely scenario but it is useful to examine it from a theoretical perspective as it can highlight the possible importance of migration to avoid kin competition or inbreeding depression (Chapter 5). Although authors have frequently cited the avoidance of kin competition and/or inbreeding as the likely mechanism driving the evolution of migration in mammals and birds, actual tests of these hypotheses are essentially nonexistent. The recently developed theoretical framework (Chapter 5) should permit more rigorous tests of these hypotheses.

The Evolution of Migration as a Response to Environmental Variability

The long-standing hypothesis that migration evolved in response the spatio-temporal heterogeneity has received both theoretical and empirical support (Chapter 5). Migration involves a suite of life history, behavioral, morphological and physiological traits that are genetically correlated. As a consequence the evolution of migration involves a complex shift in a wide set of traits. This phenomenon can serve as an ideal "model" system to investigate the influence of environmental and genetic constraints on the evolution of traits.

Phenotypic Plasticity Is Ubiquitous

Virtually every life history trait examined shows a response to some environmental factor and hence is phenotypically plastic. How much of this plasticity is adaptive is not clear; as in the case of bet-hedging it is easy to erect an adaptive explanation for practically any example of phenotypic plasticity. In some cases (Chapter 6), such as diapause induction or inducible defenses, the adaptive value may be obvious, but this does not eliminate the need to demonstrate its adaptive value.

Phenotypic Models Are Important Tools for the Analysis of Reaction Norms but a Quantitative Genetic Analysis Is Also Desirable

Phenotypic modeling is a powerful tool for predicting the optimal reaction norm given a set of constraints. However, if the environment is spatially and temporally complex, the evolution of the norm of reaction could be genetically constrained. Under theses conditions, the optimum predicted using a phenotypic modeling approach would be incorrect (Chapter 6). The optimal trait value or reaction norm might also be influenced by the presence of individuals from earlier episodes of selection (Chapter 6). By their very nature reaction norms are likely to be under continual selection. In some cases this variation can be ignored, whereas in others, it cannot. In the latter case a genetic model is required.

REFERENCES

Aastveit, A. H. 1990. Use of bootstrapping for estimation of standard deviation and confidence intervals of genetic variance- and covariance-components. Biometrics 32:515–527.

Abell, A. J. 1999. Variation in clutch size and offspring size relative to environmental conditions in the lizard *Sceloporus virgatus*. Journal of Herpetology 33:173–180.

Abrams, P. A., O. Leimar, S. Nylin and C. Wiklund. 1996. The effect of flexible growth rates on optimal sizes and development times in a seasonal environment. American Naturalist 147:381–395.

Ackerman, J. D. and A. M. Montalvo. 1990. Short- and long-term limitations to fruit production in a tropical orchid. Ecology 71:263–272.

Adams, P. B. 1980. Life history patterns in marine fishes and their consequences for fisheries management. Fisheries Bulletin 78:1–12.

Adler, F. R. and R. Karban. 1994. Defended fortresses of moving targets? Another model of inducible defenses inspired by military metaphors. American Naturalist 144:813–832.

Alatalo, R. V., L. Gustafsson and A. Lundberg. 1990. Phenotypic selection on heritable size traits: environmental variance and genetic response. American Naturalist 135:464–471.

Aleksiuk, M. 1977. Sources of mortality in concentrated garter snake populations. Canadian Field-Naturalist 91:70–72.

Alexander, R. D. 1974. The evolution of social behavior. Annual Review of Ecology and Systematics 5:325–383.

Alexander, R. D. and R. S. Bigelow. 1960. Allochronic speciation in field crickets, and a new species, *Acheta veletis*. Evolution 14:334–346.

Allander, K. 1997. Reproductive investment and parasite susceptibility in the Great Tit. Functional Ecology 11:358–364.

Allander, K. and G. F. Bennett. 1995. Retardation of breeding onset in Great Tits (*Parus major*) by blood parasites. Functional Ecology 9:677–682.

Allard, R. W. 1965. Genetic systems associated with colonizing ability in predominately self-pollinated species. Pp. 49-75 *in* H. Baker and G. Stebbins, (ed.) *The Genetics of Colonizing Species*. Academic Press, New York.

Allee, A., E. Emerson, O. Park, T. Park and K. P. Schmidt. 1949. *Principles of Animal Ecology*. Saunders, Philadelphia.

Ananeva, N. B. and S. M. Shammakov. 1986. Ecologic strategies and relative clutch mass in some species of lizard fauna in the USSR. Soviet Journal of Ecology (English translation Ekologiya) 16:241–247.

Anderson, D. J. 1990. Evolution of obligate siblicide in boobies. I. A test of the insurance-egg hypothesis. American Naturalist 135:334–350.

Anderson, J. L. and R. Boonstra. 1979. Some aspects of reproduction in the vole *Microtus townsendii*. Canadian Journal of Zoology 57:18–24.

Anderson, W. W. 1966. Genetic divergence in M. Vetukhiv's experimental populations of *Drosophila pseudoobscura*. Genetical Research 7:255–266.

Anderson, W. W. 1969. Polymorphism resulting from the mating advantage of rare male genotypes. Proceedings of the National Academy of Sciences 64:190–197.

Anderson, W. W. 1971. Genetic equilibrium and population growth under density-regulated selection. American Naturalist 105:489–498.

Anderson, W. W. 1973. Genetic divergence in body size among experimental populations of *Drosophila pseudoobscura* kept at different temperatures. Evolution 27:278–284.

Andersson, M., C. G. Wiklund and H. Rundgren. 1980. Parental defence of offspring: a model and an example. Animal Behaviour 28:536–542.

Andrewartha, H. G. and L. C. Birch. 1954. *The Distribution and Abundance of Animals*. University of Chicago Press, Chicago.

Andrews, R. M. and A. S. Rand. 1974. Reproductive effort in anoline lizards. Ecology 55:1317–1327.

Ansell, A. D. 1960. Observations on predation of *Venus strialuta* (da Costa) by *Natica alderi* Forbes. Proceedings of the Malacological Society of London 34:157–164.

Antolin, M. F. 1992b. Sex ratio variation in a parasitic wasp II. Diallel cross. Evolution 46:1511–1524.

Antonovics, J. 1968. Evolution in closely adjacent plant populations. V. Evolution of self-fertility. Heredity 23:219–238.

Aparicio, J. M. 1997. Costs and beneits of surplus offspring in the lesser kestrel (*Falco naumanni*). Behavioral Ecology and Sociobiology 41:129–137.

Appleton, R. D. and A. R. Palmer. 1988. Water-borne stimuli released by predatory crabs and damaged prey induce more predator-resistant shells in a marine gastropod. Proceedings of the National Academy of Sciences, USA 85:4387–4391.

Arendt, J. D. 1997. Adaptive intrinsic growth rates: an integration across taxa. Quarterly Review of Biology 72:149–177.

Armstrong, M. J. and P. A. Shelton. 1990. Clupeoid life-history styles in variable environments. Environmental Biology of Fishes 28:77–85.

Armstrong, R. A. and M. E. Gilpin. 1977. Evolution in a time-varying environment. Science 195:591–592.

Armstrong, T. and R. J. Robertson. 1988. Parental investment based on clutch value: nest desertion in response to partial clutch size in dabbling ducks. Animal Behaviour 36:941–943.

Arnold, J. and W. W. Anderson. 1983. Density-regulated selection in a heterogeneous environment. American Naturalist 121:656–668.

Arnold, S. J. 1992. Constraints on phenotypic evolution. American Naturalist 140 (Suppl.):s85–s107.

Arnold, T. W. 1988. Life histories of North American Game birds: a reanalysis. Canadian Journal of Zoology 66:1906–1912.

Arnqvist, G. 1989. Multiple mating in a water strider: mutual benefits or intersexual conflict? Animal Behaviour 38:749–756.

Arvesen, J. N. and T. H. Schmitz. 1970. Robust procedures for variance component problems using the jackknife. Biometrics 26:677–686.

Asmussen, M. A. 1983. Density-dependent selection incorporating intraspecific competition. II. A diploid model. Genetics 103:335–350.

Aspi, J. and A. Hoikkala. 1993. Laboratory and natural heritabilities of male courtship song characters in *Drosophila montana* and *D. littoralis*. Heredity 70:400–406.

Atchley, W. R. 1984. Ontogeny, timing of development, and genetic variance-covariance structure. American Naturalist 123:519–540.

Atkinson, D. and M. Begon. 1987. Reproductive variation and adult size in two co-occurring grasshopper species. Ecological Entomology 12:119–127.

Auslander, D., J. Guckenheimer and G. Oster. 1978. Random evolutionarily stable strategies. Theoretical Population Biology 13:276–293.

Austad, S. N. 1993. Retarded senescence in an insular population of Virginia opossums (*Didelphis virginiana*). Journal of Zoology, London 229:695–708.

Austad, S. N. 1997. Postresproductive survival. Pp. 161–174 *in* K. W. Wachter and C. E. Finch, (eds.) *Between Zeus and the Salmon*. National Academy Press, Washington.

Ayala, F. J., L. Serra and A. Prevosti. 1989. A grand experiment in evolution the Drosophila-subobscura colonization of the Americas. Genome 31:246–255.

Ayala, J. and C. A. Campbell. 1974. Frequency dependent selection. Annual Review of Ecology and Systematics 5:115–138.

Badyaev, A. V. and G. E. Hill. 2000. The evolution of sexual dimorphism in the house finch: I. Population divergence in morphological covariance structure. Evolution 54:1784–1794.

Badyaev, A. V. and T. E. Martin. 2000. Individual variation in growth trajectories: phenotypic and genetic correlations in ontogeny of the house finch (*Carpodacus mexicanus*). Journal of Evolutionary Biology 13:290–301.

Bagenal, T. B. 1955a. The growth rate of the long rough dab, *Hippoglossoides platessoides* (Fabr.). Journal of the Marine Biological Association, U.K. 34:297–311.

Bagenal, T. B. 1955b. The growth rate of the long rough dab, *Hippoglossoides platessoides* (Fabr.). Journal of the Marine Biological Association, U.K. 34:643–647.

Bagenal, T. B. 1966. A short review of fish fecundity. Pp. 89–111 *in* S. D. Gerking, (ed.) *The Biological Basis of Freshwater Fish Production*. Blackwell Scientific Publication, Oxford.

Bagenal, T. B. 1971. The interaction of the size of fish eggs, the date of spawning and the production cycle. Journal of Fish Biology 3:207–219.

Ballinger, R. E. 1973. Comparative demography of two viviparous iguanid lizards (*Sceloporus jarrovi* and *Sceloporus poinsetti*). Ecology 54:269–283.

Baltz, D. M. 1984. Life history variation among female surfperches (Perciformes: Embiotocidae). Environmental Biology of Fishes 10:159–171.

Baptist, R. and A. Robertson. 1976. Asymmetrical responses to automatic selection for body size in *Drosphila melanogaster*. Theoretical and Applied Genetics 47:209–213.

Barclay, H. J. and P. T. Gregory. 1981. An experimental test of models predicting life-history characteristics. American Naturalist 117:944–961.

Barclay, H. J. and P. T. Gregory. 1982. An experimental test of life history evolution using *Drosophila melanogaster* and *Hyla regilla*. American Naturalist 120:26–40.

Barrett, S. C. H. and B. C. Husband. 1990. The genetics of plant migration and colonization. Pp. 254–277 *in* A. H. D. Brown, M. T. Clegg, A. L. Kahler and B. S. Weir, (eds.) *Plant Population Genetics, Breeding, and Genetic Resources*. Sinauer Associates Inc., Sunderland, MA.

Barton, N. H. 1990. Pleiotropic models of quantitative variation. Genetics 124:773–782.

Barton, N. H. and M. Turelli. 1989. Evolutionary quantitative genetics: how little do we know? Annual Review of Genetics 23:337–370.

Baughman, J. F. 1991. Do protandrous males have increased mating success? The case of *Euphydryas editha*. American Naturalist 23:314–322.

Bauwens, D. and C. Thoen. 1981. Escape tactics and vulnerability to predation associated with reproduction in the lizard *Lacerta vivipara*. Journal of Animal Ecology 50:733–743.

Beck, S. D. 1980. *Insect Photoperiodism*. 2nd edition, Academic Press, New York

Becker, W. A. 1992. *Manual of Quantitative Genetics*. Academic Enterprises, Pulman, WA.

Begon, M. and G. A. Parker. 1986. Should egg size and clutch size decrease with age? Oikos 47:293–302.

Beissinger, S. R. 1990. Experimental brood manipulations and the monoparental threshold in Snail Kites. American Naturalist 136:20–38.

Belk, M. C. 1995. Variation in growth and age at maturity in bluegill sunfish: genetic or environmental effects? Journal of Fish Biology 47:237–247.

Belk, M. C. 1998. Predator-induced delayed maturity in bluegill sunfish (*Lepomis macrochirus*): Variation among populations. Oecologia 113:203–209.

Bell, A. E. and M. J. Burris. 1973. Simultaneous selection for two correlated traits in *Tribolium*. Genetical Research 21:24–46.

Bell, G. 1980. The costs of reproduction and their consequences. American Naturalist 116:45–76.

Bell, G. 1984. Measuring the cost of reproduction. I. The correlation structure of the life table of plankton rotifer. Evolution 38:300–313.

Bell, G. and V. Kofopanou. 1985. The cost of reproduction. Pp. 83–131 *in* R. Dawkins, (ed.) *Oxford Surveys of Evolutionary Biology*. Oxford University Press, Oxford.

Bell, P. D. 1979. Acoustic attraction of herons by crickets. Journal of the New York Entomological Society 87:126–127.

Bellinger, R. G. and R. L. Pienkowski. 1987. Developmental polymorphism in the red-legged grasshopper *Melanoplus femurrubrum* (DeGeer) (Orthoptera: Acridoidae). Environmental Entomology 16:120–125.

Bennet, P. M. and P. H. Harvey. 1987. Active and resting metabolism in birds: allometry, phylogeny and ecology. Journal of Zoology 213:327–363.

Benrey, B. and R. F. Denno. 1997. The slow-growth-high-mortality hypothesis: a test using the cabbage butterfly. Ecology 78:987–999.

Benton, R. A. and A. Grant. 1999. Optimal reproductive effort in stochastic, density-dependent environments. Evolution 53:677–688.

Benton, T. G. and A. Grant. 2000. Evolutionary fitness in ecology: comparing measures of fitness in stochastic, density-dependent environments. Evolutionary Ecology Research 2:769–789.

Berenbaum, M. R. and A. R. Zangerl. 1999. Coping with life as a menu option:inducible defenses of the wild parsnip. Pp. 10–32 *in* R. Tollrian and C. D. Harvell, (eds.) *The Ecology and Evolution of Inducible Defenses*. Princeton University Press, Princeton.

Berrigan, D. and J. C. Koella. 1994. The evolution of reaction norms: simple models for age and size at maturity. Journal of Evolutionary Biology. 7:549–566.

Berthold, P. 1973. Relationships between migratory restlessness and migration distance in six *Sylvia* species. Ibis 115:594–99.

Berthold, P. 1988. Evolutionary aspects of migratory behavior in European warblers. Journal of Evolutionary Biology 1:195–209.

Berthold, P. 1995. Microevolution of migratory behaviour illustrated by the blackcap *Sylvia atricapilla*: 1993 Witherby Lecture. Bird Study 42:89–100.

Berthold, P., A. J. Helbig, G. Mohr and U. Querner. 1992. Rapid microevolution of migratory behaviour in a wild bird species. Nature 360:668–669.

Berthold, P. and F. Pulido. 1994. Heritability of migratory activity in a natural bird population. Proceedings of the Royal Society of London, B 257:311–315.

Berthold, P. and U. Querner. 1981. Genetic basis of migratory behavior in European warblers. Science 212:77–79.

Bertness, M. D. 1981a. Predation, physical stress, and the organisation of a tropical hermit crab community. Ecology 62:411–425.

Bertness, M. D. 1981b. Pattern and plasticity in tropical hermit crab growth and reproduction. American Naturalist 117:754–773.

Bertness, M. D. 1981c. The influence of shell-type on hermit crab growth rate and clutch size. Crustaceana 40:197–205.

Bertschy, K. A. and M. G. Fox. 1999. The influence of age-specific survivorship on pumkinseed sunfish life histories. Ecology 80:2299–2313.

Berven, K. A. 1982a. The genetic basis of altitudinal variation in the wood frog *Rana sylvatica*. I. An experimental analysis of life history traits. Evolution 36:962–983.

Berven, K. A. 1982a. The genetic basis of altitudinal variation in the wood frog *Rana sylvatica*. II. An experimental analysis of larval development. Oecologia 52:360–369.

Berven, K. A. 1982b. The genetic basis of altitudinal variation in the wood frog *Rana Sylvatica*. II. An experimental analysis of larval development. Oecologia 52: 360–369.

Berven, K. A. 1987. The heritable basis of variation in larval developmental patterns within populations of the wood frog (*Rana sylvatica*). Evolution 41:1088–1097.

Berven, K. A. and D. E. Gill. 1983. Interpreting geographic variation in life-history traits. American Zoologist 23:85–97.

Berven, K. A., D. E. Gill and S. J. Smith-Gill. 1979. Countergradient selection in the green frog, *Rana clamitans*. Evolution 33:609–623.

Betran, E., M. Santos and A. Ruiz. 1998. Antagonistic pleiotropic effect of second-chromosome inversions on body size and early life-history traits in *Drosophila buzzatii*. Evolution 52:144–154.

Beverton, R. J. H. 1963. Maturation, growth and mortality of clupeid and engraulid stocks in relation to fishing. Journal du Conseil Permanent International pour l'Exploration de la Mer 154:44–67.

Beverton, R. J. H. 1987. Longevity in fish: some ecological and evolutionary considerations. Pp. 161–185 *in* A. D. Woodhead and K. H. Thompson, (eds.) *Evolution of Longevity in Animals*. Plenum Press, New York.

Beverton, R. J. H. 1992. Patterns of reproductive strategy paramters in some marine teleost fishes. Journal of Fish Biology 41 (suppl. B):137–160.

Beverton, R. J. H. and S. J. Holt. 1959. A review of the lifespans and mortality rates of fish in nature, and their relation to growth and other physiological characteristics. Pp. 142–177 *in* G. E. W. Wolstenholme and M. O'Connor, (ed.) *CIBA Foundation Colloquia on Ageing*, Volume 5. J. A. Churchill Ltd., London.

Bezzi, M. 1916. Riduzione e scomparsa delle ali negli insetti ditteri. Natura (Milano) 7:85–182.

Biebach, H. 1983. Genetic determination of partial migration in the European robin (*Erithacus rubecula*). Auk 100:601–606.

Biermann, G. C. and R. J. Robertson. 1983. Residual reproductive value and parental investment. Animal Behaviour 31:311–312.

Bijlsma, R., J. Bundgaard and W. F. van Putten. 1999. Environmental dependence of inbreeding depression and purging in *Drosophila melanogaster*. Journal of Evolutionary Biology 12:1125–1137.

Bijlsma, R., N. J. Ouberg and R. van Treuren. 1994. On genetic erosion and population extinction in plants: A case study in *Scabiosa columbaria* and *Salvia pratensis*. Pp. 255–271 *in* V. Loeschcke, J. Tomiuk and S. K. Jain, (eds.) *Conservation Genetics*. Birkauser Verlag Basel, Switzerland.

Birch, L. C. 1960. The genetic factor in population ecology. American Naturalist 94:5–24.

Bjorklund, M. 1996. The importance of evolutionary constraints in ecological time scales. Evolutionary Ecology 10:423–431.

Blackburn, T. M. 1991. The interspecific relationship between egg size and clutch size in birds. Auk 108:973–977.

Blakley, N. and S. R. Goodner. 1978. Size-dependent timing of metamorphosis in milkweed bugs (*Oncopeltus*) and its life history implications. Biological Bulletin 155:499–510.

Blanckenhorn, W. U. 1998. Adaptive phenotypic plasticity in growth, development, and body size in the yellow dung fly. Evolution 52:1394–1407.

Blanckenhorn, W. U. and D. J. Fairbairn. 1995. Life history adaptation along a latitudinal cline in the water strider *Aquarius remigis* (Heteroptera: Gerridae). Journal of Evolutionary Biology 8:21–41.

Blouin, M. 1992. Genetic correlations among morphometric traits and rates of growth and differentiation in the green tree frog, *Hyla cinera*. Evolution 46:735–744.

Boag, P. T. and P. R. Grant. 1981. Intense natural selection in a population of Darwin's finches (Geospizinae) in the Galapagos. Science 214:82–85.

Bohren, B. B., W. G. Hill and A. Robertson. 1966. Some observations on asymmetrical correlated responses to selection. Genetical Research 7:44–57.

Bohren, B. B. and H. E. McKean. 1961. Relative efficiencies of heritability estimates based on regression of offspring on parent. Biometrics 17:481–491.

Bolton, M., D. Houston and P. Monaghan. 1992. Nutritional constraints on egg formation in the lesser black gull: an experimental study. Journal of Animal Ecology 61:521–532.

Bomze, I. M., P. Schuster and K. Sigmund. 1983. The role of Mendelian genetics in strategic models of animal behaviour. Journal of Theoretical Biology 101:19–38.

Bondari, K., R. Willham and A. Freeman. 1978. Estimates of direct and maternal genetic correlations for pupa weight and family size in *Tribolium*. Journal of Animal Science 47:358–365.

Booth, D. T. 1998. Egg size, clutch size, and reproductive effort of the Australian broad-shelled river turtle, *Chelodina expansa*. Journal of Herpetology 32:592–596.

Boots, M. and M. Begon. 1993. Trade-offs with resistance to a granulosis virus in the Indian meal moth, examined by laboratory evolution experiment. Functional Ecology 7:528–534.

Borror, D. L. and D. M. DeLong. 1964. *An Introduction to the Study of Insects*. Holt, Rinehart and Winston, New York.

Bortolotti, G. R., K. L. Wiebe and W. M. Iko. 1991. Cannibalism of nestling American kestrel by their parents and siblings. Canadian Journal of Zoology 69:1447–1453.

Boscher, J. 1981. Reproductive effort in *Allium porrum*: relation to the length of the juvenile phase. Oikos 37:328–334.

Bostock, S. J. and R. A. Benton. 1979. The reproductive strategies of five perennial compositae. Journal of Ecology 67:91–107.

Botkin, D. B. and R. S. Miller. 1974. Mortality rates and survival of birds. American Naturalist 108:181–192.

Boucher, D. H. 1977. On wasting parental effort. American Naturalist 111:786–788.

Bourguet, D. 1999. The evolution of dominance. Heredity 83:1–4.

Box, J. F. 1978. R. A. Fisher: *The Life of a Scientist*. Wiley, New York.

Boyce, M. S. 1984. Restitution of *r*- and *K*-selection as a model of density-dependent natural selection. Annual Review of Ecology and Systematics 15: 427–447.

Boyce, M. S. and C. M. Perrins. 1987. Optimizing Great Tit clutch size in a fluctuating environment. Ecology 68:142–153.

Bradford, M. J. and D. A. Roff. 1995. Genetic and phenotypic sources of life history variation along a cline in voltinism in the cricket *Allonemobius socius*. Oecologia 103:319–326.

Bradford, M. J. and D. A. Roff. 1997. An empirical model of diapause strategies in the cricket *Allonemobius socius*. Ecology 78:442–451.

Bradshaw, W. E. 1976. Geography of photoperiodic response in a diapausing mosquito. Nature 262: 384–386.

Bradshaw, W. E. and C. M. Holzapfel. 1983. Life cycle strategies in *Wyeomyia smithii*: seasonal and geographic adaptations. Pp. 167–185 *in* V. K. Brown and I. Hodek, (eds.) *Diapause and Life Cycle Strategies in Insects*. Dr. W. Junk Publishers, The Hague.

Bradshaw, W. E. and C. M. Holzapfel. 1990. Evolution of phenology and demography in the pitcher plant mosquito, *Wyeomyia smithii*. Pp. 47–67 *in* F. Gilbert, (ed.) *Insect Life Cycles: Genetics, Evolution, and Co-ordination*. Springer-Verlag, London.

Bradshaw, W. E. and C. M. Holzapfel. 1996. Genetic constraints to life-history evolution in the pitcher-plant mosquito, *Wyeomyia smithii*. Evolution 50:1176–1181.

Bradshaw, W. E. and L. P. Lounibos. 1977. Evolution of dormancy and its photoperiodic control in pitcher-plant mosquitoes. Evolution 31:546–547.

Bradshaw, W. E., M. Holzapfel, C. A. Kleckner and J. J. Hard. 1997. Heritability of deveopment time and protandry in the pitcher-plant mosquito, *Wyeomyia smithii*. Ecology 78:969–976.

Brantley, R. K., J. Tseng and A. H. Bass. 1993. The ontogeny of inter- and intrasexual vocal muscle dimorphisms in a sound-producing fish. Brain Behavioral Evolution 42:336–349.

Brennan, J. M. and D. J. Fairbairn. 1995. Clinal variation in morphology among eastern populations of the waterstrider, *Aquarius remigis* Say (Hemiptera: Gerridae). Biological Journal of the Linnean Society 54:151–171.

Brinkhof, M. W. G., A. J. Cave, F. J. Hage and S. Verhulst. 1993. Timing of reproduction and fledging success in the Coot *Fulica atra*: evidence for a causal relationship. Journal of Animal Ecology 62:577–587.

Brinkhurst, R. O. 1959. Alary polymorphism in the Gerroidea. Journal of Animal Ecology 28:211–230.

Brinkhurst, R. O. 1963. Observations on wing-polymorphism in the Heteroptera. Proceedings of the Royal Entomological Society of London (A) 38: 15–22.

Britton, M. M. 1966. Reproductive success and survival of the young in *Peromyscus*. M. Sc. Thesis, University of British Columbia, Vancouver.

Brodie, E. D. J. and D. R. J. Formanowicz. 1983. Prey size preference of predators: differential vulnerability of larval anurans. Herpetologica 39:67–75.

Brody, M. S. and L. R. Lawlor. 1984. Adaptive variation in offspring size in the terrestrial isopod, *Armadillidium vulgare*. Oecologia 61:55–59.

Bronmark, C. and J. G. Miner. 1992. Predator-induced phenotypical change in body morphology in crucian carp. Science 258:1348–1350.

Bronmark, C. and L. B. Pettersson. 1994. Chemical cues from piscivores induce a change in morphology in crucian carp. Oikos 70:396–402.

Brooks, J. L. 1968. The effects of prey size selection by lake planktivores. Systematic Zoologist 17:273–291.

Brooks, J. L. and S. I. Dodson. 1965. Predation, body size and composition of the plankton. Science 150:28–35.

Brousseau, D. J. and J. A. Baglivo. 1988. Life tables for two field populations of soft-shell clam, *Mya arenaria*, (Mollusca: Pelecypoda) from Long Island Sound. Fishery Bulletin 86:567–579.

Browne, R. A. 1982. The costs of reproduction in brine shrimp. Ecology 63:43–47.

Browne, R. A. and W. D. Russell-Hunter. 1978. Reproductive effort in molluscs. Oecologia 37:23–27.

Browne, R. A., S. E. Sallee, D. S. Grosch, W. O. Segreti and S. M. Parser. 1984. Partitioning genetic and environmental components of reproduction and lifspan in *Artemia*. Ecology 65:949–960.

Brues, C. T. 1902. The structure and significance of vestigial wings among insects. Biological Bulletin 4:170–190.

Bryant, D. M. 1979. Reproductive costs in the house martin (*Delichon urbica*). Journal of Animal Ecology 48:655–675.

Bryant, D. M. and P. Tatner. 1991. Intraspecies variation in avian energy expenditure: correlates and constraints. Ibis 133:236–245.

Bryant, E. H. 1971. Life history consequences of natural selection: Cole's result. American Naturalist 105:75–76.

Bryant, E. H. 1976. A comment on the role of environmental variation in maintaining polymorphisms in natural populations. Evolution 30:188–190.

Bryant, E. H. 1977. Morphometric adaptation of the housefly, *Musca domestica* L., in the United States. Evolution 31:580–596.

Bull, J. J. 1987. Evolution of phenotypic variance. Evolution 41:303–315.

Bull, J. J., R. C. Vogt and C. J. McCoy. 1982. Sex determining temperatures in turtles: a geographic comparison. Evolution 36:326–332.

Bulmer, M. G. 1971a. The stability of equilibria under selection. Heredity 27:157–162.

Bulmer, M. G. 1971b. Stable equilibria under the two island model. Heredity 27:321–330.

Bulmer, M. G. 1974. Density-dependent selection and character displacement. The American Naturalist 108:45–58.

Bulmer, M. G. 1983. Models for the evolution of protandry in insects. Theoretical Population Biology 23:314–322.

Bulmer, M. G. 1985a. *The Mathematical Theory of Quantitative Genetics*. Claredon Press, Oxford.

Bulmer, M. G. 1985b. Selection for iteroparity in a variable environment. American Naturalist 126: 63–71.

Bulmer, M. G. 1989. Maintenance of genetic variability by mutation-selection balance: a child's guide through the jungle. Genome 31:761–767.

Bulmer, M. G. and C. M. Perrins. 1973. Mortality in the great tit *Parus major*. Ibis 115:277–281.

Bumpus, H. C. 1899. The elimination of the unfit as illustrated by the introduced sparrow, *Passer domesticus*. Biological Lectures, Marine Biology Laboratory, Woods Hole, MA. 209–226.

Bunnell, F. L. and D. E. N. Tait. 1981. Population dynamics of bears-implications. Pp. 75–98 *in* C. W. Fowler and T. D. Smith, (ed.) *Dynamics of Large Mammal Populations*. John Wiley and Sons, New York.

Burger, R. 1989. Linkage and the maintenance of heritable variation by mutation-selection balance. Genetics 121:175–184.

Burger, R. and R. Lande. 1994. On the distribution of the mean and variance of a quantitative trait under mutation-selection-drift balance. Genetics 138: 901–912.

Burger, R., G. P. Wagner and F. Stettinger. 1989. How much heritable variation can be maintained in finite populations by mutation-selection balance? Evolution 43:1748–1766.

Burk, T. 1982. Evolutionary significance of predation on sexually signalling males. Florida Entomologist 65:90–104.

Busack, C. and G. Gall. 1983. An initial description of the quantitative genetics of growth and reproduction in the mosquito fish. Aquaculture 32:123–140.

Butterworth, B. B. 1961. A comparative study of growth and development of the kangaroo rats, *Dipodomys deserti* Stephens and *Dipodomys merriami* Mearns. Growth 25:127–139.

Byers, G. W. 1969. Evolution of wing reduction in crane flies (Diptera: Tipulidae). Evolution 23:346–354.

Byrd, J. W., A. I. Houston and P. D. Sozou. 1991. Ydenberg's model of fledging time- A comment. Ecology 72:1893–1896.

Cade, W. 1975. Acoustically orienting parasitoids: fly phonotaxis to cricket song. Science 190:1312–1313.

Calder, W. A. I. 1976. Aging in vertebrates: Allometric considerations of spleen size and life span. Federation Proceedings 35:96–97.

Calder, W. A. I. 1983. Body size, mortality and longevity. Journal of theoretical Biology 102:135–144.

Caldwell, J. P., J. H. Thorp and T. O. Jervey. 1980. Predator-prey relationships among larval dragonflies, salamanders and frogs. Oecologia 46:285–289.

Calef, G. W. 1973. Natural mortality of tadpoles in a population of *Rana aurora*. Ecology

Calhoun. 1947. The role of temperature and natural selection in relation to the variations in the size of the English Sparrow in the United States. American Naturalist 81:203–228.

Calow, P. 1978. The evolution of life-cycle strategies in fresh-water gastropods. Malacologia 17:351–364.

Calow, P. and A. S. Woollhead. 1977. The relationship between ration, reproductive effort and age-specific mortality in the evolution of life-history strategies - some observations on freshwater triclads. Journal of Animal Ecology 46:765–781.

Cameron, G. N. 1973. Effect of litter size on posnatal growth and survival in the desert wood rat. Journal of Mammalogy 54:489–493.

Capy, P., J. R. David, R. Allemand, P. Hyytia and J. Rouault. 1983. Genetic properties of North African *Drosophila melanogaster* and comparison with European and Afrotropical populations. Genetics, Selection, Evolution. 15:185–200.

Carey, J. R. and C. Gruenfelder. 1997. Population biology of the elderly. Pp. 127–160 *in* K. W. Wachter and C. E. Finch, (eds.) *Between Zeus and the Salmon*. National Academy Press, Washington.

Carlisle, T. R. 1982. Brood success in variable environments implications for parental care allocation. Animal Behaviour 30:824–836.

Carlisle, T. R. 1985. Parental response to brood size in cichlid fish. Animal Behavior 33:234–238.

Carlquist, S. 1965. *Island Life*. Natural History Press, New York.

Carriere, Y., J.-P. Deland, D. A. Roff and C. Vincent. 1994. Life-history costs associated with the evolution of insecticide resistance. Proceedings of the Royal Society of London, B 258:35–40.

Carriere, Y. and D. A. Roff. 1995a. Change in genetic architecture resulting from the evolution of insecticide resistance: a theoretical and empirical analysis. Heredity 75:618–629.

Carriere, Y. and D. A. Roff. 1995b. The evolution of offspring size and number: a test of the Smith-Fretwell model in three species of crickets. Oecologia 102:389–396.

Carriere, Y., D. A. Roff and J.-P. Deland. 1995. The joint evolution of diapause and isecticide resistance: a test of an optimality model. Ecology 76:35–40.

Carriere, Y., A. Simons and D. A. Roff. 1996. The effect of post-diapause egg development on survival, growth and body size in *Gryllus pennsylvanicus*. Oikos 75:463–470.

Carroll, S. P. and C. Boyd. 1992. Host race radiation in the soapberry bug: natural history with a history. Evolution 46:1052–1069.

Carroll, S. P., H. Dingle and S. P. Klassen. 1997. Genetic differentiation of fitness-associated traits among rapidly evolving populations of the soapberry bug. Evolution 51:1182–1188.

Carroll, S. P., S. P. Klassen and H. Dingle. 1998. Rapidly evolving adaptations to host ecology and nutrition in the soapberry bug. Evolutionary Ecology 12:955–968.

Carroll, S. P. and J. E. Loye. 1987. Specialization of *Jadera* species (Hemiptera: Rhopalidae) on seeds of the Sapindaceae and coevolution of defense and attack. Annals of the Entomological Society of America 80:373–378.

Carson, H. L., D. E. Hardy, H. T. Spieth and W. S. Stone. 1970. The evolutionary biology of the Hawaiian Drosophilidae. Pp. 437–543 *in* M. K. Hecht and W. C. Steere, (ed.) *Essays in Evolution and Genetics in Honor of Theodosius Dobzhansky*. Appleton-Century-Crofts, New York.

Caswell, H. 1980. On the equivalence of maximizing reproductive value and fitness. Ecology 61:19–24.

Caswell, H. 1982a. Optimal life histories and the maximization of reproductive value: a general theorem for complex life cycles. Ecology 63:1218–1222.

Caswell, H. 1982b. Life history theory and the equilibrium status of populations. American Naturalist 120:317–339.

Caswell, H. 1989. *Matrix Population Models*. Sinauer Associates, Inc., Sunderland, MA.

Caswell, H. and A. Hastings. 1980. Fecundity, developmental time, and population growth rate: an analytical solution. Theoretical Population Biology 17:71–79.

Caswell, H., R. J. Naiman and R. Morin. 1984. Evaluating the consequences of reproduction in complex salmonid life cycles. Aquaculture 43:123–134.

Caswell, H. and P. A. Werner. 1978. Transient behavior and life history analysis of teasel (*Dipsacus sylvestris* Hnds.). Ecology 59:53–66.

Caughley, G. 1966. Mortality patterns in mammals. Ecology 47:906–918.

Charlesworth, B. 1971. Selection in density-regulated populations. Ecology 52:469–474.

Charlesworth, B. 1972. Selection in populations with overlapping generations. III Conditions for genetic equilibrium. Theoretical Population Biology 3: 377–395.

Charlesworth, B. 1974. Selection in populations with overlapping generations. VI. Rates of change of gene frequency and population growth rate. Theoretical Population Biology 6:108–132.

Charlesworth, B. 1980. *Evolution in Age Structured Populations*. Cambridge University Press, Cambridge.

Charlesworth, B. 1990. Optimization models, quantitative genetics, and mutation. Evolution 44:520–538.

Charlesworth, B. 1993. Natural selection on multivariate traits in age-structured populations. Proceedings of the Royal Society of London, B 251:47–52.

Charlesworth, B. 1994. Evolution in Age Structured Populations. Cambridge University Press

Charlesworth, B. and J. T. Giesel. 1972a. Selection in populations with overlapping generations. II. Relations between gene frequency and demographic variables. American Naturalist 106:388–401.

Charlesworth, B. and J. T. Giesel. 1972b. Selection in populations with overlapping generations. IV. Fluctuations in gene frequency with density dependent selection. American Naturalist 106:402–411.

Charlesworth, B. and J. A. Leon. 1976. The relation of reproductive effort to age. American Naturalist 110:449–459.

Charlesworth, B. and J. A. Williamson. 1975. The probability of survival of a mutant gene in an age-structured population and implications for the evolution of life-histories. Genetical Research 26:1–10.

Charnov, E. L. 1989a. Phenotypic evolution under Fisher's Fundamental Theorem of Natural Selection. Heredity 62:113–116.

Charnov, E. L. 1989b. Natural selection on age at maturity in shrimp. Evolutionary Ecology 3:236–239.

Charnov, E. L. 1990. On evolution of age at maturity and the adult lifespan. Journal of Evolutionary Biology 3:139–144.

Charnov, E. L. 1991a. Pure numbers, invariants, and symmetry in the evolution of life histories. Evolutionary Ecology 5:339–342.

Charnov, E. L. 1991b. Evolution of life history variation among female mammals. Proceedings of the National Academy of Scienccss, USA 88:1134–1137.

Charnov, E. L. 1993. *Life History Invariants*. Oxford University Press, Oxford.

Charnov, E. L. and D. Berrigan. 1990. Dimensionless numbers and life history evolution: age at maturity versus the adult lifespan. Evolutionary Ecology 4:273–275.

Charnov, E. L. and D. Berrigan. 1991a. Evolution of life history parameters in animals with indeterminate growth, particularly fish. Evolutionary Ecology 5:63–68.

Charnov, E. L. and D. Berrigan. 1991b. Dimensionless numbers and the assembly rules for life histories. Philosophical Transactions of the Royal Society of London (B) 332:41–48.

Charnov, E. L. and S.W. Skinner. 1985. Complementary approaches to the understanding of parasitoid oviposition decisions. Ecological Entomology 14:383–391.

Charnov, E. L. and J. R. Krebs. 1973. On clutch size and fitness. Ibis 116:217–219.

Charnov, E. L. and W. M. Schaffer. 1973. Life history consequences of natural selection: Cole's result revisited. American Naturalist 107:791–793.

Charnov, E. L. and S. W. Skinner. 1984. Evolution of host selection and clutch size in parasitoid wasps. Florida Entomologist 67:5–21.

Charnov, E. L., T. F. Turner and K. O. Winemiller. 2001. Reproductive constraints and the evolution of life histories. Nature with indeterminate growth. Proceedings of the National Academy of Sciences (USA) 98:9460–9464.

Chen, S., S. Watanabe and K. Takagi. 1988. Growth analysis on fish population in the senescence with special reference to an estimation of age at end of reproductive span and life span. Nippon Suisan Gakkaishi 54:1567–1572.

Chen, S. B. and S. Watanabe. 1989. Age dependence of natural mortality coefficient in fish population dynamics. Nippon Suisan Gakkaishi 55:205–208.

Chen, X. 1993. Comparison of inbreeding and outbreeding in hermaphroditic *Arianta arbustorum* (L.) (land snail). Heredity 71:456–461.

Cheung, T. K. and R. J. Parker. 1974. Effect of selection on heritability and genetic correlation of two quantitative traits in mice. Canadian Journal of Genetics and Cytology 16:599–609.

Cheverud, J. M. 1984. Evolution by kin selection: a quantitative genetic model illustrated by maternal performance in mice. Evolution 38:766–777.

Cheverud, J. M. 1988. A comparison of genetic and phenotypic correlations. Evolution 42:958–968.

Chiang, H. C. and A. C. Hodson. 1950. An analytical study of population growth in *Drosophila melanogaster*. Ecological Monographs 20:173–206.

Childs, J. E. 1991. And the cat shall lie down with the rat. Natural History 6:16–19.

Chippindale, A. K., D. T. Hoang, P. M. Service and M. R. Rose. 1994. The evolution of development in *Drosophila melanogaster* selected for postponed senescence. Evolution 48:1880–1899.

Choo, J. 1975. Genetic studies on walking behavior in *Drosophila melanogaster* I. Selection and hybridization analysis. Canadian Journal of Genetics and Cytology 17:535–542.

Christiansen, F. B. 1974. Sufficient conditions for protected polymorphism in a subdivided populations. American Naturalist 108:157–166.

Christiansen, F. B. 1985. Selection and population regulation with habitat variation. American Naturalist 126:418–429.

Christiansen, F. B. and T. M. Fenchel. 1977. *Theories of Populations in Biological Communities*. Springer Verlag, Berlin.

Clancy, K. M. and P. W. Price. 1987. Rapid herbivore growth enhances enemy attack: sublethal plant defenses remain a paradox. Ecology 68:736–738.

Clark, A. G. 1990. Genetic components of variation in energy storage in *Drosophila melanogaster*. Evolution 44:637–650.

Clark, C. W. and C. D. Harvell. 1992. Inducible defenses and the allocation of resources: a minimal model. American Naturalist 139:521–539.

Clark, R. 1984. J.B.S. *The Life and Work of J.B.S. Haldane*. Oxford University Press, Oxford.

Clarke, A. 1993. Reproductive trade-offs in caridean shrimps. Functional Ecology 7:411–19.

Clarke, B. 1964. Frequency-dependent selection for the dominance of rare polymorphic genes. Evolution 18:364–369.

Clarke, B. 1972. Density-dependent selection. American Naturalist 106:1–13.

Clarke, B. and P. O'Donald. 1964. Frequency-dependent selection. Heredity 19:201–206.

Clayton, G. A., G. R. Knight, J. A. Morris and A. Robertson. 1957. An experimental check on quantitative genetical theory. I. Short-term response to selection. Journal of Genetics 55:131–151.

Clayton, G. A. and A. Robertson. 1957. An experimental check on quantitative genetical theory: II. The long-term effects of selection. Journal of Genetics 55:152–170.

Clifford, H. F. and H. Boerger. 1974. Fecundity of mayflies (ephemeroptera), with special reference to mayflies of a brown-water stream of Alberta, Canada. Canadian Entomologist 106:1111–1119.

Clobert, C., J. Nichols, J. D. Danchin and A. Dhondt. (eds) 2001. *Causes, Consequences and Mechanisms of Dispersal at the Individual, Population and Community Level*. Oxford University Press, Oxford.

Clobert, J., V. Bauchau, A. A. Dhondt and C. Vansteenwegen. 1987. Survival of breeding female starlings in relation to brood size. Acta Oecologica Oecoogia Generalis 8:427–434.

Clutton-Brock, T. H. 1991. The Evolution of Parental Care. Princeton University Press, Princeton, New Jersey.

Clutton-Brock, T. H., F. E. Guinness and S. D. Albon. 1983. The costs of reproduction in red deer hinds. Journal of Animal Ecology 52:367–383.

Cody, M. L. and J. Overton. 1996. Short-term evolution of reduced dispersal in island plant populations. Journal of Ecology 84:53–61.

Cohen, D. 1966. Optimizing reproduction in a randomly varying environment. Journal of Theoretical Biology 12:119–129.

Cohen, D. 1971. Maximising final yield when growth is limited by time or by limiting resources. Journal of Theoretical Biology 33:299–307.

Cohen, D. 1976. The optimal timing of reproduction. American Naturalist 110:801–807.

Cohen, D. and S. A. Levin. 1991. Dispersal in patchy environments—the effects of temporal and spatial structure. Theoretical Population Biology 39:63–99.

Cole, L. C. 1954. The population consequences of life history phenomena. Quarterly Review of Biology 29:103–137.

Coleman, R. M., M. R. Gross and R. C. Sargent. 1985. Parental investment decision rules: a test in bluegill sunfish. Behavioral Ecology and Sociobiology 18:59–66.

Comins, H. N. 1982. Evolutionarily stable strategies for localized dispersal in two dimensions. Journal of Theoretical Biology 94:579–606.

Comins, H. N., W. D. Hamilton and R. M. May. 1980. Evolutionarily stable dispersal strategies. Journal of Theoretical Biology 82:205–230.

Connell, J. H. 1970. A predator-prey system in the marine intertidal region. I. *Balanus glandula* and several predatory species of *Thais*. Ecological Monographs 40:49–78.

Connell, J. H. 1972. Community interactions on rocky intertidal shores. Annual Review of Ecology and Systematics 3:169–192.

Conover, D. O. 1990. The relation between capacity for growth and length of growing season: evidence for and implications of countergradient variation. Transactions of the American Fisheries Society 119:416–430.

Cooch, E. G. and R. E. Ricklefs. 1994. Do variable environments significantly influence optimal reproductive effort in birds? Oikos 69:447–459.

Cook, D. 1988. Sexual selection in dung beetles. II. Female fecundity as an estimate of male reproductive success in relation to horn size, and alternative behavioral strategies in *Onthophagus binodis* Thunberg (Scarabaeidae: Onthophagini). Australian Journal of Zoology 36:521–532.

Cooke, F., P. D. Taylor, C. M. Francis and R. F. Rockwell. 1990. Directional selection and clutch size in birds. American Naturalist 136:261–267.

Cooper, P. A., R. H. Benno, M. E. Hahn and J. K. Hewitt. 1991. Genetic analysis of cerebellar foliation patterns in mice (*Mus musculus*). Behavior Genetics 21:405–420.

Cooper, W. S. and R. H. Kaplan. 1982. Adaptive 'coin-flipping': a decision-theoretic examination of natural selection for random individual variation. Journal of Theoretical Biology 94:135–151.

Corey, S. 1981. Comparative fecundity and reproductive strategies in seventeen species of the *Cumacea* (Crustacea: Peracarida). Marine Biology 62:65–72.

Courtney, S. P. 1984. The evolution of egg clustering by butterflies and other insects. American Naturalist 123:276–281.

Coyne, J. A. and E. Beecham. 1987. Heritability of two morphological characters within and among natural populations of *Drosphila melanogaster*. Genetics 117:727–737.

Creaser, E. P. Jr. 1973. Reproduction of the bloodworm (*Glycera dibranchiata*) in the Sheepscot Estuary, Maine. Journal of the Fisheries Research Board of Canada 30:161–166.

Crespi, B. J. 1988a. Alternative male mating tactics in a thrips: effects of sex ratio variation and body size. American Midland Naturalist 119:83–92.

Crespi, B. J. 1988b. Risks and benefits of lethal male fighting in the colonial, polygynous thrips *Hoplothrips karnyi* (Insects: Thysanoptera). Behavioral Ecology and Sociobiology 22:293–301.

Crespi, B. J. 1989. Causes of assortative mating in arthropods. Animal Behaviour 38:980–1000.

Cressman, R. and A. T. Dash. 1987. Density dependence and evolutionary stable strategies. Journal of Theoretical Biology 126:393–406.

Crnokrak, P. and D. A. Roff. 1998. The genetic basis of the trade-off between calling and wing morph in males of the cricket, *Gryllus firmus*. Evolution 52:1111–1118.

Crnokrak, P. and D. A. Roff. 1999. Inbreeding depression in the wild. Heredity 83:260–270.

Crnokrak, P. and D. A. Roff. 2000. The trade-off to macroptery in the cricket *Gryllus firmus*: a path analysis in males. Journal of Evolutionary Biology 13:396–408.

Croot, C. and L. Margolis (eds.). 1991. *Pacific Salmon Life Histories*. UBC Press, Vancouver.

Crow, J. F. 1993. Mutation, mean fitness, and genetic load. Pp. 3–42 *in* D. Futuyma and J. Antonovics, (eds.) *Oxford Surveys in Evolutionary Biology*. Oxford University Press, Oxford.

Crow, J. F. and M. Kimura. 1970. *An Introduction to Population Genetics Theory*. Harper & Row, New York.

Curio, E. and K. Regelmann. 1985. The behavioural dynamics of great tits (*Parus major*) approaching a predator. Zeitschrift fuer Tierpsychologie 69:3–18.

Curtsinger, J. W. 1976a. Stabilizing selection in *Drosophila melanogaster*. Journal of Heredity 67:59–60.

Curtsinger, J. W. 1976b. Stabilizing or directional selection on egg lengths? : a rejoinder. Journal of Heredity 67:246–247.

Curtsinger, J. W., P. M. Service and T. Prout. 1994. Antagonistic pleiotropy reversal of dominance and genetic polymorphism. American Naturalist 144:210–28.

Cushing, B. S. 1985. Estrous mice and vulnerability to weasel predation. Ecology 66:1976–1978.

Daan, S., C. Deerenberg and C. Dijkstra. 1996. Increased daily work precipitates natural death in the kestrel. Journal of Animal Ecology 65:539–544.

Daan, S., C. Dijkstra, R. H. Drent and T. Meijer. 1989. Food supply and the annual timing of avian reproduction. Proceedings of the XIXth International Ornithological Congress, Ottawa, 1986:392–407.

Daan, S., C. Dijkstra and J. M. Tinbergen. 1990. Family planning in the kestrel (*Falco tinnunculus*): the ultimate control of covariation of laying date and clutch size. Behaviour 114:83–116.

Daborn, G. 1975. Life history and energy relations of the giand fairy shrimp *Branchinecta gigas* Lynch 1937 (Crustacea: Anostraca). Ecology 56:1025–1039.

Danilevsky, A. S. 1965. *Photoperiodism and Seasonal Development of Insects*. Oliver and Boyd, London.

Danks, H. V. 1987. Insect dormancy: an ecological perspective. Biological Survey Canadian Monograph 1:1–439.

Danks, H. V. (ed.). 1994. Insect *Life-cycle Polymorphism: Theory, Evolution and Ecological Consequences for*

Seasonality and Diapause Control. Kluwer, Dordrecht, The Netherlands.

Darlington, P. J. J. 1943. Carabidae of mountains and islands: data on the evolution of isolated faunas, and on atrophy of wings. Ecological Monographs 13:38–61.

Darwin, C. 1876. *On the Origin of Species by Means of Natural Selection, or the Preservation of Favoured Races in the Struggle for Life*. Pickering and Chatto, London.

David, J. R., P. Gibert, E. Gravot, G. Petavy, J. Morin, D. Karan and B. Moreteau. 1997. Phenotypic plasticity and developmental temperature in *Drosophila*: Analysis and significance of reaction norms of morphological traits. Journal of Thermal Biology 22:441–451.

Davis, D. E. 1951. The analysis of population by banding. Bird-banding 22:103–107.

Davis, M. A. 1986. Geographic patterns in the flight ability of a monophagous beetle. Oecologia 69: 407–412.

Dawkins, R. 1976. *The Selfish Gene*. Oxford University Press, Oxford.

Dawkins, R. and T. R. Carlisle. 1976. Parental investment, mate desertion and a fallacy. Nature 262: 131–132.

Dawson, P. S. 1965a. Genetic homeostasis and developmental rate in *Tribolium*. Genetics 51:873–885.

Dawson, P. S. 1965b. Estimation of components of phenotypic variance for development rate in *Tribolium*. Heredity 20:403–417.

Dawson, P. S. 1975. Directional versus stabilizing selection for developmental time in natural and laboratory populations of flour beetles. Genetics 80:773–783.

Day, T. and P. D. Taylor. 1997. Von Bertalanffy's growth equation should not be used to model age and size at maturity. American Naturalist 149:381–393.

Dayton, P. K. 1971. Competition, disturbance, and community organisation: the provision and subsequent utilization of space in a rocky intertidal community. Ecological Monographs 41:351–389.

de Jong, G. 1982. Fecundity selection and maximization of equilibrium number. Netherlands Journal of Zoology 32:572–585.

de Jong, G. 1984. Selection and numbers in models of life histories. Pp. 87–101 *in* K. Wohrman and V. Loeschke, (eds.) Population Biology and Evolution. Springer-Verlag, Berlin.

de Jong, G. 1990a. Quantitative genetics of reaction norms. Journal of Evolutionary Biology 3:447–468.

de Jong, G. 1990b. Genotype-by-environment interaction and the genetic covariance between environments: multilocus genetics. Genetica 81:171–177.

de Jong, G. 1995. Phenotypic plasticity as a product of selection in a variable environment. American Naturalist 145:493–512.

de Jong, G. 1999. Unpredictable selection in a structures population leads to local genetic differentiation in evolved reaction norms. Journal of Evolutionary Biology 12:839–851.

de Jong, G. and A. J. van Noordwijk. 1992. Acquisition and allocation of resources: genetic (co)variances, selection, and life histories. American Naturalist 139:749–770.

de Jong, G. and S. C. Stearns. 1991. Phenotypic plasticity and the expression of genetic variation. Pp. 707–718 *in* E. C. Dudley, (ed.) *The Unity of Evolutionary Biology*, vol II. Dioscorides Press, Portland, Oregon.

Deakin, M. A. B. 1966. Sufficient conditions for genetic polymorphism. American Naturalist 100:690–691.

DeBenedictis, P. 1978. Frequency-dependent selection: what is the problem? Evolution 32:915–916.

Deerenberg, C., V. Appanius, S. Daan and N. Bos. 1997. Reproductive effort decreases antibody responsiveness. Procedings of the Royal Society of London B. 264:1021–1029.

Deerenberg, C., C. H. de Kogel and G. F. J. Overkamp. 1996. Costs of reproduction in the Zebra finch *Taeniopygia guttata*: manipulation of brood size in the laboratory. Journal of Avian Biology 27:321–326.

Deevey, E. S. 1947. Life tables for natural populations of animals. Quarterly Review of Biology 22:283–314.

DeFries, J. C. and R. W. Touchberry. 1961. A "maternal effect" on body weight in *Drosophila*. Genetics 46:1261–1266.

Delpuech, J., B. Moreteau, J. Chiche, E. Pla, J. Vouidibio and J. R. David. 1995. Phenotypic plasticity and reaction norms in temperature and tropical populations of *Drosophila melanogaster*: ovarian size and developmental temperature. Evolution 49:670–675.

Demas, G. E., V. Chefer, M. I. Talan and R. J. Nelson. 1997. Metabolic costs of mounting an antigen-stimulated immune response in adult and aged C57BL/6J mice. American Journal of Physiology 273:1631–1637.

Demetrius, L. 1977. Adaptedness and fitness. American Naturalist 111:1163–1168.

Den Boer, P. J. 1968. Spreading of risk and stabilization of animal numbers. Acta Biotheoretica 18:165–194.

Den Boer, P. J. 1970. On the significance of dispersal power for populations of carabid beetles (Coleoptera, Carabidae). Oecologia 4:1–28.

Den Boer, P. J. 1971. On the dispersal power of carabid beetles and its possible significance. Miscellaneous Papers, Landbouwhogeschool, Wageningen 8:119–137.

Den Boer, P. J., T. H. P. v. Huizen, W. d. Boer-Daanje, B. Aukema and C. F. M. d. Bieman. 1980. Wing polymorphism and dimorphism in ground beetles as stages in an evolutionary process (Coleoptera: Carabidae). Entomologia Generalis 6:107–134.

den Bosch, H. A. J. and R. G. Bout. 1998. Relationships between maternal size, egg size, clutch size, and hatchling size in European lacertid lizards. Journal of Herpetology 32:410–417.

Deng, H.-W. and T. T. Kibota. 1995. The importance of the environmental variance-covariance structure in predicting evolutionary responses. Evolution 49:572–574.

Denno, R. F. 1978. The optimum population strategy for planthoppers (Homoptera: Delphacidae) in stable marsh habitats. Canadian Entomologist 110: 135–142.

Denno, R. F. 1979. The relation between habitat stability and the migratory tactics of planthoppers. Miscellaneous Publications of the Entomological Society of America 11:41–49.

Denno, R. F. and E. E. Grissell. 1979. The adaptiveness of wing-dimorphism in the salt marsh-inhabiting planthopper, *Prokelisia marginata* (Homoptera: Delphacidae). Ecology 60:221–236.

Denno, R. F., M. J. Raupp, D. W. Tallamy and C. F. Reichelderfer. 1980. Migration in heterogeneous environments: differences in habitat selection between the wing forms of the dimorphic planthopper, *Prokelisia marginata* (Homoptera: Delphacidae). Ecology 61:859–867.

Denno, R. F., G. K. Roderick, K. L. Olmstead and H. G. Dobel. 1991. Density-related migration in planthoppers (Homoptera: Delphacidae): the role of habitat persistence. American Naturalist 138:1513–1541.

Denno, R. F., G. K. Roderick, M. A. Peterson, A. F. Huberty, H. G. Dobel, M. D. Eubanks, J. E. Losey and G. A. Langellotto. 1996. Habitat persistence underlies intraspecific variation in the dispersal strategies of planthoppers. Ecological Monographs 66:389–408.

Derr, J. A. 1980. The nature of variation in life history characters of *Dysdercus bimaculatus* (Heteroptera: Pyrrhocoridae), a colonizing species. Evolution 34:548–557.

Desharnais, R. A. and R. F. Costantino. 1983. Natural selection and density-dependent population growth. Genetics 105:1029–1040.

Dhondt, A. A., F. Adriaensen, E. Matthysen and B. Kempenaers. 1990. Nonadaptive clutch sizes in tits. Nature 348:723–725.

Dickerson, G. E. 1947. Composition of hog cargasses as influenced by heritable differences in rate and economy of gain. Iowa Agricultural Experimental Station, Research Bulletin 354:489–524.

Dickerson, G. E. 1955. Genetic slippage in response to selection for multiple objectives. Cold Spring Harbour Symposium on Quantitative Biology 20:213–223.

Dickins, D. W. and R. A. Clark. 1987. Games theory and siblicide in the kittiwake gull, *Rissa tridactyla*. Journal of Theoretical Biology 125:301–305.

Dijkstra, C., A. Bijlsma, S. Daan, T. Meijer and M. Zijltra. 1990. Brood size manipulations in the kestrel (*Falco tinnunculus*): effects on offspring and parent survival. Journal of Animal Ecology 59:269–285.

Dijkstra, L. J. 1986. Optimal selection and exploitation of hosts in the parasitic wasp *Colpoclyeus florus* (Hymenoptera: Eulophidae). Netherlands Journal of Zoology 36:177–301.

Dingle, H. 1979. Adaptive variation in the evolution of insect migration. Pp. 64–87 *in* R. L. Rabb and G. G. e. Kennedy, (eds.) *Movements of Highly Mobile Insects: Concepts and Methodology.* North Carolina State University Press, Raleigh, North Carolina.

Dingle, H. 1980. Ecology and the evolution of migration. Pp. 1–101 *in* J. S. A. Gauthreaux, (ed.) *Animal Migration, Orientation and Navigation.* Academic Press, New York, New York.

Dingle, H. 1984. Behavior, genes, and life histories: complex adaptations in uncertain environments. Pp. 169–184 *in* C. N. Slobodchikoff, W. S. Gaud. and P. W. Price, (eds.) A New Ecology: Novel Approaches to Interactive Systems. John Wiley & Sons, Inc.

Dingle, H. 1985. Migration and Life Histories. Contributions to Marine Science 27 (supplement):27–44.

Dingle, H. 1996. *Migration.* Oxford University Press, New York.

Dixon, A. F. G., R. Hundu and P. Kindlmann. 1993. Reproductive effort and maternal age in iteroparous insects using aphids as a model group. Functional Ecology 7:267–272.

Dobson, A. P. 1990. Survival rates and their relationship to life-history traits in some common British birds. Current Ornithology 7:115–146.

Dobzhansky, T. 1950. Evolution in the tropics. American Scientist 38:209–221.

Dobzhansky, T., B. Spassky and J. Sved. 1969. Effects of selection and migration on geotactic and phototactic behaviour of *Drosophila*. II. Proceedings of the Royal Society, B 173:191–207.

Dodson, S. 1989. Predator-induced reaction norms. BioScience 39:447–452.

Doebeli, M. and G. D. Ruxton. 1997. Evolution of dispersal rates in metapopulation models: branching and cyclic dynamics in phenotype space. Evolution 51:1730–1741.

Doebelli, M. 1995. Dispersal and dynamics. Theoretical Population Biology 47:82–106.

Dominey, W. 1980. Female mimicry in male bluegill sunfish—a genetic polymorphism? Nature 284: 546–548.

Dominey, W. J. 1984. Alternative mating tactics and evolutionarily stable strategies. American Zoologist 24:385–396.

Dorn, L. A. and T. Mitchell-Olds. 1991. Genetics of *Brassica campestris*. 1. Genetic constraints on evolution of life-history characters. Evolution 45:371–379.

Doward, E. F. 1962. Comparative biology of the white booby and brown booby *Sula* spp at Ascension. Ibis 103b:74–220.

Drent, R. and S. Daan. 1980. The prudent parent: energetic adjustments in avian biology. Ardea 68:225–252.

Drickhamer, L. C. 1974. A ten-year summary of reproductive data for free-ranging *Macaca mulatta*. Folia Primatologica 21:61–80.

Duarte, C. M. and M. Alcaraz. 1989. To produce many small or few large eggs: a size-independent reproductive tactic of fish. Oecologia 80:401–404.

Dudash, M. R. 1990. Relative fitness of selfed and outcrossed progeny in a self-compatible, protandrous species, *Sabatia angularis* L.(Gentianaceae): a comparison in three environments. Evolution 44:1129–1139.

Dudley, J. W. 1977. 76 generations of selection for oil and protein percentage in maize. Pp. 459–473 *in* E. Pollack, O. Kempthorne. and. T. B. Bailey. (eds.). *Proceedings of the International Conference on Quantitative Genetics.* Iowa State University Press, Ames, Iowa.

Dunham, A. E. and D. B. Miles. 1985. Patterns of covariation in life history traits of squamate reptiles: the effects of size and phylogeny reconsidered. American Naturalist 126:231–257.

Dunham, A. E., D. B. Miles and D. N. Reznick. 1988. Life-history patterns in squamate reptiles. Pp. 441–521 *in* C. Gans and R. B. Huey, (eds.) *Biology of the Reptilia.* Alan R. Liss, Inc., New York.

Dupont-Nivet, M., J. Mallard, J. C. Bonnet and J. M. Blanc. 1997. Quantitative genetics of growth traits in the edible snail, *Helix aspersa* Müller. Genetics Selection Evolution 29:571–587.

Dutilleul, P. and C. Potvin. 1995. Among-environment heteroscedasticity and genetic autocorrelation: Implications for the study of phenotypic plasticity. Genetics 139:1815–29.

Easton, D. M. 1997. Gompertz growth in number dead confirms medflies and nematodes show excess oldster survival. Experimental Gerontology 32:719–726.

Eberhard, W. G. 1977. Aggressive chemical mimicry by a bolas spider. Science 198:1173–1175.

Eberhard, W. G. 1980. Horned Beetles. Scientific American March:166–182.

Eberhard, W. G. 1980. The natural history and behavior of the bolas spider *Mastophora dizzydeani* sp. n. (Araneidae). Psyche 87:143–169.

Eberhard, W. G. 1982. Beetle horn dimorphism: making the best of a bad lot. American Naturalist 119:420–426.

Eckert, C. G. and S. C. H. Barrett. 1994. Inbreeding depression in partially self-fertilizing *Decodon verticillatus* (Lythraceae): population-genetic and experimental analyses. Evolution 48:952–964.

Edley, M. T. and R. Law. 1988. Evolution of life histories and yields in experimental populations of *Daphnia magna.* Biological Journal of the Linnean Society 34:309–326.

Eisen, E. J. 1972. Long-term selection response for 12-day litter weight in mice. Genetics 72:129–142.

Eisen, E. J., J. E. Legates and O. W. Robison. 1970. Selection for 12-day litter weight in mice. Genetics 64:511–532.

Eisenberg, J. F. 1988. *Reproduction in Polyprotodont Marsupials and Similae-sized Eutherians with a Speculation Concerning the Evolution of Litter Size in Mammals.* Yale University Press, New Haven.

Elgar, M. A. 1990. Evolutionary compromise between a few large and many small eggs: comparative evidence in teleost fish. Oikos 59:283–287.

Elgar, M. A. and C. P. Catterall. 1989. Density-dependent natural selection. Trends in Ecology and Evolution 4:95–96.

Ellner, P. S., J. Hairston, N. G., C. M. Kearns and D. Babai. 1999. The roles of fluctuating selection and long-term diapause in microevolution of diapause timing in a freshwater copepod. Evolution 53: 111–122.

Ellner, S. 1996. Environmental fluctuations and the maintenance of genetic diversity in age or stage-structured populations. Bulletin of Mathematical Biology 58:103–27.

Ellner, S. and J. Hairston, N. G. 1994. Role of overlapping generations in maintaining genetic variation in a fluctuating environment. American Naturalist 143:403–417.

Ellner, S. and A. Sasaki. 1996. Patterns of genetic polymorphism maintained by fluctuating selection with overlapping generations. Theoretical Population Biology 50:31–65.

Elmes, G. W. 1991. Mating strategy and isolation between the two forms, macrogyna and microgyna, of *Myrmica ruginodis* (Hymenoptera: Formicidae). Ecological Entomology 16:411–423.

Endler, J. 1980. Natural selection on color patterns in *Poecilia reticulata.* Evolution 34:319–364.

Endler, J. A. 1986. *Natural Selection in the Wild.* Princeton University Press, Princeton, New Jersey.

Endler, J. A. 1987. Predation, light intensity and courtship behaviour in *Poecilia reticulata* (Pisces: Poeciliidae). Animal Behaviour 35:1376–1385.

Enfield, F. D. 1980. Long term effects of selection: the limits to response. Pp. 69–86 *in* A. Robertson, (ed.) *Selection Experiments in Laboratory and Domestic Animals.* Commonwealth Agricultural Bureau, Slough, U.K.

Enfield, F. D., R. E. Comstock and O. Braskerud. 1966. Selection for pupa weight in *Tribolium casteneum.* I. Parameters in base populations. Genetics 54: 523–533.

Engen, S. and B. Saether. 1994. Optimal allocation of resources to growth and reproduction. Theoretical Population Biology 46:232–248.

Englert, D. C. and A. E. Bell. 1970. Selection for time of pupation in Tribolium casteneum. Genetics 64:541 552.

Englert, D. C. and A. E. Bell. 1969. Components of growth in genetically diverse populations of *Tribolium casteneum.* Canadian Journal of Genetics and Cytology 11:896–907.

Engstrom, G., L.-E. Liljedahl and T. Bjorklund. 1992. Expression of genetic and environmental variation during ageing 2. Selection for increased lifespan in *Drosophila melanogaster.* Theoretical and Applied Genetics 85:26–32.

Enquist, B. J., G. B. West, E. L. Charnov and J. H. Brown. 1999. Allometric scaling of production and life-history variation in vascular plants. Nature 401:907–911.

Erasmus, J. E. 1962. Part-period selection for egg production. Proceedings of the 12th World Poultry Congress, Sydney 1962:17–18.

Eshel, I. 1982. Evolutionarily stable strategies and viability selection in Mendelian populations. Theoretical Population Biology 22:204–217.

Ewens, W. J. 1989. An interpretation and proof of the fundamental theorem of natural selection. Theoretical Population Biology 36:167–180.

Ewens, W. J. 1992. An optimizing principle of natural selection in evolutionary population genetics. Theoretical Population Biology 42:333–346.

Fagen, R. M. 1972. An optimal life history in which reproductive effort decreases with age. American Naturalist 106:258–261.

Fairbairn, D. and L. Desranleau. 1987. Flight threshold, wing muscle histolysis, and alary polymorphism: Correlated traits for dispersal tendency in the Gerridae. Ecological Entomology 12:13–24.

Fairbairn, D. J. 1977. Why breed early? A study of reproductive tactics in *Peromyscus*. Canadian Journal of Zoology 55:862–871.

Fairbairn, D. J. 1986. Does alary polymorphism imply dispersal dimorphism in the waterstrider, *Gerris remigis*? Ecological Entomology 11:355–368.

Fairbairn, D. J. 1988. Sexual selection for homogamy in the Gerridae: an extension of Ridley's comparative approach. Evolution 42:1212–1222.

Fairbairn, D. J. 1990. Factors influencing sexual size dimorphism in temperate waterstriders. American Naturalist 136:61–86.

Fairbairn, D. J. 1994. Wing dimorphism and the migratory syndrome: Correlated traits for migratory tendency in wing dimorphic insects. Researches in Population Ecology 36:157–163.

Fairbairn, D. J. and T. C. Butler. 1990. Correlated traits for migration in the Gerridae (Hemiptera, Heteroptera): A field test. Ecological Entomology 15:131–142.

Fairbairn, D. J. and D. A. Roff. 1990. Genetic correlations among traits determining migratory tendency in the sand cricket, *Gryllus firmus*. Evolution 44:1787–1795.

Fairbairn, D. J. and D. E. Yadlowski. 1997. Coevolution of traits determining migratory tendency: correlated response of a critical enzyme, juvenile hormone esterase, to selection on wing morphology. Journal of Evolutionary Biology 10:495–513.

Fairbairn, D. J. F. 1993. Costs of loading associated with mate-carrying in the waterstrider, *Aquarius remigis*. Behavioral Ecology 4:224–231.

Fairbairn, D. J. F. and J. Reeve. 2001 (in press). Natural selection *in* C. W. Fox, D. A. Roff and D. J. Fairbairn, (eds.) *Evolutionary Ecology: Concepts and Case Studies.* Oxford University Press, Oxford.

Fairfull, R., L. Haley and J. Castell. 1981. The early growth of artificially reared American lobsters. Part 1: Genetic paramters and environments. Theoretical and Applied Genetics 60:269–273.

Falconer, D. S. 1952. The problem of environment and selection. American Naturalist 86:293–298.

Falconer, D. S. 1960. Selection of mice for growth on high and low planes of nutrition. Genetical Research 1:91–113.

Falconer, D. S. 1965. Maternal effects and selection response. Pp. 763–774 *in* S. J. Geerts, (ed.) *Genetics Today. Proceedings of the XI International Congress of Genetics.* Volume 3. Pergamon Press, Oxford, U.K.

Falconer, D. S. 1981. *Introduction to Quantitative Genetics*, Second Edition. Longmans, London and New York.

Falconer, D. S. 1989. *Introduction to Quantitative Genetics.* Longmans, New York.

Feijen, H. R. and G. G. M. Schulten. 1981. Egg parasitoids (Hymen., Trichogrammatidae) of *Diopsis macrophthalma* in Malawi. Netherlands Journal of Zoology 31:381–417.

Felsenstein, J. 1976. The theoretical population genetics of variable selection and migration. Annual Review of Genetics 10:253–280.

Felsenstein, J. 1977. Multivariate normal genetic models with finite number of loci. Pp. 227–246 *in* E. Pollack, O. Kempthorne and T. B. Bailey, (eds.) Proceedings of the the International Conference on Quantitative Genetics. Iowa State University Press, Ames, Iowa.

Ferguson, G. W. and S. F. Fox. 1984. Annual variation of survival advantage of large juvenile side-blotched lizards *Uta stansburiana*: its causes and evolutionary significance. Evolution 38:342–349.

Ferguson, I. M. and D. J. F. Fairbairn. 2000. Sex-specific selection and sexual size dimorphism in the waterstrider *Aquarius remigis*. Journal of Evolutionary Biology 13:160–170.

Ferriere, R. and M. Gatto. 1995. Lyapunov exponents and the mathematics of invasion in oscillatory or chaotic populations. Theoretical Population Biology 48:126–171.

Festa-Bianchet. 1989. Individual differences, parasites, and the costs of reproduction for bighorn ewes (*Ovis canadensis*). Journal of Animal Ecology 58:785–795.

Ffrench-Constant, R. H. 1994. The molecular and population genetics of cyclodiene insecticide resistance. Insect Biochemistry and Molecular Biology 24:335–45.

Findlay, C. S. and F. Cooke. 1983. Genetic and environmental components of clutch size variance in a wild population of lesser snow geese (*Anser caerulescens caerulescens*). Evolution 37:724–734.

Fisher, R. A. 1918. The correlation between relatives on the supposition of Mendelian inheritance. Transactions of the Royal Society of Edinburgh 52:399–433.

Fisher, R. A. 1930. *The Genetical Theory of Natural Selection*. Claredon Press, Oxford.

Fitch, H. S. 1957. Aspects of reproduction and development in the prairie vole (*Microtus ochrogaster*). University of Kansas, Publications of the Museum of Natural History 10:129–161.

Fleischer, R. C. and R. F. Johnston. 1984. The relationship between winter climate and selection on body size of house sparrows. Canadian Journal of Zoology 62:405–410.

Fleiss, J. L. 1981. *Statistical methods for rates and proportions*. John Wiley & Sons.

Fleming, T. H. and R. J. Rauscher. 1978. On the evolution of litter size in *Peromyscus leucopus*. Evolution 32:45–55.

Flux, J. E. C. and M. M. Flux. 1982. Artificial selection and gene flow in wild starlings, *Sturnus vulgaris*. Naturwissensch 69:96–97.

Foley, P. 1992. Small population genetic variability at loci under stabilizing selection. Evolution 46:763–774.

Forbes, L. S. 1991. Insurance offspring and brood reduction in a variable environment: the cost and benefits of pessimism. Oikos 62:325–332.

Forbes, L. S. 1994. The good the bad and the ugly: Lack's brood reduction hypothesis and experimental design. Journal of Avian Biology 25: 338–243.

Forbes, M. R. L. 1993. Parasitism and host reproductive effort. Oikos 67:444–450.

Ford, L. S. and R. C. Ydenberg. 1992. Sibling rivalry in a variable environment. Theoretical Population Biology 41:135–160.

Ford, N. B. and R. A. Seigel. 1989a. Relationships among body size, clutch size, and egg size in three species of oviparous snakes. Herpetologica 45:75–83.

Ford, N. B. and R. A. Siegel. 1989b. Phenotypic plasticity in reproductive traits: evidence from a viviparous snake. Ecology 70:1768–1774.

Fordham, R. H. 1971. Field populations of deermice with supplemental food. Ecology 52:138–146.

Fox, C., D. A. Roff and D. J. F. Fairbairn. 2001. *Evolutionary Ecology: Concepts and Case Studies*. Oxford University Press, New York.

Fox, C. W. 1993. The influence of maternal age and mating frequency on egg size and offspring performance in *Callosobruchus maculatus* (Coleoptera: Bruchidae). Oecologia 96:139–146.

Fox, C. W. 1993. A quantitative genetic analysis of oviposition preference and larval performance on two hosts in the bruchid beetle, *Callosobruchus maculatus*. Evolution 47:166–175.

Fox, C. W. 1994. The influence of egg size on offspring performance in the seed beetle, *Callosobruchus maculatus*. Oikos 71:321–325.

Fox, C. W. and M. E. Czesak. 2000. Evolutionary ecology of progeny size in arthropods. Annual Review of Entomology 45:341–369.

Fox, C. W., M. E. Czesak, T. A. Mousseau and D. A. Roff. 1999. The evolutionary genetics of an adaptive maternal effect: egg size plasticity in a seed beetle. Evolution 53:552–560.

Fox, C. W., L. A. McLennan and T. Mousseau. 1995. Male body size affects female lifetime reproductive success in a seed beetle. Animal Behaviour 50: 281–284.

Fox, C. W. and T. Mousseau. 1996. Larval host plant affects the fitness consequences of egg size in the seed beetle *Stator limbatus*. Oecologia 107:541–548.

Fox, C. W., M. S. Thakar and T. A. Mousseau. 1997. Egg size plasticity in a seed beetle: an adaptive maternal effect. American Naturalist 149:149–163.

Frahm, R. R. and K. Kojima. 1966. Comparison of selection responses on body weight under divergent larval density conditions in *Drosophila pseudoobscura*. Genetics 54:625–637.

Frank, S. A. 1986a. Dispersal polymorphisms in subdivided populations. Journal of Theoretical Biology 122:303–309.

Frank, S. A. 1986b. Hierarchical selection theory and sex ratios. 1. General solutions for structured populations. Theoretical Population Biology 29:312–342.

Frank, S. A. 1993. A model of inducible defense. Evolution 47:325–327.

Frank, S. A. and M. Slatkin. 1990. Evolution in variable environments. American Naturalist 136:244–260.

Frank, S. A. and M. Slatkin. 1992. Fisher's fundamental theorem of natural selection. Trends in Ecology and Evolution 7:92–95.

Frankham, R. 1977. Optimum selection intensities in artificial selection programmes: an experimental evaluation. Genetical Research 30:115–119.

Frankham, R. 1990. Are responses to artificial selection for reproductive fitness characters consistently asymmetrical? Genetical Research 56:35–42.

Fraser, A. S. 1960. Simulation of genetic systems by automatic digital computers. VI. Epistasis. Australian Journal of Biological Science 13:150–162.

Fry, J. D. 1992. The mixed-model analysis of variance applied to quantitative genetics: biological meaning of the parameters. Evolution 46:540–550.

Fukui, H. H., L. Xiu and J. W. Curtsinger. 1993. Slowing of age-specific mortality rates in *Drosophila melanogaster*. Experimental Gerontology 28:585–599.

Gadgil, M. and W. H. Bossert. 1970. Life historical consequences of natural selection. American Naturalist 104:1–24.

Gadgil, S., V. Nanjundiah and M. Gadgil. 1980. On evolutionarily stable compositions of populations of interacting genotypes. Journal of Theoretical Biology 84:737–759.

Gage, T. B. and B. Dyke. 1988. Model life tables for the larger Old World Monkeys. American Journal of Primatology 16:305–320.

Galbraith, M. G. J. 1967. Size-selective predation on *Daphnia* by rainbow trout and yellow perch. Transactions of the American Fisheries Society 96:1–10.

Gale, J. S. and M. J. Kearsey. 1968. Stable equilibria under stabilising selection in the absence of dominance. Heredity 23:553–561.

Gall, G. A. E. 1971. Replicated selection for 21-day pupa weight of *Tribolium castaneum*. Theoretical and Applied Genetics 41:164–173.

Gallardo, M. H. and N. Kohler. 1994. Demographic changes and genetic losses in populations of a subterranean rodent (*Ctenomys maulinus* Brunneus) affected by a natural catastrophe. Z. Saugetierkunde 59:358–365.

Gallardo, M. H., N. Kohler and C. Araneda. 1995. Bottleneck effects in local populations of fossorial *Ctenomys* (Rodentia, Ctenomyidae) affected by vulcanism. Heredity 74:638–646.

Gallego, A. and C. Lopez-Fanjul. 1983. The number of loci affecting a quantitative trait in *Drosophila melanogaster* revealed by artificial selection. Genetical Research 42:137–149.

Galton, F. 1889. *Natural Inheritance*. MacMillan and Co., London.

Garland, T. J. 1985. Ontogenetic and individual variation in size, shape and speed in the Australian agamid lizard *Amphibolurus nuchalis*. Journal of Zoology 207:425–439.

Garland, T. J. and P. L. Else. 1987. Seasonal, sexual, and individual variation in endurance and activity metabolism in lizards. American Journal of Physiology 252:R439–R449.

Gauthreaux, S. J. (ed.). 1980. *Animal Migration, Orientation, and Navigation*. Academic Press, New York.

Gavrilets, S. and S. M. Scheiner. 1993a. The genetics of phenotypic plasticity. VI. Theoretical predictions for directional selection. Journal of Evolutionary. Biology 6:49–68.

Gavrilets, S. and S. M. Scheiner. 1993b. The genetics of phenotypic plasticity. V .Evolution of reaction norm shape. Journal of Evolutionary Biology 6:31–48.

Geber, M. A. 1990. The cost of meristem limitation in *Polygonum arenastrum*: negative genetic correlations between fecundity and growth. Evolution 44:799–819.

Gemmil, A. W., A. Skorping and A. F. Read. 1999. Optimal timing of first reproduction in parasitic nematodes. Journal of Evolutionary Biology 12:1148–1156.

Genoud, M. and N. Perrin. 1994. Fecundity versus offspring size in the greater white-toothed shrew, *Crocidura russula*. Journal of Animal Ecology 63:328–336.

Gerard, C. and A. Theron. 1997. Age/size- and time-specific effects of *Schisosoma mansoni* on energy allocaton patterns of its snail host *Biomphalaria glabrata*. Oecologia 112:447–452.

Gibb, J. A. 1961. Bird populations. Pp. 413–446 *in* A. J. Marshall, (ed.) *Biology and Comparative Physiology of Birds*. Academic Press, New York.

Gibbons, J. W., J. L. Greene and K. K. Patterson. 1982. Variation in reproductive characteristics of aquatic turtles. Copeia 1982:776–784.

Gibson, J. B. and B. P. Bradley. 1974. Stabilising selection in constant and fluctuating environments. Heredity 33:293–302.

Giesel, J. T. 1974. Fitness and polymorphism for net frequency distribution in iteroparous populations. American Naturalist 108:321–331.

Giles, N. 1984. Implications of parental care of offspring for the anti-predator behaviour of adult male and female three-spined sticklebacks, *Gasterosteus aculeatus* L. Pp. 275–289 *in* P. G. W. and R. J. Wootton, (eds.) *Fish Reproduction, Strategies and Tactics*. Academic Press, London.

Gill, D. E. 1972. Intrinsic rate of increase, saturation densities and competitive ability. I. An experiment with *Paramecium*. American Naturalist 106:461–471.

Gill, D. E. 1974. Intrinsic rate of increase, saturation density, and competitive ability II. The evolution of competitive ability. American Naturalist 108:103–116.

Gillespie, J. H. 1973. Polymorphism in random environments. Theoretical Population Biology 4:193–195.

Gillespie, J. H. 1974. Natural selection for within-generation variance in offspring number. Genetics 76:601–606.

Gillespie, J. H. 1975. The role of migration in the genetic structure of populations in temporally and spatially varying environments I. Conditions for polymorphism. American Naturalist 109:127–135.

Gillespie, J. H. 1976. A general model to account for enzyme variation in natural populations II Characterization of the fitness function. American Naturalist 110:809–821.

Gillespie, J. H. 1977a. A general model to account for enzyme variation in natural populations. III. Multiple alleles. Evolution 31:85–90.

Gillespie, J. H. 1977b. Natural selection for variance in offspring numbers: a new evolutionary principle. American Naturalist 111:1010–1014.

Gillespie, J. H. 1978. A general model to account for enzyme variation in natural populations. V. The SAS-CFF model. Theoretical Population Biology 14:1–45.

Gillespie, J. H. and M. Turelli. 1989. Genotype-environment interactions and the maintenance of polygenic variation. Genetics 121:129–138.

Gimelfarb, A. 1986. Additive variation maintained under stabilizing selection: a two-locus model of pleiotropy for two quantitative characters. Genetics 112:717–725.

Gimelfarb, A. 1989. Genotypic variation for a quantitative character maintained under stabilizing selection without mutations: epistasis. Genetics 123:217–227.

Gjerde, B. and T. Gjedrem. 1984. Estimates of phenotypic and genetic parameters for carcass traits in Atlantic salmon and rainbow trout. Aquaculture 36:97–110.

Gjerde, B. and L. R. Schaeffer. 1989. Body traits in rainbow trout II. Estimates of heritabilities and of phenotypic correlations. Aquaculture 80:25–44.

Gliwicz, Z. M. and C. Guisande. 1992. Family planning in *Daphnia*—resistance to starvation in offspring born to mothers grown at different food levels. Oecologia 91:463–467.

Glynn, P. W. 1970. On the ecology of the Caribbean chitons *Acanthopleura granulata* Gmelin and *Chiton tuberculatus* Linne: density, mortality, feeding, reproduction and growth. Smithsonian Contributions in Zoology 66:1–21.

Godfray, H. C. J. 1986. Clutch size in a leaf-mining fly (*Pegomya nigritarsis*: Anthomyiidae). Ecological Entomology 11:75–81.

Godfray, H. C. J. 1987. The evolution of clutch size in invertebrates. Oxford Surveys in Evolutionary Biology 4:117–154.

Godfray, H. C. J. and A. B. Harper. 1990. The evolution of brood reduction by siblicide in birds. Journal of Theoretical Biology 145:163–175.

Godfray, H. C. J., L. Partridge and P. H. Harvey. 1991. Clutch size. Annual Review of Ecology and Systematics 22:409–429.

Godin, J. J. and S. A. Smith. 1988. A fitness cost to foraging in the guppy. Nature 333:69–71.

Goldsmith, S. K. 1985. Male dimorphism in *Dendrobias mandibularis* Audinet-Serville (Coleoptera: Ceramycidae). Journal of the Kansas Entomological Society 58:534–538.

Goldsmith, S. K. and J. Alcock. 1993. The mating chances of small males of the Cerambycid Beetle *Trachyderes mandibularis* differ in different environments (Coleoptera: Cerambycidae). Journal of Insect Behavior 6:351–360.

Gomez, J. M. 2000. Phenotypic selection and response to selection in *Lobularia maritima*: importance of direct and correlational components of natural selection. Journal of Evolutionary Biology 13:689–699.

Gomulkiewicz, R. and M. Kirkpatrick. 1992. Quantitative genetics and the evolution of reaction norms. Evolution 46:390–411.

Gonor, J. J. 1972. Gonad growth in the sea urchin, *Strongylocentrotus purpuratus* (Stimpson) (Echinodermata: Echinoidea) and the assumptions of gonad index methods. Journal of Experimental Marine Biology and Ecology 10:89–103.

Goodman, D. 1974. Natural selection and a cost ceiling on reproductive effort. American Naturalist 108:247–268.

Goodman, D. 1982. Optimal life histories, optimal notation, and the value of reproductive value. American Naturalist 119:803–823.

Goodman, D. 1984. Risk spreading as an adaptive strategy in iteroparous life histories. Theoretical Population Biology 25:1–20.

Goodnight, C. 1987. On the effect of founder events on epistatic genetic variance. Evolution 41:80–91.

Goodnight, C. J. 1988. Epistasis and the effect of founder events on the additive genetic variance. Evolution 42:441–454.

Goodnight, C. J. 1995. Epistasis and the increase in additive genetic variance: implications for phase 1 of Wright's shifting-balance process. Evolution 49:502–511.

Gotthard, K. 1998. Life history plasticity in the satyrine butterfly *Lasiommata petropolitana*. Journal of Evolutionary Biology 11:21-39.

Gotthard, K. 1999. Life History Analysis of Growth Strategies in Temperate Butterflies. Chapter V: Increased risk of predation as a cost of high growth rate: an experimental test in the speckled wood butterfly, *Pararge aegeria*. Department of Zoology. Stockholm, Stockholm.

Gotthard, K., S. Nylin and C. Wiklund. 1994. Adaptive variation in growth rate: life history costs and consequences in the speckled wood butterfly *Pararge aegeria*. Oecologia 99:281–289.

Gotthard, K., S. Nylin and C. Wiklund. 1999. Seasonal plasticity in two satyrine butterflies: state-dependent decision making in relation to daylength. Oikos 84.

Grafen, A. and R. Sibly. 1978. A model of mate desertation. Animal Behaviour 26:645–652.

Grant, A. 1997. Selection pressures on vital rates in density dependent populations. Proceedings of the Royal Society, B 264:303–306.

Grau, C. R. 1984. Egg formation. Pp. 33–58 *in* C. C. Whittow and R. Rahn, (eds.) *Seabird Energetics*. Plenum, New York.

Graves, J. 1991. Comments on the sample sizes use to test the effect of experimental brood enlargment on adult survival. Auk 108:967–969.

Green, J. 1954. Growth, size and reproduction in *Daphnia magma* (Crustacea: Cladocera). Proceedings of the Zoological Society of London 124:535–545.

Green, J. 1967. The distribution and variation of *Daphnia lumholtzii* (Crustacea: Cladocera) in relation to fish predation in Lake Albert, East Africa. Journal of Zoology 151:181–197.

Green, R. and P. R. Painter. 1975. Selection for fertility and development. American Naturalist 109:1–10.

Green, R. F. 1980. A note on *r*- and *K*-selection. American Naturalist 116:291–296.

Green, R. H. and K. D. Hobson. 1970. Spatial and temporal structure in a temperate intertidal community, with special emphasis on *Gemma gemma* (Pelecypoda: Mollusca). Ecology 51:999–1011.

Greene, E. 1989. A diet-induced developmental polymorphism in a caterpillar. Science 243:643–646.

Greenwood, P. J. 1980. Mating systems, philopatry and dispersal in birds and mammals. Animal Behaviour 28:1140–1162.

Greenwood, P. J. 1983. Mating systems and the evolutionary consequences of dispersal. Pp. 116–131 *in* I. R. Swingland and P. J. Greenwood, (eds.) *The Ecology of Animal Movement*. Claredon Press, Oxford.

Grill, C. P. and A. J. Moore. 1998. Effects of a larval antipredator response and larval diet on adult phenotype in an aposematic ladybird beetle. Oecologia 114:274–282.

Gromko, M. H. 1977. What is frequency-dependent selection? Evolution 31:438–442.

Gromko, M. H. 1987. Genetic constraint on the evolution of courtship behaviour in *Drosophila melanogaster*. Heredity 58:435–441.

Gromko, M. H., A. Briot, S. C. Jensen and H. H. Fukui. 1991. Selection on copulation duration in *Drosphila melanogaster*: predictability of direct response versus unpredictibility of correlated response. Evolution 45:69–81.

Gross, M. R. 1982. Sneakers, satellites and parentals: polymorphic mating strategies in North American sunfishes. Zeitschrift fuer Tierpsychologie 60:1–26.

Gross, M. R. 1985. Disruptive selection for alternative life histories in salmon. Nature 313:47–48.

Gross, M. R. 1991. Evolution of alternative reproductive strategies: frequency-dependent sexual selection in male bluegill sunfish. Philosophical Transactions of the Royal Society of London, B 332:59–66.

Gross, M. R. and E. L. Charnov. 1980. Alternative male life histories in bluegill sunfish. Proceedings of the National Academy of Sciences, USA 77:6937–6940.

Grossberg, R. K. 1988. Life-history variation within a population of the colonial ascidian *Botryllus schlosseri*. Evolution 42:900–920.

Grossman, M. and H. W. Norton. 1974. Simplification of the sampling variance of the correlation coefficients. Theoretical and Applied Genetics 44:332.

Gu, H. and J. S. F. Barker. 1995. Genetic and phenotypic variation for flight ability in the cactophilic *Drosophila* species, *D. aldrichi* and *D. buzzatii*. Entomologia Experimentalis et Applicata 76:25–35.

Guisande, C., J. Sanchez, I. Maneiro and A. Miranda. 1996. Trade-off between offspring number and offspring size in the marine copepod *Euterpina acutifrons* at different food concentrations. Marine Ecology Progress Series 143:37–44.

Gunderson, D. R. 1997. Trade-off between reproductive effort and adult survival in oviparous and viviparous fishes. Canadian Journal of Fisheries and Aquatic Sciences 54: 990–998.

Gunderson, D. R. 1980. Using *r-K* selection theory to predict natural mortality. Canadian Journal of Fisheries and Aquatic Sciences 37:2266–2271.

Gunnarsson, B. 1988. Body size and survival: implications for an overwintering spider. Oikos 52: 274–282.

Guntrip, J., R. M. Sibly and R. H. Smith. 1997. Controlling resource acquisition to reveal a life history trade-off: egg mass and clutch size in an iteroparous seed predator, *Prostephanus truncatus*. Ecological Entomology 22:264–270.

Gurney, W. S. C. and D. A. J. Middleton. 1996. Optimal resource allocation in a randomly varying environment. Functional Ecology 10:602–612.

Gustafsson, L. 1986. Lifetime reproductive success and heritabilities: empirical support for Fisher's fundamental theorem. American Naturalist 128:761–764.

Gustafsson, L., D. Nordling, M. S. Andersson, B. C. Sheldon and A. Qvarnstrom. 1994. Infectious deseases, reproductive effort and the cost of reproduction in birds. Philosophical Transactions of the Royal Society of London, B. 346:323–331.

Gwynne, D. T. and G. N. Dodson. 1983. Nonrandom provisioning by the digger wasp, *Palmedes laeviventris* (Hymenoptera: Sphecidae). Annals of the Entomological Society of America 76:434–436.

Gwynne, D. T. and K. M. O'Neill. 1980. Territoriality in digger wasps result in sex biased predation on males (Hymenoptera: Sphecidae, *Philanthus*). Journal of the Kansas Entomological Society 53:220–224.

Haccou, P. and Y. Iwasa. 1995. Optimal mixed strategies in stochastic environments. Theoretical Population Biology 47:212–243.

Hairston, N. G. J., S. Ellner and C. M. Kearns. 1996. Overlapping generations: the storage effect and he maintenance of biotic diversity. Pp. 109–145 *in* J. O. E. Rhodes, R. K. Chesser and M. H. Smith, (eds.) Population Dynamics in Ecological Space and Time. The University of Chicago Press, Chicago.

Hairston, N. G. J., and E. J. Olds. 1987. Population differences in the timing of diapause: a test of hypotheses. Oecologia 71:339–344.

Hairston, N. G. J., C. M. Kearns and S. P. Ellner. 1996. Phenotypic variation in a zooplankton egg bank. Ecology 77:2382–2392.

Hairston, N. G. J., and E. J. Olds. 1984. Population differences in the timing of diapause: adaption in a spatially heterogeneous environment. Oecologia 61:42–48.

Hairston, N. G. J., D. W. Tinkle and H. M. Wilbur. 1970. Natural selection and the parameters of population growth. Journal of Wildlife Management 34:681–690.

Hairston, N. G. J. and T. A. Dillon. 1990. Fluctuating selection and response in a population of freshwater copepods. Evolution 44:1796–1805.

Hairston, N. G. J. and C. M. Kearns. 1995. The interaction of photoperiod and temperature in diapause timing: A copepod example. Biological Bulletin 189:42–8.

Hairston, N. G. J. and W. R. J. Munns. 1984. The timing of copepod diapause as an evolutionarily stable strategy. American Naturalist 123:733–751.

Hairston, N. G. J. and E. J. Olds. 1986. Partial photoperiodic control of diapause in three populations of the freshwater copepod *Diaptomus sanguineus*. Biological Bulletin 171:135–142.

Hairston, N. G. J., W. E. Walton and K. T. Li. 1983. The causes and consequences of sex-specific mortality in a freshwater copepod. Limnology and Oceanography 28:935–947.

Haldane, J. B. S. 1954. The measurement of natural selection. Proceedings of the 9th International

Congress of Genetics (Caryologia, Supplement to Volume 6) 1:480–487.

Haldane, J. B. S. and S. D. Jayakar. 1963a. Polymorphism due to selection of varying direction. Journal of Genetics 58:237–242.

Haldane, J. B. S. and S. D. Jayakar. 1963b. Polymorphism due to selection depending on the composition of a population. Journal of Genetics 58:318-323.

Hall, C. A. S. 1988. An assessment of several of the historically most influential theoretical models used in ecology and of the data provided in their support. Ecological Modelling 43:5–31.

Hall, D. L., S. T. Threlkeld, C. W. Burns and P. H. Crowley. 1976. The size efficiency hypothesis and the size structure of zooplankton communities. Annual Review of Ecology and Systematics 7:177–208.

Hamilton, W. D. and R. M. May. 1977. Dispersal in stable habitats. Nature 269:578–581.

Hamilton, W. J. 1962. Reproductive adaptations in the red tree mouse. Journal of Mammalogy 43:486–504.

Hammond, K. and F. W. Nicholas. 1972. The sampling variance of the correlation coefficients estimated from two-fold nested and offspring-parent regression analysis. Theoretical and Applied Genetics 42:97–100.

Hanski, I. 1999. *Metapopulation Ecology*. Oxford University Press, Oxford.

Haq, S. M. 1972. Breeding of *Euterpina acutifrons*, a harpacticid copepod, with special reference to dimorphic males. Marine Biology 15:221–235.

Hard, J. J., W. E. Bradshaw and C. M. Holzapfel. 1993a. The genetic basis of photoperiodism and its evolutionary divergence among populations of the pitcher-plant mosquito, *Wyeomyia smithii*. American Naturalist 142:457–473.

Hard, J. J., W. E. Bradshaw and C. M. Holzapfel. 1993b. Genetic coordination of demography and phenology in the pitcher-plant mosquito, *Wyeomyia smithii*. Journal of Evolutionary Biology 6:707–723.

Harding, J., H. Huang and T. Byrne. 1991. Maternal, paternal, additive, and dominance components of variance in *Gerbera*. Theoretical and Applied Genetics 82:756–760.

Hardy, I. C. W., N. T. Griffiths and H. C. J. Godfray. 1992. Clutch size in a parasitoid wasp: a manipulation experiment. Journal of Animal Ecology 61: 121–129.

Harper, J. L. 1977. *Population Biology of Plants*. Academic Press, London.

Harper, J. L. and J. Ogden. 1970. The reproductive strategies of higher plants. I. The concept of strategy with special reference to *Senecio vulgaris*. Journal of Ecology 58:681–698.

Harper, J. L. and J. White. 1974. The demography of plants. Annual Review of Ecology and Systematics 5:419–463.

Harris, V. E. and J. W. Todd. 1980. Male-mediated aggregation of male, female and 5th instar southern green stink bugs and concomitant attraction of a tachinid parasite. Entomologica Experimentalis et Applicata 27:117–126.

Harrison, R. G. 1980. Dispersal polymorphism in insects. Annual Review of Ecology and Systematics 11:95–118.

Hart, J. L. 1973. Pacific fishes of Canada. Fisheries Research Board of Canada Bulletin 180:1–740.

Hart, M. W. 1995. What are the costs of small egg size for a marine invertabrate with feeding planktonic larvae? American Naturalist 146:415–426.

Harvey, G. T. 1985. Egg weight as a factor in the overwintering survival of spruce budworm [*Choristoneura fumiferana*] (Lepidoptera: Tortricidae) larvae. Canadian Entomologist 117: 1451–1462.

Harvey, G. T. 1983. Environmental and genetic effects on mean egg weight in spruce budworm (Lepidoptera: Tortricidae) and its significance in population dynamics. Canadian Entomologist 115: 1109–1117.

Harvey, P. H. and T. H. Clutton-Brock. 1985. Life history variation in primates. Evolution 39:559–581.

Harvey, P. H., M. J. Stenning and B. Campbell. 1985. Individual variation in seasonal breeding success of pied flycatchers (*Ficedula hypoleuca*). Journal of Animal Ecology 54:391–398.

Hasler, J. F. and E. M. Banks. 1975. Reproductive performance and growth in captive collared lemmings (*Dicostonyx groenlandicus*). Canadian Journal of Zoology 53:777–787.

Hastings, A. 1978. Evolutionarily stable strategies and the evolution of life history strategies: 1. density dependent models. Journal of Theoretical Biology 75:527–536.

Hastings, A. and H. Caswell. 1979. Environmental variability in the evolution of life history strategies. Proceedings of the National Academy of Sciencess, USA 76:4700–4703.

Hatchwell, B. J. 1991. An experimental study of the effects of timing of breeding on the reproductive success of Common Guillemots (*Uria aalge*). Journal of Animal Ecology 60:721–736.

Haukioja, E. and T. Hakala. 1978. Life history evolution in *Anodonta piscinalis* (Mollusca, Pelecypoda). Oecologia 35:253–266.

Haybittle. 1998. The use of the Gompertz function to relate changes in life expectancy to the standardized mortality ratio. International Journal of Epidemiology 27:885–889.

Hazel, L. N. 1943. The genetic basis for constructing selection indices. Genetics 28:476–490.

Hazel, W. N. 1977. The genetic basis of pupal color dimorphism and its maintenance by natural selection in *Papilio polyxenes* (Papilionidae: Lepidoptera). Heredity 38:227–236.

Hazel, W. N., R. S. Mock and M. D. Johnson. 1990. A polygenic model for the evolution and maintenance of conditional strategies. Proceedings of the Royal Society of London, B 242:181–187.

Heaney, V. and P. Monaghan. 1995. A within-clutch trade-off between egg production and rearing in birds. Proceedings of the Royal Society of London B 261:361–365.

Heaney, V. and P. Monaghan. 1996. Optimal allocation of effort between reproductive phases: the trade-off between incubation costs and subsequent brood rearing capacity. Proceedings of the Royal Society of London, B 263:1719–1724.

Heath, D. D., R. H. Devlin, J. W. Heath and G. K. Iwama. 1994. Genetic environmental and interaction effects on the incidence of jacking in *Oncorhynchus tshawytscha* (chinook salmon). Heredity 72:146–154.

Hedrick, P. W. 1986. Genetic polymorphism in heterogeneous environments: a decade later. Annual Review of Ecology and Systematics 17:535–566.

Hedrick, P. W., M. E. Genevan and E. P. Ewing. 1976. Genetic polymorphism in heterogeneous environments. Annual Review of Ecology and Systematics 7:1–32.

Hegmann, J. P. and H. Dingle. 1982. Phenotypic and genetic covariance structure in milkweed bug life history traits. Pp. 177–186 *in* H. Dingle and J. P. Hegmann, (eds.) Evolution and Genetics of Life Histories. Springer-Verlag, New York.

Hegner, R. E. and J. C. Wingfield. 1987. Effects of brood-size manipulations on parental investment, breeding success, and reproductive endocrinology of house sparrows. Auk 104:470–480.

Heino, M. and V. Kaitala. 1996. Optimal resource allocation between growth and reproduction in clams: why does indeterminate growth exist? Functional Ecology 10:245–251.

Heisler, I. L. 1984. A quantitative genetic model for the origin of mating preferences. Evolution 38:1283–1295.

Helbig, A. J. 1992. Population differentiation of migratory directions in birds: comparison between ringing results and orientation behaviour of hand-raised migrants. Oecologia 90:483–88.

Hemmingsen, A. M. 1956. Deep-boring oviposition of some crane-fly species (Tipulidae) of the subgenera *Vestiplex* Bezzi and *Oreomyza* Pok, and some associated phenomena. Videnskabelige Meddelelser fra Dansk Naturhistorisk Forening Kbh. 118:243–315.

Henderson, P. A., R. H. A. Holmes and R. N. Bamber. 1988. Size-selective overwintering mortality in the sand smelt, *Atherina boyeri* Risso, and its role in population regulation. Journal of Fish Biology 33:221–233.

Hendry, A. P. and M. T. Kinnison. 1999. The pace of modern life: measuring rates of micro-evolution. Evolution 53:1637–1654.

Henneman, W. W. I. 1983. Relationship among body mass, metabolic rate, and the intrinsic rate of natural increase in mammals. Oecologia 56:104–108.

Henneman, W. W. I. 1984. Intrinsic rates of natural increase of altricial and precocial eutherian mammals: the potential price of precociality. Oikos 43:363–368.

Henrich, S. and J. Travis. 1988. Genetic variation in reproductive traits in a population of *Heterandria formosa* (Pisces: Poeciliidae). Journal of Evolutionary Biology 1:275–280.

Hessen, D. O. 1985. Selective zooplankton predation by pre-adult roach (*Rutilus rutilus*): the size-selective hypothesis versus the visibility-selective hypothesis. Hydrobiologia 124:73–79.

Hickey, J. J. 1952. Survival studies of banded birds. USDA Fish and Wildlife Service, Special Scientific Report, Wildlife No. 15, Washington, DC:117p.

Hickman, J. C. 1975. Environmental unpredictability and the plastic energy allocation strategies in the annual *Polygonum cascadense* (Polygonaceae). Journal of Ecology 63:689–702.

Higgins, L. and M. A. Rankin. 1999. Nutritional requirements for web synthesis in the tetragnathid spider *Nephila clavipes*. Physiological Entomology 24:263–270.

Hill, J. 1964. Effects of correlated gene distributions in the analysis of dialel crosses. Heredity 19:27–46.

Hill, R. W. 1972. The amount of maternal care in *Peromyscus leucopus* and its thermal significance for the young. Journal of Mammalogy 53:774–790.

Hill, W. G. 1972. Estimation of realized heritabilities from selection experiments. II. selection in one direction. Biometrics 28:767–780.

Hill, W. G. and F. W. Nicholas. 1974. Estimation of heritability by both regression of offspring on parent and intra-class correlation of sibs in one experiment. Biometrics 30:447–468.

Hillesheim, E. and S. C. Stearns. 1991. The responses of *Drosophila melanogaster* to artificial selection on body weight and its phenotypic plasticity in two larval food environments. Evolution 45:1909–1923.

Hines, W. G. S. 1980. An evolutionarily stable strategy for randomly mating diploid populations. Journal of Theoretical Biology 87:379–384.

Hines, W. G. S. 1987. Can and will a sexual diploid population evolve to an ESS: the multi-locus linkage equilibrium case. Journal of Theoretical Biology 126:1–5.

Hipeau-Jacquotte, R. 1984. A new concept in the evolution of the Copepoda: *Pachypygus gibber* (Notodelphyidae), a species with two breeding males. Crustaceana Supplement 7:60–67.

Hoekstra, R. F., R. Bijlsma and A. J. Dolman. 1985. Polymorphism from environmental heterogeneity: models are only robust if the heterozygote is close in fitness to the favored homozygote in each environment. Genetical Research 45:299–314.

Hoffman. 2000. Laboratory and field heritabilities Some lessons from *Drosophila*. Pp. 200–218 *in* T. Mousseau, B. Sinervo and J. Endler, (eds.) *Adaptive Genetic Variation in the Wild*. Oxford University Press, New York.

Hogstedt, G. 1980. Evolution of clutch size in birds: adaptive variation in relation to territory quality. Science 210:1148–1150.

Hogstedt, G. 1981. Should there be a positive or negative correlation between survival of adults in a bird population and their clutch size? American Naturalist 118:568–571.

Holloway, G. J., S. R. Povey and R. M. Sibly. 1990. The effect of new environment on adapted genetic architecture. Heredity 64:323–330.

Horvitz, C. C. and D. W. Schemske. 1988. Demographic cost of reproduction in a neotropical herb: An experimental field study. Ecology 69:1741–1745.

Hospital, F. and C. Chevalet. 1993. Effects of population size and linkage on optimal selection intensity. Theoretical and Applied Genetics 86:775–780.

Houde, A. E. 1992. Sex-linked heritability of a sexually selected character in a natural population of *Poecilia reticulata* (Pisces: Poeciliidae) (guppies). Heredity 69:229–235.

Houle, D. 1989. The maintenance of polygenic variation in finite populations. Evolution 43:1767–1780.

Houle, D. 1991. Genetic covariance of fitness correlates: what genetic correlations are made of and why it matters. Evolution 45:630–648.

Houle, D. 1992. Comparing evolvability and variability of quantitative traits. Genetics 130:195–204.

Houston, A. I. and J. M. McNamara. 1992. Phenotypic plasticity as a state-dependent life-history decision. Evolutionary Ecology 6:243–253.

Houston, A. I. and J. M. McNamara. 1999. *Models of Adaptive Behaviour*. Cambridge University Press, Cambridge.

Howard, D. J. 1993. Small populations, inbreeding, and speciation. Pp. 118–142 in N. W. Thornhill, (ed.) *The Natural History of Inbreeding and Outbreeding*. University of Chicago Press, Chicago.

Howard, D. J. and D. G. Furth. 1986. Review of the *Allonemobius fasciatus* (Orthoptera: Gryllidae) complex with the description of two new species separated by electrophoresis, songs and morphometrics. Annals of the Entomological Society of America 79:472–481.

Howell, D. J. and B. S. Roth. 1981. Sexual reproduction in agaves: the benefits of bats; the costs of semelparous advertising. Ecology 62:1–7.

Hoyle, J. A. and A. Keast. 1987. The effect of prey morphology and size on handling time in a piscivore, the largemouth bass (*Micropterus salmoides*). Canadian Journal of Zoology 65:1972–1977.

Hrbacek, J. and M. Hrbackova-Esslova. 1960. Fish stock as a protective agent in the occurrence of slow-developing dwarf species and strains of the genus *Daphnia*. Intnationale Revue de Gesamten Hydrobiologie 45:355–358.

Huey, R. B., G. W. Gilchrist, M. L. Carlson, D. Berrigan and L. Serra. 2000. Rapid evolution of a geographic cline in size in an introduced fly. Science 287:308–309.

Huey, R. B. and E. R. Pianka. 1981. Ecological consequences of foraging model. Ecology 62:991–999.

Hughes, K. A. and B. Charlesworth. 1994. A genetic analysis of senescence in *Drosophila*. Nature 367:64–65.

Hughes, R. N. 1971. Ecological energetics of the keyhole limpet *Fissurella barbadensis* Gmelin. Journal of Experimental Marine Biology 6:167–178.

Hughes, R. N. and D. J. Roberts. 1980. Reproductive effort of winkles (*Littorina* spp.) with contrasted methods of reproduction. Oecologia 47:130–136.

Humphries, M. M. and S. Boutin. 2000. The determinants of optimal litter size in free-ranging red squirrels. Ecology 81:2867–2877.

Hussell, D. J. T. and T. E. Quinney. 1986. Food abundance and clutch size of tree swallows *Tachycineta bicolor*. Ibis 129:243–258.

Hutchings, J. A. 1991. Fitness consequences of variation in egg size and food abundance in brook trout *Salvelinus fontinalis*. Evolution 45:1162–1168.

Hutchings, J. A. 1993. Adaptive life histories effected by age-specific survival and growth rate. Ecology 74:673–684.

Hutchings, J. A. 1997. Life history responses to environmental variability in early life. Pp. 139–168 *in* R. C. Chambers and E. A. Trippel, (eds.) *Early Life History and Recruitment in Fish Populations*. Chapman and Hall, London.

Hutchings, J. A. and D. W. Morris. 1985. The influence of phylogeny, size and behavior on patterns of covariation in salmonid life histories. Oikos 45:118–124.

Hutchings, J. A. and R. A. Myers. 1987. Escalation of an asymmetric contest: mortality resulting from mate competition in Atlantic salmon, *Salmo salar*. Canadian Journal of Zoology 65:766–768.

Hutchings, J. A. and R. A. Myers. 1988. Mating success of alternative maturation phenotypes in male Atlantic salmon, *Salmo salar*. Oecologia 75:169–174.

Hutchings, J. A. and R. A. Myers. 1994. The evolution of alternative mating strategies in variable environments. Evolutionary Ecology 8:256–268.

Huxley, J. S. 1942. *Evolution, the Modern Synthesis*. Allen and Unwin, London.

Iason, G. R. 1990. The effects of size, age and a cost of early breeding on reproduction in female mountain hares. Holarctic Ecology 13:81–89.

Innes, D. G. and J. S. Millar. 1979. Growth of *Clethrionomys gapperi* and *Microtus pennsylvanicus* in captivity. Growth 43:208–217.

Irschick, D. J., C. C. Austin, K. Petren, R. N. Fisher, J. B. Losos and O. Ellers. 1996. A comparative analysis of clinging ability among pad-bearing lizards. Biological Journal of the Linnean Society 59:21–35.

Istock, C. A., J. Zisfein and K. J. Vavra. 1976. Ecology and evolution of the pitcher-plant mosquito. 2. The substructure of fitness. Evolution 30:535–547.

Ives, A. R. 1989. The optimal clutch size of insects when many females oviposit per patch. American Naturalist 133:671–687.

Iwasa, Y. 1991. Asynchronous pupation of univoltine insects as evolutionarily stable phenology. Researches in Population Ecology 33:213–227.

Iwasa, Y. and D. Cohen. 1989. Optimal growth schedule of a perennial plant. American Naturalist 133:480–505.

Iwasa, Y. and E. Teramoto. 1980. A criterion of life history evolution based on density dependent selection. Journal of Theoretical Biology 84:545–566.

Iwasa, Y., A. Yamauchi and S. Nozoe. 1992. Optimal seasonal timing of univoltine and bivoltine insects. Ecological Research 7:55–62.

Jackson, D. J. 1928. The inheritance of long and short wings in the weevil *Sitona hispidula*, with a discussion of wing reduction among beetles. Transactions of the Royal Society of Edinburgh 55:655–735.

Jacobsen, K., K. E. Erikstad and B. Saether. 1995. An experimental study of the costs of reproduction in the kittiwake, *Rissa tridactyla*. Ecology 76:1636–1642.

James, J. W. 1974. Genetic covariances under the partition of resources model. Appendix 1 in Sheridan and Barker (1974). Australian Journal of Biological Science 27:99–101.

Janssen, G. M., G. de Jong, E. N. G. Joosse and W. Scharloo. 1988. A negative maternal effect in springtails. Evolution 42:828–834.

Janzen, F. J. 1992. Heritable variation for sex ratio under environmental sex determination in the common snapping turtle (*Chelydra serpentina*). Genetics 131:155–161.

Jarvinen, O. 1976. Migration, extinction, and alary morphism in water-striders. Annales Academiae Scientiarum Fennicae A, IV Biologica 206:1–7.

Jarvinen, O. and K. Vepsalainen. 1976. Wing dimorphism as an adaptive strategy in water-striders (*Gerris*). Hereditas 84:61–68.

Jasieniuk, M., A. L. Brule-Babel and I. N. Morrison. 1996. The evolution and genetics of herbicide resistance in weeds. Weed Science 44:176–193.

Jenkins, N. L. and A. A. Hoffman. 1994. Genetic and maternal variation for heat resistance in *Drosophila* from the field. Genetics 137:783–789.

Jennings, H. S. and R. S. Lynch. 1928. Age, mortality, fertility and individual diversities in the rotifer *Proales sordida* Grosse. II. Life history in relation to mortality and fecundity. Journal of Experimental Zoology 51:339–381.

Jensen, A. L. 1996. Beverton and Holt life history invariants result from optimal trade-off of reproduction and survival. Canadian Journal of Fisheries and Aquatic Sciences 53:820–822.

Jensen, A. L. 1998. Simulation of relations among fish life history parameters with a bioenergetics-based population model. Canadian Journal of Fisheries and Aquatic Sciences 55:353–357.

Jensen, J. P. 1958. The relation between body size and number of eggs in marine malacostrakes.

Meddelelser fra Danmarks Fiskeri-og Havundersogelser 2:1–25.

Jimenez, J. A., K. A. Hughes, G. Alaks, L. Graham and R. C. Lacy. 1994. An experimental study of inbreeding depression in a natural habitat. Science 266:271–273.

Jinks, J. L., J. M. Perkins and E. L. Breese. 1969. A general method of detecting additive, dominance and epistatic variation for metrical traits. Heredity 24:45–57.

Johnsen, T. S. and M. Zuk. 1999. Parasites and tradeoffs in the immune response of female red jungle fowl. Oikos 86:487–492.

Johnson, C. G. 1969. *Migration and Dispersal of Insects by Flight*. Methuen, London.

Johnson, M. L. and M. S. Gaines. 1990. Evolution of dispersal: Theoretical models and empirical tests using birds and mammals. Annual Review of Ecology and Systematics 21:449–480.

Johnston, M. O. 1992. Effects of cross and self-fertilization on progeny fitness in *Lobelia cardinalis* and *L. siphilitica*. Evolution 46:688–702.

Johnston, R. F., D. M. Niles and S. A. Rowher. 1972. Hermon Bumpus and natural selection in the house sparrow *Passer domesticus*. Evolution 26:20–31.

Johnston, R. F. and R. K. Selander. 1964. House sparrows: rapid evolution of races in North America. Science 144:548–550.

Jones, L. P., R. Frankham and J. S. F. Barker. 1968. The effects of population size and selection intensity in selection for a quantitative character in *Drosophila* II. Long-term response to selection. Genetical Research 12:249–266.

Jones, R. E. 1977. Movement patterns and egg distribution in cabbage butterflies. Journal of Animal Ecology 46:195–212.

Jones, R. E. 1987. Behavioural evolution in the cabbage butterfly (*Pieris rapae*). Oecologia 72:69–76.

Jones, R. E., J. R. Hart and G. D. Bull. 1982. Temperature, size and egg production in the cabbage butterfly, *Pieris rapae*. Australian Journal of Zoology 30:223–232.

Jones, R. E. and P. M. Ives. 1979. The adaptiveness of searching and oviposition behaviour in *Pieris rapae* L. Australian Journal of Ecology 4:75–86.

Jones, R. E., V. G. Nealis, P. M. Ives and E. Scheermeyer. 1987. Seasonal and spatial variation in juvenile survival of the cabbage butterfly *Pieris rapae*: evidence for patchy density-dependence. Journal of Animal Ecology 56:723–737.

Jones, W. T. 1985. Body size and life-history variables in Heteromyids. Journal of Mammalogy 66:128–132.

Jonsson, B. and K. Hinder. 1982. Reproductive strategy of dwarf and normal Arctic charr (*Salvelinus alpinus*) from Vangsvatnet Lake, western Norway. Canadian Journal of Fisheries and Aquatic Sciences 39:1404–1413.

Joshi, A. and L. D. Mueller. 1988. Evolution of higher feeding rate in *Drosophila* due to density-dependent natural selection. Evolution 42:1090–1093.

Kachi, N. and T. Hirose. 1985. Population dynamics of *Oenothera glazioviana* in a sand-dune system with special reference to the adaptive significance of size-dependent reproduction. Journal of Ecology 73:887–901.

Kalela, O. 1957. Regulation of reproductive rate in sub-arctic populations of the vole *Clethrionomys rufocanus* (Sund.). Annales Academiae Scientiarum Fennicae, Series A 4:1–60.

Kallman, K. D. 1983. The sex determining mechanism of the poecilid fish, *Xiphophorus montezumae*, and the genetic control of the sexual maturation process and adult size. Copeia 1983:733–769.

Kallman, K. D. and V. Borkoski. 1977. A sex-linked gene controlling the onset of sexual maturity in female and male platyfish (*Xiphophorus maculatus*), fecundity in females and adult size in males. Genetics 898:79–119.

Kallman, K. D., M. P. Schreibman and V. Borkoski. 1973. Genetic control of gonadotroph differentiation in the platyfish, *Xiphophorus maculatus* (Poeciliidae). Science 181:678–680.

Kambysellis, M. P. and W. B. Heed. 1971. Studies of oogenesis in natural populations of Drosophilidae. I. Relation of ovarian development and ecological habitats of the Hawaiian species. American Naturalist 105:31–49.

Kaplan, R. H. and W. S. Cooper. 1984. The evolution of developmental plasticity in reproductive character-istics: an application of the "adaptive coin-flipping" principle. American Naturalist 123:393–410.

Karban, R. and I. T. Baldwin. 1997. *Induced Responses to Herbivory.* University of Chicago Press, Chicago.

Kari, J. S. and R. B. Huey. 2000. Size and seasonal tem-perature in free-ranging *Drosophila subobscura*. Journal of Thermal Biology 25:267–272.

Karlin, S. and R. B. Campbell. 1981. The existence of a protected polymorphism under condition of soft as opposed to hard selection in a multideme popula-tion system. American Naturalist 117:262–275.

Karlin, S. and U. Liberman. 1974. Random temporal variation in selection intensities: case of large popu-lation size. Theoretical Population Biology 6: 355–382.

Karlin, S. and U. Liberman. 1975. Random temporal variation in selection intensities: one-locus two allele model. Journal of Mathematical Biology 2:1–17.

Kasule, F. K. 1992. A quantitative genetic analysis of reproductive allocation in the cotton stainer bug, *Dysdercus fasciatus*. Heredity 69:141–149.

Kaufman, D. W. and G. A. Kaufman. 1987. Repro-duction by *Peromyscus polinotus*: number, size, and survival of young. Journal of Mammalogy 68:275–280.

Kaufman, P. K., F. D. Enfield and R. E. Comstock. 1977. Stabilizing selection for pupa weight in *Tribolium cas-teneum*. Genetics 87:327–341.

Kawasaki, T. 1980. Fundamental relations among the selections of life history in the marine teleosts. Bulletin of the Japanese Society of Scientific Fisheries 46:289–293.

Kawecki, T. J. 1993. Age and size at maturity in a patchy environment: fitness maximization versus evolutionary stability. Oikos 66:309–317.

Kawecki, T. J. and S. C. Stearns. 1993. The evolution of life histories in spatially heterogeneous environ-ments: Optimal reaction norms revisited. Evolutionary Ecology 7:155–174.

Kearsey, M. J. and J. S. Gale. 1968. Stabilising selection in the absence of dominance: an additional note. Heredity 23:617–620.

Keeley, J. E. and W. J. Bond. 1999. Mast flowering and semelparity in bamboos: the bamboo fire cycle hypothesis. American Naturalist 154:383–391.

Keightley, P. D. and W. G. Hill. 1990. Variation main-tained in quantitative traits with mutation-selection balance: pleiotropic side-effects on fitness traits. Proceedings of the Royal Society of London. B 242:95–100.

Kekic, V. and D. Marinkovic. 1974. Multiple-choice selection for light preference in *Drosophila subobscura*. Behavior Genetics 4:285–300.

Keller, B. L. and C. J. Krebs. 1970. *Microtus* population biology; III. Reproductive changes in fluctuating populations of *M. ochrogaster* and *M. pennsylvanicus* in southern Indiana. Ecological Monographs 40:263–294.

Kempthorne, O. and O. B. Tandon. 1953. The estima-tion of heritability by regression of offspring on par-ent. Biometry 9:90–100.

Kerfoot, W. C. 1974. Egg size cycle of a cladoceran. Ecology 55:1259–1270.

Kerfoot, W. C. 1977. Competition in cladoceran com-munities: the cost of evolving defenses against cope-pod predation. Ecology 58:303–313.

Kessler, A. 1971. Relation between egg production and food consumption in species of the genus *Pardosa* (Lycosidae, Araneae) under experimental conditions of food abundance and food shortage. Oecologia 8:93–109.

Kessler, S. 1969. The genetics of *Drosophila* mating behavior. II. The genetic architecture of mating speed in *Drosophila pseudobscura*. Genetics 62: 421–433.

Kilgore, D. L. 1970. The effects of northward dispersal on growth rate of young, size of young at birth, and litter size in *Sigmodon hispidus*. American Midland Naturalist 84:510–520.

Kimura, J. and S. Masaki. 1977. Brachypterism and sea-sonal adaptation in *Orgyia thyellina* Butler (Lepidoptera, Lymantridae). Kontyu 45:97–106.

King, C. E. and W. W. Anderson. 1971. Age-specific selection. II. The interaction between *r* and *K* during population growth. American Naturalist 105: 137–156.

Kingsland, S. E. 1985. *Modeling Nature*. University of Chicago Press, Chicago and London.

Kingsolver, J. G. and D. W. Schemske. 1991. Path analysis of selection. Trends in Ecology and Evolution 6:276–280.

Kinney, T. B. J. 1969. A summary of reported estimates of heritabilities and of genetic and phenotypic correlations for traits of chickens. Washington: US Department of Agriculture Agricultural Handbook 363:1–49.

Kiorboe, T. and M. Sabatini. 1995. Scaling of fecundity, growth and development in marine planktonic copepods. Marine Ecology Progress Series 120: 285–98.

Kirkpatrick, M. 1982. Sexual selection and the evolution of female choice. Evolution 36:1–12.

Kirkpatrick, M. and N. Heckman. 1989. A quantitative genetic model for growth, shape, reaction norms, and other infinite-dimensional characters. Journal of Mathematical Biology 27:429–450.

Kirkpatrick, M., G. W. Hill and R. Thompson. 1994. Estimating the covariance structure of traits during growth and ageing, illustrated with lactation in dairy cattle. Genetical Research 64:57–69.

Kirkpatrick, M. and R. Lande. 1989. The evolution of maternal characters. Evolution 43:485–503.

Kirkpatrick, M. and D. Lofsvold. 1992. Measuring selection and constraint in the evolution of growth. Evolution 46:954–971.

Kirkpatrick, M., D. Lofsvold and M. Bulmer. 1990. Analysis of the inheritance, selection and evolution of growth trajectories. Genetics 124:979–993.

Kirkwood, T. B. L., G. M. Martin and L. Partridge. 1999. Evolution, senescence, and health in old age. Pp. 219–230 *in* S. C. Stearns, (ed.) *Evolution in Health and Disease*. Oxford University Press, Oxford.

Kisdi, E. and G. Meszena. 1995. Life histories with lottery competition in a stochastic environment: ESSs which do not prevail. Theoretical Population Biology 47:191–211.

Kitahara, T., Y. Hiyama and T. Tokai. 1987. A preliminary study on quantitative relations among growth, reproduction and mortality in fishes. Researches in Population Ecology 29:85–95.

Kleckner, C. A., W. A. Hawley, W. E. Bradshaw, C. M. Holzapfel and I. J. Fisher. 1995. Protandry in *Aedes sierrensis*: the significance of temporal variation in female fecundity. Ecology 76:1242–50.

Klingenberg, C. P. and J. R. Spence. 1997. On the role of body size for life-history evolution. Ecological Entomology 22:55–68.

Klinkhamer, P. G. L., T. J. de Jong, J. A. J. Metz and J. Val. 1987. Life history tactics of annual organisms: the joint effect of dispersal and delayed germination. Theoretical Population Biology 32:127–156.

Klinkhamer, P. G. L., T. J. de Jong and E. Meelis. 1990. How to test for proportionality in the reproductive effort of plants. American Naturalist 135:291–300.

Klomp, H. 1970. The determination of clutch-size in birds: a review. Ardea 58:1–124.

Klomp, H. and B. J. Teerink. 1962. Host selection and number of eggs per oviposition in the egg-parasite *Trichogramma embyophagum* Htg. Nature 195: 1020–1021.

Klomp, H. and B. J. Teerink. 1967. The significance of oviposition rates in the egg parasite, *Trichogramma embryophagum*. Archives Neerlandaises de Zoologie 17:350–375.

Kluyver, H. N. 1963. The determination of reproductive rate in Paridae. Proceedings of the International Ornithologial Congress 13:706–716.

Knapp, S. J., J. Bridges, W. C. and M. Yang. 1989. Nonparametric confidence estimators for heritability and expected selection response. Genetics 121:891–898.

Knight, R. L. and S. A. Temple. 1983. Nest defence in the American goldfinch. Animal Behaviour 34: 887–897.

Knight, W. 1968. Asymptotic growth: an example of nonsense disguised as mathematics. Journal of the Fisheries Research Board of Canada 25:1303–1307.

Koella, J. C. and J. Offenberg. 1999. Food availability and parasite infection influence the correlated response of life history traits to selection for age at pupation in the mosquito *Aedes aegypti*. Journal of Evolutionary Biology 12:760–769.

Kojima, K. 1959. Stable equilibria for the optimum model. Proceedings of the National Academy of Sciences 45:989–993.

Kojima, K. I. 1961. Effects of dominance and size of population on response to mass selection. Genetical Research 2:177–188.

Kojola, I. 1991. Influence of age on the reproductive effort of male reindeer. Journal of Mammalogy 72:208–210.

Konarzewski, M. 1993. The evolution of clutch size and hatching asynchrony in altricial birds: the effects of environmental variability, egg failure and predation. Oikos 67:97–106.

Konig, B., J. Reister and H. Markl. 1988. Maternal care in house mice (*Mus musculus*): 2. The energy cost of lactation as a function of litter size. Journal of Zoology, London 216:195–210.

Konig, C. and P. Schmid-Hempel. 1995. Foraging activity and immunocompetence in workers of the bumble bee, *Bombus terrestris* L. Procedings of the Royal Society of London Biology, B 260:225–227.

Koots, K. R. and J. P. Gibson. 1994. How precise are genetic correlation estimates? Proceedings of the 5th World Congress on Genetics Applied to Livestock Production 18:353–356.

Koots, K. R. and J. P. Gibson. 1996. Realized sampling variances of estimates of genetic parameters and the difference between genetic and phenotypic correlations. Genetics 143:1409–1416.

Korpomaki, E., H. Hakkarainen and G. F. Bennett. 1993. Blood parasites and reproductive success of Tengmalm's owls: detrimental effects on females but not on males? Functional Ecology 7:420–426.

Koskela, E. 1998. Offspring growth, survival and reproductive success in the bank vole:a litter size manipulation experiment. Oecologia 115:379–384.

Koufopanou, V. and G. Bell. 1984. Measuring the cost of reproduction. IV Predation experiments with *Daphnia pulex*. Oecologia 64:81–86.

Kovacs, K. M. and D. M. Lavigne. 1985. Neonatal growth and organ allometry of northwest Atlantic harp seals (*Phoca groenlandica*). Canadian Journal of Zoology 63:2793–2799.

Kovacs, K. M., D. M. Lavigne and S. Innes. 1990. Mass transfer efficiency between harp seal (*Phoca groenlandica*) mothers and their pups during lactation. Journal of Zoology 223:213–221.

Kozlowski, J. 1996. Optimal allocation of resources explains interspecific life-history patterns in animals with indeterminate growth. Proceedings of the Royal Society of London, B 263:559–66.

Kozlowski, J. and S. C. Stearns. 1989. Hypotheses for the production of excess zygotes: models of bet-hedging and selective abortion. Evolution 43: 1369–1377.

Kozlowski, J. and J. Uchmanski. 1987. Optimal individual growth and reproduction in perennial species with indeterminate growth. Evolutionary Ecology 1:214–230.

Kozlowski, J. and J. Weiner. 1997. Interspcific allometries are by-products of body size optimization. American Naturalist 149:352–380.

Kozlowski, J. and R. G. Wiegert. 1986. Optimal allocation of energy to growth and reproduction. Theoretical Population Biology 29:16–37.

Kozlowski, J. and R. G. Wiegert. 1987. Optimal age and size at maturity in annuals and perennials with determinate growth. Evolutionary Ecology 1:231–244.

Kramer, D. L. 1973. Parental behaviour in the blue gourami *Trichogaster trichopterus* (Pisces, Belontiidae) and its induction during exposure to varying numbers of conspecific eggs. Behaviour XLVII:14–32.

Kramer, D. L. 2001. Foraging Behavior. Pp. 232–246 in C. W. Fox, D. A. Roff and D. J. Fairbairn, (eds.) *Evolutionary Ecology: Concepts and Case Studies*. Oxford University Press, New York.

Krohne, D. T. 1981. Intraspecific litter size variation in *Microtus californicus*: variation within populations. Journal of Mammalogy 62:29–40.

Kruuk, L. E. B., T. H. Clutton-Brock, J. Slate, J. M. Pemberton and S. Brotherstone. 2000. Heritability of fitness in a wild mammal population. Proceedings of the National Academy of Sciences 97:698–703.

Kulesza, G. 1990. An analysis of clutch-size in New World passerine birds. Ibis 132:407–422.

Kuno, E. 1981. Dispersal and the persistence of populations in unstable habitats: a theoretical note. Oecoligia 49:123–126.

Kuramoto, M. 1978. Correlations of quantitative parameters of fecundity in amphibians. Evolution 32:287–296.

Kusano, T. 1982. Postmetamorphic growth, survival, and age at first reproduction of the salamander, *Hynobius nebulosus tokyoensis* Tago in relation to a consideration on the optimal timing of first reproduction. Researches on Population Ecology 24:329–344.

Kusch, J. 1995. Adaptation of inducible defense in *Euplotes daidaleos* (Ciliophora) to prediction risks by various predators. Microbial Ecology 30:79–88.

Kusch, J. and H. W. Kuhlmann. 1994. Cost of Stenostomum-induced morphological defence in the ciliate *Euplotes octocarinatus*. Archives Hydrobiologica 130:257–267.

L'Abee-Lund, J. H., B. Jonsson, A. J. Jensen, L. M. Saettem, T. G. Heggberget, B. O. Johnsen and T. F. Naesje. 1989. Latitudinal variation in life-history characteristics of sea-run migrant brown trout *Salmo trutta*. Journal of Animal Ecology 58:525–542.

Lacey, E. P., L. Real, J. Antonovics and D. G. Heckel. 1983. Variance models in the study of life histories. American Naturalist 122:114–131.

Lachance, S. and G. J. FitzGerald. 1992. Parental care tactics of three-pined sticklebacks living in a harsh environment. Behavioral Ecology 3:360–366.

Lack, D. 1943a. The age of the blackbird. British Birds 36:166–172.

Lack, D. 1943b. The age of some more British birds. British Birds 36:193–197, 214–221.

Lack, D. 1947. The significance of clutch size 1. Intraspecific variation. Ibis 89:302–352.

Lack, D. 1948. The significance of litter size. Journal of Animal Ecology 17:45–50.

Lack, D. 1954. *The Natural Regulation of Animal Numbers*. Clarendon Press, Oxford.

Lack, D. 1966. *Population Studies of Birds*. Clarendon Press, Oxford.

Lackey, J. A. 1976. Reproduction, growth and development in the Yucatan deer mouse, *Peromyscus yucatanicus*. Journal of Mammalogy 57:638–655.

Lackey, J. A. 1978. Reproduction, growth, and development in high-latitude populations of *Peromyscus leucopus* (Rodentia). Journal of Mammalogy 59:69–83.

Lafferty, K. D. 1993. The marine snail, *Cerithidea californica*, mature at smaller sizes where parasitism is high. Oikos 68:3–11.

Lamey, T. C., R. M. Evans and J. D. Hunt. 1996. Insurances reproductive value and facultative brood reduction. Oikos 77:285–290.

Landahl, J. T. and R. B. Root. 1969. Differences in the life tables of tropical and temperate milkweed bugs, genus *Oncopeltus* (Hemiptera, Lygaeidae). Ecology 50:734–737.

Lande, R. 1975. The maintenance of genetic variation by mutation in a polygenic character with linked loci. Genetical Research 26:221–235.

Lande, R. 1976. Natural selection and random genetic drift in phenotypic evolution. Evolution 30:314–334.

Lande, R. 1979. Quantitative genetic analysis of multivariate evolution applied to brain body size allometry. Evolution 33:402–416.

Lande, R. 1981. Models of speciation by sexual selection on polygenic traits. Proceedings of the National Academy of Sciences 78:3721–3725.

Lande, R. 1982. A quantitative genetic theory of life history evolution. Ecology 63:607–615.

Lande, R. 1993. Risks of population extinction from demographic and environmental stochasticity and random catastrophes. American Naturalist 142:911–927.

Lande, R. 1995. Mutation and conservation. Conservation Biology 9:782–91.

Lande, R. and S. J. Arnold. 1983. The measurement of selection on correlated characters. Evolution 37:1210–1226.

Lande, R. and M. Kirkpatrick. 1990. Selection response in traits with maternal inheritance. Genetical Research 55:189–197.

Lande, R. and S. H. Orzack. 1988. Extinction dynamics of age-structured populations in a fluctuating environment. Proceedings of the National Academy of Sciencess, USA 85:7418–7421.

Landwer, A. J. 1994. Manipulation of egg production reveals costs of reproduction in the tree lizard (*Urosaurus ornatus*). Oecologia 100:243–249.

Langslow, D. R. 1979. Movements of blackcaps ringed in Britain and Ireland. Bird Study 26:239–52.

Larruga, J. M., J. Rozas, M. Hernandez, A. M. Gonzalez and V. M. Cabrera. 1993. Latitudinal differences in sex chromosome inversions, sex linked allozymes, and mitochindrial DNA variation in *Drosophila subobscura*. Genetica 92:67–74.

Latter, B. D. H. 1964. Selection for a threshold character in *Drosophila* I. An analysis of the phenotypic variance on the underlying scale. Genetical Research 5:198–210.

Latter, B. D. H. and A. Robertson. 1960. Experimental design in the estimation of heritability by regression methods. Biometrics 16:348–353.

Lavery, R. J. and M. H. A. Keenleyside. 1990. Parental investment of a biparental fish *Cichlasoma nigrofasciatum*, in relation to brood size and past investment. Animal Behaviour 40:1128–1137.

Law, R. 1979a. The cost of reproduction in annual meadow grass. American Naturalist 113:3–16.

Law, R. 1979b. Optimal life histories under age-specific predation. American Naturalist 114:399–417.

Law, R., A. D. Bradshaw and P. D. Putwain. 1977. Life history variation in *Poa annua*. Evolution 31:233–247.

Lawlor, L. R. 1976. Parental investment and offspring fitness in the terrestrial isopod *Armadillidium vulgare* (Latr.), (Crustacea: Oniscoidea). Evolution 30:775–785.

Lee, S. J., M. S. Witter, I. C. Cuthill and A. R. Goldsmith. 1996. Reduction in escape performance as a cost of reproduction in gravid starlings, *Sturnus vulgaris*. Proceedings of the Royal Society of London, B 263:619–24.

Leggett, W. C. and J. E. Carscadden. 1978. Latitudinal variation in reproductive characteristics of American Shad (*Alosa sapidissima*): evidence for population specific life history strategies in fish. Journal of the Fisheries Research Board of Canada 35:1469–1478.

Legner, E. F. 1991. Estimation of the number of active loci, dominance and heritability in polygenic inheritance of gregarious behavior in *Muscidifurax raptorellus* (Hymenoptera: Pteromalidae). Entomophaga 36:1–18.

Leimar, O. and U. Norberg. 1997. Metapopulation extinction and genetic variation in dispersal-related traits. Oikos 80:448–458.

Lemen, C. A. and H. K.Voris. 1981. A comparison of reproductive strategies among marine snakes. Journal of Animal Ecology 50:89–101.

Leon, J. A. 1976. Life histories as adaptive strategies. Journal of Theoretical Biology 60:301–335.

Leon, J. A. and B. Charlesworth. 1978. Ecological versions of Fisher's fundamental theorem of natural selection. Ecology 59:457–464.

Lepage, D., G. Gauthier and A. Desrochers. 1998. Larger clutch size increases fledging success and offspring quality in a precocial species. Journal of Animal Ecology 67:210–216.

Leprince, D. J. and L. D. Foil, 1993. Relationships among body size, blood meal size, egg volume, and egg production of *Tabanus fuscicostatus* (Diptera: Tabanidae). Journal of Medical Entomology 30:865–871.

Lerner, I. M. 1958. *The Genetic Basis of Selection*. Wiley, New York.

Lerner, I. M. and E. R. Dempster. 1951. Attenuation of genetic progress under continued selection in poultry. Heredity 5:75–94.

Lerner, I. M. and C. A. Gunns. 1952. Egg size and reproductive fitness. Poultry Science 31:537–544.

Lessells, C. M. 1986. Brood size in Canada geese: a manipulation experiment. Journal of Animal Ecology 55:669–689.

Levene, H. 1953. Genetic equilibrium when more than one ecological niche is available. American Naturalist 87:331–333.

Levin, S. A., D. Cohen and A. Hastings. 1984. Dispersal strategies in patchy environments. Theoretical Population Biology 26:165–191.

Levin, S. A. and H. C. Muller-Landau. 2000. The evolution of dispersal and seed size in plant communities. Evolutionary Ecology Research 2:409–435.

Levins, R. 1966. The strategy of model building in population biology. American Scientist 54:421–431.

Levins, R. 1969. The effect of random variations of different types on population growth. Proceedings of the National Academy of Sciencess, USA 62:1061–1065.

Lewontin, R. C. 1964. The interaction of selection and linkage. II. Optimum models. Genetics 50:757–782.

Lewontin, R. C. 1965. Selection for colonizing ability. Pp. 77–94 *in* H. G. Baker and G. L. Stebbins, (eds.) *The Genetics of Colonizing Species*. Academic Press, New York.

Lewontin, R. C. and D. Cohen. 1969. On population growth in a randomly varying environment. Proceedings of the National Academy of Sciencess, USA 62:1056–1060.

Licht, L. E. 1974. Survival of embryos, tadpoles and adults of the frogs *Rana aurora aurora* and *Rana pretiosa pretiosa* sympatric in southwestern British Columbia. Canadian Journal of Zoology 52:613–627.

Lidicker, W. Z. J. and R. L. (eds.) Caldwell. 1982. *Dispersal and Migration*. Hutchinson Ross Publishing Company, Stroudsburg, Pennsylvania.

Liebherr, J. K. 1988. Gene flow in ground beetles (Coleoptera: Carabidae) of differing habitat preference and flight-wing development. Evolution 42:129–137.

Lima, S. L. and L. M. Dill. 1990. Behavioral decisions made under risk of predation: a review and prospectus. Canadian Journal of Zoology 68:619–640.

Lindstedt, S. and W. A. Calder. 1976. Body size and longevity in birds. Condor 78:91–94.

Lindstedt, S. L. and W. A. Calder. 1981. Body size, physiological time, and longevity of homeothermic animals. Quarterly Review of Biology 56:1–16.

Linzey, A. R. 1970. Postnatal growth and development of *Peromyscus maniculatus nubiterrae*. Journal of Mammalogy 51:152–155.

Lively, C. M. 1986a. Competition, comparative life histories, and maintenance of shell dimorphism in a barnacle. Ecology 67:858–864.

Lively, C. M. 1986b. Canalization versus developmental conversion in a spatially variable environment. American Naturalist 128:561–572.

Lloyd, D. G. 1984. Variation strategies of plants in heterogeneous environments. Biological Journal of the Linnean Society 21:357–385.

Lloyd, J. E. 1965. Aggressive mimcry in *Photuris*: firefly femmes fatales. Science 149:653–654.

Lloyd, J. E. and S. R. Wing. 1983. Nocturnal aerial predation of fireflies by light-seeking fireflies. Science 222:634–635.

Logerwell, E. A. and M. D. Ohman. 1999. Egg-brooding, body size and predation risk in planktonic marine copepods. Oecologia 121:426–431.

Loman, J. 1977. Factors affecting clutch and brood size in the crow *Corvus cornix*. Oikos 29:294–301.

Loman, J. 1982. A model of clutch size determination in birds. Oecologia 52:253–257.

Lope, F. D., G. Gonzalez, J. J. Perez and A. P. Moller. 1993. Increased detrimental effects of ectoparasites on their bird hosts during adverse environmental conditions. Oecologia 95:234–240.

Lopez, M. A. and C. Lopez-Fanjul. 1993. Spontaneous mutation for a quantitative trait in *Drosophila melanogaster*. II. Distribution of mutant effects on the trait and fitness. Genetical Research 61:117–126.

Lopez-Fanjul, C. and W. G. Hill. 1973. Genetic differences between populations of *Drosophila melanogaster* for a quantitative trait. Genetical Research 22:69–78.

Losos, J. B. 1994. Integrative approaches to evolutionary ecology: *Anolis* lizards as model systems. Annual Review of Ecology and Systematics 25:467–493.

Losos, J. B. and D. J. Irschick. 1996. The effect of perch diameter on escape behaviour of *Anolis* lizards: laboratory predictions and field tests. Animal Behaviour 51:593–602.

Losos, J. B., K. I. Warheit and T. W. Schoener. 1997. Adaptive differentiation following experimental island colonization in *Anolis* lizards. Nature 387:70–72.

Luck, R. F., H. Podoler and R. Kfir. 1982. Host selection and egg allocation behavior by *Aphytis melinus* and *A. lingnanensis*: comparison of two facultatively gregarious parasitoids. Ecological Entomology 7:397–408.

Luckinbill, L. S. 1984. An experimental analysis of a life history theory. Ecology 65:1170–1184.

Luckinbill, L. S., R. Arking, M. J. Clare, W. C. Cirocco and S. Buck. 1984. Selection for delayed senescence in *Drosophila melanogaster*. Evolution 38:996–1003.

Ludwig, D. and S. A. Levin. 1991. Evolutionary stability of plant communities and the maintenance of multiple dispersal types. Theoretical Population Biology 40:285–307.

Ludwig, D. and L. Rowe. 1990. Life-history strategies for energy gain and predator avoidance under time constraints. American Naturalist 135:686–707.

Lunberg, S. 1985. The importance of egg hatchability and nest predation in clutch size evolution in altricial birds. Oikos 45:110–117.

Lundqvist, H. and G. Fridberg. 1982. Sexual maturation versus immaturity: different tactics with adaptive values in Baltic salmon (*Salmo salar*) male smolts. Canadian Journal of Zoology 60:1822–1827.

Lynch, C. B. 1994. Evolutionary inferences from genetic analyses of cold adaptation in laboratory and wild populations of the house mouse. Pp. 278–304 *in* C. R. B. Boake, (ed.) *Quantitative Genetic Studies of Behavioral Evolution*. University of Chicago Press, Chicago.

Lynch, M. 1977. Fitness and optimal body size in zooplankton populations. Ecology 58:763–774.

Lynch, M. 1980a. The evolution of cladoceran life histories. Quarterly Review of Biology 55:23–42.

Lynch, M. 1980b. Predation, enrichment and the evolution of cladoceran life histories: a theoretical approach. Pp. 367–376 *in* W. C. Kerfoot, (ed.) *The Evolution and Ecology of Zooplankton Communities*. University Press of New England, Hanover, NH.

Lynch, M. 1988. The rate of polygenic mutation. Genetical Research 51:137–148.

Lynch, M. and W. G. Hill. 1986. Phenotypic evolution by neutral mutation. Evolution 40:915–935.

Lynch, M. and B. Walsh. 1998. *Genetics and Analysis of Quantitative Traits*. Sinauer Associates, Sunderland, MA.

MacArthur, R. H. 1962. Some generalized theorems of natural selection. Proceedings of the National Academy of Sciencess, USA 48:1893–1897.

MacArthur, R. H. and E. O. Wilson. 1967. *The Theory of Island Biogeography*. Princeton University Press, Princeton, NJ.

Machin, D. and S. Page. 1973. Effects of reduction of litter size on subsequent growth and reproductive performance in mice. Animal Production 16:1–6.

Mackay, T. F. C. 1980. Genetic variance, fitness, and homeostasis in varying environments: an experimental check of the theory. Evolution 34:1219–1222.

Mackay, T. F. C. 1981. Genetic variation in varying environments. Genetical Research 37:79–93.

Mackay, T. F. C., R. F. Lyman, M. S. Jackson, C. Terzian and W. G. Hill. 1992. Polygenic mutation in *Drosophila melanogaster*: estimates from divergence among inbred strains. Evolution 46:300–316.

Maekawa, K. and T. Hino. 1987. Effect of cannibalism on alternative life histories in charr. Evolution 41:1120–1123.

Maekawa, K. and H. Onozato. 1986. Reproductive tactics and fertilization success of mature male Miyabe charr, *Salvelinus malma miyabei*. Environmental Biology of Fishes 15:119–129.

Magnhagen, C. and K. Vestergaard. 1991. Risk taking in relation to reproductive investments and future reproductive opportunities: field experiments on nest-guarding common gobies, *Pomatoschistus microps*. Behavioural Ecology 2:351–359.

Magnusson, W. E. L., L. J. d. Paiva, R. M. d. Rocha, C. R. Franke, L. A. Kasper and A. P. Lima. 1985. The correlates of foraging mode in a community of Brazilian lizards. Herpetologica 41:324–332.

Maigret, J. L. and M. T. Murphy. 1997. Costs and benefits of parental care in eastern kingbirds. Behavioral Ecology 8:250–259.

Mallet, J. 1989. The evolution of insecticide resistance: have the insects won? Trends in Ecology and Evolution 4:336–340.

Mangel, M. 1996. Life history invariant, age at maturity and the ferox trout. Evolutionary Ecology 10:249–263.

Mangold, J. R. 1978. Attraction of *Euphasiopteryx ochracea, Corethrella* sp. and gryllids to broadcast songs of the southern mole cricket. Florida Entomologist 61:57–61.

Mani, G. S., B. C. Clarke and P. R. Sheltom. 1990. A model of quantitative traits under frequency-dependent balancing selection. Proceedings of the Royal Society of London B 240:15–28.

Mann, R. H. K. and C. A. Mills. 1979. Demographic aspects of fish fecundity. Symposium of the Zoological Society of London 44:161–177.

Manning, A. 1961. The effects of artificial selection for mating speed in *Drosophila melanogaster*. Animal Behaviour 9:82–92.

Manton, K. 1998. Commentary. Journal of Gerontology: Biological Sciences 53A:B404–B405.

Mappes, J. and A. Kaitala. 1994. Experiments with *Elasmucha grisea* L. (Heteroptera: Acanthosomatidae): does a female parent bug lay as many eggs as she can defend? Behavioral Ecology 5:314–317.

Mappes, T., E. Koskela and H. Ylonen. 1995. Reproductive costs and litter size in the bank vole. Proceedings of the Royal Society of London B 261:19–24.

Mappes, T. and H. Ylonen. 1997. Reproductive effort of female bank voles in a risky environment. Evolutionary Ecology 11:591–598.

Marshall, L. D. 1988. Small male advantage in mating in *Parapediasia teterrella* and *Agriphila plumbifimbriella* (Lepidoptera: Pyralidae). American Midland Naturalist 119:412–419.

Marshall, L. D. 1990. Intra-specific variation in reproductive effort by female *Parapediasia teterrella* (Lepidoptera: Pyralidae) and its relations to body size. Canadian Journal of Zoology 68:44–48.

Martin, G. A. and A. E. Bell. 1960. An experimental check on the accuracy of prediction of response during selection. Pp. 178–187 *in* O. Kempthorne, (ed.) *Biometrical Genetics*. Pergamon Press, London.

Martin, R. D. and M. MacLaron. 1985. Gestation period, neonatal size and maternal investment in placental mammals. Nature 313:220–223.

Martin, T. E. 1995. Avian life history evolution in relation to nest sites, nest predation, and food. Ecological Monographs 65:101–127.

Martinez, M. L., A. E. Freeman and P. J. Berger. 1983. Age of dam and direct and maternal effects on calf livability. Journal of Dairy Science 66:1714–1720.

Maruyama, T. and P. A. Fuerst. 1985. Population bottlenecks and nonequilibrium models in population genetics: II. Number of alleles in a small population that was formed by a recent bottleneck. Genetics 111:675–689.

Masaki, S. 1967. Geographic variation and climatic adaptation in a field cricket (Orthoptera: Gryllidae). Evolution 21:725–741.

Masaki, S. 1973. Climatic adaptation and photoperiodic response in the band-legged ground cricket. Evolution 26:587–600.

Masaki, S. 1978a. Climatic adaptation and species status in the Lawn Ground Cricket. II. Body Size. Oecologia 35:343–356.

Masaki, S. 1978b. Seasonal and latitudinal adaptations in the life cycles of crickets. Pp. 72–100 *in* H. Dingle, (ed.) *Evolution of Insect Migration and Diapause*. Springer-Verlag, New York.

Masaki, S. 1996. Geographical variation of life cycle in crickets (Ensifera: Grylloidea). European Journal of Entomology 93:281–302.

Masaki, S. 1999. Seasonal adaptations of insects as revealed by latitudinal diapause clines. Entomological Science 2:539–549.

Mathews, R. W. and J. R. Mathews. 1970. Adaptive aspects of insular evolution. Science 167:909–910.

Matos, M., M. R. Rose, M. T. Rocha Pita, C. Rego and T. Avelar. 2000. Adaptation to the laboratory environment in *Drosophila subobscura*. Journal of Evolutionary Biology 13:9–19.

Mattingly, D. K. and P. A. McClure. 1982. Energetics of reproduction in large-littered cotton rats (*Sigmodon hispidus*). Ecology 63:183–195.

Mattingly, H. T. and M. J. Butler. 1994. Laboratory predation on the Trinidadian guppy: implications for the size-selective predation hypothesis and guppy life history evolution. Oikos 69:54–64.

Mauck, R. A. and T. C. Grubb. 1995. Petrel parent shunt all experimentally increased reproductive costs to their offspring. Animal Behaviour 49:999–1008.

May, R. M. 1971. Stability in model ecosystems. Proceedings of the Ecological Society of Australia 6:18–56.

May, R. M. 1973. Stability in randomly fluctuating versus deterministic environments. American Naturalist 107:621–650.

Maynard Smith, J. 1966. Sympatric speciation. American Naturalist 100:637–650.

Maynard Smith, J. 1970. Genetic polymorphism in a varied environment. American Naturalist 104:487–490.

Maynard Smith, J. 1977. Parental investment: a prospective analysis. Animal Behaviour 25:1–9.

Maynard Smith, J. 1981. Will a sexual population evolve to an ESS? American Naturalist 117:1015–1018.

Maynard Smith, J. 1982. *Evolution and the Theory of Games*. Cambridge University Press, Cambridge.

Maynard Smith, J. and R. L. Brown. 1986. Competition and body size. Theoretical Population Biology 30:166–179.

Maynard Smith, J., R. Burian, S. Kauffman, P. Alberch, J. Campbell, B. Goodwin, R. Lande, D. Raup and L. Wolpert. 1985. Developmental constraints and evolution. Quarterly Review of Biology 60:265–287.

Maynard Smith, J. and R. Hoekstra. 1980. Polymorphism in a varied environment: how robust are the models? Genetical Research 35:45–57.

Maynard Smith, J. and G. R. Price. 1973. The logic of animal conflict. Nature 246:15–18.

McCall, C., D. M. Waller and T. Mitchell-Olds. 1994. Effects of serial inbreeding on fitness components in *Impatiens capensis*. Evolution 48:818–827.

McClenaghan, L. R. J. and M. S. Gaines. 1978. Reproduction in marginal populations of the hispid cotton rat (*Sigmodon hispidus*) in northeastern Kansas. Occasional Papers of the Museum of Natural History, University of Kansas 74:1–16.

McClure, P. A. 1981. Sex-biased litter reduction in food-restricted wood rats (*Neotoma floridana*). Science 211:1058–1060.

McCollum, S. A. and J. Van Buskirk. 1996. Costs and benefits of a predator-induced polyphenism in the gray treefrog *Hyla chrysoscelis*. Evolution 50:583–593.

McCoy, E. D. and J. R. Rey. 1981. Patterns of abundance, distribution, and alary polymorphism among the salt marsh Delphacidae (Homoptera: Fulgoroidea) of Northwest Florida. Ecological Entomology 6:285–291.

McFarquhar, A. M. and F. W. Robertson. 1963. The lack of evidence for co-adaptation in crosses between geographical races of *Drosophila subobscura* Coll. Genetical Research 4:104–131.

McGillivray, W. B. 1983. Intraseasonal reproductive costs for the house sparrow (*Passer domesticus*). Auk 100:25–32.

McGinley, M. A. 1989. The influence of a positive correlation between clutch size and offspring fitness on the optimal offspring size. Evolutionary Ecology 3:150–156.

McGinley, M. A., D. H. Temme and M. A. Geber. 1987. Parental investment in offspring in variable environments: theoretical and empirical considerations. American Naturalist 130:370–398.

McGurk, M. D. 1986. Natural mortality of marine pelagic fish eggs and larvae: role of spatial patchiness. Marine Ecology - Progress Series 34:227–242.

McKenzie, J. A. and P. Batterham. 1994. The genetic, molecular and phenotypic consequences of selection for insecticide resistance. Trends in Ecology and Evolution 9:166–169.

McLaren, I. 1976. Inheritance of demographic and production parameters in the marine copepod, *Eurytemora herdmani*. Biological Bulletin 151:200–213.

McLaren, I. and C. Corkett. 1978. Unusual genetic variation in body size, development times, oil storage, and survivorship in the marine copepod *Pseudocalanus*. Biological Bulletin 155:347–359.

McLaren, I. A. 1966. Adaptive significance of large size and long life of the chaetognath *Saggitta elegans* in the Arctic. Ecology 47:852–856.

McLaughlin, R. L. 1989. Search modes of birds and lizards: evidence for alternative movement patterns. American Naturalist 133:654–670.

McMillan, I., M. Fitz-Earle and D. S. Robson. 1970a. Quantitative genetics of fertility. I. Lifetime egg production of *Drosophila melanogaster* - Theoretical. Genetics 65:349–353.

McMillan, I., M. Fitz-Earle and D. S. Robson. 1970b. Quantitative genetics of fertility. II Lifetime egg production of *Drosophila melanogaster*—Experimental. Genetics 65:355–369.

McNamara, J. M. 1993. State-dependent life-history equations. Acta Biotheoretica 41:165–74.

McPeek, M. A. and R. D. Holt. 1992. The evolution of dispersal in spatially and temporally varying environments. American Naturalist 140:1010–1027.

Meats, A. 1971. The relative importance to population increase of fluctuations in mortality, fecundity and time variables of the reproductive schedule. Oecologia 6:223–237.

Mellors, W. K. 1975. Selective predation on ephippial *Daphnia* and the resistance of ephippial eggs to digestion. Ecology 56:974–980.

Menge, B. A. 1973. Effect of predation and environmental patchiness on the body size of a tropical pulmonate limpet. Velinger 16:87–92.

Menge, B. A. 1974. Effect of wave action and competition on brooding and reproductive effort in the sea star *Leptasterias hexactis*. Ecology 55:84–93.

Menu, F. 1993. Strategies of emergence in the chestnut weevil *Curculio elephas* (Coleoptera: Curculionidae). Oecologia 96:383–390.

Menu, F. and D. Debouzie. 1993. Coin-flipping plasticity and prolonged diapause in insects: Example of the chestnut weevil *Curculio elephas* (Coleoptera: Curculionidae). Oecologia 93:367–373.

Merila, J. and M. Bjorklund. 1999. Population divergence and morphometric integration in the greenfinch (*Carduelis chloris*) - evolution against the trajectory of least resistance? Journal of Evolutionary Biology 12:103–112.

Merritt, E. S. 1962. Selection for egg production in geese. Proceedings of the 12 World Poultry Congress 1962:85–87.

Mertz, D. B. 1971. The mathematical demography of the California condor population. American Naturalist 105:437–453.

Mertz, D. R. 1975. Senescent decline in flour beetle strains selected for early adult fitness. Physiological Zoology 48:1–23.

Messina, F. J. 1989. Genetic basis of variable oviposition behavior in *Callosobruchus maculatus* (Coleoptera: Bruchidae). Annals of the Entomological Society of America 82:792–796.

Meyer, A. 1990. Morphometrics and allometry in the trophically polymorphic cichlid fish, *Cichlasoma citrinellum*: Alternative adaptations and ontogenetic changes in shape. Journal of Zoology 221:237–260.

Michener, G. R. 1989. Reproductive effort during gestation and lactation by Richardson's ground squirrels. Oecologia 78:77–86.

Michod, R. E. 1979. Evolution of life histories in response to age-specific mortality factors. American Naturalist 113:531–550.

Millar, J. S. 1975. Tactics of energy partitioning in breeding *Peromyscus*. Canadian Journal of Zoology 53:967–976.

Millar, J. S. 1977. Adaptive features of mammalian reproduction. Evolution 31:370–386.

Millar, J. S. 1978. Energetics of reproduction in *Peromyscus leucopus*: the cost of lactation. Ecology 59:1055–1061.

Millar, J. S. 1981. Pre-partum reproductive characteristics of eutherian mammals. Evolution 35:1149–1163.

Millar, J. S. 1984. The role of design constraints in the evolution of mammalian reproductive rates. Acta Zoologica Fennica 171:133–136.

Millar, J. S. and G. J. Hickling. 1991. Body size and the evolution of mammalian life histories. Functional Ecology 5:588–593.

Millar, J. S. and R. M. Zammuto. 1983. Life histories of mammals: an analysis of life tables. Ecology 64:631–635.

Miller, K. M. and T. H. Carefoot. 1989. The role of spatial and size refuges in the interaction between juvenile barnacles and grazing limpets. Journal of Experimental Marine Biology and Ecology 134:157–174.

Miller, P. S. 1994. Is inbreeding depression more severe in a stressful environment? Zoo Biology 13:195–208.

Minchella, D. J. and P. T. Loverde. 1981. A cost of increased early reproductive effort in the snail *Biomphalaria glabrata*. American Naturalist 118:876–881.

Misra, R. K. and E. C. R. Reeve. 1964. Clines in body dimensions in populations of *Drosophilia subobscura*. Genetical Research 5:240–56.

Mitchell, R. 1975. The evolution of oviposition tactics in the bean weevil, *Callosobruchus maculatus*. Ecology 56:696–702.

Mitchell-Olds, T. 1991. Quantitative genetic changes in small populations. Pp. 634–638 *in* E. C. Dudley, (ed.) *The Unity of Evolutionary Biology*, vol II. Dioscorides Press, Portland, Oregon.

Mitchell-Olds, T. and J. J. Rutledge. 1986. Quantitative genetics in natural plant populations: a review of the theory. American Naturalist 127:379–402.

Mitchell-Olds, T. and D. M. Waller. 1985. Relative performance of selfed and outcrossed progeny in *Impatiens capensis*. Evolution 39:533–544.

Mittelbach, G. G. 1981. Patterns of invertebrate size and abundance in aquatic habitats. Canadian Journal of Fisheries and Aquatic Sciences 38:896–904.

Miyatake, T. and M. Yamagishi. 1993. Active quality control in mass reared melon flies. Quantitative genetic aspects. Pp. 201–213. Management of insect pests: nuclear and related molecular and genetic techniques. International Atomic Energy Agency.

Mock, D. W., H. Drummond and C. H. Stinson. 1990. Avian Siblicide. American Scientist 78:438–449.

Mock, D. W. and G. A. Parker. 1986. Advantages and disadvantages of egret and heron brood reduction. Evolution 40:459–470.

Møller, A. P. 1990. Effects of parasitism by a haematophagous mite on reproduction in the barn swallow. Ecology 71:2345–2357.

Møller, A. P. 1993. Ectoparasites enhance the cost of reproduction in their hosts. Journal of Animal Ecology 62:309–322.

Moller, H., R. H. Smith and R. M. Sibly. 1989. Evolutionary demography of a bruchid beetle. I.

Quantitative genetical analysis of the female life history. Functional Ecology 3:673–681.

Monaghan, P., M. Bolton and D. C. Houston. 1995. Egg production constraints and the evolution of avian clutch size. Proceedings of the Royal Society of London, B 259:189–191.

Montalvo, A. M. and J. D. Ackerman. 1987. Limitations to fruit production in *Ionopsis utricularioide* (Orchidaceae). Biotropica 19:24–31.

Montgomerie, R. D. and P. J. Weatherhead. 1988. Risks and rewards of nest defence by parent birds. Quarterly Review of Biology 63:167–187.

Moran, N. 1992. The evolutionary maintenance of alternative phenotypes. American Naturalist 139:971–989.

Morand, S. 1996. Life-history traits in parasitic nematodes: a comparative approach for the search of invariants. Functional ecology 10:210–218.

Moreno, J. 1993. Physiological mechanisms underlaying reproductive trade-offs. Etologia 3:41–56.

Moreno, J., J. J. Sanz and E. Arriero. 1999. Reproductive effort and T-lymphocyte cell-mediated immunocompetence in female pied flycatchers *Ficedula hypoleuca*. Proceedings of the Royal Society of London, B 266:1105–1109.

Moreteau, B., J. Morin, P. Gibert, G. Petavy, E. Pla and J. R. David. 1997. Evolutionary changes of nonlinear reaction norms according to thermal adaptation: a comparison of two *Drosophila* species. C. R. Acadamie des Sciences, Paris, Sciences de la vie 320:833–841.

Morin, J. P., B. Moreteau, G. Petavy, R. Parkash and J. R. David. 1997. Reaction norms of morphological traits in *Drosophila*: adaptive shape changes in a stenotherm circumtropical species. Evolution 51:1140–1148.

Morris, D. W. 1986. Proximate and ultimate controls of life-history variation: the evolution of litter size in white-footed mice (*Peromyscus leucopus*). Evolution 40:169–181.

Morris, D. W. 1992. Optimum brood size: tests of alternative hypotheses. Evolution 46:1848–1861.

Morris, J. A. 1963. Continuous selection for egg production using short term records. Australian Journal of Agricultural Research 14:909–925.

Motro, U. 1982a. Optimal rates of dispersal. I. Haploid populations. Theoretical Population Biology 21:394–411.

Motro, U. 1982b. Optimal rates of dispersal. II. Diploid populations. Theoretical Population Biology 21:412–429.

Mountford, M. D. 1968. The significance of litter size. Journal of Animal Ecology 37:363–367.

Mousseau, T. 2000. Intra- and interpopulation genetic variation: explaining the past and predicting the future. Pp. 219–250 *in* T. Mousseau, B. Sinervo and J. Endler, (eds.) *Adaptive Genetic Variation in the Wild.* Oxford University Press, New York.

Mousseau, T. and C. W. Fox (editors). 1998. *Maternal Effects as Adaptations.* Oxford University Press, New York.

Mousseau, T., K. Ritland and D. D. Heath. 1998. A novel method for estimating heritabilities using molecular markers. Heredity 80:218–224.

Mousseau, T. A. 1991. Geographic variation in maternal-age effects on diapause in a cricket. Evolution 45:1053–1059.

Mousseau, T. A. and H. Dingle. 1991. Maternal effects in insect life histories. Annual Review of Entomology 36:511–534.

Mousseau, T. A. and D. A. Roff. 1987. Natural selection and the heritability of fitness components. Heredity 59:181–198.

Mousseau, T. A. and D. A. Roff. 1989. Adaptation to seasonality in a cricket: patterns of phenotypic and genotypic variation in body size and diapause expression along a cline in season length. Evolution 43:1483–1496.

Mueller, L. D. 1988a. Density-dependent population growth and natural selection in food-limited environments: the *Drosophila* model. American Naturalist 132:786–809.

Mueller, L. D. 1988b. Evolution of competitive ability in *Drosophila* by density-dependent natural selection. Proceedings of the National Academy of Sciencess, USA 85:4383–4386.

Mueller, L. D. 1990. Density-dependent natural selection does not increase efficiency. Evolutionary Ecology 4:290–297.

Mueller, L. D. 1991. Ecological determinants of life-history evolution. Philosophical Transactions of the Royal Society of London B 332:25–30.

Mueller, L. D. and F. J. Ayala. 1981. Trade-off between r-selection and K-selection in *Drosophila* populations. Proceedings of the National Academy of Sciencess, USA 78:1303–1305.

Mueller, L. D., P. Guo and F. J. Ayala. 1991. Density-dependent natural selection and trade-offs in life history traits. Science 253:433–435.

Mueller, L. D. and V. F. Sweet. 1986. Density-dependent natural selection in *Drosophila*: evolution of pupation height. Evolution 40:1354–1356.

Muir, W., W. E. Nyquist and S. Xu. 1992. Alternative partitioning of the genotype-by-environment interaction. Theoretical and Applied Genetics 84:193–200.

Mullin, T. J., E. K. Morgenstern, Y. S. Park and D. P. Fowler. 1992. Genetic parameters from a clonally replicated test of black spruce (*Picea mariana*). Canadian Journal of Forest Research 22:24–36.

Murphy, E. C. 1985. Bergmann's rule, seasonality, and geographic variation in body size of house sparrows. Evolution 39:1327–1334.

Murphy, G. I. 1968. Pattern in life history and the environment. American Naturalist 102:391–403.

Murphy, M. T. 2000. Evolution of clutch size in the Eastern kingbird: tests of alternative hypotheses. Ecological Mongraphs 70:1–20.

Murray, J. and B. Clarke. 1967. Inheritance of shell size in *Partula*. Heredity 23:189–198.

Myers, J. H. 1976. Distribution and dispersal in populations capable of resource depletion. Oecologia 23:255–269.

Myers, P. and L. L. Master. 1983. Reproduction by *Peromyscus maniculatus*: size and compromise. Journal of Mammalogy 64:1–18.

Myers, R. A. 1986. Game theory and the evolution of Atlantic salmon (*Salmo salar*) age at maturation. Special Publication of Canadian Fisheries and Aquatic Sciences 89:53–61.

Myers, R. A. and R. W. Doyle. 1983. Predicting natural mortality rates and reproduction—mortality tradeoffs from fish life history data. Canadian Journal of Fisheries and Aquatic Sciences 40:612–620.

Myers, R. A. and J. A. Hutchings. 1986. Selection against parr maturation in Atlantic Salmon. Aquaculture 53:313–320.

Naevdal, G., M. Holm, D. Moller and O. D. Osthus. 1976. Variation in growth rate and age at sexual maturity in Atlantic salmon. International Council for the Exploration of the Sea. 1976/E, 40:10pp.

Nager, R. G., L. F. Keller and A. J. van Noordwijk. 2000. Understanding natural selection on traits that are influenced by environmental conditions. Pp. 95–115 *in* T. Mousseau, B. Sinervo and J. Endler, (eds.) *Adaptive Genetic Variation in the Wild*. Oxford University Press, New York.

Nager, R. G., C. Ruegger and A. J. van Noordwijk. 1997. Nutrient or energy limitation of egg formation: A feeding experiment in great tits. Journal of Animal Ecology 66:495–507.

Nager, R. G. and A. J. van Noordwijk. 1995. Proximate and ultimate aspects of phenotypic plasticity in timing of great tit breeding in a heterogeneous environment. American Naturalist 146:454–474.

Nakamura, S. 1995. Optimal clutch size for maximizing reproductive success in a parasitoid fly, *Exorista japonica* (Diptera: Tachinidae). Applied Entomology and Zoology 30:425–431.

Nakasuji, F. 1982. Seasonal changes in native host plants of a migrant skipper, *Parnara guttata* Bremer et Grey (Lepidoptera: Hesperiidae). Applied Entomology and Zoology 17:146–148.

Nakasuji, F. 1987. Egg size of skippers (Lepidoptera: Hesperiidae) in relation to their host specificity and to leaf toughness of host plants. Ecological Research 2:175–183.

Nakasuji, F. and M. Kimura. 1984. Seasonal polymorphism of egg size in a migrant skipper, *Parnara guttata guttata* (Lepidoptera: Hesperiidae). Kontyu 52:253–259.

Namkoong, G. and H. R. Gregorius. 1985. Conditions for protected polymorphisms in subdivided plant populations. 2. Seed versus pollen migration. American Naturalist 125:521–534.

Naylor, A. F. 1964. Natural selection through maternal influence. Heredity 19:509–511.

Nei, M., T. Maruyama and R. Chakraborty. 1975. The bottleneck effect and genetic variability in populations. Evolution 29:1–10.

Newman, R. A. 1988. Genetic variation for larval anuran (*Scaphiopus couchii*) development time in an uncertain environment. Evolution 42:763–773.

Ni, I. 1978. Comparative fish population studies. Ph.D. Thesis, University of British Columbia, Vancouver, Canada.

Nice, M. M. 1937. Studies in the life history of the song sparrow. Part 1. Transactions of the Linnaean Society of New York 4:1–247.

Nicholas, F. W. 1980. Size of population required for artificial selection. Genetical Research 35:85–105.

Nickell, C. D. and J. E. Grafius. 1969. Analysis of a vegatative response to selection for high yield in winter barley, *Hordeum vulgare* L. Crop Science 9:447–451.

Nijhout, H. F. 1975. A threshold size for metamorphosis in the tobacco hornworm, *Manduca sexta* (L.). Biological Bulletin 149:214–225.

Nijhout, H. F. 1979. Stretch-induced moulting in *Oncopeltus fasciatus*. Journal of Insect Physiology 25:277–281.

Nijhout, H. F. and C. M. Williams. 1974a. Control of moulting and metamorphosis in the tobacco hornworm, *Manduca sexta* (L): cessation of juvenile hormone secretion as a trigger for pupation. Journal of Experimental Biology 61:493–501.

Nijhout, H. F. and C. M. Williams. 1974b. Control of moulting and metamorphosis in the tobacco hornworm, *Manduca sexta* (L.): growth of the last-instar larva and the decision to pupate. Journal of Experimental Biology 61:481–491.

Nilsson, P. A., C. Bronmark and L. B. Pettersson. 1995. Benefits of a predator-induced morphology in crucian carp. Oecologia 104:291–296.

Nomura, T. and K. Yonezawa. 1990. Genetic correlations among life history characters of adult females in the Azuki bean weevil, *Callosobruchus chinensis* (L.) (Coleoptera: Bruchidae). Applied Entomology and Zoology 25:423–430.

Noordwijk, van A. J. and G. de Jong. 1986. Acquisition and allocation of resources: their influence on variation in life history tactics. American Naturalist 128:137–142.

Nordling, D., M. Andersson, S. Zohari and L. Gustafsson. 1998. Reproductive effort reduces specific immune response and parasite resistance. Proceedings of the Royal Society of London B 265:1291–1298.

Nordskog, A. W. 1977. Success and Failure of quantitative genetic theory in poultry. Pp. 569–586 *in* E. Pollack, O. Kempthorne and T. B. Bailey, (eds.). Proceedings of the International Conference on

Quantitative Genetics. Iowa State University Press, Iowa.

Nordskog, A. W. and M. Festing. 1962. Selection and correlated responses in the fowl. Proceedings of the 12th World Poultry Congress, Sydney 1962:25–29.

Norris, K., M. Anwar and A. F. Read. 1994. Reproductive effort influences the prevalence of haematozoan parasites in great tits. Journal of Animal Ecology 63:601–610.

Nunney, L. 1996. The response to selection for fast larval development in *Drosophila melanogaster* and its effect on adult weight: an example of a fitness trade-off. Evolution 50:1193–1204.

Nur, N. 1983. On parental investment during the breeding season. Animal Behaviour 31:309–311.

Nussbaum, R. A. 1981. Seasonal shifts in clutch-size and egg-size in the side-blotched lizard, *Uta stansburiana*, Baird and Girard. Oecologia 49:8–13.

Nylin, S. 1992. Seasonal plasticity in life history traits: growth and development in *Polygonia c-album* (Lepidoptera: Nymphalidae). Biological Journal of the Linnean Society 47:301–323.

Nylin, S., K. Gotthard and C. Wiklund. 1996. Reaction norms for age and size at maturity in *Lasiommata* butterflies: Predictions and tests. Evolution 50:1351–1358.

Nylin, S. and L. Svard. 1991. Latitudinal patterns in the size of European butterflies. Holarctic Ecology 14:192–202.

O'Brien, W. J., D. Kettle and H. Riessen. 1979. Helmets and invisible armour: structures reducing predation from tactile and visual planktivores. Ecology 60:287–294.

O'Brien, W. J., N. A. Slade and G. L. Vinyard. 1976. Apparent size as the determinant of prey selection by bluegill sunfish (*Lepomis macrochirus*). Ecology 57:1304–1310.

O'Connor, R. J. 1978. Brood reduction in birds: selection for fratricide, infanticide or suicide? Animal Behaviour 26:79–96.

Ohsumi, S. 1979. Interspecies relationships among some biological parameters in cetaceans and estimation of the natural mortality coefficient of the southern hemisphere minke whale. Report of the International Whaling Commission 29:397–406.

Olivieri, I., Y. Michalakis and P. H. Gouyon. 1995. Metapopulation genetics and the evolution of dispersal. American Naturalist 146:202–228.

Olsen, P. D., R. B. Cunningham and C. F. Donnelly. 1994. Avian egg morphometrics: Allometric models of egg volume clutch volume and shape. Australian Journal of Zoology 42:307–321.

Olsen, P. D., R. B. Cunningham and C. F. Donnelly. 1994. Is there a trade-off between egg size and clutch size in altricial and precocial nonpasserines? A test of a model of the relationship between egg and clutch size. Australian Journal of Zoology 42:323–328.

Olson, S. L. 1973. Evolution of the rails of the South Atlantic Islands (Aves: Rallidae). Smithsonian Contributions to Zoology 152: 1–53.

Olsson, G. 1960. Some relations between number of seeds per pod, seed size and oil content and the effects of selection for these characters in *Brassica* and *Sinapis*. Hereditas 46:29–70.

Olsson, M., R. Shine and E. Bak-Olsson. 2000. Locomotor impairment of gravid lizards: is the burden physical or physiological? Journal of Evolutionary Biology 13:263–268.

Oosthuizen, E. and N. Daan. 1974. Egg fecundity and maturity of North Sea Cod, *Gadus morhua*. Netherlands Journal of Sea Research 8:378–397.

Oppliger, A., P. Christe and H. Richner. 1997. Clutch size and malalarial parasites in female great tits. Behavioral Ecology 8:148–152.

Oppliger, A., H. Richner and P. Christe. 1994. Effect of an actoparasite on lay date, nest-site choice, desertion, and hatching success in the great tit (*Parus major*). Behavioral ecology 5:130–134.

Orengo, D. J. and A. Prevosti. 1999. Wing-size heritability in a natural population of *Drosophila subobscura*. Heredity 1999:100–106.

Orozco, F. 1976. A dynamic study of genotype environment interaction with egg laying of *Tribolium casteneum*. Heredity 37:157–171.

Orzack, S. H. 1997. Life-history evolution and extinction *in* S. Tuljapurkar and H. Caswell, (eds.) *Structured-Population Models in Marine, Terrestrial and Freshwater Systems*. Chapman and Hall, NY.

Orzack, S. H. and S. Tuljapurkar. 1989. Population dynamic in variable environments. VII. The demography and evolution of iteroparity. American Naturalist 133:901–923.

Ots, I. and P. Horak. 1996. Great tits, *Parus major*, trade health for reproduction. Proceedings of the Royal Society of London B 263:1443–1447.

Packard, G. C. 1967. House sparrows: Evolution of populations from the Great Plains and Colorado Rockies. Systematic Zoologist 16:73–89.

Padilla, D. K. and S. C. Adolph. 1996. Plastic inducible morphologies are not always adaptive: The importance of time delays in a stochastic environment. Evolutionary Ecology 10:105–117.

Padley, D. 1985. Do the life history parameters of passerines scale to metabolic rate independently of body size? Oikos 45:285–287.

Paine, R. T. 1965. Natural history, limiting factors and energetics of the opisthobranch *Navanax inermis*. Ecology 46:603–619.

Paine, R. T. 1976. Size-limited predation: an observational and experimental approach with the *Mytilus - Pisaster* interaction. Ecology 57:858–873.

Pak, G. A. and E. R. Oatman. 1982. Biology of *Trichogramma brevicapillum*. Entomologica Experimentalis Applicata 32:61–67.

Palmer, A. R. 1990. Predation size, prey size, and the scaling of vulnerability: hatching gastropods vs barnacles. Ecology 71:759–775.

Palmer, J. O. and H. Dingle. 1986. Direct and correlated responses to selection among life-history traits in mikweed bugs (*Oncopeltus fasciatus*). Evolution 40:767–777.

Parker, G. A. and M. Begon. 1986. Optimal egg size and clutch size: effects of environmental and maternal phenotype. American Naturalist 128:573–592.

Parker, G. A. and S. P. Courtney. 1984. Models of clutch size in insect oviposition. Theoretical Population Biology 26:27–48.

Parker, G. A. and J. Maynard Smith. 1990. Optimality theory in evolutionary biology. Nature 348:27–33.

Parker, G. A. and D. W. Mock. 1987. Parent-offspring conflict over clutch size. Evolutionary Ecology 1:161–174.

Parker, G. A., D. W. Mock and T. C. Lamey. 1989. How selfish should stronger sibs be? American Naturalist 133:846–868.

Parker, R. J., L. D. McGilliard and J. L. Gill. 1969. Genetic correlation and response to selection in simulated populations I. Additive model. Theoretical and Applied Genetics 39:365–370.

Parker, R. J., L. D. McGilliard and J. L. Gill. 1970a. Genetic correlation and response to selection in simulated populations II. Model of complete dominance. Theoretical and Applied Genetics 40:106–110.

Parker, R. J., L. D. McGilliard and J. L. Gill. 1970b. Genetic correlation and response to selection in simulated populations III. Correlated response to selection. Theoretical and Applied Genetics 40:157–162.

Parker, R. R. and P. A. Larkin. 1959. A concept of growth in fishes. Journal of the Fishes Research Board of Canada 16:721–745.

Parkes, A. S. 1926. The growth of young mice according to size of litter. Annals of Applied Biology 13:374–394.

Parry, G. D. 1981. The meanings of *r*- and *K*-selection. Oecologia 48:260–264.

Parsons, P. A. 1964. Egg lengths in *Drosophila melanogaster* and correlated responses to selection. Genetica 35:175–181.

Partridge, L., B. Barrie, K. Fowler and V. French. 1994. Evolution and development of body size and cell size in *Drosophila melanogaster* in response to temperature. Evolution 48:1269–1276.

Partridge, L. and M. Farquhar. 1981. Sexual activity reduces life span of male fruit flies. Nature 294:580–582.

Partridge, L. and K. Fowler. 1992. Direct and correlated response to selection on age at reproduction in *Drosophila melanogaster*. Evolution 46:76–91.

Partridge, L. and P. H. Harvey. 1985. Costs of reproduction. Nature 316:20.

Partridge, L. L. 1997. Evolutionary biology and age-related mortality. Pp. 78–95 *in* K. W. Wachter and C. E. Finch, (eds.) *Between Zeus and the Salmon*. National Academy Press, Washington.

Partridge, L. L. and J. A. Coyne. 1997. Bergmann's rule in ectotherms: is it adaptive? Evolution 5:632–635.

Pascual, M., F. J. Ayala, A. Prevosti and L. Serra. 1993. Colonization of North America by *Drosophila subobscura*. Zeitschrift fuer Zoollogische Systematik und Evolutionsforschung 31:216–226.

Pashley, D. P. 1988. Quantitative genetics, development, and physiological adaptation in host straints of fall armyworm. Evolution 42:93–102.

Pauly, D. 1978. A preliminary compilation of fish length growth parameters. Berichte aus dem Institute fur Meereskunde an der Christian-Albrechts-Universitat Kiel 55:1–200.

Pauly, D. 1980. On the interrelationships between natural mortality, growth parameters, and mean environmental temperature in 175 fish stocks. Journal du Conseil Permanent International pour l'Exploration de la Mer 39:175–192.

Pearl, R. 1940. *Introduction to Medical Biometry and Statistics*. Saunders, Philadelphia.

Pearl, R. and C. R. Doering. 1923. A comparison of the mortality of certain organisms with that of man. Science 57:209–212.

Pearl, R. and J. R. Miner. 1935. Experimental studies on the duration of life. XIV. The comparative mortality of certain lower organisms. Quarterly Review of Biology 10:60–79.

Pease, C. M. and J. J. Bull. 1988. A critique of methods for measuring life history trade-offs. Journal of Evolutionary Biology 1:293–303.

Perrin, N. and J. Goudet. 2001. Inbreeding, kinship, and the evolution of natal dispersal *in* C. Clobert, J. Nichols, J. D. Danchin and A. Dhondt, (eds.) *Causes, Consequences and Mechanisms of Dispersal at the Individual, Population and Community Level*. Oxford University Press, Oxford.

Perrin, N. and V. Mazalov. 1999. Dispersal and inbreeding avoidance. American Naturalist 154:282–292.

Perrin, N. and V. Mazalov. 2000. Local competition, inbreeding, and the evolution of sex-biased dispersal. American Naturalist 155:116–127.

Perrin, N. and J. F. Rubin. 1990. On dome-shaped norms of reaction for size-at-age at maturity in fishes. Functional Ecology 4:53–57.

Perrin, N., R. M. Sibly and N. K. Nichols. 1993. Optimal growth strategies when mortality and production rates are size-dependent. Evolutionary Ecology 7:576–592.

Perrins, C. M. 1970. The timing of birds' breeding seasons. Ibis 112:242–255.

Perrins, C. M. and P. J. Jones. 1974. The inheritance of clutch size in the great tit (*Parus major* L.). Condor 76:225–229.

Peterman, R. M. 1990. Statistical power analysis can improve fisheries research and management. Canadian Journal of Fisheries and Aquatic Sciences 47:2–15.

Peters, R. H. 1983. *The Ecological Implications of Body Size*. Cambridge University Press, Cambridge.

Peterson, B. 1950. The relation between size of mother and number of eggs and young in some spiders. Experientia 6:96–98.

Peterson, C. C., K. A. Nagy and J. Diamond. 1990. Sustained metabolic scope. Proceedings of the National Academy of Sciences 87:2324–2328.

Peterson, I. and J. S.Wroblewski. 1984. Mortality rate of fishes in the pelagic ecosystem. Canadian Journal of Fisheries and Aquatic Sciences 41:1117–1120.

Pettersson, L. B. and C. Bronmark. 1997. Density-dependent costs of an inducible morphological defense in crucian carp. Ecology 76:1805–1815.

Pettifor, R. A. 1993a. Brood-manipulation experiments. I. The number of offspring surviving per nest in blue tits (*Parus caeruleus*). Journal of Animal Ecology 62:131–144.

Pettifor, R. A. 1993b. Brood-manipulation experiments: II. A cost of reproduction in blue tits (*Parus caeruleus*). Journal of Animal Ecology 62:145–159.

Pettifor, R. A., C. M. Perrins and R. H. McCleery. 1988. Individual optimization of clutch size in great tits. Nature 336:160–162.

Pfennig, D. W. 1992. Polyphenism in spadefoot toad tadpoles as a locally adjusted evolutionarily stable strategy. Evolution 46:1408–1420.

Phillips, J. C. 1928. Wild birds introduced or transplanted in North America. United States Department of Agriculture Technical Bulletin 61:1–63.

Pianka, E. R. 1970. On *r*- and *K*-selection. American Naturalist 104:592–597.

Pianka, E. R. and W. S. Parker. 1975. Age specific reproductive tactics. American Naturalist 109:453–464.

Pimm, S. L. and A. Redfearn. 1988. The variability of population densities. Nature 334: 613–614.

Pinero, D., J. Sarukhan and P. Alberdi. 1982. The costs of reproduction in a tropical palm *Astrocaryum mexicanum*. Journal of Ecology 70:473–481.

Pinkowski, B. C. 1975. Growth and development of Eastern Bluebirds. Bird-Banding 46:273–289.

Pitelka, L. F. 1977. Energy allocation in annual and perennial lupines (*Lupinus*: Leguminosae). Ecology 58:1055–1065.

Pitt, T. K. 1966. Sexual maturity and spawning of the American plaice, *Hippoglossoides platessoides* (Fabr.), from Grand Bank and Newfoundland areas. Journal of the Fisheries Research Board of Canada 23:651–672.

Pitt, T. K. 1967. Age and growth of American plaice (*Hippoglossoides platessoides*) in the Newfoundland area on the northwest Atlantic. Journal of the Fisheries Research Board of Canada 24:1077–1099.

Platenkamp, G. A. J. and R. G. Shaw. 1992. Environmental and genetic constraints on adaptive population differentiation in *Anthoxanthum odoratum*. Evolution 46:341–352.

Poizat, G., E. Rosecchi and A. J. Crivelli. 1999. Empirical evidence of a trade-off between reproductive effort and expectation of future reproduction in female three-spined sticklebacks. Proceedings of the Royal Society of London, B 266:1543–1548.

Polak, M. and W. T. Starmer. 1998. Parasite-induced risk of mortality elevates reproductive effort in male *Drosophila*. Proceedings of the Royal Society of London B 265:2197–2201.

Pollack, E. and O. Kempthorne. 1970. Malthusian parameters in genetic populations. Part I. Haploid and selfing models. Theoretical Population Biology 1:315–345.

Pollack, E. and O. Kempthorne. 1971. Malthusian parameters in genetic populations. Part II. Random mating populations in infinite habitats. Theoretical Population Biology 2:351–390.

Pomeroy, D. 1990. Why fly? The possible benefits for lower mortality. Biological Journal of the Linnean Society 40:53–65.

Pomiankowski, A. and Y. Iwasa. 1993. Evolution of multiple sexual preferences by Fisher's runaway process of sexual selection. Proceedings of the Royal Society of London B 253:173–181.

Pomiankowski, A. N. 1988. The evolution of female mate preferences for male genetic quality. Pp. 136–184 *in* P. H. Harvey and L. Partridge (eds.). Oxford Surveys in Evolutionary Biology. Oxford University Press, Oxford.

Pontier, D., J. M. Gaillard and D. Allaine. 1993. Maternal investment per offspring and demographic tactics in placental mammals. Oikos 66:424–430.

Post, J. R. and D. O. Evans. 1989. Size-dependent overwinter mortality of young-of-the-year yellow perch (*Perca flavescens*): laboratory, in situ enclosure, and field experiments. Canadian Journal of Fisheries and Aquatic Sciences 46:1958–1968.

Poulin, R. 1995. Clutch size and egg size in free-living and parasitic copepods: A comparative analysis. Evolution 49:325–336.

Powell, J. A. 1976. A remarkable new genus of brachypterous moth from coastal sand dunes in California (Lepidoptera: Gelechioidea, Scythrididae). Annals of the Entomological Society of America 69:325–339.

Powles, P. M. 1958. Studies of reproduction and feeding of Atlantic cod (*Gadus callarias* L.) in the Southwestern Gulf of St. Lawrence. Journal of the Fisheries Research Board of Canada 15:1383–1402.

Pressley, P. H. 1981. Parental effort and the evolution of nest-guarding tactics in the threespine stickleback, *Gasterosteus aculeatus* L. Evolution 35:282–295.

Prevosti, A. 1955. Geographical variability in quantitative traits in populations of *Drosophila subobscura*. Quantitative Biology 20:294–299.

Prevosti, A., G. Ribo, L. Serra, M. Aguade, J. Balana, M. Monclus and F. Mestres. 1988. Colonization of America by *Drosophila subobscura*: Experiment in natural populations that supports the adaptive role of chromosomal-inversion polymorphism. Proceedings of the National Academy of Sciences 85:5597–5600.

Preziosi, R. F. and D. J. Fairbairn. 1997. Sexual size dimorphism and selection in the wild in the water-strider *Aquarius remigis*: lifetime fecundity selection on female total length and its components. Evolution 51:467–474.

Preziosi, R. F. and D. J. Fairbairn. 2000. Lifetime selection on adult body size and components of body size in a waterstrider: opposing selection and maintenance of sexual size dimorphism. Evolution 54:558–566.

Preziosi, R. F., D. J. Fairbairn, D. A. Roff and J. M. Brennan. 1996. Body size and fecundity in the water-strider *Aquarius remigis*: a test of Darwin's fecundity advantage hypothesis. Oecologia 108:424–431.

Price, D. K. and N. T. Burley. 1993. Constraints on the evolution of attractive traits: genetic (co)variance of zebra finch bill colour. Heredity 71:405–412.

Price, G. R. 1972a. Fisher's 'fundamental theorem' made clear. Annals of Human Genetics 36:129–140.

Price, T., M. Kirkpatrick and S. J. Arnold. 1988. Directional selection and the evolution of breeding date in birds. Science 240:798–799.

Price, T. and L. Liou. 1989. Selection on clutch size in birds. American Naturalist 134:950–959.

Price, T. and D. Schluter. 1991. On the low heritability of life-history traits. Evolution 45:853–861.

Price, T. D. 1984. The evolution of sexual size dimorphism in Darwin's finches. American Naturalist 123:500–518.

Price, T. D., C. R. Brown and M. B. Brown. 2000. Evaluation of selection on cliff swallows. Evolution 54:1824–1827.

Price, T. D. and P. R. Grant. 1984. Life history traits and natural selection for small body size in a population of Darwin's finches. Evolution 38:483–494.

Primack, R. B. 1979. Reproductive effort in annual and perennial species of *Plantago* (Plantaginaceae). American Naturalist 114:51–62.

Primack, R. B. and P. Hall. 1990. Costs of reproduction in the pink lady's slipper orchid: a four-year experimental study. American Naturalist 136:638–656.

Pritchard, G. 1965. Prey capture by dragonfly larvae (Odonata: Anisoptera). Canadian Journal of Zoology 43:271–289.

Promislow, D. E. L. 1991. Senscence in natural populations of mammals: a comparative study. Evolution 45:1869–1887.

Promislow, D. E. L. and P. H. Harvey. 1990. Living fast and dying young: a comparative analysis of life-history variation among mammals. Journal of Zoology 220:417–437.

Prout, T. 1962. The effects of stabilising selection on the time development in *D. melanogaster*. Genetical Research 3:364–382.

Prout, T. 1980. Some relationships between density-dependent selection and density dependent population growth. Evolutionary Biology 13:1–68.

Prout, T. and J. S. F. Barker. 1989. Ecological aspects of the heritability of body size in *Drosophila buzzatii*. Genetics 123:803–813.

Provine, W. B. 1971. *Origins of Theoretical Population Genetics*. University of Chicago Press, Chicago.

Provine, W. B. 1986. *Sewall Wright and Evolutionary Biology*. University of Chicago Press, Chicago.

Pulido, F., P. Berthold and A. J. van Noordwijk. 1996. Frequency of migrants and migratory activity are genetically correlated in a bird population: Evolutionary implications. Proceedings of the National Academy of Sciences 93:14642–14647.

Qualls, F. J. and R. Shine. 1997. Geographic variation in cost of reproduction in the scincid lizard *Lampropholis guichenoti*. Functional Ecology 11: 757–763.

Rabb, R. L. and G. G. Kennedy (eds.). 1979. Movement of Highly Mobile Animals: Concepts and Methodology in Research. North Carolina State University

Radwan, J. 1993. The adaptive significance of male polymorphism in the acarid mite *Caloglyphus berlesei*. Behavioral Ecology and Sociobiology 33:201–208.

Radwan, J. 1995. Male morph determination in two species of acarid mites. Heredity 74:669–673.

Rahn, H., P. R. Sotherland and C. V. Paganelli. 1985. Interrelationships between egg mass and adult body mass and metabolism among passerine birds. Journal of Ornithology 126:263–271.

Ramsay, S. L. and D. C. Houston. 1997. Nutritional constraints on egg production in the blue tit: a supplementary feeding study. Journal of Animal Ecology 66:649–657.

Rand, D. A., H. B. Wilson and J. M. McGlade. 1994. Dynamics and evolution: evolutionarily stable attractors, invasion exponents and phenotypic dynamics. Philosophical Transactions of the Royal Society of London, Series B, Biological Sciences 343: 261–283.

Randolph, P. A., J. C. Randolph, K. Mattingly and M. M. Foster. 1977. Energy costs of reproduction in the cotton rat, *Sigmodon hispidus*. Ecology 58:31–45.

Rapoport, A. 1985. Applications of game-theoretic concepts in biology. Bulletin of Mathematical Biology 47:161–192.

Rathie, K. A. and J. S. F. Barker. 1968. Effectiveness of regular cycles of intermittent artificial selection for a quantitative character in *Drosophila melanogaster*. Australian Journal of Biological Science 21:1187–213.

Read, A. F. and P. H. Harvey. 1989. Life history differences among the eutherian radiations. Journal of Zoology 219:329–353.

Reddingius, J. and P. J. den Boer. 1970. Simulation experiments illustrating stabilization of animal numbers by spreading of risk. Oecologia 5:240–284.

Reed, J. and N. C. Stenseth. 1984. On evolutionarily stable strategies. Journal of Theoretical Biology 108:491–508.

Reeve, E. C. and F. W. Robertson. 1953. Studies in quantitative inheritance. II. Analysis of a strain of *Drosophila melanogaster* selected for long wings. Journal of Genetics 51:276–316.

Reeve, E. C. R. 1955. The variance of the genetic correlation coefficient. Biometrics 11:357–374.

Reeve, J. and D. J. Fairbairn. 2001. Predicting the evolution of sexual size dimorphism. Journal of Evolutionary Biology 14:244–254.

Reimchen, T. E. 1988. Inefficient predators and prey injuries in a population of giant stickleback. Canadian Journal of Zoology 66:2036–2044.

Reiss, M. J. 1985. The allometry of reproduction: why larger species invest relatively less in their offspring. Journal of Theoretical Biology 113:529–544.

Reiter, J. and B. J. Le Boeuf. 1991. Life history consequences of variation in age at primiparity in northern elephant seals. Behavioral Ecology and Sociobiology 28:153–160.

Reusch, T. and W. U. Blackenhorn. 1998. Quantitative genetics of the dung fly *Sepsis cynipsea*: Cheverud's conjecture revisited. Heredity 81:111–119.

Reznick, D. 1982b. The impact of predation on life history evolution in Trinidadian guppies: genetic basis of observed life history patterns. Evolution 36:1236–1250.

Reznick, D. 1985. Costs of reproduction: an evaluation of the empirical evidence. Oikos 44:257–267.

Reznick, D. and J. A. Endler. 1982. The impact of predation on life history evolution in Trinidadian guppies (*Poecilia reticulata*). Evolution 36:160–177.

Reznick, D. A., H. Bryga and J. A. Endler. 1990. Experimentally induced life-history evolution in a natural population. Nature 346:357–359.

Reznick, D. N. 1982a. Genetic determination of offspring size in the guppy (*Poecilia reticulata*). American Naturalist 120:181–188.

Reznick, D. N. and H. Bryga. 1987. Life-history evolution in guppies (*Poecilia reticulata*): 1. Phenotypic and genetic changes in an introduction experiment. Evolution 41:1370–1385.

Reznick, D. N., F. H. Shaw, F. H. Rodd and R. G. Shaw. 1997. Evaluation of the rate of evolution in natural populations of guppies (*Poecilia reticulata*). Science 275:1934–1937.

Rhodes, C. P. and D. M. Holdich. 1982. Observations on the fecundity of the freshwater crayfish, *Austropotamobius pallipes* (Lereboullet) in the British Isles. Hydrobiologia 89:231–236.

Richner, H., P. Christe and A. Oppliger. 1995. Paternal investment affects malaria prevalence. Proceedings of the National Academy of Sciences 92:1192–1194.

Richner, H., A. Oppliger and P. Christe. 1993. Effect of an ectoparasite on reproduction in great tits. Journal of Animal Ecoogy 62:703–710.

Ricklefs, R. 1982. A comment on the optimization of body size in *Drosophila* to Roff's life history model. American Naturalist 120:686–688.

Ricklefs, R. E. 1974. Energetics of reproduction in birds. Pp. 152–292 *in* J. R. A. Paynter, (ed.) *Avian Energetics*. Nuttall Ornithological Club, Cambridge, Massachusetts.

Ricklefs, R. E. 1977. On the evolution of reproductive strategies in birds: reproductive effort. American Naturalist 111:453–478.

Ricklefs, R. E. 1998. Evolutionary theories of aging: Confirmation of a fundamental prediction, with implications for the genetic basis and evolution of life span. American Naturalist 152:24–45.

Ridley, M. 1983. *The Explanation of Organic Diversity*. Clarendon Press, Oxford.

Riechert, S. E. and P. Hammerstein. 1983. Game theory in the ecological context. Annual Review of Ecology and Systematics 14:377–409.

Riessen, H. P. 1992. Cost-benefit model for the induction of an antipredator defense. American Naturalist 140:349–362.

Riessen, H. P. and W. G. Sprules. 1990. Demographic costs of antipredator defenses in *Daphnia pulex*. Ecology 71:1536–1546.

Rijnsdorp, A. D., F. van Lent and K. Groeneveld. 1983. Fecundity and the energetics of reproduction and growth of North Sea Plaice (*Pleuronectes platessa* L.). International Council for the Exploration of the Sea C.M. 1983/g:31:1–12.

Riley, J. G. 1979. Evolutionary equilibrium strategies. Journal of Theoretical Biology 76:109–123.

Rinder, T. E., A. M. Collins and M. A. Brown. 1983. Heritabilities and correlations of honeybee response to *Nosema apis* longevity and alarm response to isopentylacetate. Apidologie 14:79–86.

Ripley, S. D. 1977. *Rails of the World*. Godine, Boston.

Riska, B. 1986. Some models for development, growth and morphometric correlation. Evolution 40:1303–1311.

Riska, B., W. R. Atchley and J. J. Rutledge. 1984. A genetic analysis of targeted growth in mice. Genetics 107:79–101.

Riska, B., T. Prout and M. Turelli. 1989. Laboratory estimates of heritabilities and genetic correlations in nature. Genetics 123:865–871.

Ritland, K. 1983. The joint evolution of seed dormancy and flowering time in annual plants living in variable environment. Theoretical Population Biology 24:213–243.

Ritland, K. 1996. A marker-based method for inferences about quantitative inheritance in natural populations. Evolution 50:1062–1073.

Ritland, K. 2000. Detecting inheritance with inferred relatedness in nature. Pp. 187–199 *in* T. Mousseau, B. Sinervo and J. Endler, (eds.) *Adaptive Genetic Variation in the Wild*. Oxford University Press, New York.

Roach, D. A. 1986. Life history variation in *Geranium carolinianum*. 1. Covariation between characters at different stages of the life cycle. American Naturalist 128:47–57.

Roach, D. A. and R. D.Wulff. 1987. Maternal effects in plants. Annual Review of Ecology and Systematics 18:209–235.

Roberts, R. C. 1966a. The limits to artificial selection for body weight in the mouse: I. The limits attained in earlier experiments. Genetical Research 8:347–360.

Roberts, R. C. 1966b. The limits to artificial selection for body weight in the mouse. II. The genetic nature of the limits. Genetical Research 8:361–377.

Robertson, A. 1959a. Experimental design in the evaluation of genetic parameters. Biometrics 15:219–226.

Robertson, A. 1959b. The sampling variance of the genetic correlation coefficient. Biometrics 15:469–485.

Robertson, A. 1960a. A theory of limits in artificial selection. Proceedings of the Royal Society of London B 153:234–249.

Robertson, A. 1960b. Experimental design on the measurement of heritabilities and genetic correlations. Pp. 101–106 *in* O. Kempthorne, (ed.) *Biometrical Genetics*. Pergamon Press, London.

Robertson, A. 1970a. Some optimum problems in individual selection. Theoretical Population Biology 1:120–127.

Robertson, A. 1970b. A theory of limits in artificial selection with many linked loci. Pp. 246–288 *in* K. Kojima, (ed.) *Mathematical Topics in Population Genetics*. Springer-Verlag, Berlin.

Robertson, A. and W. G. Hill. 1983. Population and quantitative genetics of many linked loci in finite populations. Proceedings of the Royal Society of London B 219:253–264.

Robertson, A. E. 1980. Selection Experiments in Laboratory and Domestic Animals. Proceedings of a symposium held at Harrogate, U. K., 21–22 July 1979. Commonwealth Agricultural Bureau, Slough, U.K.

Robertson, F. W. 1955. Selection response and the properties of genetic variation. Cold Spring Harbor Symposium in Quantitative Genetics 20:166–177.

Robertson, F. W. 1957. Studies in quantitative inheritance XI. Genetic and environmental correlation between body size and egg production in *Drosophila melanogaster*. Genetics 55:428–443.

Robertson, F. W. 1960a. The ecological genetics of growth in *Drosophila* 1. Body size and developmental time on different diets. Genetical Research 1:288–304.

Robertson, F. W. 1960b. The ecological genetics of growth in *Drosophila* 3. Growth and competitive ability of strains selected on different diets. Genetical Research 1:333–350.

Robertson, F. W. 1963. The ecological genetics of growth in *Drosophila* 6. The genetic correlation between the duration of the larval period and body size in relation to larval diet. Genetical Research 4:74–92.

Robertson, F. W. and E. Reeve. 1952. Studies in quantitative inheritance I. The effects of selection of wing and thorax length in *Drosophila melanogster*. Journal of Genetics 50:414–448.

Robertson, F. W., M. Shook, G. Takei and H. Gaines. 1968. Observations on the biology and nutrition of *Drosophila disticha*, Hardy, and indigenous Hawaiian species. Studies in Genetics, IV. Research Reports 4:279–299.

Robertson, R. J. and G. C. Biermann. 1979. Parental investment strategies determined by expected benefits. Zeitschrift fuer Tierpsychologie 50:124–128.

Roby, D. D. and R. E. Ricklefs. 1986. Energy expenditure in adult least auklets and diving petrels during the chick-rearing period. Physiological Zoology 59:661–678.

Roff, D. A. 1974a. Spatial heterogeneity and the persistence of populations. Oecologia 15:245–258.

Roff, D. A. 1974b. The analysis of a population model demonstrating the importance of dispersal in a heterogeneous environment. Oecologia 15:259–275.

Roff, D. A. 1975. Population stability and the evolution of dispersal in a heterogeneous environment. Oecologia 19:217–237.

Roff, D. A. 1980a. A motion for the retirement of the Von Bertalanffy function. Canadian Journal of Fisheries and Aquatic Sciences 37:127–129.

Roff, D. A. 1980b. Optimizing development time in a seasonal environment: the "ups and downs" of clinal variation. Oecologia 45:202–208.

Roff, D. A. 1981a. On being the right size. American Naturalist 118:405–422.

Roff, D. A. 1981b. Reproductive uncertainty and the evolution of iteroparity: why don't flatfish put all their eggs in one basket? Canadian Journal of Fisheries and Aquatic Sciences 38:968–977.

Roff, D. A. 1982. Reproductive strategies in flatfish: a first synthesis. Canadian Journal of Fisheries and Aquatic Sciences 39:1686–1698.

Roff, D. A. 1983a. An allocation model of growth and reproduction in fish. Canadian Journal of Fisheries and Aquatic Sciences 40:1395–1404.

Roff, D. A. 1983b. Development rates and the optimal body size in *Drosophila*: a reply to Ricklefs. American Naturalist 122:570–575.

Roff, D. A. 1983c. Phenological adaptation in a seasonal environment: a theoretical perspective. Pp. 253–270 *in* V. K. Brown and I. Hodek (eds.). Diapause and Life Cycle Strategies in Insects. Dr. W. Junk, The Hague.

Roff, D. A. 1984a. The cost of being able to fly: a study of wing polymorphism in two species of crickets. Oecologia 63:30–37.

Roff, D. A. 1984b. The evolution of life history parameters in teleosts. Canadian Journal of Fisheries and Aquatice Sciences 41:984–1000.

Roff, D. A. 1986. The evolution of wing dimorphism in insects. Evolution 40:1009–1020.

Roff, D. A. 1990a. Selection for changes in the incidence of wing dimorphism in *Gryllus firmus*. Heredity 65:163–168.

Roff, D. A. 1990b. Understanding the evolution of insect life cycles: the role of genetical analysis. Pp. 5–27 *in* F. Gilbert, (ed.) *Genetics, Evolution and Coordination of Insect Life Cycles*. Springer-Verlag, New York.

Roff, D. A. 1990c. The evolution of flightlessness in insects. Ecological Monographs 60:389–421.

Roff, D. A. 1992a. *The Evolution of Life Histories: Theory and Analysis*. Chapman and Hall, New York.

Roff, D. A. 1992b. The evolution of alternative life histories: quantitative genetics and the evolution of wing dimorphism in insects. Bulletin of the Society of Population Ecology 49:28–35.

Roff, D. A. 1994a. The evolution of dimorphic traits: predicting the genetic correlation between environments. Genetics 136:395–401.

Roff, D. A. 1994b. Habitat persistence and the evolution of wing dimorphism in insects. American Naturalist 144:772–798.

Roff, D. A. 1994c. The evolution of flightlessness: is history important ? Evolutionary Ecology 8:639–657.

Roff, D. A. 1994d. Evidence that the magnitude of the trade-off in a dichotomous trait is frequency dependent. Evolution 48:1650–1656.

Roff, D. A. 1996a. The evolution of threshold traits in animals. Quarterly Review of Biology 71:3–35.

Roff, D. A. 1996b. The evolution of genetic correlations: an analysis of patterns. Evolution 50:1392–1403.

Roff, D. A. 1997a. *Evolutionary Quantitative Genetics*. Chapman and Hall, New York.

Roff, D. A. 1998a. Evolution of threshold traits: the balance between directional selection, drift and mutation. Heredity 80:25–32.

Roff, D. A. 1998b. The maintenance of phenotypic and genetic variation in threshold traits by frequency-dependent selection. Journal of Evolutionary Biology 11:513–529.

Roff, D. A. 1998c. Effects of inbreeding on morphological and life history traits of the sand cricket, *Gryllus firmus*. Heredity 81:28–37.

Roff, D. A. 2000. Trade-offs between growth and reproduction: an analysis of the quantitative genetic evidence. Journal of Evolutionary Biology 13:434–445.

Roff, D. A. 2001. The threshold model as a general purpose normalizing transformation. Heredity 86:Pt. 4 404–411.

Roff, D. A. and M. J. Bradford. 2000. A quantitative genetic analysis of phenotypic plasticity of diapause induction in the cricket *Allonemobius socius*. Heredity 84:193–200.

Roff, D. A. and M. DeRose. 2001. The evolution of trade-offs: effects of inbreeding on fecundity relationships in the cricket, *Gryllus firmus*. Evolution 55:111–121.

Roff, D. A. and D. J. Fairbairn. 1991. Wing dimorphisms and the evolution of migratory polymorphisms among the insecta. American Zoologist 31:243–251.

Roff, D. A. and D. J. Fairbairn. 2001. The genetic basis of dispersal and migration and its consequences for the evolution of correlated traits *in* C. Clobert, J. Nichols, J. D. Danchin and A. Dhondt, (eds.) *Causes, Consequences and Mechanisms of Dispersal at the Individual, Population and Community Level*. Oxford University Press, Oxford.

Roff, D. A. and T. A. Mousseau. 1987. Quantitative genetics and fitness: lessons from *Drosophila*. Heredity 58:103–118.

Roff, D. A. and R. Preziosi. 1994. The estimation of the genetic correlation: the use of the jackknife. Heredity 73:544–548.

Roff, D. A. and A. M. Simons. 1997. The quantitative genetics of wing dimorphism under laboratory and 'field' conditions in the cricket *Gryllus pennsylvanicus*. Heredity 78:235–240.

Roff, D. A., G. Stirling and D. J. Fairbairn. 1997. The evolution of threshold traits: a quantitative genetic analysis of the physiological and life history correlates of wing dimorphism in the sand cricket. Evolution 51:1910–1919.

Roff, D. A., S. Mostowy and D. J. Fairbairn. 2002 (in press). The evolution of trade-offs: testing predictions on response to selection and environmental variation. Evolution

Roff, D. A., J. Tucker, G. Stirling and D. J. Fairbairn. 1999. The evolution of threshold traits: effects of selection on fecundity and correlated response in wing dimorphism in the sand cricket. Journal of Evolutionary Biology 12:535–546.

Rohwer, F. C. 1988. Inter- and intraspecific relationships between egg size and clutch size in waterfowl. Auk 105:161–176.

Ronce, O. and I. Olivier. 1997. Evolution of reproductive effort in a metapopulation with local extinctions and ecological succession. American Naturalist 150:220–249.

Roosenburg, W. M. and K. C. Kelley. 1996. The effect of egg size and incubation temperature on growth in the turtle, *Malaclemys terrapin*. Journal of Herpetology 30:198–204.

Rose, M. R. 1982. Antagonistic pleitropy, dominance, and genetic variation. Heredity 48:63–78.

Rose, M. R. 1984. Artificial selection on a fitness-component in *Drosophila melanogaster*. Evolution 38:516–526.

Rose, M. R. 1985. Life history evolution with antagonistic pleiotropy and overlapping generations. Theoretical Population Biology 28:342–358.

Rose, M. R. 1991. *Evolutionary Biology of Aging*. Oxford University Press, New York.

Rose, M. R. 1997. Toward an evolutionary demography. Pp. 96–107 *in* K. W. Wachter and C. E. Finch, (eds.) *Between Zeus and the Salmon*. National Academy Press, Washington.

Rose, M. R. and B. Charlesworth. 1981. Genetics of life history in *Drosophila melanogaster*. II Exploratory selection experiments. Genetics 97:187–196.

Roskam, J. C. and P. M. Brakefield. 1996. A comparison of temperature-induced polyphenism in African *Bicyclus* butterflies from a seasonal savannah-rain-forest ecotone. Evolution 50:2360–2372.

Ross, K. G., E. L. Vargo, L. Keller and J. C. Trager. 1993. Effect of a founder event on variation in the genetic sex-determining system of the fire ant *Solenopsis invicta*. Genetics 135:843–854.

Roughgarden, J. 1971. Density-dependent natural selection. Ecology 52:453–468.

Roush, R. T. and J. A. McKenzie. 1987. Ecological genetics of insecticide and acaricide resistance. Annual Review of Entomology 32:361–380.

Rowe, L. and D. Ludwig. 1991. Size and timing of metamorphosis in complex life cycles: time constraints and variation. Ecology 72:413–427.

Rowe, L., D. Ludwig and D. Schluter. 1994. Time, condition, and the seasonal decline of avian clutch size. American Naturalist 143:698–722.

Rowell, G. A. and W. H. Cade. 1993. Simulation of alternative male reproductive behavior: calling and satellite behavior in field crickets. Ecological Modelling 65:265–280.

Ruano, R. G., F. Orozco and C. Lopez-fanjul. 1975. The effect of different selection intensities on selection response in egg-laying of *Tribolium castaneum*. Genetical Research 25:17–27.

Rutherford, J. C. 1973. Reproduction, growth and mortality of the holothurian *Cucumaria pseudocurata*. Marine Biology 22:167–176.

Rutledge, J. J., E. J. Eisen and J. E. Legates. 1973. An experimental evaluation of genetic correlation. Genetics 75:709–726.

Ryan, M. J., Tuttle, M. D. and A. S. Rand. 1982. Bat predation and sexual advertisement in a Neotropical anuran. American Naturalist 119:136–139.

Ryan, M. J., Tuttle, M. D. and L. K. Taft. 1981. The costs and benefits of frog chorusing behavior. Behavioral Ecology and Sociobiology 8:273–278.

Sabat, A. M. 1994. Costs and benefits of parental effort in a brood-guarding fish (*Ambloplites rupestris*, Centrarchidae). Behavioral Ecology 5:195–201.

Sacher, G. A. 1959. Relation of lifespan to brain weight. Pp. 115–141 *in* G. E. W. Wolstenholme and O. M. O'Connor, (ed.) *The Lifespan of Animals*. Little Brown, Boston.

Sadleir, R. M. F. S. 1974. The ecology of the deer mouse *Peromyscus maniculatus* in a coastal coniferous forest. II. Reproduction. Canadian Journal of Zoology 52:119–131.

Saether, B. E. 1987. The influence of body weight on the covariation between reproductive traits in European birds. Oikos 48:79–88.

Saether, B. E. 1988. Pattern of covariation between life-history traits of European birds. Nature 331:616–617.

Saether, B. E. 1989. Survival rates in relation to body weight in European birds. Ornis Scandinavica 20:13–21.

Sakaluk, S. K. and J. J. Belwood. 1984. Gecko phonotaxis to cricket calling song: a case of satellite predation. Animal Behavior 32:659–662.

Salthe, S. N. 1969. Reproductive modes and the number and size of ova in the urodeles. American Midland Naturalist 81:467–490.

Sandercock, B. K. 1994. The effect of manipulated brood size on parental defence in a precocial bird, the willow ptarmigan. Journal of Avian Biology 25:281–286.

Sang, J. H. 1962. Selection for rate of larval development using *Drosophila melanogaster* cultured axenically on deficient diets. Genetical Research 3:90–109.

Sang, J. H. and G. A. Clayton. 1957. Selection for larval development time in *Drosophila*. Journal of Heredity 48:265–270.

Sargeant, A. B., S. H. Allen and R. T. Eberhard. 1984. Red fox predation on breeding ducks in midcontinental North America. Wildlife Monographs 89: 1–41.

Sargent, P. C. and M. R. Gross. 1985. Parental investment decision rules and the Concorde fallacy. Behavioral Ecology and Sociobiology 17:43–45.

Sauer, J. R. and N. A. Slade. 1987. Size-based demography of vertebrates. Annual Review of Ecology and Systematics 18:71–90.

Saxton, A., W. Hershberger and R. Iwamoto. 1984. Smoltification in the net pen culture of coho salmon: quantitative genetic analysis. Transactions of the American Fisheries Society 113:339–347.

Schaffer, W. M. 1974a. Selection for optimal life histories: the effects of age structure. Ecology 55:291–303.

Schaffer, W. M. 1974b. Optimal reproductive effort in fluctuating environments. American Naturalist 108:783–790.

Schaffer, W. M. 1979a. Equivalence of maximizing reproductive value and fitness in the case of reproductive strategies. Proceedings of the National Academy of Sciences, USA 76:3567–3569.

Schaffer, W. M. 1979b. The theory of life-history evolution and its application to Atlantic Salmon. Symposium of the Zoological Society of London 44:307–326.

Schaffer, W. M. 1981. On reproductive value and fitness. Ecology 62:1683–1685.

Schaffer, W. M. and M. L. Rosenzweig. 1977. Selection for optimal life histories II. multiple equilibria and the evolution of alternative reproductive strategies. Ecology 58:60–72.

Scharloo, W. 1984. Genetics of adaptive reactions. Pp. 5–15 *in* K. Wöhrman and V. Loeschcke, (ed.) *Population Biology and Evolution*. Springer-Verlag, Berlin.

Scheiner, S. M. 1993. Genetics and evolution of phenotypic plasticity. Annual Review of Ecology and Systematics 24:35–68.

Scheiner, S. M. 1998. The genetics of phenotypic plasticity. VII. Evolution in a spatially-structured environment. Journal of Evolutionary Biology 11:303–320.

Scheiner, S. M. and C. J. Goodnight. 1984. The comparison of phenotypic plasticity and genetic variation in populations of the grass *Danthonia spicata*. Evolution 38:845–855.

Scheiner, S. M., R. J. Mitchell and H. S. Callahan. 2000. Using path analysis to measure natural selection. Journal of Evolutionary Biology 13:423–433.

Schemske, D. W. 1983. Breeding system and habitat effects on fitness components in three neotropical *Costus* (Zingiberaceae). Evolution 37:523–539.

Schlichting, C. D. 1986. The evolution of phenotypic plasticity in plants. Annual Review of Ecology and Systematics 17:667–693.

Schlichting, C. D. and M. Pigliucci. 1998. *Phenotypic Evolution: A Reaction Norm Perspective*. Sinauer Associates, Inc., Sunderland, MA.

Schluter, D. 1988. Estimating the form of natural selection on a quantitative trait. Evolution 42:849–861.

Schluter, D. 1996. Adaptive radiation along genetic lines of least resistance. Evolution 50:1766–1774.

Schluter, D. 2000. *The Ecology of Adaptive Radiation*. Oxford University Press, Oxford.

Schmalhausen, I. I. 1949. *Factors of Evolution: The Theory of Stabilizing Selection*. 1986 reprint. University of Chicago Press, Chicago.

Schmitt, J. and D. W. Ehrhardt. 1990. Enhancement of inbreeding depression by dominance and suppression in *Impatiens capensis*. Evolution 44:269–278.

Schneider, J. M. and Y. Lubin. 1997. Does high adult mortality explain semelparity in the spider *Stegodyphus lineatus* (Eresidae)? Oikos 79:92–100.

Schoen, D. J., G. Bell and M. Lechowicz. 1994. The ecology and genetics of fitness in forest plants. IV. Quantitative genetics of fitness components in *Impatiens pallida* (Balsaminaceae). American Journal of Botany 81:232–239.

Schuler, L. 1985. Selection for fertility in mice—the selection plateau and how to overcome it. Theoretical and Applied Genetics 70:72–79.

Schultz, D. L. 1989. The evolution of phenotypic variance with iteroparity. Evolution 43:473–475.

Schultz, E. T. and D. O. Conover. 1997. Latitudinal differences in somatic energy storage: adaptive responses to seasonality in an estuarine fish (Atherinidae: *Menidia menidia*). Oecologia 109:516–529.

Schultz, E. T. and D. O. Conover. 1999. The allometry of energy reserve depletion: test of a mechanism for size-dependent winter mortality. Oecologia 119:474–483.

Schultz, E. T., D. O. Conover and A. Ehtisham. 1998. The dead of winter: size-dependent variation and genetic differences in seasonal mortality among Atlantic silverside (Atherinidae: *Menidia menidia*) from different latitudes. Canadian Journal of Fisheries and Aquatic Sciences 55:1149–1157.

Schwaegerle, K. E. and D. A. Levin. 1990. Quantitative genetics of seed size variation in *Phlox*. Evolutionary Ecology 4:143–148.

Schwaegerle, K. E. and D. A. Levin. 1991. Quantitative genetics of fitness traits in a wild population of *Phlox*. Evolution 45:169–177.

Schwarzkopf, L., M. W. Blows and M. J. Caley. 1999. Life-history consequences of divergent selection on egg size in *Drosophila melanogaster*. American Naturalist 154:333–341.

Schwarzkopf, L. and R. Shine. 1992. Costs of reproduction in lizards: escape tactics and susceptibility to predation. Behavioural Ecology and Sociobiology 31:17–25.

Scott, H. 1935. General conclusions regarding the insect fauna of the Seychelles and adjacent islands. Transactions of the Linnean Society of London. Zoology XIX:307–391.

Scott, W. B. and E. J. Crossman. 1973. Freshwater fishes of Canada. Fisheries Research Board of Canada Bulletin 184:1–966.

Seed, R. A. and R. A. Brown. 1978. Growth as a strategy of survival in two marine bivalves, *Cerastoderma* and *Modiolus modiolus*. Journal of Animal Ecology 47:283–292.

Seigel, R. A. and H. S. Fitch. 1984. Ecological patterns of relative clutch mass in snakes. Oecologia 61:293–301.

Seigel, R. A., H. S. Fitch and N. B. Ford. 1986. Variation in relative clutch mass in snakes among and within species. Herpetologica 42:179–185.

Seigel, R. A., M. M. Huggins and N. B. Ford. 1987. Reduction in locomotor ability as a cost of reproduction in gravid snakes. Oecologia 73:481–485.

Semlitsch, R. D. 1985. Reproductive strategy of a facultatively paedomorphic salamander Ambystoma talpoideum. Oecologia 65:305–313.

Sen, B. K. and A. Robertson. 1964. An experimental examination of methods for the simultaneous selection of two characters, using *Drosophila melanogaster*. Genetics 50:199–209.

Sharp, P. M. 1984. The effect of inbreeding on competitive male-mating ability in *Drosophila melanogaster*. Genetics 106:601–612.

Shaw, R. G. and J. D. Fry. 2001. Estimating and interpreting genetic covariances across environments *in* P. M. Brakefield and P. H. Van Tienderen, (eds.) *Evolution of Phenotypic Plasticity*. Oxford University Press, Oxford.

Sheldon, B. C. and S. Verhulst. 1996. Ecological immunology: costly parasite defences and trade-offs in evolutionary ecology. Trends in Ecology and Evolution 11: 317–321.

Sheridan, A. K. 1988. Agreement between estimated and realised genetic parameters. Animal Breeding Abstracts 56:877–889.

Sheridan, A. K. and J. S. F. Barker. 1974. Two-trait selection and the genetic correlation. II. Changes in the genetic correlation during two-trait selection. Australian Journal of Biological Sciences 27:89–101.

Shine, R. 1980. 'Costs' of reproduction in reptiles. Oecologia 46:92–100.

Shine, R. and E. L. Charnov. 1992. Patterns of survival, growth, and maturation in snakes and lizards. American Naturalist 139:1257–1269.

Shine, R. and J. B. Iverson. 1995. Patterns of survival, growth and maturation in turtles. Oikos 72:343–348.

Shine, R. and L. Schwarzkopf. 1992. The evolution of reproductive effort in lizards and snakes. Evolution 46:62–75.

Shukla, S. and K. R. Khanna. 1992. Genetical study for earliness in *Papaver somniferum* L. Indian Journal of Genetics 52:33–38.

Sibly, R. and P. Calow. 1983. An integrated approach to life-cycle evolution using selective landscapes. Journal of Theoretical Biology 102:527–547.

Sibly, R. and P. Calow. 1985. Classification of habitats by selection pressures: a synthesis of life cycle and *r/K* theory. Pp. 75–90 *in* R. M. Sibly and R. H. Smith, (ed.) *Behavioural Ecology*. Blackwell, Oxford.

Sibly, R., P. Calow and R. H. Smith. 1988. Optimal size of seasonal breeders. Journal of Theoretical Biology 133:13–21.

Sibly, R. and K. Monk. 1987. A theory of grasshopper life cycles. Oikos 48:186–194.

Sibly, R. M., L. Linton and P. Calow. 1991. Testing life-cycle theory by computer simulation - II bet-hedging revisited. Computers in Biology and Medicine 21:357–367.

Siemens, D. H. and T. Mitchell-Olds. 1998. Evolution of pest-induced defenses in *Brassica* plants: test of theory. Ecology 79:632–646.

Sih, A. 1980. Optimal behavior: can foragers balance two conflicting demands. Science 210:1041–1043.

Sih, A. 1982. Foraging strategies and the avoidance of predation by an aquatic insect *Notonecta hoffmanni*. Ecology 63:786–796.

Sih, A., J. Krupa and S. Travers. 1990. An experimental study on the effects of predation risk and feeding regime on the mating behavior of the water strider. American Naturalist 135:284–290.

Sih, A. and R. D. Moore. 1993. Delayed hatching of salamander eggs in response to enhanced larval predation risk. American Naturalist 142:947–60.

Siikamaki, P., O. Ratti, M. Hovi and G. F. Bennett. 1997. Association between heamatozoan infections and reproduction in the Pied Flycatcher. Functional Ecology 11:176–183.

Siler, W. 1979. A competing-risk model for animal mortality. Ecology 60:750–757.

Simons, A. M., Y. Carriere and D. A. Roff. 1998. The quantitative genetics of growth in a field cricket. Journal of Evolutionary Biology 11:721–734.

Simons, A. M. and M. Ó. Johnston. 1997. Developmental instability as a bet-hedging strategy. Oikos 80:401–406.

Simons, A. M. and D. A. Roff. 1994. The effect of environmental variability on the heritabilities of traits of a field cricket. Evolution 48:1637–1649.

Simons, A. M. and D. A. Roff. 1996. The effect of a variable environment on the genetic correlation structure in a field cricket. Evolution in press

Simpson, M. R. 1995. Covariation of spider egg and clutch size: the influence of foraging and parental care. Ecology 76:795–800.

Sinervo, B. 1990. The evolution of maternal investment in lizards: an experimental and comparative analysis of egg size and its effects on offspring performance. Evolution 44:279–294.

Sinervo, B. and D. F. DeNardo. 1996. Costs of reproduction in the wild: path analysis of natural selection and experimental tests of causation. Evolution:1299–1313.

Sinervo, B. and P. Doughty. 1996. Interactive effects of offspring size and timing of reproduction on offspring reproduction: Experimental, maternal, and quantitative genetic aspects. Evolution:1314–1327.

Sinervo, B., P. Doughty, R. B. Huey and K. Zamudio. 1992. Allometric engineering: a causal analysis of natural selection on offsrping size. Science 258:1927–1930.

Sinervo, B., R. Hedge and S. C. Adolph. 1991. Decreased sprint speed as a cost of reproduction in the lizard *Sceloporus occidentalis*: variation among populations. Journal of experimental biology 155:323–326.

Sinervo, B. and P. Licht. 1991a. Hormonal and physiological control of clutch size, and egg shape in side-blotched lizards (*Uta stansburiana*): Constraints on the evolution of lizard life histories. Journal of Experimental Zoology 257:252–264.

Sinervo, B. and P. Licht. 1991b. Proximate constraints on the evolution of egg size, number, and total clutch mass in lizards. Science 252:1300–1302.

Singh, M. and R. C. Lewontin. 1966. Stable equilibria under optimizing selection. Proceedings of the National Academy of Sciences 56:1345–1348.

Skinner, S. W. 1985. Clutch size as an optimal foraging problem for insects. Behavioral Ecology and Sociobiology 17:231–238.

Skogland, T. 1989. Natural selection of wild reindeer life history traits by food limitation and predation. Oikos 55:101–110.

Skorping, A., A. F. Read and A. E. Keymer. 1991. Life history covariation in intestinal nematodes of mammals. Oikos 60:365–372.

Skulason, S., S. S. Snorrason, D. L. G. Noakes, M. M. Ferguson and H. J. Malmquist. 1989. Segregation in spawning and early life history among polymorphic Arctic charr, *Salvelinus alpinus*, in Thingvallavatn, Iceland. Journal of Fish Biology 35 (suppl. A):225–232.

Slatkin, M. 1974. Hedging one's evolutionary bets. Nature 250:704–705.

Slatkin, M. 1978. Spatial patterns in the distribution of polygenic characters. Journal of Theoretical Biology 70:213–228.

Slatkin, M. 1979. The evolutionary response to frequency- and density-dependent interactions. American Naturalist 114:384–398.

Slatkin, M. and S. A. Frank. 1990. The quantitative genetic consequences of pleiotropy under stabilizing and directional selection. Genetics 125:207–213.

Slatkin, M. and R. Lande. 1976. Niche width in a fluctuating environment-density independent model. American Naturalist 110:31–55.

Slobodkin, L. B. 1966. *Growth and Regulation of Animal Populations*. Holt, Reinhart, and Winston, New York.

Smith, A. P. and T. P. Young. 1982. The cost of reproduction in *Senecio leriodendron*, a giant rosette species of Mt Kenya. Oecologia 55:243–247.

Smith, C. C. and S. D. Fretwell. 1974. The optimal balance between size and number of offspring. American Naturalist 108:499–506.

Smith, H. G. 1989. Larger clutches take longer to incubate. Ornis Scandinavica 20:156–158.

Smith, R. H. and C. M. Lessells. 1985. Oviposition, ovicide and larval competition in granivorous insects. Pp. 423–448 *in* R. M. and R. H. Smith, (eds.) *Behavioural Ecology:Ecological Consequences of Adaptive Behaviour*. Blackwell, Oxford.

Smith, W. and J. J. McManus. 1975. The effects of litter size on the bioenergetics and water requirements of lactating *Mus musculus*. Comparative Biochemistry and Physiology 51:111–115.

Smith-Gill, S. J. 1983. Developmental plasticity: developmental conversion versus phenotypic modulation. American Zoologist 23:47–55.

Snell, T. W. 1977. Clonal selection: competition among clones. Archiv fuer Hydrobiologie 8:202–204.

Snell, T. W. 1978. Fecundity, development time, and population growth rate. Oecologia 32:119–125.

Snell, T. W. and D. G. Burch. 1975. The effects of density on resource partitioning in *Chamaesyce hirta* (Euphorbiaceae). Ecology 56:742–746.

Snyder, R. J. 1991. Quantitative genetic analysis of life histories in two freshwater populations of the threespine stickleback. Copeia 1991:526–529.

Sokal, R. R. 1970. Senescence and genetic load: evidence from *Tribolium*. Science 167:1733–1734.

Sokal, R. R. and F. J. Rohlf. 1995. *Biometry* 2d (ed.) W. H. Freeman and Co., San Francisco.

Solbrig, O. T. and B. B. Simpson. 1974. Components of regulation of dandelions in Michigan. Journal of Ecology 62:473–486.

Sondhi, K. C. 1961. Selection for a character with a bounded distribution of phenotypes in *Drosophila subobscura*. Journal of Genetics 57:193–221.

Sondhi, K. C. 1962. The evolution of a pattern. Evolution 16:186–191.

Soper, R. S., G. E. Shewell and D. Tyrrell. 1976. *Colidonamyia auditrix* Nov. sp. (Diptera: Sarcophagidae), a parasite which is attracted by the mating song of its host, *Okanagana rimosa* (Homoptera: Cicadae). Canadian Entomologist 108:61–68.

Sorci, G., J. Clobert and Y. Michalikis. 1996. Cost of reproduction and cost of parasitism in the common lizard, *Lacerta vivipara*. Oikos 76:121–130.

Sorenson, D. A. and W. G. Hill. 1982. Effect of short term directional selection on genetic variability: experiments with *Drosophila melanogaster*. Heredity 48:27–33.

Sorenson, D. A. and W. G. Hill. 1983. Effects of disruptive selection on genetic variance. Theoretical and Applied Genetics 65:173–180.

Southwood, T. R. E. 1962. Migration of terrestrial arthropods in relation to habitat. Biological Reviews 37:171–214.

Spight, T. M. and J. Emlen. 1976. Clutch size of two marine snails with a changing food supply. Ecology 57:1162–1178.

Spiller, D. A., J. B. Losos and T. W. Schoener. 1998. Impact of a catastrophic hurricane on island populations. Science 281:695–697.

Spinage, C. A. 1972. African ungulate life tables. Ecology 53:645–652.

Stafford, J. 1956. The wintering of Blackcaps in the British Isles. Bird Study 3:251–257.

Stancyk, S. E. and G. S. Moreira. 1988. Inheritance of male dimorphism in Brazilian populations of *Euterpina acutifrons* (Dana) (Copepoda: Harpacticoida). Journal of Experimental Marine Biology and Ecology 120:125–144.

Stanley, W. F. 1941. The effect of temperature upon wing size in *Drosophila*. Journal of Experimental Zoology 69:459–495.

Stearns, S. C. 1976. Life-history tactics: a review of the ideas. Quarterly Review of Biology 51:3–46.

Stearns, S. C. 1977. The evolution of life history traits: a critique of the theory and a review of the data. Annual Review of Ecology and Systematics 8:145–171.

Stearns, S. C. 1989. Trade-offs in life-history evolution. Functional Ecology 3:259–268.

Stearns, S. C. 1992. *The Evolution of Life Histories*. Oxford University Press, New York.

Stearns, S. C. and J. C. Koella. 1986. The evolution of phenotypic plasticity in life-history traits: predictions of reaction norms for age and size at maturity. Evolution 40:893–913.

Stein, V. W. 1977. Die beziehung zwischen biotop-alter und auftreten der kurz-flugeligkeit bei populationen dimorpher russelkafer-arten (Col., Curculionidae). Z. ang. Ent. 83:37–39.

Stenhouse, S. L., N. G. Hairston and A. E. Cobey. 1983. Predation and competition in *Ambystoma* larvae: field and laboratory experiments. Journal of Herpetology 17:210–220.

Stenning, M. J. 1996. Hatching asynchrony, brood reduction and other rapidly reproducing hypotheses. Trends in Research in Ecology and Evolution 11:243–246.

Stephens, D. W. and J. R. Krebs. 1986. *Foraging Theory*. Princeton University Press, Princeton.

Stewart, R. E. A. 1986. Energetics of age-specific reproductive effort in female harp seals, *Phoca groenlandica*. Journal of Zoology A208:503–517.

Stibor, H. 1992. Predator induced life-history shifts in a freshwater cladoceran. Oecologia 92:162–165.

Stibor, H. and J. Luning. 1994. Predator-induced phenotypic variation in the pattern of growth and reproduction in *Daphnia hyalina* (Crustacea: Cladocera). Functional Ecology 8:97–101.

Stinson, C. H. 1979. On the selective advantage of fratricide in raptors. Evolution 33:1219–1225.

Stirling, G., D. J. F. Fairbairn, S. Jensen and D. A. Roff. 2001. Does a negative genetic correlation between wing morph and early fecundity imply a functional constraint in *Gryllus firmus*. Journal of Evolutionary Biology 3:157–177.

Suh, D. S. and T. Mukai. 1991. The genetic structure of natural populations of *Drosophila melanogaster*. XXIV. Effects of hybrid dysgenesis on the components of genetic variance of viability. Genetics 127:545–552.

Sullivan, T. P. 1976. Demography and dispersal in island and mainland populations of the deer mouse, *Peromyscus maniculatus*. University of British Columbia, Vancouver.

Sultan, S. E. 1987. Evolutionary implications of phenotypic plasticity in plants. Evolutionary Biology 21:127–178.

Sutherland, W. J. 1988. The heritability of migration. Nature 334:471–472.

Sutherland, W. J., A. Grafen and P. H. Harvey. 1986. Life history correlations and demography. Nature 320:88.

Svendsen, G. 1964. Comparative reproduction and development in two species of mice in the genus *Peromyscus*. Transactions of the Kansas Academy of Sciences 67:527–538.

Svensson, J.-E. 1992. The influence of visibility and escape ability on sex-specific susceptibility to fish predation in *Eudiaptomus gracilis* (Copepoda, Crustacea). Hydrobiologia 234:143–150.

Svensson, J.-E. 1995. Predation risk increases with clutch size in a copepod. Functional Ecology 9:774–777.

Swihart, R. K. 1984. Body size, breeding season length, and life history tactics of lagomorphs. Oikos 43:282–290.

Swingland, I. R. and P. J. Greenwood. 1983. *The Ecology of Animal Movement*. Claredon Press, Oxford.

Taborsky, M., B. Hudde and P. Wirtz. 1987. Reproductive behaviour and ecology of *Symphodus* (*Crenilabrus*) *ocellatus*, a European wrasse with four types of male behaviour. Behaviour 102:82–118.

Tait, D. E. N. 1980. Abandonment as a reproductive tactic - The example of grizzly bears. American Naturalist 115:800–808.

Takada, T. 1995. Evolution of semelparous and iteroparous perennial plants: comparison between the density-indepdendent and density-dependent dynamics. Journal of Theoretical Biology 173:51–60.

Takagi, M. 1985. The reproductive strategy of the gregarious parasitoid *Pteromalus puparum* (Hymenoptera: Pteromalidae). I. Optimal number of eggs in a single host. Oecologia 68:1–6.

Tallamy, D. W. 1982. Age specific maternal defense in *Gargaphia solani* (Hemiptera: Tinigidae). Behavioral Ecology and Sociobiology 11:7–11.

Tallamy, D. W. and R. F. Denno. 1982. Life history trade-offs in *Gargaphia solani* (Hemiptera: Tingidae): the cost of reproduction. Ecology 63:616–620.

Tallis, G. M. 1987. Ancestral covariance and the Bulmer effect. Theoretical and Applied Genetics 73:815–820.

Tanaka, S. 1986. Developmental characteristics of two closely related species of *Allonemobius* and their hybrids. Oecologia 69:388–394.

Tanaka, S. and V. J. Brookes. 1986. Altitudinal adaptation of the life cycle in *Allonemobius fasciatus* DeGeer (Orthoptera: Gryllidae). Canadian Journal of Zoology 61:1986–1990.

Tanaka, T., K. Tokuda and S. Kotero. 1970. Effects of infant loss on the interbirth interval of Japanese monkeys. Primates 11:113–117.

Tanaka, Y. 1989. Genetic variance and covariance patterns of larval development in the small white butterfly *Pieris rapae crucivora* Boisduval. Researches in Population Ecology 31:311–324.

Tanaka, Y. 1991a. Heritability estimates of life history traits in small white butterflly *Pieris rapae crucivora*. Researches in Population Ecology 33:323–329.

Tanaka, Y. 1991b. Genetic variation in age-specific fecundity of the adzuki bean weevil *Callosobruchus chinensis* (Coleoptera: Bruchidae). Applied Entomology and Zoology 26:263–265.

Tanaka, Y. 1993. A genetic mechanism for the evolution of senescence in *Callosobruchus chinensis* (the azuki bean weevil). Heredity 70:318–321.

Tantawy, A. O. 1956. Selection for long and short wing lengths in Drosophila melanogaster with different systems of mating. Genetica 28:231–262.

Tantawy, A. O. and M. R. El-Helw. 1966. Studies on natural populations of Drosophila. V. Correlated response to selection in *Drosophila melanogaster*. Genetics 53:97–110.

Tantawy, A. O. and M. R. El-Helw. 1970. Studies on natural populations of Drosophila. IX. Some fitness components and their heritabilities in natural and mutant populations of *Drosophila melanogaster*. Genetics 64:79–91.

Tantawy, A. O. and F. A. Rakha. 1964. Studies on natural populations of Drosophila. IV. Genetic variances of and correlations between four characters in *D. melanogaster* and *D. simulans*. Genetics 50:1349–1355.

Taper, M. L. and T. J. Case. 1992. Models of character displacement and the theoretical robustness of taxon cycles. Evolution 46:317–333.

Tatar, M. 2001(in press). Senescence *in* C. W. Fox, D. A. Roff and D. J. Fairbairn, (eds.) *Evolutionary Ecology: Concepts and Case Studies*. Oxford University press, New York.

Tatar, M., J. R. Carey and J. W. Vaupel. 1993. Long-term cost of reproduction with and without accelerated senescence in *Callosobruchus maculatus*: analysis of age-specific mortality. Evolution 47:1302–1312.

Tatar, M., D. E. L. Promislow, A. A. Khazaeli and J. W. Curtsinger. 1996. Age-specific patterns of genetic variance in *Drosophila melanogaster*. II. Fecundity and its genetic covariance with age-specific mortality. Genetics 143:849–858.

Tauber, M. J., C. A. Tauber and S. Masaki. 1986. *Seasonal Adaptations of Insects*. Oxford University Press., New York.

Taylor, A. D. 1988a. Host effects on larval competition in the gregarious parasitoid *Bracon hebetor*. Journal of Animal Ecology 57:163–172.

Taylor, B. E. and W. Gabriel. 1992. To grow or not to grow: optimal resource allocation for *Daphnia*. American Naturalist 139:248–266.

Taylor, C. E. 1976. Genetic variation in heterogeneous environments. Genetics 83:887–894.

Taylor, C. E. 1986a. Genetics and the evolution of resistance to insecticide. Biological Journal of the Linnean Society 27:103–112.

Taylor, F. 1980. Optimal switching to diapause in relation to the onset of winter. Theoretical Population Biology 18:125–133.

Taylor, F. 1986b. The fitness functions associated with diapause induction in arthropods II. The effects of fecundity and survivorship on the optimum. Theoretical Population Biology 30:93–110.

Taylor, F. 1986c. The fitness functions associated with diapause induction in arthropods. 1. The effects of age structure. Theoretical Population Biology 30: 76–92.

Taylor, F. and J. B. Spalding. 1986. Geographical patterns in the photoperiodic induction of hibernal diapause. Pp. 66–85 *in* F. Taylor and R. Karban, (eds.) *The Evolution of Insect Life Cycles*. Springer-Verlag, New York.

Taylor, F. and J. B. Spalding. 1988. Fitness functions for alternative developmental pathways in the timing of diapause induction. American Naturalist 131:678–699.

Taylor, H. M., R. S. Gourby, C. E. Lawrence and R. S. Kaplan. 1974. Natural selection of life history attributes: an analytical approach. Theoretical Population Biology 5:104–122.

Taylor, M. and R. Feyereisen. 1996. Molecular biology and evolution of resistance to toxicants. Molecular Biology and Evolution 13:719–734.

Taylor, P. 1991. Optimal life histories with age dependent tradeoff curves. Journal of Theoretical Biology 148:33–48.

Taylor, P. D. 1988b. An inclusive fitness model for dispersal of offspring. Journal of Theoretical Biology 130:363–378.

Tedin, O. 1925. Verebung, variation, und systematik in der gatung *Camelina*. Hereditas 6:275–386.

Temme, D. H. and E. L. Charnov. 1987. Brood size adjustment in birds: economical tracking in a temporally varying environment. Journal of Theoretical Biology 126:137–147.

Thessing, A. and J. Ekman. 1994. Selection on the genetical and environmental components of tarsal growth in juvenile willow tits (*Parus montanus*). Journal of Evolutionary Biology 7:713–726.

Thoday, J. M. 1953. Components of fitness. Symposium of the Society for Experimental Biology 7:96–113.

Thoday, J. M. 1959. Effects of disruptive selection. I. Genetic flexibility. Heredity 13:187–203.

Thomas, F., A. T. Teriokhin, F. Renaud, T. De Meeus and J. F. Guegan. 2000. Human longevity at the cost of reproductive success: evidence from global data. Journal of Evolutionary Biology 13:409–414.

Thompson, S. D. 1987. Body size, duration of parental care and the intrinsic rate of natural increase in eutherian and metatherian mammals. Oecologia 71:201–209.

Thornhill, R. 1980. Mate choice in *Hylobittacus apicalis* (Insecta: Mecoptera) and its relation to some models of female choice. Evolution 34:519–538.

Tilley, S. G. 1968. Size-fecundity relationships in five desmognathine salamanders. Evolution 22:806–816.

Tilley, S. G. 1972. Aspects of parental care and embryonic development in *Desmognathus ochrophaeus*. Copeia 1972:532–540.

Tinbergen, J. M. and M. W. Dietz. 1994. Parental energy expenditure during brood rearing in the great tit (*Parus major*) in relation to body mass, temperature, food availability and clutch size. Functional Ecology 8:563–572.

Tinbergen, J. M., J. H. van Balen and H. M. van Eck. 1985. Density-dependent survival in an isolated great tit population: Kluyvers data reanalyzed. Ardea 73:38–48.

Tinkle, D. W. 1967. The life and demography of the side-blotched lizard. Miscellaneous Publications of the Museum of Zoology, University of Michigan 132:1–182.

Tinkle, D. W. 1969. The concept of reproductive effort and its relation to the evolution of life histories in lizards. American Naturalist 103:501–516.

Tinkle, D. W. 1972. The dynamics of a Utah population of *Sceloporus undulatus*. Herpetologica 28:351–359.

Tinkle, D. W. 1973. A population analysis of the sage-bush lizard, *Sceloporus graciosus*, in southern Utah. Copeia 1973:284–295.

Tinkle, D. W. and R. E. Ballinger. 1972. *Sceloporus undulatus*: a study of the intraspecific comparative demography of a lizard. Ecology 53:570–584.

Tinkle, D. W. and J. W. Gibbons. 1977. The distribution and evolution of viviparity in reptiles. Miscellaneous

Publications of the Museum of Zoology, University of Michigan 154:1–55.

Tinkle, D. W., H. M. Wilbur and S. G. Tilley. 1970. Evolutionary strategies in lizard reproduction. Evolution 24:55–74.

Tollrian, R. 1995. Predator-induced morphological defense: costs, life history shifts, and maternal effects in *Daphnia pulex*. Ecology 76:1691–1705.

Tollrian, R. and C. D. Harvell (eds.). 1999. *The Ecology and Evolution of Inducible Defenses*. Princeton University Press, Princeton.

Townsend, D. S. 1986. The costs of male parental care and its evolution in a neotropical frog. Behavioral Ecology and Sociobiology 19:187–195.

Townsend, D. S., M. M. Stewart and F. H. Pough. 1984. Male parental care and its adaptive significance in a neotropical frog. Animal Behaviour 32:421–431.

Trail, P. W. 1987. Predation and antipredator behaviour at Guianan Cock-of-the-Rock leks. Auk 104:496–507.

Travis, J. 1983a. Variation in development patterns of larval anurans in temporary ponds 1. Persistent variations within a *Hyla gratiosa* population. Evolution 37:496–512.

Travis, J. 1983b. Variation in growth and survival of *Hyla gratiosa* larvae in experimental enclosures. Copeia 1983:232–237.

Travis, J., S. B. Emerson and M. Blouin. 1987. A quantitative genetic analysis of larval life-history traits in *Hyla crucifer*. Evolution 41:145–156.

Trivers, R. L. 1972. Parental investment and sexual selection. Pp. 136–179 *in* B. G. Campbell, (ed.) *Sexual Selection and the Descent of Man* 1871–1971. Aldine, Chicago.

Trouve, S., P. Sasal, J. Jourdane, F. Renaud and S. Moran. 1998. The evolution of life-history traits in parasitic and free-living platyhelminthes: a new perspective. Oecologia 115:370–378.

Trussell, G. C. 1996. Phenotypic plasticity in an intertidal snail: the role of a common crab predator. Evolution 50:448–454.

Tucic, N., D. Cvetkovic and D. Milanovic. 1988. The genetic variation and covariation among fitness components in *Drosophila melanogaster* females and males. Heredity 60:55–60.

Tucic, N., D. Cvetkovic, V. Stojikovic and D. Bejakovic. 1990. The effects of selection for early and late reproduction on fecundity and longevity in bean weevil (*Acanthoscelides obtectus*). Genetica 80:221–227.

Tucic, N., I. Gliksman, D. Seslija, D. Milanovic, S. Mikuljanac and O. Stojkovic. 1996. Laboratory evolution of longevity in the bean weevil (*Acanthoscelides obtectus*). Journal of Evolutionary Biology 94:485–503.

Tucic, N., I. Gliksman, D. Seslija, O. Stojkovic and D. Milanovic. 1998. Laboratory evolution of life-history traits in the bean weevil (*Acanthoscelides obtectus*): the effects of selection on developmental time in populations with different previous history. Evolution 52:1713–1725.

Tucic, N., M. Milosevic, I. Gliksman, D. Milanovic and I. Aleksic. 1991. The effect of larval density on genetic variation and covariation among life-history traits in the bean weevil (*Acanthoscelides obtectus* Say). Functional ecology 5:525–534.

Tuljapurkar, S. 1989. An uncertain life: demography in random environments. Theoretical Population Biology 35:227–294.

Tuljapurkar, S. 1990. Population Dynamics in Variable Environments. Springer-Verlag, Berlin.

Tuljapurkar, S. D. 1982. Population dynamics in variable environments. II. Evolutionary dynamics of *r*-selection. Theoretical Population Biology 21:114–140.

Turelli, M. 1977. Random environments and stochastic calculus. Theoretical Population Biology 12:140–178.

Turelli, M. 1984. Heritable genetic variation via mutation-selection balance: Lerch's zeta meets the abdominal bristle. Theoretical Population Biology 25:138–193.

Turelli, M. 1985. Effects of pleiotropy on predictions concerning mutation-selection balance for polygenic traits. Genetics 111:165–195.

Turelli, M. and N. H. Barton. 1990. Dynamics of polygenic characters under selection. Theoretical Population Biology 38:1–57.

Turelli, M. and D. Petry. 1980. Density-dependent selection in a random environment: an evolutionary process that can maintain stable population dynamics. Proceedings of the National Academy of Sciences (US) 77:7501–7505.

Tuttle, M. D. and M. J. Ryan. 1981. Bat predation and the evolution of frog vocalizations in the neotropics. Science 214:677–678.

Tyndale-Biscoe, M. and R. D. Hughes. 1968. Change in the female reproductive system as age indicators in the bushfly *Musca retustissima* Wlk. Bulletin of Entomological Research 59:129–141.

Ueno, H. 1994. Genetic estimations for body size characters, development period and development rate in a coccinelid beetle *Harmonia axyridis*. Researches on Population Ecology 36:121–4.

Unwin, M. J., M. T. Kinnison and T. P. Quinn. 1999. Exceptions to semelparity: postmaturation survival, morphology, and energetics of male chinook salmon (*Oncorhynchus tshawytscha*). Canadian Journal of Fisheries and Aquatic Sciences 56:1172–1181.

Vahl, O. 1981. Age specific residual reproductive value and reproductive effort in the Iceland scallop, *Chlamys islandica* (D.F. Muller). Oecologia 51:53–56.

Van Damme, R., D. Bauwens and R. F. Verheyen. 1989. Effect of relative clutch mass on sprint speed in the lizard *Lacerta vivipara*. Journal of Herpetology 23:459–461.

Van Dijk, T. S. 1979. On the relationship between reproduction, age and survival in two carabid beetles: *Calathus melanocephalus* L. and *Pterostichus caerulescens* L. (Coleoptera, Carabidae). Oecologia 40:63–80.

Van Dijken, F. R. and W. Scharloo. 1979. Divergent selection on locomotor activity in *Drosophila melanogaster*. I. Selection response. Behaviour Genetics 9:543–553.

Van Donk, E., M. Lurling and W. Lampert. 1999. Consumer-induced changes in phytoplankton: inducibility, costs, benefits and the impacts of grazers. Pp. 89–103 *in* R. Tollrian and C. D. Harvell, (eds.) *The Ecology and Evolution of Inducible Defenses*. Princeton University Press, Princeton.

Van Jaarsveld, A. S., P. R. K. Richardson and M. D. Anderson. 1995. Post-natal growth and sustained lactational effort in the aardwolf: life-history implications. Functional Ecology 9:492–497.

Van Noordwijk, A. J. and G. de Jong. 1986. Acquisition and allocation of resources: their influence on variation in life history tactics. American Naturalist 128:137–142.

Van Tienderen, P. H. 1991. Evolution of generalists and specialists in spatially heterogeneous environments. Evolution 45:1317–1331.

Van Tienderen, P. H. and H. P. Koelewijn. 1994. Selection on reaction norms, genetic correlations and constraints. Genetical Research 64:115–125.

Van Valen, L. 1971. Group selection and the evolution of dispersal. Evolution 25:591–598.

Van Valen, L. and G. W. Mellin. 1967. Selection in natural populations. 7. New York babies (Fetal life study). Annals of Human Genetics 31:109–127.

Van Vleck, L. D. 1968. Selection bias in estimation of the genetic correlation. Biometrics 24: 951–962.

Van Vleck, L. D. and C. R. Henderson. 1961. Empirical sampling estimates of genetic correlations. Biometrics 17:359–371.

VanderWerf, E. 1992. Lack's clutch size hypotheses: an explanation of the evidence using meta-analysis. Ecology 73:1699–1705.

Vanfleteren, J. R., A. D. Vreese and B. P. Braeckman. 1998. Two-Parameter Logistic and Weibull Equations provide better fits to survival data from isogenic populations of *Caenorhabditis elegans* in axenic culture than does the Gompertz Model. Journals of Gerontology, Biological Sciences and Medical Sciences 53:393.

Venable, D. and J. S. Brown. 1988. The selective interactions of dispersal, dormancy, and seed size as adaptations for reducing risk in variable environments. American Naturalist 131:360–384.

Vepsalainen, K. 1978. Wing dimorphism and diapause in *Gerris*: determination and adaptive significance. Pp. 218–253 *in* H. Dingle, (ed.) *Evolution of Insect Migration and Diapause*. Springer-Verlag, Berlin.

Verhulst, S., J. H. Van Balen and J. M. Tinbergen. 1995. Seasonal decline in reproductive success of the great tit: variation in time or quality? Ecology 76:2392–2403.

Vermeij, G. J. 1974. Marine faunal dominance and molluscan shell form. Evolution 28:656–664.

Vet, L. E. M., A. Datema, A. Janssen and H. Snellen. 1994. Clutch size in a larval-pupal endoparasitoid: consequences for fitness. Journal of Animal Ecology 63:807–815.

Vet, L. E. M., A. Datema, K. van Welzen and H. Snellen. 1993. Clutch size in larval-pupal endoparasitiod 1. Variation across and within host species. Oecologia 95:410–415.

Via, S. 1984a. The quantitative genetics of polyphagy in an insect herbivore. 1. Genotype-environment interaction in larval performance on different host plant species. Evolution 38:881–895.

Via, S. 1984b. The quantitative genetics of polyphagy in an insect herbivore. II. Genetic correlations in larval performance within and among host plants. Evolution 38:896–905.

Via, S. 1987. Genetic constraints on the evolution of phenotypic plasticity. Pp. 47–71 *in* V. Loeschcke, (ed.) *Genetic Constraints on Adaptive Evolution*. Springer-Verlag, Berlin.

Via, S., R. Gomulkiewicz, G. d. Jong, S. M. Scheiner, C. D. Schlichting and P. H. v. Tienderen. 1995. Adaptive phenotypic plasticity: consensus and controversy. Trends in Ecology and Evolution 5:212–217.

Via, S. and R. Lande. 1985. Genotype-environment interaction and the evolution of phenotypic plasticity. Evolution 39:505–522.

Vickers, G. T. and C. Cannings. 1987. On the definition of an evolutionarily stable strategy. Journal of Theoretical Biology 129:349–353.

Vincent, T. L. and J. S. Brown. 1988. The evolution of ESS theory. Annual Review of Ecology and Systematics 19:423–443.

Vincent, T. L. and H. R. Pulliam. 1980. Evolution of life history strategies for an asexual annual plant model. Theoretical Population Biology 17:215–231.

Vinegar, M. B. 1975. Demography of the striped plateau lizard, *Sceloporus virgatus*. Ecology 56:172–182.

Vinuela, J. 1997. Adaptation vs. constraint: intraclutch egg-mass variation in birds. Journal of Animal Ecology 66:781–792.

Visman, V., S. Pesant, J. Dion, B. Shipley and R. H. Peters. 1996. Joint effects of maternal and offspring sizes on clutch mass and fecundity in plants and animals. Ecoscience 3:173–82.

Vitt, L. J. and J. D. Congdon. 1978. Body shape, reproductive effort, and relative clutch mass in lizards: resolution of a paradox. American Naturalist 112:595–608.

Vollestad, L. A. and J. H. L'Abee-Lund. 1994. Evolution of the life history of Arctic charr *Salvelinus alpinus*. Evolutionary Ecology 8:315–27.

Vollestad, L. A., J. H. L'Abee-Lund and H. Saegrov. 1993. Dimensionless numbers and life history variation in brown trout. Evolutionary Ecology 7: 207–218.

Vuorinen, I., M. Rajasilta and J. Salo. 1983. Selective predation and habitat shift in a copepod species-support for the predation hypothesis. Oecologia 59:62–64.

Waage, J. K. and S. M. Ng. 1984. The reproductive strategy of a parasitic wasp. 1. Optimal progeny and sex allocation in *Trichogramma evanescens*. Journal of Animal Ecology 53:401–416.

Waddington, C. H. 1942. Canalization of development and the inheritance of acquired characters. Nature 150:563–5.

Waddington, C. H. 1957. *The Strategy of the Genes*. George Allen & Unwin Ltd, London.

Wade, M. J. 1985. Soft selection, hard selection, kin selection and group selection. American Naturalist 125:61–73.

Wagner, G. P. 1989. Multivariate mutation-selection balance with constrained pleiotropic effects. Genetics 122:223–234.

Walker, T. J. 1964. Experimental demonstration of a cat locating orthopteran prey by the prey's call. Florida Entomologist 47:163–165.

Wallace, A. R. 1880. *Island Life: or, the Phenomena and Causes of Insular Faunas and Floras, Including a Revision and Attempted Solution of the Problem of Geological Climates*. MacMillan, London.

Wallace, B. 1968. Polymorphism, population size, and genetic load. Pp. 87–108 *in* R. C. Lewontin, (ed.). *Population Biology and Evolution*. Syracuse University Press, Syracuse.

Wallace, B. 1975. Hard and soft selection revisited. Evolution 29:465–473.

Wallace, H. R. 1953. Notes on the biology of *Coranus subapterus* de Geer (Hemiptera: Reduviidae). Proceedings of the Royal Entomological Society 28:100–110.

Wallinga, J. H. and H. Bakker. 1978. Effect of long-term selection for litter size in mice on lifetime reproduction rate. Journal of Animal Science 46:1563–1571.

Walsh, J. B. 1984. Hard lessons from soft selection. American Naturalist 124:518–526.

Ware, D. M. 1975a. Growth, metabolism, and optimal swimming speed of a pelagic fish. Journal of the Fisheries Research Board of Canada 32:33–41.

Ware, D. M. 1975b. Relation between egg size, growth and natural mortality of larval fish. Journal of the Fisheries Research Board of Canada 32:2503–2512.

Ware, D. M. 1977. Spawning time and egg size of Atlantic mackerel, *Scomber scombrius*, in relation to the plankton. Journal of the Fisheries Research Board of Canada 34:2308–2315.

Ware, D. M. 1978. Bioenergetics of pelagic fish: theoretical change in swimming speed and ration with body size. Journal of the Fisheries Research Board of Canada 35:220–228.

Ware, D. M. 1980. Bioenergetics of stock and recruitment. Canadian Journal of Fisheries and Aquatic Sciences 37:1012–1024.

Warshaw, S. J. 1972. Effects of alewives (*Alosa pseudoharengus*) on the zooplankton of Lake Wononskopomac, Connecticut. Limnology and Oceanography 17:816–825.

Watanabe, T. K. and W. W. Anderson. 1976. Selection for geotaxis in *Drosophila melanogaster*: heritability, degree of dominance, and correlated responses to selection. Behavior Genetics 6:71–86.

Watkinson, A. and J. White. 1985. Some life history consequences of modular construction in plants. Philosophical Transactions of the Royal Society of London (B) 313:31–51.

Wauters, L. A., Y. Hutchinson, D. T. Parkin and A. A. Dhondt. 1994. The effects of habitat fragmentation on demography and the loss of genetic variation in the red squirrel. Procedings of the Royal Society of London (Biology) 255:107–111.

Weber, K. E. 1990. Increased selection response in larger populations. I. Selection for wing-tip height in *Drosophila melanogaster* at three population sizes. Genetics 125:579–84.

Weber, K. E. and L. T. Diggins. 1990. Increased selection response in larger populations. II. Selection for ethanol vapor resistance in *Drosophila melanogaster* at two population sizes. Genetics 125:585–597.

Weeks, S. and G. K. Meffe. 1996. Quantitative genetic and optimality analyses of life-history plasticity in the Eastern Mosquitofish, *Gambusia holbrooki*. Evolution 50:1358–1365.

Weigensberg, I., Y. Carriere and D. A. Roff. 1998. Effects of male genetic contribution and paternal investment on egg and hatchling size in the cricket *Gryllus firmus*. Journal of Evolutionary Biology 11:135–146.

Weigensberg, I. and D. A. Roff. 1996. Natural heritabilities: Can they be reliably estimated in the laboratory. Evolution 50:2149–2157.

Weis, A. F., P. W. Price and M. Lynch. 1983. Selective pressures on clutch size in the gall maker *Asteromyia carbonifera*. Ecology 64:688–695.

Weissburg, M. 1986. Risky business: on the ecological relevance of risk-sensitive foraging. Oikos 46:261–262.

Weisser, W. W., C. Braendle and N. Minoretti. 1999. Predator-induced morphological shift in the pea aphid. Proceedings of the Royal Society of London, B 266:1175–1181.

Wells, K. D. 1977. The social behaviour of anuran amphibians. Animal Behaviour 25:666–693.

Wells, L. 1970. Effects of alewife predation on zooplankton populations in Lake Michigan. Limnology and Oceanography 15:556–565.

Werner, E. E. 1974. The fish size, prey size and handling time relation in several sunfishes and some implications. Journal of the Fisheries Research Board of Canada 31:1531–1536.

Werner, E. E. 1986. Amphian metamorphosis: growth rate, predation risk, and the optimal size at transformation. American Naturalist 128:319–341.

Werner, E. E. 1988. Size, scaling, and the evolution of complex life cycles. Pp. 60–81 *in* B. Ebenman and L.

Persson, (eds.) *Size-Structured Populations*. Springer-Verlag, Berlin.

Werner, E. E. and B. R. Anholt. 1993. Ecological consequences of the trade-off between growth and mortality rates mediated by foraging activity. American Naturalist 142:242–272.

Werner, E. E., J. F. Gilliam, D. J. Hall and G. G. Mittelbach. 1983. An experimental test of the effects of predation on habitat use in fish. Ecology 64:1540–1548.

Werner, P. A. 1975. Predictions of fate from rosette size in teasel, *Dipsacus fullonum* L. Oecologia 20:197–201.

Werner, P. A. 1976. Ecology of plant populations in successional environments. Systematic Botanist 1:246–268.

Wesselingh, R. A. and T. J. de Jong. 1995. Bidirectional selection on threshold size for flowering in *Cynoglossum officinale* (hound's-tongue). Heredity 74:415–424.

Wesselingh, R. A., T. J. D. Jong, P. G. L. Klinkhamer, M. J. V. Dijk and E. G. M. Schlatmann. 1993. Geographical variation in threshold size for flowering in *Cynoglossum officinale*. Acta Botanica Neerlandica 42:81–91.

Western, D. 1979. Size, life history and ecology in mammals. African Journal of Ecology 17:185–204.

Wheeler, A. 1969. *The Fish of the British Isles and North-West Europe*. Macmillan, London.

Wheelwright, N. T., J. Leary and C. Fitzgerald. 1991. The costs of reproduction in tree swallows (*Tachycineta bicolor*). Canadian Journal of Zoology 69:2540–2547.

Whittaker, R. H. and D. Goodman. 1979. Classifying species according to their demographic strategy 1. Population fluctuations and environmental heterogeneity. American Naturalist 113:185–200.

Wiackowski, K. and M. Szkarlat. 1996. Effects of food availability on predator-induced morphological defence in the ciliate *Euplotes octocarinatus* (Protista). Hydrobiologia 321:47–52.

Wicklow, B. J. 1997. Signal-induced defensive phenotypic changes in ciliated protists: Morphological and ecological implications for predator and prey. Journal of Eukaryotic Microbiology 44:176–188.

Wigglesworth, V. B. 1934. The physiology of ecdysis in *Rhodnius prolixus* (Hemiptera). II. Factors controlling moulting and metamorphosis. Quarterly Journal of Microscopical Science 77:191–222.

Wiklund, C. and J. Forsberg. 1991. Sexual size dimorphism in relation to female polygamy and protandry in butterflies: a comparative study of Swedish Pieridae and Satyridae. Oikos 60:373–381.

Wiklund, C., S. Nylin and J. Forsberg. 1991. Sex-related size variation in growth rate as a result of selection for large size and protandry in a bivoltine butterfly, *Pieris napi*. Oikos 60

Wiklund, C., P. O. Wickman and S. Nylin. 1992. A sex difference in the propensity to enter direct/diapause development: a result of selection for protandry? Evolution 46:519–528.

Wiklund, C. G. 1990. Offspring protection by Merlin *Falco clumbarius* females: the importance of brood size and expected offspring survival for defense of young. Behavioral Ecology and Sociobiology 26:217–223.

Wilbur, H. M. 1977. Propagule size, number and dispersion pattern in *Ambystoma* and *Asclepias*. American Naturalist 111:43–68.

Wilbur, H. M. 1980. Complex life cycles. Annual Reviews in Ecology and Systematics 11:67–93.

Wilbur, H. M., D. W. Tinkle and J. P. Collins. 1974. Environmental certainty, trophic level, and resource availability in life history evolution. American Naturalist 108:805–817.

Wilbur, H. M. and J. P. Collins. 1973. Ecological aspects of amphibian metamorphosis. Science 182:1305–1314.

Willham, R. L. 1963. The covariance between relatives for characters composed of components contributed by related individuals. Biometrics 19:18–27.

Willham, R. L. 1972. The role of maternal effects in animal breeding: III. Biometrical aspects of maternal effects in animals. Journal of Animal Science 35:1288–1293.

Williams, G. C. 1966. Natural selection, the costs of reproduction and a refinement of Lack's principle. American Naturalist 100:687–690.

Williams, J. 1988. Field metabolism of tree swallows during the breeding season. Auk 105:706–714.

Willis, J. H. and H. A. Orr. 1993. Increased heritable variation following population bottlenecks: The role of dominance. Evolution 47:949–957.

Wilmoth. 1997. In search of limits. Pp. 38–64 *in* K. W. Wachter and C. E. Finch, (eds.) *Between Zeus and the Salmon*. National Academy Press, Washington.

Wilson, A. M. and K. Thompson. 1989. A comparative study of reproductive allocation in 40 British grasses. Functional Ecology 3:297–302.

Wilson, D. L. 1994. The analysis of survival (mortality) data: Fitting Gompertz, Weibull, and logistic functions. Mechanisms of Ageing and Development 74:15–39.

Wilson, D. L. 1994. Evolution of clutch size in insects. II. A test of static optimality models using the beetle *Callosobruchus maculatatus* (Coleoptera: Bruchidae). Journal of Evolutionary Biology 7:365–86.

Wilson, K. and C. M. Lessells. 1994. Evolution of clutch size in insects. I. A review of static optimality models. Journal of Evolutionary Biology 7:339–363.

Wilson, M. E., T. P. Gordon and I. S. Bernstein. 1978. Timing of births and reproductive success in rhesus monkey social groups. Journal of Medical Primatology 7:202–212.

Wilson, S. P., H. D. Goodale, W. H. Kyle and E. F. Godfrey. 1971. Long term selection for body weight in mice. Journal of Heredity 62:228–234.

Windig, J. J. 1997. The calculation and significance testing of genetic correlations across environments. Journal of Evolutionary Biology 10:853–874.

Windig, J. J., P. M. Brakefield, N. Reitsma and J. G. M. Wilson. 1994. Seasonal polyphenism in the wild: survey of wing patterns in five species of *Bicyclus* butterflies in Malawi. Ecological Entomology 19:285–298.

Winfield, I. J. and C. R. Townsend. 1983. The cost of copepod reproduction: increased susceptibility to fish predation. Oecologia 60:406–411.

Winkler, D. 1987. A general model for parental care. American Naturalist 130:526–553.

Winkler, D. W. 1991. Parental investment decision rules in Tree Swallows: parental defense, abandonment, and the so-called Concorde fallay. Behavioral Ecology 2:133–142.

Wodinsky, J. 1977. Hormonal inhibition of feeding and death in *Octopus*: control by optic gland secretion. Science 198:948–951.

Wollaston, T. V. 1854. *Insect Maderensia*. John Van Voorst, London.

Woltereck, R. 1909. Weitere experimentelle Untersuchungen uber Artveranderung, speziell uber das Wesen quantitativer Artunterschiede bei Daphniden. Verhandlungen der Deutschensch Tsch. Zoologischen Gessellschaft 1909:110–172.

Woodring, J. P. 1983. Control of moulting in the house cricket, *Acheta domesticus*. Journal of Insect Physiology 29:461–464.

Woombs, M. and J. Laybourn-Parry. 1984. Growth, reproduction and longevity in nematodes from sewage treatment plants. Oecologia 64:168–172.

Wootton, R. J. 1979. Energy costs of egg production and environmental determinants of fecundity in teleost fishes. Symposium of the Zoological Society of London 44:133–159.

Wootton, R. J. 1984. *A Functional Biology of Sticklebacks*. Croom Helm, London.

Wright, S. 1929. Fisher's theory of dominance. The American Naturalist 63:274–279.

Wright, S. 1934. Physiological and evolutionary theories of dominance. The American Naturalist 68:24–53.

Wright, S. 1948. On the roles of directed and random changes in gene frequency in the genetics of natural populations. Evolution 2:279–294.

Wright, S. and O. N. Eaton. 1929. The persistence of differentiation among inbred families of guinea pigs. Technical Bulletin of the U.S. Department of Agriculture 103:1–45.

Wyatt, T. 1973. The biology of *Oikopleura dioica* and *Fritillaria borealis* in the Southern Bight. Marine Biology 22:137–158.

Yamada, S. B., S. A. Navarrete and C. Needham. 1998. Predation indurced changes in behavior and growth rate in three populations of the intertidal snail,*Littorina sitkana* (Philippi). Journal of Experimental Marine Biology and Ecology 220:213–226.

Yamada, Y. 1962. Genotype by environment interaction and genetic correlation of the same trait under different environments. Japanese Journal of Genetics 37:498–509.

Yamada, Y. and A. E. Bell. 1969. Selection for larval growth in *Tribolium* under two levels of nutrition. Genetical Research 13:175–195.

Ydenberg, R. C. 1989. Growth-mortality trade-offs, parent-offspring conflict, and the evolution of juvenile life histories in the avian family, Alcidae. Ecology 70:1494–1506.

Ydenberg, R. C. and D. F. Bertram. 1989. Lack's clutch size hypothesis and brood enlargement studies of colonial seabirds. Colonial Waterbirds 12:134–137.

Ylonen, H. and H. Ronkainen. 1994. Breeding supression in the bank vole as antipredatory adaptation in a predictable environment. Evolutionary Ecology 8:658–666.

Yodzis, P. 1981. Concerning the sense in which maximizing fitness is equivalent to maximizing reproductive value. Ecology 62:1681–1682.

Yokoi, Y. 1989. An analysis of age- and size-dependent flowering: a critical-production model. Ecological Research 4:387–397.

Yoo, B. H. 1980a. Long-term selection for a quantitative character in large replicate populations of *Drosophila melanogaster* 1. Response to selection. Genetical Research 35:1–17.

Yoo, B. H. 1980b. Long-term selection for a quantitative character in large replicate populations of *Drosophila melanogaster* Part 3: The nature of residual genetic variability. Theoretical and Applied Genetics 57:25–32.

Yoo, B. H., F. W. Nicholas and K. A. Rathie. 1980. Long-term selection for a quantitative character in large replicate populations of *Drosophila melanogaster* Part 4: Relaxed and reversed selection. Theoretical and Applies Genetics 57:113–117.

Yoshimura, J. and C. W. Clark. 1991. Individual adaptations in stochastic environments. Evolutionary Ecology 5:173–192.

Young, B. E. 1996. An experimental analysis of small clutch size in tropical house wrens. Ecology 77:472–488.

Young, S. S. Y. and H. Weiler. 1960. Selection for two correlated traits by independent culling levels. Journal of Genetics 57:329–338.

Young, T. P. 1981. A general model of comparative fecundity for semelparous and iteroparous life histories. American Naturalist 118:27–36.

Yule, G. U. 1902. Mendel's laws and their probable relations to intra-racial heredity. New Phytologist 1:193–207, 222–238.

Yule, G. U. 1906. On the theory of inheritance of quantitative compound characters on the basis of Mendel's laws - a preliminary note. Report of the 3rd International Conference of Genetics:140–142.

Zach, R. 1988. Growth curve analysis: a critical reevaluation. Auk 105:208–210.

Zar, J. H. 1999. *Biostatistical Analysis* 4th (ed.) Prentice-Hall, Englewood Cliffs, N.J.

Zaret, T. M. 1980. *Predation in Freshwater Communities.* Yale University Press, New Haven, Connecticut.

Zaret, T. M. and W. C. Kerfoot. 1975. Fish predation on *Bosmina longirostris*: body-size selection versus visibility selection. Ecology 56:232–237.

Zeng, Z.-B. 1988. Long-term correlated response, interpopulation covariation, and interspecific allometry. Evolution 42:363–374.

Zeveloff, S. I. and M. S. Boyce. 1980. Parental investment and mating systems in mammals. Evolution 34:973–982.

Zhivotovsky, L. A. and M. W. Feldman. 1992. On the difference between mean and optimum of quantitative characters under selection. Evolution 46:1574–1578.

Zimmerer, E. J. and K. D. Kallman. 1989. Genetic basis for alternative reproductive tactics in the pygmy swordtail, *Xiphophorus nigrensis*. Evolution 43:1298–1307.

Ziolko, M. and J. Kozlowski. 1983. Evolution of body size: an optimization model. Mathematical Bioscience 64:127–143.

Zuk, M. 1996. Trade–offs in parasitology, evolution and behavior. Parasitology Today 12:46-47.

Zuk, M. 1999. Parasites and tradeoffs in the immune response of female red jungle fowl. Oikos 86:487-492.

Zuk, M. and T. S. Johnsen. 1998. Seasonal changes in the relationship between ornamentation and immune response in red jungle fowl. Proceedings of the Royal Society of London B 265:1631-1635.

Zuk, M. and G. T. Kolluru. 1998. Exploitation of sexual signals by predators and parasitoids. Quarterly Review of Biology 73:415–438.

Zuk, M., L. W. Simmons and J. T. Rotenberry. 1995. Acoustically-orienting parasitoids in calling and silent males of the field cricket *Teleogryllus oceanicus*. Ecological Entomology 20:380–383.

Zwann, B., R. Bijlsma and R. F. Hoekstra. 1995. Artificial selection for developmental time in *Drosophila melanogaster* in relation to the evolution of aging: direct and correlated responses. Evolution 49:635–648.

INDEX

Numbers in **bold-faced italic** refer to information in an illustration, table, or box.

517

Antarctic petrel
  handicapping studies in, *415*
Antechinus
  example of semelparity, 188
Anthoxanthum odoratum
  costs of reproduction in, 144
Antopercus, 215
Apera spica-venti
  source and founder popula-
    tions, *38*
Aphaereta minuta
  clutch size and site quality, 253
  effects of host size on clutch
    size in, 421, *422, 423*
Aphytis lingnanensis
  clutch size and site quality, 253
Aphytis nerii
  clutch size and site quality, 253
Aquarius remigis, 170
  costs of reproduction in, 133
  field and lab body sizes, 375
  phenotypic correlations in, *259*
  stabilizing selection in, 82, 83
Arctic char, 384
  covariation in traits, *278*
Arithmetic mean, 293
  of *r*, 69
  of the finite rate of increase, 70
  vs geometric mean, 294
Armadillidium vulgare
  egg size plasticity in, 442
Artemia
  fecundity, 182
Artificial selection
  and reduction in heritability, 25
Asplanchna brightwelli
  and *r*- and *k*-selection, 78
Asplanchna spp
  costs of inducible defense in,
    444
Asteronyia carbonifera
  and clutch size, 251, 255
Astrocaryum mexicanum
  reproductive effort and age, 235
Asymmetry of response, 40–44
Ateledrosophila, 215
Atlantic mackerel
  egg size plasticity in, 442
Atlantic salmon
  and evolution of semelparity,
    189
  modelling dimorphism in, 229
Atlantic silverside
  winter mortality in, 386
Atlantisia, 355
Autoregulatory morphogenesis
  defined, 360

Balloon vine, 167
Bamboo
  example of semelparity, 188
Band-legged ground cricket
  seasonal adjustment of growth,
    380, *381*
Bang-bang life history
  defined, 374
Barn swallow
  costs of reproduction in, 139
Bet-hedging, 10
  clutch size reduction, 318, *322*
  defined, 293
  dimorphic variation, 300–301
  general solution for, 296–298
  inducible defense, 448
Biomphalaria glabrata
  adaptive reaction norms in, *453*
  induced changes in, 452, *455,
    456*
Bitis caudalis
  foraging mode and re, 240
Bittacus strigosus
  reproduction and survival in,
    128
Bivariate distribution, 56
Black eagle
  and brood reduction, *324*
Blackcap
  rapid evolution in, 162, 164–166,
    165
Blackcap, *4*
Blue tit
  costs of reproduction, *136, 138*
  handicapping studies in, *415*
Blue-footed booby
  and brood reduction, *324*
Bluegill sunfish
  dimorphic behavior, 227
  modelling dimorphism in, 229
Blue-winged teal
  and desertion, 415
Bolas spider
  reproduction and survival in,
    127
Bootstrap, 50
Bosmina longirostris
  and *r*- and *K*-selection, 78
  egg size plasticity in, 442
  size-specific survival, 210
Botryllus schlosseri
  example of semelparity, 188
Brachionous calyciflorus
  costs of inducible defense in,
    444
Brassica campestris
  development time and size, *203*
  selection on seed size in, 260

Brassica napus
  selection on seed size in, 260
Brassica rapa
  costs of inducible defense in,
    445
Bream, 132
Breeding values, 51
Brood fitness function
  defined, 412
Brook trout, 222
Brown booby
  and brood reduction, 323
Bumblebees, 132
Bupalus, 253, 254

Cabbage butterfly
  fecundity, 182
Calathea ovandensis
  costs of reproduction, 141
California condor, 187
Callosobruchus chinensis
  genetic correlations in, *145*
Callosobruchus maculates
  clutch size, 252
  clutch size and site quality, 253
  costs of reproduction in, 115
  development time and size, *203*
  genetic correlations in, 146
  genetic parameters, size and
    fecundity, *207*
  optimal clutch size in, 424–429
Caloglyphus berlesei
  costs and benefits of dimor-
    phism, *226*
Camelina
  selection on seed size in, 260
Canalization
  defined, 360
Canalizing selection, 171
Carabidae
  evolution of flightlessness in,
    354
Carassius carassius
  costs of inducible defense in,
    445
Cardiospermum coruidum, 167, *168*
Cardiospermum halicacabum, 167,
    168
Carduelis tristis
  and parental defense, *418*
Carex olivacea, 438
Carrying capacity, 78
Cavia porcellus
  costs of reproduction, *131*
Cerastoderma
  size-specific survival, 208
Ceratophyllus gallinae, 415
Cercidium floridum, 433, *434, 435*